Synthesis of Natural Products Using Engineered Plants and Microorganisms

Synthesis of Natural Products Using Engineered Plants and Microorganisms

Yongjun Wei
Lingbo Qu
Xiao-Jun Ji

MDPI

Basel • Beijing • Wuhan • Barcelona • Belgrade • Novi Sad • Cluj • Manchester

Editors

Yongjun Wei
School of Pharmaceutical Sciences
Zhengzhou University
Zhengzhou
China

Lingbo Qu
School of Chemical Engineering
Zhengzhou University
Zhengzhou
China

Xiao-Jun Ji
State Key Laboratory of Materials-Oriented Chemical Engineering
Nanjing Tech University
Nanjing
China

Editorial Office
MDPI AG
Grosspeteranlage 5
4052 Basel, Switzerland

This is a reprint of articles from the Special Issue published online in the open access journal *Molecules* (ISSN 1420-3049) (available at: www.mdpi.com/journal/molecules/special_issues/F8CG4N1697).

For citation purposes, cite each article independently as indicated on the article page online and as indicated below:

Lastname, A.A.; Lastname, B.B. Article Title. *Journal Name* **Year**, *Volume Number*, Page Range.

ISBN 978-3-7258-2606-3 (Hbk)
ISBN 978-3-7258-2605-6 (PDF)
doi.org/10.3390/books978-3-7258-2605-6

Contents

About the Editors

Yongjun Wei

Dr. Wei completed his bachelor's degree in biology at Shaanxi Normal University before pursuing his Ph.D. in Microbiology from the Chinese Academy of Sciences (CAS) in 2014. Following his doctoral studies, he conducted a post-doctoral fellowship at Chalmers University of Technology from 2014 to 2016. After that, he worked as an assistant researcher at CAS for a year before joining Zhengzhou University's faculty in 2018. Currently working at Zhengzhou University, Dr. Wei has published over 90 peer-reviewed papers and holds more than 20 patents. Furthermore, he serves as an editorial board member or youth editor for over 20 different journals. His research primarily focuses on the microbial production of plant bioactive natural products.

Lingbo Qu

Lingbo Qu has a Ph.D. in Pharmacy and is a Professor. His primary research areas include the safety testing and evaluation of food and pharmaceuticals. He has led or participated in over 20 national and provincial projects, winning four awards for Scientific and Technological Progress from Henan Province, including one first prize and three second prizes. He has applied for more than 40 patents for inventions and published over 300 academic papers.

Xiao-Jun Ji

Dr. Xiao-Jun Ji is currently a Professor of Bioengineering and Biochemical Engineering at Nanjing Tech University (NanjingTech). He received his B.Sc. and PhD from NanjingTech in 2005 and 2009 with honors and conducted visiting research in the Systems and Synthetic Biology Lab led by Professor Jens Nielsen at Chalmers University of Technology, Sweden, during 2016 and 2017. His recent research focuses on bio-manufacturing pharmaceutical and nutritional chemicals using non-conventional yeast through metabolic engineering and the emerging synthetic biology tools. To date, he has published more than 150 peer-reviewed papers, including Trends in *Biotechnology, Biotechnology Advances, ACS Synthetic Biology, Biotechnology and Bioengineering, AIChE Journal*, etc., and co-edited two books (Elsevier). Currently, he serves as the Associate Editor of *Biotechnology for Biofuels and Bioproducts, Frontiers in Bioengineering and Biotechnology*, etc. He served as the advisory board member of the *European Journal of Lipid Science and Technology*. He has got many awards, such as the Fok Ying-Tung Foundation Young Scholars Award (2014), the Hou Te-Pang Chemical Youth Award of the Chemical Industry and Engineering Society of China (CIESC) (2016), the National Technological Invention Award of China (2018), the Excellent Young Scholars of the National Natural Science Foundation of China (NSFC) (2019), the Newton Advanced Fellowship of the Royal Society (2020), the Green Chemistry Emerging Investigator of the Royal Society of Chemistry (2024), etc.

molecules

Editorial

Synthesis of Natural Products Using Engineered Plants and Microorganisms

Yongjun Wei [1,2,*], Lingbo Qu [2,*] and Xiaojun Ji [3,*]

1 School of Pharmaceutical Sciences, Zhengzhou University, Zhengzhou 450001, China
2 Zhengzhou Research Base, State Key Laboratory of Cotton Bio-Breeding and Integrated Utilization, Laboratory of Synthetic Biology, Zhengzhou University, Zhengzhou 450001, China
3 State Key Laboratory of Materials-Oriented Chemical Engineering, College of Biotechnology and Pharmaceutical Engineering, Nanjing Tech University, Nanjing 211816, China
* Correspondence: yongjunwei@zzu.edu.cn (Y.W.); qlb@zzu.edu.cn (L.Q.); xiaojunji@njtech.edu.cn (X.J.)

Microorganisms and plants, particularly medicinal herbs, are abundant sources of diverse natural products, many of which are bioactive molecules with significant pharmaceutical or health benefits, and include artemisinin [1,2], ginsenosides [3], and *Ganoderic* triterpenes [4]. However, the contents of these bioactive compounds in their natural sources are often low [5]. The advancement of omics technologies and synthetic biology has opened new avenues for producing these valuable molecules through metabolically engineered plants and microorganisms [6]. Notable examples include the scaled-up production of artemisinin, rare ginsenosides, and various other natural products, which have potential applications in drugs, cosmetics, functional foods, animal feed, and other health-related products [7]. There is growing interest in utilizing cutting-edge synthetic biology technologies to further enhance the utilization of these natural products through the engineering of both plants and microorganisms [8].

This Special Issue is a collection of relevant research articles and reviews that highlight recent advancements in the production of bioactive natural products using engineered plants or microorganisms. Zhao, et al. summarized the artemisinin biosynthesis pathway, discussed advances in heterologous production using *Escherichia coli* and *Saccharomyces cerevisiae*, and explored metabolic engineering strategies in *Artemisia annua* and other plants to enhance artemisinin yields [9]. Zhang, et al. examined the pharmacological activities, biosynthesis pathways, extraction methods, and advances in the biotransformation and microbial synthesis of *Epimedium* flavonoids. Moreover, they discussed strategies to boost the industrial-scale production and utilization of *Epimedium* flavonoids [10]. Xu, et al. described the extraction, content determination, bioactivities, and product development of *Lycium ruthenicum* Murray's bioactive substances, and discussed their potential medicinal and economic value [11]. Drożdżyński, et al. explored how plants and their endophytes adapt to harsh environments, focusing on the bioactive metabolites produced by them in synanthropic ruderal plants for potential use in agriculture and medicine [12]. Su, et al., using transcriptome and metabolome tools, revealed that the adaptation of *Citrus reticulata* 'Chachi' to a saline environment enhanced its high production of naringin and narirutin [13].

The function of bioactive natural products in plants and microorganisms are discussed in this Special Issue. Ma, et al. investigated the triterpenoids and ergostane steroids in *Ganoderma luteomarginatum*, which is distributed in Hainan province, China, and their evaluation of several natural products indicated that certain natural products displayed potential anti-tumor effects [14]. Alanazi, et al. found that *Syzygium aromaticum* (clove) oil had the ability to inhibit methicillin-resistant *Staphylococcus aureus* infection [15]. Sun, et al. identified and evaluated the anti-inflammatory constituents in Ardisiae Japonicae Herba (AJH) extracts, and found that ethyl acetate extract (EA) displayed significant anti-inflammatory activity and contained key pharmacodynamic components like bergenin,

Citation: Wei, Y.; Qu, L.; Ji, X. Synthesis of Natural Products Using Engineered Plants and Microorganisms. *Molecules* **2024**, *29*, 5054. https://doi.org/10.3390/molecules29215054

Received: 11 October 2024
Accepted: 18 October 2024
Published: 26 October 2024

quercetin, and epigallocatechingallate, crucial for treating acute lung injury (ALI) [16]. Zhang, et al. found that ginsenoside Rb1, compound K, and 20(S)-protopanaxadiol had the ability to attenuate high-fat-diet-induced hyperlipidemia in rats by modulating the gut microbiota and bile acid metabolism [17].

Sun, et al. successfully increased the yield of 1,4-diaminobutane to 272 mg/L from engineered *E. coli* via the optimization of the synthesis of the cofactors PLP and NADPH [18]. Caballero Cerbon, et al. highlighted that eliminating L-cysteine desulfhydrases is key to improving L-cysteine production in engineered microbes, and outlined several strategies to tackle this issue [19]. Li, et al. demonstrated that biological nano selenium (SeNPs) and sodium selenite (Na_2SeO_3) treatments enhanced the organic selenium content in cowpea pods, and identified that a certain ABC transporter family played a crucial role in selenium absorption and transport [20]. Tu, et al. isolated *Klebsiella aerogenes* TL3 from termite guts, and demonstrated that it could degrade lignin anaerobically through a novel mechanism involving C-O and C-C bond cleavage, side chain oxidation, demethylation, and aromatic ring destruction [21].

Despite significant achievements in the synthesis of natural products using engineered plants and microorganisms, many bioactive natural products remain undeveloped [22]. Therefore, it is essential to leverage interdisciplinary tools, such as synthetic biology and engineering biology, to promote the development of related industries [23]. Through close collaboration between academia and industry, it is anticipated that a greater number of bioactive natural products could be discovered and utilized via engineered plants and microorganisms, thereby promoting technological innovation and market expansion in related industries.

Author Contributions: Conceptualization, Y.W., L.Q. and X.J.; writing—original draft preparation, Y.W.; writing—review and editing, L.Q. and X.J. All authors have read and agreed to the published version of the manuscript.

Funding: This study was supported by the National Natural Science Foundation of China (No. 32371485 and 32111530179), Henan Province Outstanding Youth Foundation (No. 242300421097), Muyuan Laboratory (No. 12106022401), Zhengzhou Research Base, National Key Laboratory of Cotton Bio-breeding and Integrated Utilization (No. CBIU2024002 and CB2024A30), and the State Key Laboratory of Materials-Oriented Chemical Engineering (No. SKL-MCE-23B04).

Conflicts of Interest: The authors declare no conflicts of interest.

References

1. Ro, D.K.; Paradise, E.M.; Ouellet, M.; Fisher, K.J.; Newman, K.L.; Ndungu, J.M.; Ho, K.A.; Eachus, R.A.; Ham, T.S.; Kirby, J.; et al. Production of the antimalarial drug precursor artemisinic acid in engineered yeast. *Nature* **2006**, *440*, 940–943. [CrossRef] [PubMed]
2. Paddon, C.J.; Keasling, J.D. Semi-synthetic artemisinin: A model for the use of synthetic biology in pharmaceutical development. *Nat. Rev. Microbiol.* **2014**, *12*, 355–367. [CrossRef] [PubMed]
3. Yan, X.; Fan, Y.; Wei, W.; Wang, P.; Liu, Q.; Wei, Y.; Zhang, L.; Zhao, G.; Yue, J.; Zhou, Z. Production of bioactive ginsenoside compound K in metabolically engineered yeast. *Cell Res.* **2014**, *24*, 770–773. [CrossRef] [PubMed]
4. Yuan, W.; Jiang, C.; Wang, Q.; Fang, Y.; Wang, J.; Wang, M.; Xiao, H. Biosynthesis of mushroom-derived type II ganoderic acids by engineered yeast. *Nat. Commun.* **2022**, *13*, 7740. [CrossRef]
5. Zhang, X.; Miao, Q.; Pan, C.; Yin, J.; Wang, L.; Qu, L.; Yin, Y.; Wei, Y. Research advances in probiotic fermentation of Chinese herbal medicines. *Imeta* **2023**, *2*, e93. [CrossRef]
6. Choi, K.R.; Jang, W.D.; Yang, D.; Cho, J.S.; Park, D.; Lee, S.Y. Systems Metabolic Engineering Strategies: Integrating Systems and Synthetic Biology with Metabolic Engineering. *Trends Biotechnol.* **2019**, *37*, 817–837. [CrossRef]
7. Ma, J.; Gu, Y.; Marsafari, M.; Xu, P. Synthetic biology, systems biology, and metabolic engineering of Yarrowia lipolytica toward a sustainable biorefinery platform. *J. Ind. Microbiol. Biotechnol.* **2020**, *47*, 845–862. [CrossRef]
8. Chen, R.; Gao, J.; Yu, W.; Chen, X.; Zhai, X.; Chen, Y.; Zhang, L.; Zhou, Y.J. Engineering cofactor supply and recycling to drive phenolic acid biosynthesis in yeast. *Nat. Chem. Biol.* **2022**, *18*, 520–529. [CrossRef] [PubMed]
9. Zhao, L.; Zhu, Y.; Jia, H.; Han, Y.; Zheng, X.; Wang, M.; Feng, W. From Plant to Yeast—Advances in Biosynthesis of Artemisinin. *Molecules* **2022**, *27*, 6888. [CrossRef]
10. Zhang, X.; Tang, B.; Wen, S.; Wang, Y.; Pan, C.; Qu, L.; Yin, Y.; Wei, Y. Advancements in the Biotransformation and Biosynthesis of the Primary Active Flavonoids Derived from Epimedium. *Molecules* **2023**, *28*, 7173. [CrossRef]

11. Xu, M.-L.; He, Y.-F.; Xie, L.; Qu, L.-B.; Xu, G.-R.; Cui, C.-X. Research Progress on Active Ingredients and Product Development of *Lycium ruthenicum* Murray. *Molecules* **2024**, *29*, 2269. [CrossRef]
12. Drożdżyński, P.; Rutkowska, N.; Rodziewicz, M.; Marchut-Mikołajczyk, O. Bioactive Compounds Produced by Endophytic Bacteria and Their Plant Hosts—An Insight into the World of Chosen Herbaceous Ruderal Plants in Central Europe. *Molecules* **2024**, *29*, 4456. [CrossRef] [PubMed]
13. Su, J.; Peng, T.; Bai, M.; Bai, H.; Li, H.; Pan, H.; He, H.; Liu, H.; Wu, H. Transcriptome and Metabolome Analyses Provide Insights into the Flavonoid Accumulation in Peels of Citrus reticulata 'Chachi'. *Molecules* **2022**, *27*, 6476. [CrossRef] [PubMed]
14. Ma, Q.; Zhang, S.; Yang, L.; Xie, Q.; Dai, H.; Yu, Z.; Zhao, Y. Lanostane Triterpenoids and Ergostane Steroids from Ganoderma luteomarginatum and Their Cytotoxicity. *Molecules* **2022**, *27*, 6989. [CrossRef] [PubMed]
15. Alanazi, A.K.; Alqasmi, M.H.; Alrouji, M.; Kuriri, F.A.; Almuhanna, Y.; Joseph, B.; Asad, M. Antibacterial Activity of *Syzygium aromaticum* (Clove) Bud Oil and Its Interaction with Imipenem in Controlling Wound Infections in Rats Caused by Methicillin-Resistant Staphylococcus aureus. *Molecules* **2022**, *27*, 8551. [CrossRef]
16. Sun, S.; Liu, X.; Zhao, D.; Zheng, L.; Han, X.; Tian, Y.; Feng, S. Discovery of the Active Compounds of the Ethyl Acetate Extract Site of *Ardisia japonica* (Thunb.) Blume for the Treatment of Acute Lung Injury. *Molecules* **2024**, *29*, 770. [CrossRef]
17. Zhang, K.-X.; Zhu, Y.; Song, S.-X.; Bu, Q.-Y.; You, X.-Y.; Zou, H.; Zhao, G.-P. Ginsenoside Rb1, Compound K and 20(S)-Protopanaxadiol Attenuate High-Fat Diet-Induced Hyperlipidemia in Rats via Modulation of Gut Microbiota and Bile Acid Metabolism. *Molecules* **2024**, *29*, 1108. [CrossRef]
18. Sun, T.; Zhao, Y.; Wang, J.; Kang, W.; Sun, X.; Sun, Y.; Chu, M.; Liu, Z.; Lu, F.; Li, M. Increasing 1,4-Diaminobutane Production in Escherichia coli by Optimization of Cofactor PLP and NADPH Synthesis. *Molecules* **2024**, *29*, 3094. [CrossRef]
19. Caballero Cerbon, D.A.; Gebhard, L.; Dokuyucu, R.; Ertl, T.; Härtl, S.; Mazhar, A.; Weuster-Botz, D. Challenges and Advances in the Bioproduction of L-Cysteine. *Molecules* **2024**, *29*, 486. [CrossRef]
20. Li, L.; Xiong, Y.; Wang, Y.; Wu, S.; Xiao, C.; Wang, S.; Cheng, S.; Cheng, H. Effect of Nano-Selenium on Nutritional Quality of Cowpea and Response of ABCC Transporter Family. *Molecules* **2023**, *28*, 1398. [CrossRef]
21. Tu, Z.; Geng, A.; Xiang, Y.; Zayas-Garriga, A.; Guo, H.; Zhu, D.; Xie, R.; Sun, J. Lignin Degradation by Klebsiella aerogenes TL3 under Anaerobic Conditions. *Molecules* **2024**, *29*, 2177. [CrossRef] [PubMed]
22. Bureau, J.A.; Oliva, M.E.; Dong, Y.; Ignea, C. Engineering yeast for the production of plant terpenoids using synthetic biology approaches. *Nat. Prod. Rep.* **2023**, *40*, 1822–1848. [CrossRef] [PubMed]
23. Brooks, S.M.; Alper, H.S. Applications, challenges, and needs for employing synthetic biology beyond the lab. *Nat. Commun.* **2021**, *12*, 1390. [CrossRef] [PubMed]

molecules

MDPI

Review

Advancements in the Biotransformation and Biosynthesis of the Primary Active Flavonoids Derived from *Epimedium*

Xiaoling Zhang [1,2,3], Bingling Tang [1,2], Sijie Wen [1,2], Yitong Wang [1,2], Chengxue Pan [1,*], Lingbo Qu [4], Yulong Yin [5] and Yongjun Wei [2,*]

1 School of Pharmaceutical Sciences, Zhengzhou University, Zhengzhou 450001, China
2 Laboratory of Synthetic Biology, School of Pharmaceutical Sciences, Zhengzhou University, Zhengzhou 450001, China
3 Key Laboratory of Food Safety Quick Testing and Smart Supervision Technology for State Market Regulation, Zhengzhou 450003, China
4 College of Chemistry, Zhengzhou University, Zhengzhou 450001, China
5 Institute of Subtropical Agriculture, Chinese Academy of Sciences, Changsha 410081, China
* Correspondence: pancx@zzu.edu.cn (C.P.); yongjunwei@zzu.edu.cn (Y.W.); Tel.: +86-18937160698 (Y.W.)

Abstract: *Epimedium* is a classical Chinese herbal medicine, which has been used extensively to treat various diseases, such as sexual dysfunction, osteoporosis, cancer, rheumatoid arthritis, and brain diseases. Flavonoids, such as icariin, baohuoside I, icaritin, and epimedin C, are the main active ingredients with diverse pharmacological activities. Currently, most *Epimedium* flavonoids are extracted from *Epimedium* plants, but this method cannot meet the increasing market demand. Biotransformation strategies promised huge potential for increasing the contents of high-value *Epimedium* flavonoids, which would promote the full use of the *Epimedium* herb. Complete biosynthesis of major *Epimedium* flavonoids by microbial cell factories would enable industrial-scale production of *Epimedium* flavonoids. This review summarizes the structures, pharmacological activities, and biosynthesis pathways in the *Epimedium* plant, as well as the extraction methods of major *Epimedium* flavonoids, and advancements in the biotransformation and complete microbial synthesis of *Epimedium* flavonoids, which would provide valuable insights for future studies on *Epimedium* herb usage and the production of *Epimedium* flavonoids.

Keywords: *Epimedium*; flavonoids; pharmacological activities; extraction methods; biotransformation; biosynthesis

check for updates

Citation: Zhang, X.; Tang, B.; Wen, S.; Wang, Y.; Pan, C.; Qu, L.; Yin, Y.; Wei, Y. Advancements in the Biotransformation and Biosynthesis of the Primary Active Flavonoids Derived from *Epimedium*. *Molecules* **2023**, *28*, 7173. https://doi.org/10.3390/molecules28207173

Academic Editor: George Grant

Received: 21 September 2023
Revised: 12 October 2023
Accepted: 17 October 2023
Published: 19 October 2023

1. Introduction

The *Epimedium* genus, belonging to the Berberidaceae family, contains 68 species worldwide, with 58 of them (85.3%) distributed in China [1]. China is the center of geographical distribution and varieties of *Epimedium*. Over 15 *Epimedium* species have a long history of use in Traditional Chinese Medicine (TCM) and are believed to have kidney-nourishing and *Yang*-reinforcing properties [2]. Tao Hongjing, a renowned medical scientist, learned from shepherds that male sheep consuming a certain plant experienced significantly increased penile erections and mating frequency. Tao believed that this plant could enhance "*Yang*" energy, and named it "Yin-Yang-Huo" in Chinese [3]. It was later discovered that this was an *Epimedium* plant.

Epimedium was first mentioned over 2000 years ago in the "Shen Nong Ben Cao Jing". It was later listed as a medium-grade herb in the "Ben Cao Gang Mu" by Li Shizhen during the Ming Dynasty [4]. In the *Chinese Pharmacopoeia* (2020 edition), *Epimedii Folium* (EF) refers to the dried leaves of four *Epimedium* plants, namely *E. brevicornum* Maxim, *E. sagittatum* (Sieb. et Zucc.) Maxim, *E. pubescens* Maxim, and *E. koreanum* Nakai [5]. EF is a classical herbal medicine. Alone, or combined within diverse prescriptions, it has been used to treat various diseases, including sexual dysfunction [6,7], osteoporosis [8], cancer [9], rheumatoid arthritis [10], and brain diseases [11]. Additionally, EF has been used in functional food production, and is available in alcoholic health beverages, health tea, and medicated gruel and noodle diets [2,12]. Published reviews have primarily concentrated on the compositions and molecular structures of the active components of EF, as well as the pharmacological activities of EF flavonoids [3,8–11]. In this review, we aim to emphasize the recent advancements in the biotransformation and biosynthesis of EF's active flavonoids.

2. The Pharmacological Activities of Major *Epimedium* Flavonoids

More than 379 compounds have been detected in EF, including flavonoids, lignans, organic acids, terpenoids, dihydrophenanthrene derivatives, alkaloids, and other constituents [13]. Flavonoids, such as epimedin A, epimedin B, epimedin C, icariin, baohuoside I (also known as icariside II), icariside I, and icaritin have been recognized as major phytochemical and pharmacological active ingredients (Figure 1) [13,14]. These compounds differ in varying degrees of glycosylation at the C-3 and C-7 of icaritin [2,14]. There were great variations among the flavonoid contents in *Epimedium* from different species, collection and/or storage times and/or locations [15,16]. Icariin was the most abundant component in *E. brevicornum* Maxim and *E. koreanum* Nakai, followed by epimedin B, epimedin C, and epimedin A. However, epimedin C was the most abundant component in *E. sagittatum* (Sieb. et Zucc.) Maxim, *E. pubescens* Maxim, and *E. wushanense* T.S. Ying, followed by icariin, epimedin B, and epimedin A [15,16]. The average proportions of the total contents of epimedin A, B, C, and icariin to the 15 investigated flavonoid contents were 85.6%, 82.3%, 68.8%, 74.9%, and 69.8% in *E. brevicornum* Maxim, *E. koreanum* Nakai, *E. sagittatum* (Sieb. et Zucc.) Maxim, *E. pubescens* Maxim, and *E. wushanense* T.S. Ying, respectively [15,16]. In another study, epimedin A, B, C, and icariin accounted for over 52% of the total flavonoid contents in *E. brevicornum* Maxim [17]. In the *Chinese Pharmacopoeia* (2020 edition), the total amount of epimedin A, B, C and icariin was identified as the quality control indicator for the EF herb [5]. The major flavonoids in EF exhibit significant and diverse pharmacological activities (Figure 1).

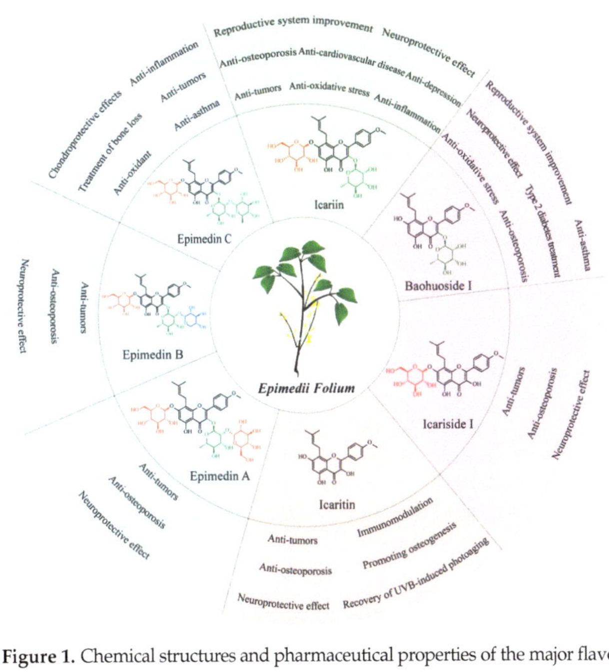

Figure 1. Chemical structures and pharmaceutical properties of the major flavonoids from *Epimedii Folium*.

2.1. Icariin and Its Pharmaceutical Effects

Icariin, the major bioactive component in EF (about 1%) [17], has been found to possess various pharmacological effects. These include improved reproductive system function, a neuroprotective effect, an anti-osteoporosis effect, protective effects from cardiovascular disease, an anti-inflammation effect, an anti-oxidative stress effect, an anti-depressive effect, and an anti-tumor effect [18–21]. In ancient China, EF was commonly used to treat sexual dysfunction [3]. Icariin can enhance erectile function in spontaneously hypertensive rats by reducing endothelial microparticle levels in the blood and inhibiting platelet activation [22]. In male mice, icariin can improve sexual function through the PI3K/AKT/eNOS/NO signaling pathway [23]. In the female reproductive system, icariin promotes estrogen biosynthesis in human ovarian granulosa-like KGN cells, and upregulates the expression of aromatase, which is responsible for the conversion of androgens to estrogens in vertebrates [24]. Additionally, icariin has exhibited protective effects in various nervous system disorders, including Alzheimer's disease, Parkinson's disease, and depressive disorder [20,25–27]. Moreover, icariin is regarded as a potential drug for osteoporosis treatment. Recent studies have demonstrated that icariin could prevent bone loss in ovariectomized rat models by modulating gut microbiota and regulating metabolite alterations [28] or by activating autophagy [29], as well as protect against iron overload-induced bone loss via suppressing oxidative stress [30].

2.2. Baohuoside I and Its Pharmaceutical Effects

Baohuoside I, although presents in low contents (<0.15%) in the raw material of EF compared to icariin, exhibits a wider range of pharmacological activities [31,32]. Baohuoside I has better bioavailability in vivo than icariin, as it is more easily absorbed by the capillaries of intestinal epithelial cells because of its lower polarity [33]. Cheng et al. found that 91.2% of icariin was converted to baohuoside I after oral administration in rats [34]. Similarly, human intestinal microflora metabolized most icariin to baohuoside I in a short time before absorption in the human intestine [35]. Baohuoside I has been proved to have a significant therapeutic effect on various diseases, such as sexual dysfunction, osteoporosis, and cancers [9,36–38]. For improving erectile dysfunction, baohuoside I could facilitate the differentiation of adipose-derived stem cells into Schwann cells and preserve the erectile function of bilateral cavernous nerve injury (BCNI) in rats [36,39]. The anti-osteoporotic activity of baohuoside I was suggested to be associated with its ability to induce bone marrow stromal cell differentiation into osteoblasts while inhibiting adipocyte formation, regulating immune functions, and providing antioxidant activity [40]. Baohuoside I could inhibit osteoclastogenesis and protect against ovariectomy-induced bone loss in mice, surpassing the effects of icariin [38]. Current studies have shown that baohuoside I exhibits promising anti-tumor effects on lung cancer cells [41], melanoma cells [42], breast cancer cells [43], prostate cancer cells [44], and osteosarcoma cells [45]. Furthermore, baohuoside I has shown its potential application in type 2 diabetes treatment [46], neuroprotection [47], and asthma inhibition [48].

2.3. Icaritin and Its Pharmaceutical Effects

Icaritin is a flavonoid aglycone in EF [14], which can be generated by hydrolytic reactions that remove the glycone parts of icariin, baohuoside I, and icariside I [49,50]. Icaritin possesses diverse pharmacological activities [51], including protection of neurons against amyloid-induced neurotoxicity [52], promotion of differentiation from embryonic stem cells into cardiomyocytes [53], anti-osteoporosis effects and osteogenesis promotion [54,55], immunomodulation [56,57], and recovery of UVB-induced photoaging of human keratinocytes [58]. Moreover, icaritin is considered as a promising candidate for the treatment of various cancers [59,60], including hepatocellular carcinoma [61–63], breast cancer [64], lung cancer [65], ovarian cancer [66], endometrial cancer [67], human oral squamous cell carcinoma [68], and multiple hematological malignancies [59,69–71]. In the treatment of hepatocellular carcinoma, icaritin can suppress cell growth and promote cell apoptosis via down-regulating alpha-fetoprotein gene expression in hepatocellular carcinoma [62] and inducing anti-tumor immune responses [61,63]. In 2022, an icaritin soft capsule was marketed as a small molecule immunomodulatory drug, providing a solution for patients with advanced hepatocellular carcinoma with poor prognosis, and significantly improving the life quality of patients with hepatocellular carcinoma [72,73].

2.4. Epimedin C and Its Pharmaceutical Effects

Epimedin C is a trioglycoside ingredient in EF, with the highest content among all flavonol glycosides in certain *Epimedium* species, such as *E. brevicornu* Maxim [74], *E. wushanense* T.S. Ying, and *E. sagittatum* Maxim [75]. Epimedin C is considered as the quality control standard for evaluating the quality of *E. wushanense* T.S. Ying in the *Chinese Pharmacopoeia* (2020 edition) [5]. The pharmacological activities of epimedin C mainly include treatment of bone loss, anti-oxidant effects, and anti-inflammation. Epimedin C has shown significant anti-inflammatory and chondroprotective effects by increasing the expression of extracellular matrix components in osteoarthritis chondrocytes [76]. Epimedin C could alleviate the suppressive impact of dexamethasone on the osteogenesis of larval zebrafish and MC3T3-E1 cells via triggering the PI3K/AKT/RUNX2 signaling pathway [77]. Notably, epimedin C has a stronger anti-osteoporosis effect than icariin at the same dose on dexamethasone-induced osteoporosis in a mouse model [78]. Furthermore, epimedin C has been found to protect against H_2O_2-induced peroxidation injury by enhancing the function of endothelial progenitor human umbilical vein endothelial cells, which plays an important role in repairing endothelial cell vascular injury [79]. In an ovalbumin-induced murine asthma model, epimedin C was demonstrated to dose-dependently decrease the protein levels of p52 and RelB, and the phosphorylation of ERK1/2, and p38 MAPK, which are pivotal in the development of Th9 cells and Treg cells, thereby inhibiting airway inflammation [80].

2.5. Other Flavonoids and Their Pharmaceutical Effects

Other flavonoids presented in EF include epimedin A and B, icariside I, and sagittatoside A, B, and C [2,13]. Their contents in EF are very low, and limited pharmacological research is available on them. However, similar to the major flavonoids described above, icariside I, epimedin A, and epimedin B also exhibit anti-osteoporosis, neuroprotective, and anti-cancer effects. Epimedin A has shown excellent efficacy against senile osteoporosis [8], and in vitro and in vivo experiments demonstrated that a complex epimedin A drug significantly enhances bone regeneration [81]. In addition, epimedin A could ameliorate 2,4-dinitrofluorobenzene (DNFB)-induced allergic contact dermatitis in mice, due to its ability to suppress the NF-κB/NLRP3 pathway, enhance the Nrf2 pathway, and modulate local inflammation [82]. Diao et al. provided evidence that epimedin B ameliorates osteoporosis in male mice via regulating PI3K-Akt, MAPK, and PPAR signaling pathways [83]. Additionally, epimedin B can exert a neuroprotective effect against Parkinson's disease in an MPTP-induced mouse model [84]. Chen et al. suggested that icariside I performed tumor immunotherapy activity by blocking the kynurenine-AhR pathway and tumor immune escape [85]. Icariside I could significantly inhibit B16F10 melanoma growth in vivo through regulation of gut microbiota and host immunity [85]. Moreover, icariside I also effectively ameliorated estrogen deficiency-induced osteoporosis in an ovariectomy mouse model [86].

In addition to the various beneficial effects of EF flavonoids, it is important to note that EF can potentially cause drug-induced liver injury (Table 1). In clinical applications, there are increasing evidences indicate that Zhuangguguanjie pills and Xianlinggubao capsules have toxic effects, leading to liver injury in humans [87,88]. Both medicines contain EF as their major components, and are used to treat rheumatism, bone pain, arthritis, osteoporosis, and other diseases. Recently, animal studies have indicated that EF extracts can cause liver toxicity in mice and rats, with the severity of hepatotoxic effects increasing with higher dosages and prolonged exposure [89,90]. However, the exact compound(s) and the underlying mechanisms contributing to the observed liver toxicity remain unclear. Zhang et al. suggested that icariside I and sagittatoside A are the most relevant compounds related to the hepatotoxicity of EF extracts [91]. Epimedin C has been reported to have potential hepatotoxicity. Song et al. revealed that mRNA methylation might be associated with epimedin C-induced liver injury by the UPLC-MS/MS method [92]. When treated with the normal human liver cell line (HL-7702) and human hepatocellular carcinoma cell line (HepG2), 2″-O-Rhamnosyl icariside II, baohuoside I, and baohuoside II showed significant dose-toxic effects, and baohuoside I was more likely to be involved in the hepatotoxicity of EF [93]. Therefore, the hepatotoxicity of EF, like other TCMs, is probably due to the combined effects of multiple components. Further investigations are needed to fully understand the hepatotoxicity mechanism in order to avoid EF-induced liver injury.

Table 1. The published mechanisms of the hepatotoxicity effects of *Epimedium* flavonoids.

Epimedium Flavonoids	Research Systems	Mechanisms	Reference
Alcohol extracts of *E. koreanum* Nakai and *E. wushanense* T.S. Ying	SD rats	Compared with the normal group, animal groups treated with EF extracts showed severer hepatotoxicity, which was positively correlated with the dose and course. Additionally, the females experienced more significant damage compared to the males.	[90]
Icariside I and sagittatoside A	HL-7702 and HepG2 cells	Icariside I could destroy the cell structure and cause oxidative stress. Sagittatoside A could cause oxidative stress and damage to mitochondria.	[91]
Epimedin C	Male Balb/c mice	Epigenetic modification changed in mouse liver after epimedin C treatment with a test dose, and the m^6A and m^5C may be associated with epimedin C-induced liver injury.	[92]
Baohuoside I	HL-7702 and HepG2 cells	The toxicity mechanism(s) of baohuoside I may be involved in increasing oxidative stress and inducing apoptosis.	[93]
E. koreanum Nakai ethanol extract	Male Sprague Dawley rats	The mechanism of hepatotoxicity of *E. koreanum* Nakai was probably related to the induction of ferroptosis in hepatocytes.	[94]

3. The Biosynthetic Pathway of Prenylated Flavonoids in EF

The biosynthetic pathway of *Epimedium* flavonoids has been explored in *E. sagittatum* Maxim and *E. pubescens* Maxim [95–97]. It can be divided into three phases (Figure 2): phase 1 involves the phenylpropanoid pathway, phase 2 refers to the core pathway, and phase 3 involves further enzymatic modification pathways [97]. The starting precursors for the phenylpropanoid pathway are phenylalanine and tyrosine, which are produced via the shikimate pathway [98]. In plant cells, glucose metabolism produces phosphoenolpyruvate (PEP) and erythrose-4-phosphate, and they are catalyzed by seven enzymes to convert to chorismite, which is the common precursor for the synthesis of phenylalanine and tyrosine [98,99].

In phase 1, phenylalanine is converted to cinnamic acid catalyzed by phenylalanine ammonia lyase (PAL); cinnamic acid is then catalyzed by cinnamate-4-hydroxylase (C4H) to produce *p*-coumaric acid, and tyrosine is converted to *p*-coumaric acid by tyrosine ammonia lyase (TAL) [100]. Then, 4-coumarate CoA ligase (4CL) converts *p*-coumaric acid to 4-coumaroyl-CoA. Another precursor, malonyl-CoA, is mainly derived from acetyl-CoA by acetyl-CoA carboxylase (ACC), and acetyl-CoA is often produced from the classic glucose metabolic pathway. In phase 2, one molecule of 4-coumaryl-CoA and three molecules of malonyl-CoA are condensed by chalcone synthase (CHS) to form naringenin chalcone (a C6-C3-C6 backbone unit for all flavonoids) [97,100]. Subsequently, naringenin chalcone is converted to naringenin by chalcone isomerase (CHI) to complete ring closing. Naringenin is transformed to dihydroflavonol by flavanone 3-hydroxylase (F3H). Flavonol synthase (FLS) further converts dihydroflavonol to flavonols. In phase 3, isopentenyl is added to flavonols by prenyltransferase (PT), and further modifications are carried out by various post-modification enzymes, such as UDP-glycosyltransferase (UGT) and O-methyltransferase (OMT), to produce a series of icaritin flavonoids, including epimedin A, epimedin B, epimedin C, icariin, baohuoside I, icariside I, and icaritin [96,97].

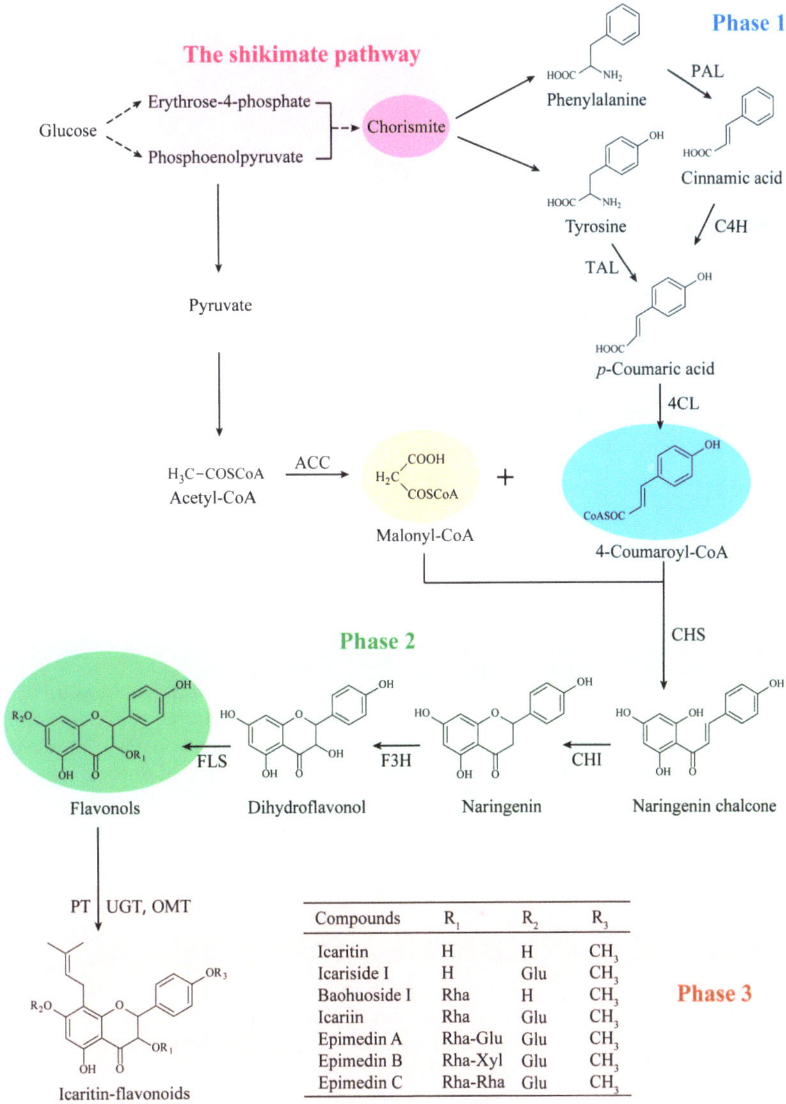

Compounds	R_1	R_2	R_3
Icaritin	H	H	CH_3
Icariside I	H	Glu	CH_3
Baohuoside I	Rha	H	CH_3
Icariin	Rha	Glu	CH_3
Epimedin A	Rha-Glu	Glu	CH_3
Epimedin B	Rha-Xyl	Glu	CH_3
Epimedin C	Rha-Rha	Glu	CH_3

Figure 2. The biosynthetic pathway of prenylated flavonoids recovered from EF. Phase 1: the phenylpropanoid pathway; phase 2: the core pathway; phase 3: further enzymatic modification pathways. PAL: phenylalanine ammonia lyase, C4H: cinnamate-4-hydroxylase, TAL: tyrosine ammonia lyase, 4CL: 4-coumarate CoA ligase, ACC: acetyl-CoA carboxylase, CHS: chalcone synthase, CHI: chalcone isomerase, F3H: flavanone 3-hydroxylase, FLS: flavonol synthase, PT: prenyltransferase, UGT: UDP-glycosyltransferase, OMT: O-methyltransferase, Glu: Glucose, Rha: Rhamnose, Xyl: Xylose.

4. Extraction Methods of *Epimedium* Flavonoids

Currently, commercially available *Epimedium* flavonoids are extracted from *Epimedium* plants. Several techniques have been developed for isolating flavonoids from *Epimedium* (Figure 3), including hot water extraction, alcohol extraction, ultrasonic extraction, microwave-assisted extraction, and ultra-high-pressure extraction. Among these techniques, hot water extraction and alcohol extraction have been implemented in industrial production, while others are at the lab-scale stage.

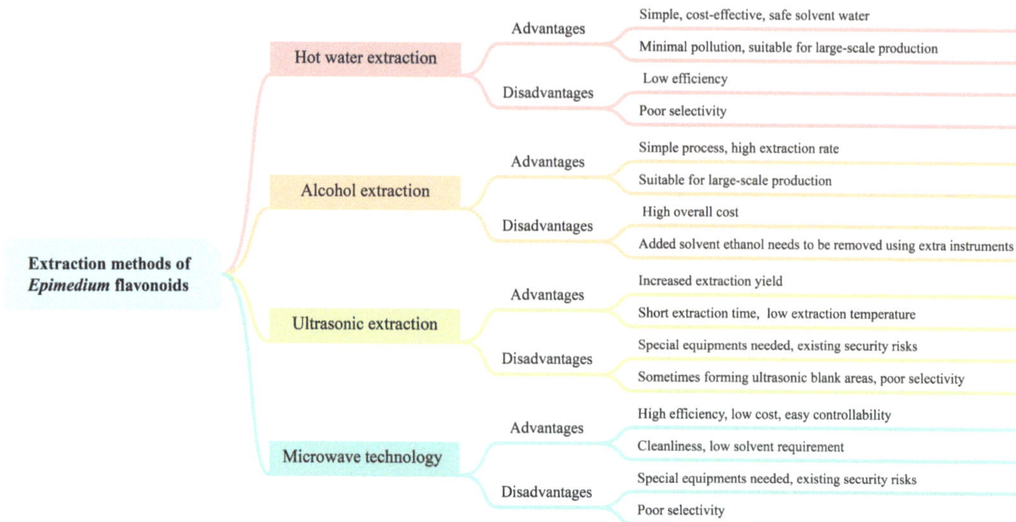

Figure 3. The advantages and disadvantages of different extraction methods of *Epimedium* flavonoids.

4.1. Hot Water Extraction

Hot water extraction is a traditional method used for decocting Chinese herbs. In this method, the crushed herbs are immersed in water in a container for an appropriate amount of time, then heated and gently boiled for a certain period of time. The liquid is subsequently filtered, and the process of decoction is repeated 2–3 times. The decocted liquids from each iteration are mixed and concentrated to achieve the desired flavonoid concentration. Wang et al. optimized the hot water extraction procedure with an orthogonal test [101]. The results showed that the optimized extraction procedure was 2% Na_2CO_3, 15 times the water volume of the weight of dried material, with three 1.5 h extractions. The final extracting ratio of the total flavonoids was 97.92%. Other new technologies, such as microwave technology, have been used to enhance hot water extraction. Compared to the conventional hot water extraction method, microwave-assisted extraction offers higher extraction efficiency and is time-saving [102]. The hot water extraction process is simple and cost-effective, and utilizes water as a safe solvent. The whole process generates minimal pollution. Therefore, hot water extraction is suitable for the large-scale production of flavonoids. However, the efficiency of hot water extraction for flavonoid extraction is low, and it lacks selectivity in capturing specific flavonoids.

4.2. Alcohol Extraction

The alcohol extraction method is the most commonly used technique for extracting flavonoids, adopted by the *Chinese Pharmacopoeia* (2020 edition) [5]. In this method, ethanol is generally employed as the extraction solvent. The process of the alcohol extraction method is relatively simple, and well-suited for industrial applications. However, a large amount of ethanol is added to the extraction reactor, which subsequently needs to be removed using extra instruments. As a result, the overall cost of this method is higher compared to hot water extraction. Zhang et al. demonstrated that the extraction rate of icariin using the alcohol extraction method was significantly higher than that of the water extraction method [75]. The optimal extraction parameters were determined to be 50% ethanol, 1:10 solid–liquid ratio, 60 °C extraction temperature, 2 h extraction time, and two extraction cycles. In addition, an ultrasonic-assisted ethanol extraction procedure has shown to increase the extraction yield of epimedin A, epimedin B, epimedin C, and icariin from *Herba Epimedii*, when compared to the conventional ethanol boiling extraction method [75].

4.3. Other Extraction Methods

Ultrasonic extraction utilizes the effects of strong vibrations, cavitation, and thermal energy generated by ultrasound to extract the active components of plants into solvents. Ultrasonic extraction is regarded as a powerful tool for extracting flavonoids from plant biomass, offering several advantages, such as increased extraction yield, shorter extraction time, and lower extraction temperature [103]. Microwave technology utilizes the ability to generate heat within cells and vaporize water to break down the cell walls, allowing for better release of active ingredients in plant cells. The microwave technique presents numerous benefits, including high efficiency, low energy consumption, short processing time, low cost, cleanliness, easy controllability, and low solvent requirement [104]. Both ultrasonic extraction and the microwave technique are often used to assist common extraction methods, such as water extraction and alcohol extraction, to improve the efficiency of extracting *Epimedium* flavonoids [75,102,105,106]. Furthermore, ultra-high-pressure extraction has also been utilized for extracting flavonoids from *E. sagittatum*. Compared to heating extraction and ultrasonic-assisted extraction, ultra-high-pressure extraction presents distinctive advantages in superior extraction yield and a higher percentage of marker compounds [107].

5. Biotransformation and Biosynthesis of *Epimedium* Flavonoids

The natural resources of medicinal *Epimedium* species have experienced a significant decline over the past several decades, primarily due to over-harvesting and curtailment of habitats [108]. *Epimedium* is a low-growing perennial plant that is commonly propagated by rhizome division because of its low seed viability [109]. *Epimedium* needs specific growth conditions and thrives in acidic soil with shade and relatively high humidity [2]. The growth conditions directly affect the contents of the major components present in *Epimedium* plants, such as *Epimedium* flavonoids. The cultivation of large-scale *Epimedium* plants is challenging [110], as selecting high-quality and stable *Epimedium* species, advancing seedling propagation technology, and enhancing farmers' planting income are difficult [111]. Thus, improving the use efficiency of the existing wild resources of *Epimedium* and exploring new alternative methods for producing *Epimedium* flavonoids to meet the huge market demand are necessary and of great interest.

5.1. Biotransformation of Epimedium Flavonoids by Enzymes

Epimedium flavonoids mainly consist of epimedin A, epimedin B, epimedin C, icariin, baohuoside I, and icaritin. These compounds share a similar structure with the common aglycone skeleton icaritin, while differing in the type and number of lined sugar groups at the C-3 or C-7 positions [2,14]. Notably, icariin and baohuoside I are considered as the most effective components [18,33,112]. However, the contents of icariin and baohuoside I in *Epimedium* plants are extremely rare [17,32]. Therefore, converting other flavonoid components (epimedin A, epimedin B, and epimedin C) into icariin and baohuoside I through enzymatic hydrolysis of the terminal extra sugar groups, provides a feasible strategy to increase their contents in *Epimedium* extracts (Figure 4). Additionally, enzymatic hydrolysis shows enormous potential in the preparation of icariin and baohuoside I due to its notable selectivity, mild reaction conditions, high efficiency, and environmental friendliness [113,114].

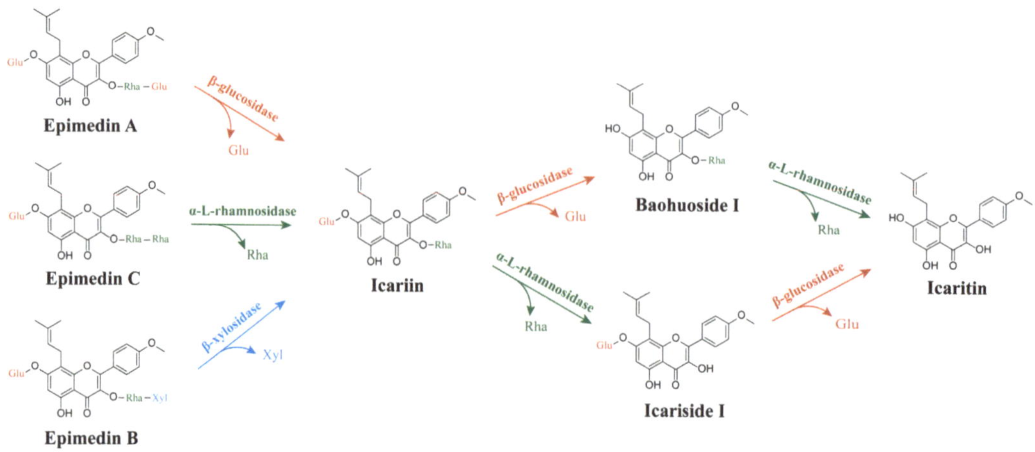

Figure 4. Schematic diagram of the transformation of high glycosylation flavonoids (epimedin A, epimedin B, and epimedin C) into low glycosylation flavonoids (icariin, baohuoside I, icariside I, and icaritin) by different enzymes. This conversion involves the enzymatic hydrolysis of the terminal sugar groups attached to the flavonoid molecules, resulting in the formation of the desired low glycosylation flavonoids.

The transformation of *Epimedium* flavonoids includes the hydrolysis of three types of sugar moieties: the glucose group, rhamnose group, and xylose group (Figures 1 and 4). Three types of enzymes, namely glucosyl hydrolase, rhamnosyl hydrolase, and xylosyl hydrolase, are required to hydrolyze the corresponding sugar moieties (Table 2). High glycosylation flavonoids gradually hydrolyze sugar groups to form low glycosylation flavonoids (Figure 4). For example, epimedin C can be converted to icariin by removing its terminal rhamnose using α-L-rhamnosidase. To produce baohuoside I from icariin, β-glucosidase is employed to hydrolyze the glucose attached to the 7-O position of icariin. The aglycone icaritin can be generated by releasing the rhamnose moiety from the 3-O position of baohuoside I by α-L-rhamnosidase.

Previous studies have identified multiple enzymes capable of facilitating the conversion of *Epimedium* flavonoids (Table 2). Xie et al. discovered a novel thermostable GH78 family α-L-rhamnosidase (PodoRha) from *Paenibacillus odorifer*, which exhibited high specific activity in cleaving the outer α-1,2-rhamnopyranosyl moieties of epimedin C to produce icariin [114]. Another thermostable glucose-tolerant GH1 β-glucosidase (IagBgl1) derived from the hyperthermophile *Ignisphaera aggregans* was found to efficiently produce baohuoside I from icariin [32]. Based on these two enzymes, Xie et al. built a two-step conversion system consisting of PodoRha and IagBgl1 to transform epimedin C into baohuoside I with a conversion rate of 98% under optimized conditions [114]. A fungal α-L-rhamnosidase, AmRha, could hydrolyze the α-1,2-rhamnoside bond between two rhamnoses in epimedin C, achieving the production of icariin with a molar conversion rate of 92.3% in vitro [112]. When AmRha cooperated with β-glucosidase/β-xylosidase Dth3 [115], epimedin A, epimedin B, epimedin C, and icariin in the total *Epimedium* flavonoids could be converted to baohuoside I efficiently and simultaneously [112]. In addition, snailase, a kind of β-D-glucose hydrolase, was reported to contain more than 20 kinds of enzymes, including cellulase, pectinase, hemicellulose, and β-glucuronidase [116]. Snailase was applied to hydrolyze the total *Epimedium* flavonoids to the hydrolysate with enhanced anti-tumor activity [117].

Table 2. Three types of functional enzymes were used to transform *Epimedium* flavonoids.

Enzyme Type	Enzyme Name	Enzyme Source	Enzyme Characterization	Enzyme Functions	References
Rhamnosyl hydrolase	α-L-rhamnosidase (AmRha)	*Aspergillus mulundensis*	107.27 kDa, glycoside hydrolase (GH) 78 family; the optimal activity was achieved at 65 °C and pH 5.5, wide application temperature range, high level of enzyme catalytic ability and stability in the range of pH 5.0–7.5, acting on α-1,2-rhamnoside and α-1,6-rhamnoside bonds directly connected with glucose.	Catalyzes the bioconversion of epimedin C to icariin	[112]
	α-L-rhamnosidase	*Papiliotrema laurentii* ZJU-L07	100 kDa; the optimal activity was achieved at 55° C and pH 7.0, sensitive to temperature, stable at a pH range of 5.5–9.0 with an activity of over 80%, higher selectivity to cleave the α-1,2 glycosidic linkage between glucoside and rhamnoside and the α-1,2 glycosidic linkage between rhamnoside and rhamnoside.	Produces icariin from epimedin C	[118]
	α-L-rhamnosidase (PodoRha)	*Paenibacillus odorifer*	The molecular weight of the monomer was 100.12 kDa, the native recombinant PodoRha was a trimer, GH78 family; the optimal activity was achieved at 45 °C and pH 6.5, a broad range of activity within a pH range of 5.0–8.5, excellent thermostability at 40 °C and 35 °C, high specificity on α-1,2-glycoside in epimedin C.	Converts epimedin C into icariin	[114]
	α-L-rhamnosidase (DthRha)	*Dictyoglomus thermophilum* DSM3960	106.96 kDa, GH78 family; the optimal activity was achieved at 95 °C and pH 6.5, stable within the pH range of 4.5–7.5, residual activities exceeded 50% after incubation at 85 °C for 3 h and exceeded 90% after incubation at 75 °C for 3 h. Efficient hydrolyzation of the α-L-rhamnosidic bond of *Epimedium* flavonoids.	Converts epimedin C into icariin, converts icariin into icariside I, and converts baohuoside I into icaritin	[50]
	α-L-rhamnosidase (^{syn}AnRhaE)	*Aspergillus nidulans*	95 kDa; the optimal activity was achieved at 55 °C and pH 4.5, stable in an acidulous pH range below 55 °C, high specificity on α-1,2 rhamnoside glycosidic bond in epimedin C.	Converts epimedin C into icariin	[119]
	α-L-rhamnosidase (SPRHA2)	*Novosphingobium* sp. GX9	120 kDa, GH106 family, when combined with PBGL, the optimal temperature for the reaction was 55 °C, and the highest activity was observed in 200 mM borate saline buffer at pH 8.5.	Catalyzes icariin into icariside I, converts baohuoside I into icaritin	[14]
	α-L-rhamnosidase (BtRha)	*Bacteroides thetaiotaomicron* VPI-5482	83.3 kDa, GH78 family; the optimal activity was achieved at 55 °C and pH 6.5, high selectivity to cleave the α-1,2 and α-1,6 glycosidic bond between rhamnoside and rhamnoside, rhamnoside and glycoside, respectively.	Transforms epimedin C to icariin	[120]
	α-L-rhamnosidase (Rhase-I)	*Talaromyces stollii* CLY-6	140 kDa, GH106 family; the optimal activity was achieved at 45 °C and pH 4.5, high thermal stability at a temperature lower than 50 °C and superior stability in an acidic environment (pH 2.0–5.0), be activated by Ca^{2+} and Mg^{2+}, efficiently cleaving both the outer and inner rhamnosidic bonds of epimedin C.	Converts epimedin C into icariin, and converts icariin into icariside I	[121]
	α-L-rhamnosidase (AtRha)	*Aspergillus terreus* CCF3059	96.9 kDa, GH78 family; the optimal activity was achieved at 60 °C and pH 6.5, excellent thermal stability and pH stability, hydrolyzed icariin containing the α-1 rhamnoside linkage.	Hydrolyzes icariin to icariside I	[122]

Table 2. *Cont.*

Enzyme Type	Enzyme Name	Enzyme Source	Enzyme Characterization	Enzyme Functions	References
Glucosyl hydrolase	β-glucosidase (Tpebgl3)	*Thermomotoga petrophila* DSM 13,995	81.24 kDa, GH3 family; the optimal activity was achieved at 90 °C and pH 5.0, the thermostability of the enzyme was improved by Ca^{2+}, good stability at high temperatures and organic solvents.	Produces baohuoside I from icariin	[31,123]
	β-glucosidase (IagBgl1)	*Ignisphaera aggregans*	The molecular weight of the monomer was 56.36 kDa, the native recombinant IagBgl1 was a trimer, GH1 family; the optimal activity was achieved at 95 °C and pH 6.5, thermostable and glucose-tolerant, retained more than 70% after incubation at 90 °C for 4 h, high catalytic activity towards icariin.	Produces baohuoside I from icariin	[32]
	β-glucosidase (PBGL)	*Paenibacillus cookii* GX-4	84 kDa, GH3 family, when combined with SPRHA2, the optimal temperature for the reaction was 55 °C, and the highest activity was observed in 200 mM borate saline buffer at pH 8.5.	Converts icariin into baohuoside I, converts icariside I to icaritin	[14]
	β-1,3-glucanase (*Ct*Lam55)	*Chaetomium thermophilum*	82.7 kDa, GH55 family; the optimal activity was achieved at 60 °C and pH 5.0, thermostable at 50 °C, exo-β-1,3-glucanase activity.	Hydrolyzes icariin to baohuoside I	[113,124]
	β-glucosidase (Dth3)	*Dictyoglomus thermophilum* DSM3960	88.7 kDa, GH3 family; the optimal activity was achieved at 90 °C and pH 5.5, highly tolerant to glucose.	Converts epimedin A into baohuoside I, converts icariin into baohuoside I	[112,115]
	β-glucosidase	*Trichoderma viride*	60 kDa; the optimal activity was achieved at 41 °C and pH 4.0.	Prepares baohuoside I from icariin	[125]
	dextranase	-	The optimal activity was achieved at 40 °C and pH 5.4, sensitive to pH.	Hydrolyzes icariin to baohuoside I	[126]
	cellulase	-	The optimum conditions for the cellulase were 50 °C and pH 5.0.	Transforms icariin to baohuoside I	[127]
Xylosyl hydrolase	β-xylosidase (BbXyl)	*Bifidobacterium breve* K-110	70 kDa, GH43 family; the optimal activity was achieved at 45 °C and pH 5.5, the residual activity was more than 80% after being incubated at 45 °C for 4 h, showed over 70% of the maximum activity at a pH from 4.5 to 7.0, higher catalytic efficiency and selection specificity.	Converts epimedin B into icariin	[128]
	β-xylosidase (Dt-2286)	*Dictyoglomus turgidum*	85.1 kDa, GH3 family; the optimal activity was achieved at 98 °C and pH 5.0, excellent thermostable/haloduric/organic solvent-tolerance, a multifunctional enzyme with β-xylosidase, α-arabinofuranoside, α-arabinopyranoside and β-glucosidase activities.	Converts epimedin B into sagittatoside B, and converts sagittatoside B into icariside I	[129]
	β-xylosidase (Dth3)	*Dictyoglomus thermophilum* DSM3960	88.7 kDa, GH3 family; the optimal activity was achieved at 90 °C and pH 5.5, activity was not affected by xylose in high concentration.	Converts epimedin B into baohuoside I	[112,115]

-: Not determined.

5.2. Biotransformation of *Epimedium* Flavonoids by Whole-Cell Catalysis

Icariin, baohuoside I, and icaritin are the most effective components in *Epimedium* [31,118,119,122], and considered as high-value *Epimedium* flavonoids. Enzyme transformation is the preferred method for the preparation of icariin, baohuoside I, and icaritin, due to its high selectivity and catalytic efficiency, fewer by-products, mild reaction conditions, and convenient purification of the products [114,130]. However, enzymes, as a kind of biological macromolecule catalyst, have limitations in the large-scale use in industry, such as easily reduced activity and the high cost associated with enzyme expression and purification [131]. To address these challenges, enzymes are usually combined with carriers to improve their operational stability and reusability [131]. For example, Lu et al. successfully prepared an immobilized enzyme (4LP-Tpebgl3@Na-Y) by fusing the 4LP linker to thermostable β-glucosidase (Tpebgl3) on Na-Y zeolite. This immobilized enzyme exhibited enhanced tolerance of organic solvent and glucose, higher activity, and more reuse cycles, as compared to free enzymes during the production of baohuoside I from icariin [31]. Whole-cell catalysis is an alternative strategy that eliminates enzyme isolation and purification processes, thereby reducing costs (Figure 5). In this strategy, the cells provide a natural and protective environment for the enzymes [132]. Therefore, hydrolysis of flavonoid glycosides by whole-cell catalysis has the potential to be a low-cost strategy while retaining high efficiency.

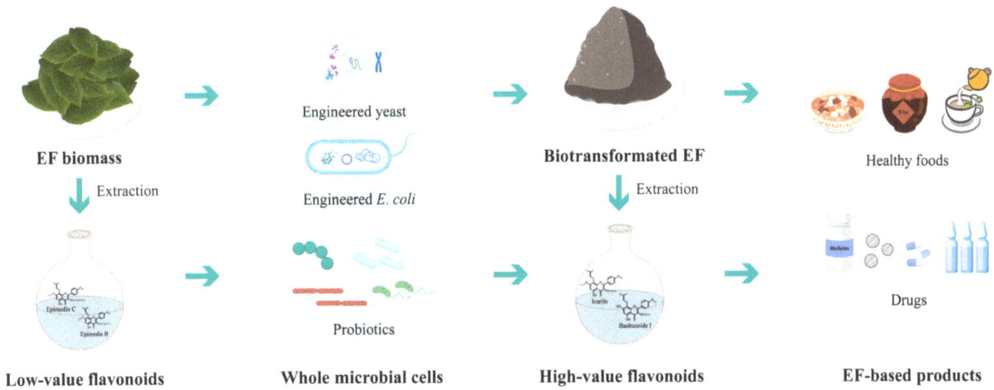

Figure 5. The work flow for biotransformation of *Epimedium* flavonoids by whole-cell catalysis, and their industrial applications.

Several studies have used whole microbial cells to transform *Epimedium* flavonoids. The α-L-rhamnosidase ^{syn}AnRhaE from *A. nidulans* was expressed in *Escherichia coli*, and the recombinant *E. coli* could efficiently transform epimedin C to icariin [119]. Another α-L-rhamnosidase SPRHA2 and β-glucosidase PBGL were expressed in *E. coli*, respectively. The mixture of SPRHA2 cells and PBGL cells was used to transform icariin into icaritin, and the yield rate was 95.23%. Moreover, crude icariin extracts were also efficiently hydrolyzed by the recombinant *E. coli* strain that co-expressed SPRHA2 and PBGL [14]. Additionally, the recombinant *E. coli* strain transformed with α-L-rhamnosidase BtRha and β-xylosidase BbXyl could convert the total flavonoids of *Epimedium* into icariin [128]. Furthermore, the engineered *Komagataella phaffii* GS115 strain expressed with α-L-rhamnosidase AmRha could transform epimedin C to icariin efficiently. This strain also largely removed the feedback inhibition of L-rhamnose and enabled a five-fold increase in the concentration of raw *Epimedium* flavonoids [112]. Whole-cell catalysis, as an intermediate approach between fermentation and in vitro enzyme catalysis, offers several advantages over the use of purified or immobilized enzymes. These include significantly reduced catalyst costs, increased stability due to residual cell wall compounds, and no need for external cofactor addition [133]. Overall, these findings highlight the potential of whole-cell catalysis as a promising approach for future industrial production of high-value *Epimedium* flavonoids.

5.3. Complete Biosynthesis of Epimedium Flavonoids

Biotransformation processes have demonstrated the ability to convert low-value flavonoids into high-value flavonoids. However, the production of natural products from plant biomass requires large amounts of arable land and labor-intensive cultivation. Metabolic engineering and synthetic biology methods have emerged as alternative, simple, and eco-friendly strategies for the production of high-value natural herb products [134,135]. Currently, microbial cell factories have been built to produce diverse natural products, such as artemisinic acid [136], opioids [137], and glycyrrhetinic acid [135].

The biosynthetic pathways of specific *Epimedium* flavonoids have been elucidated, and successful production of icaritin has been achieved. Wang et al. utilized a metabolic engineering strategy to build yeast strains by introducing 11 heterologous genes and modifying 12 native yeast genes. After relocating the methyltransferase GmOMT2 into yeast mitochondria or co-culturing the engineered yeast with an *E. coli* strain expressing GmOMT2, they realized the biosynthesis of icaritin from glucose, and the yields of icaritin were 7.2 and 19.7 mg/L [110].

The biosynthesis of other *Epimedium* flavonoids, such as monoglycoside flavonoids of baohuoside I and icariside I, diglycoside flavonoid of icariin, and trioglycoside flavonoids of epimedin A, B, and C, requires specific glycosyltransferase, rhamnosyltransferase, xylosyltransferase, and other modifying enzymes [97]. Some of these enzymes are not well characterized, and their compatibility with eukaryotic chasses and the sequential reactions involved in *Epimedium* flavonoid biosynthesis have not yet been determined. Despite these challenges, the application of metabolic engineering and synthetic biology holds great potential for advancing the biosynthesis of *Epimedium* flavonoids (Figure 6). Future research efforts focused on characterizing enzymes, improving compatibility with yeasts and other microbial hosts, and optimizing biosynthetic flavonoid pathways will contribute to the de novo production of *Epimedium* flavonoids in microbial cell factories.

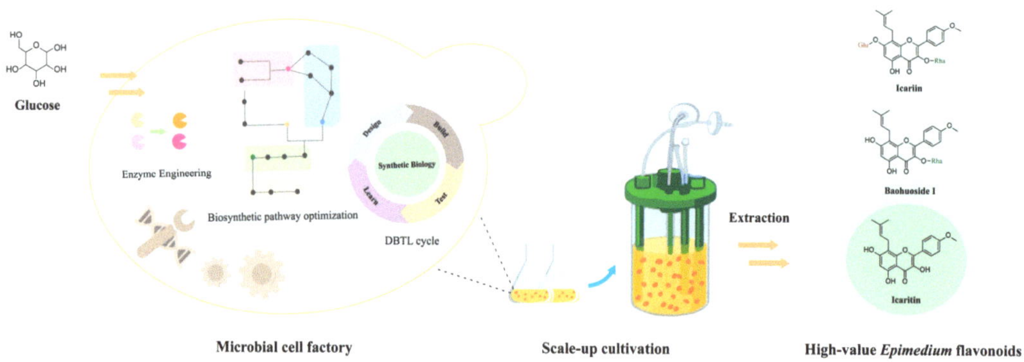

Figure 6. Complete biosynthesis of *Epimedium* flavonoids in a microbial cell factory.

6. Perspective

EF, an important TCM, has been safely used for more than 2000 years to treat various diseases, such as sexual dysfunction, osteoporosis, rheumatoid arthritis, cardiovascular disease, nervous system diseases, and tumors [138–143]. Modern evidences have confirmed that the major active pharmacological ingredients of EF, including epimedin A, epimedin B, epimedin C, icariin, baohuoside I, and icaritin, have corresponding pharmacological activities.

To enhance the production of bioactive high-value *Epimedium* flavonoids, multiple methods have been developed, including biotransformation by enzymes or whole cells [113,115], chemical synthesis [144], and biological synthesis by microbial cell factories [107]. Among these methods, whole-cell catalysis is regarded as a low-cost, stable, highly efficient, and environmentally friendly strategy. Actually, the treatment of Chinese herbs with microorganisms is a traditional technique; TCM fermentation was recorded thousands of years ago in China [145]. Fermenting Chinese herbs with microorganisms, especially probiotics, could promote the release of effective ingredients, reduce toxicities, generate new bioactive substances, enhance bioavailability, and improve pharmacological activities [146]. Fermenting EF with *Lactobacillus plantarum* has shown to increase the relative contents of flavonoid aglycones, monoglycosides, and phenolic acids, while the relative contents of multilevel and secondary flavonoid glycosides decreased, and the antioxidation of fermentation products was significantly increased [147].

Currently, synthetic biology is a powerful tool for the production of complex and high-value natural products using microbial cell factories [136]. However, the recovery of the key enzymes and building of the desirable engineered microorganisms remain challenging tasks. Omics technologies have been widely used to identify new functional genes, while metabolic engineering and metabolic flux analyses help to build complete and efficient natural-product synthetic pathways in microbial cell factories. These advanced technologies would greatly promote the efficient production of Chinese herb-derived bioactive ingredients in the near future, including *Epimedium* flavonoids.

Author Contributions: Y.W. (Yongjun Wei), C.P. and L.Q. conceived the study. X.Z., B.T. and S.W. drafted the manuscript, and X.Z., B.T., S.W. and Y.W. (Yitong Wang) prepared the figures and tables. Y.W. (Yongjun Wei), L.Q., C.P. and Y.Y. revised and polished the manuscript. All authors have read and agreed to the published version of the manuscript.

Funding: This work was supported by the National Natural Science Foundation of China (No. 32101003, 32371485, and 32111530179) and Clinical Laboratories, Shenyou Bio.

Institutional Review Board Statement: Not applicable.

Informed Consent Statement: Not applicable.

Data Availability Statement: Not applicable.

Conflicts of Interest: The authors declare no conflict of interest.

References

1. Xu, Y.Q.; Jiang, Y.; Huang, H.; Li, R.Q.; Li, F.Q.; Liu, Y.; Huang, X.F. Taxonomic study of *Epimedium* L.: Status, issues and prospect. *Guihaia* **2020**, *40*, 601–617.
2. Ma, H.; He, X.; Yang, Y.; Li, M.; Hao, D.; Jia, Z. The genus *Epimedium*: An ethnopharmacological and phytochemical review. *J. Ethnopharmacol.* **2011**, *134*, 519–541. [CrossRef] [PubMed]
3. Niu, Y.; Lin, G.; Pan, J.; Liu, J.; Xu, Y.; Cai, Q.; Wang, T.; Luan, Y.; Chen, Y.; Feng, Y.; et al. Deciphering the myth of icariin and synthetic derivatives in improving erectile function from a molecular biology perspective: A narrative review. *Transl. Androl. Urol.* **2022**, *11*, 1007–1022. [CrossRef] [PubMed]
4. Seyedi, Z.; Amiri, M.S.; Mohammadzadeh, V.; Hashemzadeh, A.; Haddad-Mashadrizeh, A.; Mashreghi, M.; Qayoomian, M.; Hashemzadeh, M.R.; Simal-Gandara, J.; Taghavizadeh Yazdi, M.E. Icariin: A promising natural product in biomedicine and tissue engineering. *J. Funct. Biomater.* **2023**, *14*, 44. [CrossRef] [PubMed]
5. Chinese Pharmacopoeia Commission. *Pharmacopoeia of the People's Republic of China*; China Medical Science Press: Beijing, China, 2020; Volume 1.
6. Punyawudho, B.; Puttilerpong, C.; Wirotsaengthong, S.; Aramwit, P. A randomized, double-blind, placebo-controlled crossover study of Cappra®for the treatment of mild or mild to moderate erectile dysfunction in Thai male. *Afr. J. Tradit. Complement. Altern. Med.* **2012**, *10*, 310–315. [CrossRef]
7. Zhang, H.; Sun, Z.X.; Men, B.; Fu, X.J.; Chen, J.S. Clinical and mechanism research on functional erectile dysfunction treated with moxibustion and qiangshen shugan qiwei decoction. *Zhongguo Zhen Jiu* **2021**, *41*, 1325–1330.
8. Wang, L.; Li, Y.; Guo, Y.; Ma, R.; Fu, M.; Niu, J.; Gao, S.; Zhang, D. *Herba Epimedii*: An ancient Chinese herbal medicine in the prevention and treatment of osteoporosis. *Curr. Pharm. Des.* **2016**, *22*, 328–349. [CrossRef]
9. Chen, M.; Wu, J.; Luo, Q.; Mo, S.; Lyu, Y.; Wei, Y.; Dong, J. The anticancer properties of *Herba Epimedii* and its main bioactive components icariin and icariside II. *Nutrients* **2016**, *8*, 563. [CrossRef]
10. Zhang, L.B.; Yan, Y.; He, J.; Wang, P.P.; Chen, X.; Lan, T.Y.; Guo, Y.X.; Wang, J.P.; Luo, J.; Yan, Z.R.; et al. *Epimedii Herba*: An ancient Chinese herbal medicine in the prevention and treatment of rheumatoid arthritis. *Front. Chem.* **2022**, *10*, 1023779. [CrossRef]

11. Cho, J.H.; Jung, J.Y.; Lee, B.J.; Lee, K.; Park, J.W.; Bu, Y. *Epimedii Herba*: A promising herbal medicine for neuroplasticity. *Phytother. Res.* **2017**, *31*, 838–848. [CrossRef]
12. Zhang, H.F.; Yang, X.H. Application of *Herba Epimedii* in food industry: Current status and prospect. *Sci. Technol. Food Ind.* **2010**, *31*, 390–393.
13. Qian, H.Q.; Wu, D.C.; Li, C.Y.; Liu, X.R.; Han, X.K.; Peng, Y.; Zhang, H.; Zhao, B.Y.; Zhao, Y. A systematic review of traditional uses, phytochemistry, pharmacology and toxicity of *Epimedium koreanum* Nakai. *J. Ethnopharmacol.* **2024**, *318*, 116957. [CrossRef] [PubMed]
14. Lin, Y.; Chen, W.W.; Ding, B.; Guo, M.; Liang, M.; Pang, H.; Wei, Y.T.; Huang, R.B.; Du, L.Q. Highly efficient bioconversion of icariin to icaritin by whole-cell catalysis. *Microb. Cell Fact.* **2023**, *22*, 64. [CrossRef] [PubMed]
15. Chen, X.J.; Ji, H.; Zhang, Q.W.; Tu, P.F.; Wang, Y.T.; Guo, B.L.; Li, S.P. A rapid method for simultaneous determination of 15 flavonoids in *Epimedium* using pressurized liquid extraction and ultra-performance liquid chromatography. *J. Pharm. Biomed. Anal.* **2008**, *46*, 226–235. [CrossRef]
16. Chen, X.J.; Guo, B.L.; Li, S.P.; Zhang, Q.W.; Tu, P.F.; Wang, Y.T. Simultaneous determination of 15 flavonoids in *Epimedium* using pressurized liquid extraction and high-performance liquid chromatography. *J. Chromatogr. A* **2007**, *1163*, 96–104. [CrossRef]
17. Gao, Y.; Shi, W.; Tu, C.; Li, P.; Zhao, G.; Xiao, X.; Wang, J.; Bai, Z. Immunostimulatory activity and structure-activity relationship of epimedin B from *Epimedium brevicornu* Maxim. *Front. Pharmacol.* **2022**, *13*, 1015846. [CrossRef] [PubMed]
18. He, C.; Wang, Z.; Shi, J. Pharmacological effects of icariin. *Adv. Pharmacol.* **2020**, *87*, 179–203.
19. Li, Z.; Li, D.; Chen, R.; Gao, S.; Xu, Z.; Li, N. Cell death regulation: A new way for natural products to treat osteoporosis. *Pharmacol. Res.* **2023**, *187*, 106635. [CrossRef]
20. Wang, S.; Ma, J.; Zeng, Y.; Zhou, G.; Wang, Y.; Zhou, W.; Sun, X.; Wu, M. Icariin, an up-and-coming bioactive compound against neurological diseases: Network pharmacology-based study and literature review. *Drug Des. Devel. Ther.* **2021**, *15*, 3619–3641. [CrossRef]
21. Zeng, Y.; Xiong, Y.; Yang, T.; Wang, Y.; Zeng, J.; Zhou, S.; Luo, Y.; Li, L. Icariin and its metabolites as potential protective phytochemicals against cardiovascular disease: From effects to molecular mechanisms. *Biomed. Pharmacother.* **2022**, *147*, 112642. [CrossRef]
22. Li, X.; Yang, H.F.; Chen, Y.; Pei, L.J.; Jiang, R. Effect of the icariin on endothelial microparticles, endothelial progenitor cells, platelets, and erectile function in spontaneously hypertensive rats. *Andrology* **2022**, *10*, 576–584. [CrossRef]
23. Ding, J.; Tang, Y.; Tang, Z.; Zu, X.; Qi, L.; Zhang, X.; Wang, G. Icariin improves the sexual function of male mice through the PI3K/AKT/eNOS/NO signalling pathway. *Andrologia* **2018**, *50*, e12802. [CrossRef] [PubMed]
24. Yang, L.; Lu, D.; Guo, J.; Meng, X.; Zhang, G.; Wang, F. Icariin from *Epimedium brevicornum* Maxim promotes the biosynthesis of estrogen by aromatase (CYP19). *J. Ethnopharmacol.* **2013**, *145*, 715–721. [CrossRef] [PubMed]
25. Angeloni, C.; Barbalace, M.C.; Hrelia, S. Icariin and its metabolites as potential protective phytochemicals against Alzheimer's Disease. *Front. Pharmacol.* **2019**, *10*, 271. [CrossRef]
26. Wang, G.Q.; Li, D.D.; Huang, C.; Lu, D.S.; Zhang, C.; Zhou, S.Y.; Liu, J.; Zhang, F. Icariin reduces dopaminergic neuronal loss and microglia-mediated inflammation in vivo and in vitro. *Front. Mol. Neurosci.* **2017**, *10*, 441. [CrossRef] [PubMed]
27. Wei, K.; Xu, Y.; Zhao, Z.; Wu, X.; Du, Y.; Sun, J.; Yi, T.; Dong, J.; Liu, B. Icariin alters the expression of glucocorticoid receptor, FKBP5 and SGK1 in rat brains following exposure to chronic mild stress. *Int. J. Mol. Med.* **2016**, *38*, 337–344. [CrossRef] [PubMed]
28. Wang, S.; Wang, S.; Wang, X.; Xu, Y.; Zhang, X.; Han, Y.; Yan, H.; Liu, L.; Wang, L.; Ye, H.; et al. Effects of icariin on modulating gut microbiota and regulating metabolite alterations to prevent bone loss in ovariectomized rat model. *Front. Endocrinol.* **2022**, *13*, 874849. [CrossRef]
29. Liang, X.; Hou, Z.; Xie, Y.; Yan, F.; Li, S.; Zhu, X.; Cai, L. Icariin promotes osteogenic differentiation of bone marrow stromal cells and prevents bone loss in OVX mice via activating autophagy. *J. Cell Biochem.* **2019**, *120*, 13121–13132. [CrossRef] [PubMed]
30. Jing, X.; Du, T.; Chen, K.; Guo, J.; Xiang, W.; Yao, X.; Sun, K.; Ye, Y.; Guo, F. Icariin protects against iron overload-induced bone loss via suppressing oxidative stress. *J. Cell Physiol.* **2019**, *234*, 10123–10137. [CrossRef]
31. Lu, S.; Zou, K.; Guo, B.; Pei, J.; Wang, Z.; Xiao, W.; Zhao, L. One-step purification and immobilization of thermostable β-glucosidase on Na-Y zeolite based on the linker and its application in the efficient production of baohuoside I from icariin. *Bioorg. Chem.* **2022**, *121*, 105690. [CrossRef]
32. Xie, J.; Xu, H.; Jiang, J.; Zhang, N.; Yang, J.; Zhao, J.; Wei, M. Characterization of a novel thermostable glucose-tolerant GH1 β-glucosidase from the hyperthermophile *Ignisphaera aggregans* and its application in the efficient production of baohuoside I from icariin and total epimedium flavonoids. *Bioorg. Chem.* **2020**, *104*, 104296. [CrossRef] [PubMed]
33. Yan, H.; Song, J.; Jia, X.; Zhang, Z. Hyaluronic acid-modified didecyldimethylammonium bromide/ d-a-tocopheryl polyethylene glycol succinate mixed micelles for delivery of baohuoside I against non-small cell lung cancer: In vitro and in vivo evaluation. *Drug Deliv.* **2017**, *24*, 30–39. [CrossRef] [PubMed]
34. Cheng, T.; Zhang, Y.; Zhang, T.; Lu, L.; Ding, Y.; Zhao, Y. Comparative pharmacokinetics study of icariin and icariside II in rats. *Molecules* **2015**, *20*, 21274–21286. [CrossRef] [PubMed]
35. Wu, H.; Kim, M.; Han, J. Icariin metabolism by human intestinal microflora. *Molecules* **2016**, *21*, 1158. [CrossRef]
36. Ge, P.; Guo, Y.; Shen, J. Icariside II facilitates the differentiation of ADSCs to SCs via let-7i/STAT3 axis to preserve erectile function. *Biol. Res.* **2019**, *52*, 54. [CrossRef]

37. Khan, M.; Maryam, A.; Qazi, J.I.; Ma, T. Targeting apoptosis and multiple signaling pathways with icariside II in cancer cells. *Int. J. Biol. Sci.* **2015**, *11*, 1100–1112. [CrossRef]
38. Ma, M.; Fan, A.Y.; Liu, Z.; Yang, L.Q.; Huang, J.M.; Pang, Z.Y.; Yin, F. Baohuoside I inhibits osteoclastogenesis and protects against ovariectomy-induced bone loss. *Front. Pharmacol.* **2022**, *13*, 874952. [CrossRef]
39. Zheng, T.; Zhang, T.; Zhang, W.; Lv, K.; Jia, D.; Yang, F.; Sun, Y.; Lian, J.; Wang, R. Icariside II facilitates the differentiation of ADSCs to schwann cells and restores erectile dysfunction through regulation of miR-33/GDNF axis. *Biomed. Pharmacother.* **2020**, *125*, 109888. [CrossRef]
40. Xi, Y.; Jiang, T.; Yu, J.; Xue, M.; Xu, N.; Wen, J.; Wang, W.; He, H.; Ye, X. Preliminary studies on the anti-osteoporosis activity of Baohuoside I. *Biomed. Pharmacother.* **2019**, *115*, 108850. [CrossRef]
41. Kong, Q.; Ma, M.; Zhang, L.; Liu, S.; He, S.; Wu, J.; Liu, B.; Dong, J. Icariside II potentiates the anti-PD-1 antitumor effect by reducing chemotactic infiltration of myeloid-derived suppressor cells into the tumor microenvironment via ROS-mediated inactivation of the SRC/ERK/STAT3 signaling pathways. *Phytomedicine* **2023**, *110*, 154638. [CrossRef]
42. Peng, Y.G.; Zhang, L. Baohuoside-I suppresses cell proliferation and migration by up-regulating miR-144 in melanoma. *Pharm. Biol.* **2018**, *56*, 43–50. [CrossRef] [PubMed]
43. Wang, S.; Wang, N.; Huang, X.; Yang, B.; Zheng, Y.; Zhang, J.; Wang, X.; Lin, Y.; Wang, Z. Baohuoside i suppresses breast cancer metastasis by downregulating the tumor-associated macrophages/C-X-C motif chemokine ligand 1 pathway. *Phytomedicine* **2020**, *78*, 153331. [CrossRef] [PubMed]
44. Li, S.; Zhan, Y.; Xie, Y.; Wang, Y.; Liu, Y. The impact of icariside II on human prostate cancer cell proliferation, mobility, and autophagy via PI3K-AKT-mTOR signaling pathway. *Drug Des. Devel. Ther.* **2020**, *14*, 4169–4178. [CrossRef] [PubMed]
45. Choi, H.J.; Eun, J.S.; Kim, D.K.; Li, R.H.; Shin, T.Y.; Park, H.; Cho, N.P.; Soh, Y. Icariside II from *Epimedium koreanum* inhibits hypoxia-inducible factor-1alpha in human osteosarcoma cells. *Eur. J. Pharmacol.* **2008**, *579*, 58–65. [CrossRef] [PubMed]
46. Kim, D.H.; Jung, H.A.; Sohn, H.S.; Kim, J.W.; Choi, J.S. Potential of icariin metabolites from *Epimedium koreanum* Nakai as antidiabetic therapeutic agents. *Molecules* **2017**, *22*, 986. [CrossRef]
47. Gao, J.; Ma, C.; Xia, D.; Chen, N.; Zhang, J.; Xu, F.; Li, F.; He, Y.; Gong, Q. Icariside II preconditioning evokes robust neuroprotection against ischaemic stroke, by targeting Nrf2 and the OXPHOS/NF-κB/ferroptosis pathway. *Br. J. Pharmacol.* **2023**, *180*, 308–329. [CrossRef]
48. Zhou, Y.; Huang, X.; Yu, H.; Shi, H.; Chen, M.; Song, J.; Tang, W.; Teng, F.; Li, C.; Yi, L.; et al. TMT-based quantitative proteomics revealed protective efficacy of Icariside II against airway inflammation and remodeling via inhibiting LAMP2, CTSD and CTSS expression in OVA-induced chronic asthma mice. *Phytomedicine* **2023**, *118*, 154941. [CrossRef]
49. Wang, Z.; Liu, C.; Yu, H.; Wu, B.; Huai, B.; Zhuang, Z.; Sun, C.; Xu, L.; Jin, F. Icaritin preparation from icariin by a special *Epimedium* flavonoid-glycosidase from *Aspergillus* sp. y848 strain. *J. Microbiol. Biotechnol.* **2022**, *32*, 437–446. [CrossRef]
50. Zhang, S.; Luo, J.; Dong, Y.; Wang, Z.; Xiao, W.; Zhao, L. Biotransformation of the total flavonoid extract of epimedium into icaritin by two thermostable glycosidases from *Dictyoglomus thermophilum* DSM3960. *Process Biochem.* **2021**, *105*, 8–18. [CrossRef]
51. Huong, N.T.; Son, N.T. Icaritin: A phytomolecule with enormous pharmacological values. *Phytochemistry* **2023**, *213*, 113772. [CrossRef]
52. Wang, Z.; Zhang, X.; Wang, H.; Qi, L.; Lou, Y. Neuroprotective effects of icaritin against beta amyloid-induced neurotoxicity in primary cultured rat neuronal cells via estrogen-dependent pathway. *Neuroscience* **2007**, *145*, 911–922. [CrossRef] [PubMed]
53. Zhu, D.Y.; Lou, Y.J. Inducible effects of icariin, icaritin, and desmethylicaritin on directional differentiation of embryonic stem cells into cardiomyocytes in vitro. *Acta Pharmacol. Sin.* **2005**, *26*, 477–485. [CrossRef]
54. Gao, L.; Zhang, S.Q. Antiosteoporosis effects, pharmacokinetics, and drug delivery systems of icaritin: Advances and prospects. *Pharmaceuticals* **2022**, *15*, 397. [CrossRef] [PubMed]
55. Wei, Q.; Wang, B.; Hu, H.; Xie, C.; Ling, L.; Gao, J.; Cao, Y. Icaritin promotes the osteogenesis of bone marrow mesenchymal stem cells via the regulation of sclerostin expression. *Int. J. Mol. Med.* **2020**, *45*, 816–824. [CrossRef] [PubMed]
56. Bi, Z.; Zhang, W.; Yan, X. Anti-inflammatory and immunoregulatory effects of icariin and icaritin. *Biomed. Pharmacother.* **2022**, *151*, 113180. [CrossRef]
57. Liao, J.; Liu, Y.; Wu, H.; Zhao, M.; Tan, Y.; Li, D.; Long, H.; Dai, Y.; Yung, S.; Chan, T.M.; et al. The role of icaritin in regulating Foxp3/IL17a balance in systemic lupus erythematosus and its effects on the treatment of MRL/lpr mice. *Clin. Immunol.* **2016**, *162*, 74–83. [CrossRef]
58. Hwang, E.; Lin, P.; Ngo, H.T.T.; Gao, W.; Wang, Y.S.; Yu, H.S.; Yi, T.H. Icariin and icaritin recover UVB-induced photoaging by stimulating Nrf2/ARE and reducing AP-1 and NF-κB signaling pathways: A comparative study on UVB-irradiated human keratinocytes. *Photochem. Photobiol. Sci.* **2018**, *17*, 1396–1408. [CrossRef]
59. Yang, X.J.; Xi, Y.M.; Li, Z.J. Icaritin: A novel natural candidate for hematological malignancies therapy. *Biomed. Res. Int.* **2019**, *2019*, 4860268. [CrossRef]
60. Zhang, C.; Sui, X.; Jiang, Y.; Wang, X.; Wang, S. Antitumor effects of icaritin and the molecular mechanisms. *Discov. Med.* **2020**, *29*, 5–16.
61. Fan, Y.; Li, S.; Ding, X.; Yue, J.; Jiang, J.; Zhao, H.; Hao, R.; Qiu, W.; Liu, K.; Li, Y.; et al. First-in-class immune-modulating small molecule Icaritin in advanced hepatocellular carcinoma: Preliminary results of safety, durable survival and immune biomarkers. *BMC Cancer* **2019**, *19*, 279. [CrossRef]

62. Li, H.; Liu, Y.; Jiang, W.; Xue, J.; Cheng, Y.; Wang, J.; Yang, R.; Zhang, X. Icaritin promotes apoptosis and inhibits proliferation by down-regulating AFP gene expression in hepatocellular carcinoma. *BMC Cancer* **2021**, *21*, 318. [CrossRef]
63. Tao, H.; Liu, M.; Wang, Y.; Luo, S.; Xu, Y.; Ye, B.; Zheng, L.; Meng, K.; Li, L. Icaritin induces anti-tumor immune responses in hepatocellular carcinoma by inhibiting splenic myeloid-derived suppressor cell generation. *Front. Immunol.* **2021**, *12*, 609295. [CrossRef]
64. Wang, X.; Zheng, N.; Dong, J.; Wang, X.; Liu, L.; Huang, J. Estrogen receptor-α36 is involved in icaritin induced growth inhibition of triple-negative breast cancer cells. *J. Steroid Biochem. Mol. Biol.* **2017**, *171*, 318–327. [CrossRef] [PubMed]
65. Zheng, Q.; Liu, W.W.; Li, B.; Chen, H.J.; Zhu, W.S.; Yang, G.X.; Chen, M.J.; He, G.Y. Anticancer effect of icaritin on human lung cancer cells through inducing S phase cell cycle arrest and apoptosis. *J. Huazhong Univ. Sci. Technolog. Med. Sci.* **2014**, *34*, 497–503. [CrossRef] [PubMed]
66. Gao, L.; Chen, M.; Ouyang, Y.; Li, R.; Zhang, X.; Gao, X.; Lin, S.; Wang, X. Icaritin induces ovarian cancer cell apoptosis through activation of p53 and inhibition of Akt/mTOR pathway. *Life Sci.* **2018**, *202*, 188–194. [CrossRef]
67. Jin, Y.B.; Liang, X.C.; Cai, J.H.; Wang, K.; Wang, C.Y.; Wang, W.H.; Chen, X.L.; Bao, S. Mechanism of action of icaritin on uterine corpus endometrial carcinoma based on network pharmacology and experimental evaluation. *Front. Oncol.* **2023**, *13*, 1205604. [CrossRef] [PubMed]
68. Yang, J.G.; Lu, R.; Ye, X.J.; Zhang, J.; Tan, Y.Q.; Zhou, G. Icaritin reduces oral squamous cell carcinoma progression via the inhibition of STAT3 signaling. *Int. J. Mol. Sci.* **2017**, *18*, 132. [CrossRef]
69. Wu, T.; Wang, S.; Wu, J.; Lin, Z.; Sui, X.; Xu, X.; Shimizu, N.; Chen, B.; Wang, X. Icaritin induces lytic cytotoxicity in extranodal NK/T-cell lymphoma. *J. Exp. Clin. Cancer Res.* **2015**, *34*, 17. [CrossRef]
70. Zhu, J.; Li, Z.; Zhang, G.; Meng, K.; Kuang, W.; Li, J.; Zhou, X.; Li, R.; Peng, H.; Dai, C.; et al. Icaritin shows potent anti-leukemia activity on chronic myeloid leukemia in vitro and in vivo by regulating MAPK/ERK/JNK and JAK2/STAT3/AKT signalings. *PLoS ONE* **2011**, *6*, e23720. [CrossRef]
71. Zhu, S.; Wang, Z.; Li, Z.; Peng, H.; Luo, Y.; Deng, M.; Li, R.; Dai, C.; Xu, Y.; Liu, S.; et al. Icaritin suppresses multiple myeloma, by inhibiting IL-6/JAK2/STAT3. *Oncotarget* **2015**, *6*, 10460–10472. [CrossRef]
72. Liu, X.; Yang, F.; Jia, D.; Dong, X.; Zhang, Y.; Wu, Z. Case report: A case study on the treatment using icaritin soft capsules in combination with lenvatinib achieving impressive PR and stage reduction in unresectable locally progressive pancreatic cancer and a literature review. *Front. Genet.* **2023**, *14*, 1167470. [CrossRef]
73. Tang, X.; Zhang, Y.; Dong, X.; Jiang, G.; Hong, D.; Liu, X. The synergy of gene targeting drug icaritin soft capsule with immunomodulator and TACE brings new hope for drug combination in patients with advanced liver cancer: A case report and literature review. *Cancer Manag. Res.* **2023**, *15*, 707–717. [CrossRef] [PubMed]
74. Guo, Y.; Wang, X.; Gao, J. Simultaneous preparation and comparison of the osteogenic effects of epimedins A–C and icariin from *Epimedium brevicornu*. *Chem. Biodivers.* **2018**, *15*, e1700578. [CrossRef] [PubMed]
75. Zhang, H.F.; Yang, T.S.; Li, Z.Z.; Wang, Y. Simultaneous extraction of epimedin A, B, C and icariin from *Herba Epimedii* by ultrasonic technique. *Ultrason. Sonochem.* **2008**, *15*, 376–385. [CrossRef] [PubMed]
76. Ziadlou, R.; Barbero, A.; Martin, I.; Wang, X.; Qin, L.; Alini, M.; Grad, S. Anti-inflammatory and chondroprotective effects of vanillic acid and epimedin C in human osteoarthritic chondrocytes. *Biomolecules* **2020**, *10*, 932. [CrossRef]
77. Xu, Y.; Chen, S.; Huang, L.; Han, W.; Shao, Y.; Chen, M.; Zhang, Y.; He, R.; Xie, B. Epimedin C alleviates glucocorticoid-induced suppression of osteogenic differentiation by modulating PI3K/AKT/RUNX2 signaling pathway. *Front. Pharmacol.* **2022**, *13*, 894832. [CrossRef] [PubMed]
78. Liu, Y.L.; Huang, M.; Feng, J.; Xia, P.; Wang, Y.; Wei, X.; Qiu, L. Effects of icariin and epimedium C on microstructure of bone tissue in glucocorticoid osteoporosis model mice based on Micro-CT technique. *Drug Eval. Res.* **2020**, *43*, 1733–1739.
79. Wei, D.H.; Deng, J.L.; Shi, R.Z.; Ma, L.; Shen, J.M.; Hoffman, R.; Hu, Y.H.; Wang, H.; Gao, J.L. Epimedin C protects H_2O_2-induced peroxidation injury by enhancing the function of endothelial progenitor HUVEC populations. *Biol. Pharm. Bull.* **2019**, *42*, 1491–1499. [CrossRef]
80. Huang, M.; Wei, Y.; Dong, J. Epimedin C modulates the balance between Th9 cells and Treg cells through negative regulation of noncanonical NF-κB pathway and MAPKs activation to inhibit airway inflammation in the ovalbumin-induced murine asthma model. *Pulm. Pharmacol. Ther.* **2020**, *65*, 102005. [CrossRef]
81. Liu, Y.; Bi, Y.; Chai, L.; Song, L.; Huang, J.; Wang, Q.; Li, Y.; Zhou, K. Development of epimedin A complex drugs for treating the osteoporosis. *J. Mater. Sci. Mater. Med.* **2021**, *32*, 17. [CrossRef]
82. Balaha, M.F.; Ahmed, N.J.; Almalki, Z.S.; Alahmari, A.K.; Alshehri, A.M.; Soliman, G.A.; Hamad, A.M. Epimedin A ameliorates DNFB-induced allergic contact dermatitis in mice: Role of NF-κ B/NLRP3-driven pyroptosis, Nrf2/HO-1 pathway, and inflammation modulation. *Life Sci.* **2022**, *302*, 120653. [CrossRef] [PubMed]
83. Diao, X.; Wang, L.; Zhou, Y.; Bi, Y.; Zhou, K.; Song, L. The mechanism of Epimedin B in treating osteoporosis as revealed by RNA sequencing-based analysis. *Basic Clin. Pharmacol. Toxicol.* **2021**, *129*, 450–461. [CrossRef]
84. Zhang, M.; Hu, Z.F.; Dong, X.L.; Chen, W.F. Epimedin B exerts neuroprotective effect against MPTP-induced mouse model of Parkinson's disease: GPER as a potential target. *Biomed. Pharmacother.* **2022**, *156*, 113955. [CrossRef]
85. Chen, G.; Huang, J.; Lei, H.; Wu, F.; Chen, C.; Song, Y.; Cao, Z.; Zhang, C.; Zhang, C.; Ma, Y.; et al. Icariside I—A novel inhibitor of the kynurenine-AhR pathway with potential for cancer therapy by blocking tumor immune escape. *Biomed. Pharmacother.* **2022**, *153*, 113387. [CrossRef]

86. Chen, C.; Wu, M.; Lei, H.; Cao, Z.; Wu, F.; Song, Y.; Zhang, C.; Qin, M.; Zhang, C.; Du, R.; et al. A novel prenylflavonoid icariside I ameliorates estrogen deficiency-induced osteoporosis via simultaneous regulation of osteoblast and osteoclast differentiation. *ACS Pharmacol. Transl. Sci.* **2023**, *6*, 270–280. [CrossRef] [PubMed]
87. Cheng, J.; Cai, H. Adverse reactions to Zhuangguguanjie Wan and cause analysis. *Advers. Drug React. J.* **2000**, *2*, 15–19.
88. Du, Q.; Wang, Z.; Yun, N.R.; Huang, Y.H.; Xu, Q.; Wang, B.H. Literature analysis of 185 cases of ADR induced by Xianling Gubao capsule. *China Pharm.* **2017**, *28*, 3785–3787.
89. Wang, Q.; Zhang, P.Y.; Yuan, X.M.; Bi, Y.N.; Zhou, K.; Zhang, Y. Long-term toxicity of different extracts of *Epimedium brevicornu* maxim in mice. *Chin. J. Pharmacovigil.* **2018**, *15*, 65–69.
90. Zhang, L.; Zhang, J.X.; Fan, Q.Y.; Su, Z.Q.; Chen, C.; Peng, L.; Wang, T. Hepatoxicity of *Epimedii folium* in rat model based on uniform design and regression analysis. *Chin. J. Exp. Tradit. Med. Formulae.* **2018**, *24*, 189–197.
91. Zhang, L.; Xu, A.L.; Yang, S.; Zhao, B.S.; Wang, T. In vitro screening and toxic mechanism exploring of leading components with potential hepatotoxicity of *Herba Epimedii* extracts. *Toxicol. In Vitro* **2020**, *62*, 104660. [CrossRef]
92. Song, Z.; Li, Z.; Wen, X.; Liu, R.; Tian, X. UPLC-MS/MS method for simultaneously determining nucleosides and methyl-nucleosides in liver mRNA of Epimedin C-induced liver injury mouse model. *Chin. Med.* **2021**, *16*, 91. [CrossRef] [PubMed]
93. Zhang, L.; Wang, T.; Zhao, B.S.; Zhang, J.X.; Yang, S.; Fan, C.L.; Li, P. Effect of 2″-O-rhamnosyl icariside II, baohuoside I and baohuoside II in *Herba Epimedii* on cytotoxicity indices in HL-7702 and HepG2 Cells. *Molecules* **2019**, *24*, 1263. [CrossRef] [PubMed]
94. Li, P.; Zhang, L.; Guo, Z.; Kang, Q.; Chen, C.; Liu, X.; Ma, Q.; Zhang, J.; Hu, Y.; Wang, T. *Epimedium koreanum* Nakai-induced liver injury-A mechanistic study using untargeted metabolomics. *Front. Pharmacol.* **2022**, *13*, 934057. [CrossRef] [PubMed]
95. Huang, W.; Zeng, S.; Xiao, G.; Wei, G.; Liao, S.; Chen, J.; Sun, W.; Lv, H.; Wang, Y. Elucidating the biosynthetic and regulatory mechanisms of flavonoid-derived bioactive components in *Epimedium sagittatum*. *Front. Plant. Sci.* **2015**, *6*, 689. [CrossRef] [PubMed]
96. Shen, G.; Luo, Y.; Yao, Y.; Meng, G.; Zhang, Y.; Wang, Y.; Xu, C.; Liu, X.; Zhang, C.; Ding, G.; et al. The discovery of a key prenyltransferase gene assisted by a chromosome-level *Epimedium pubescens* genome. *Front. Plant Sci.* **2022**, *13*, 1034943. [CrossRef] [PubMed]
97. Xu, C.; Liu, X.; Shen, G.; Fan, X.; Zhang, Y.; Sun, C.; Suo, F.; Guo, B. Time-series transcriptome provides insights into the gene regulation network involved in the icariin-flavonoid metabolism during the leaf development of *Epimedium pubescens*. *Front. Plant Sci.* **2023**, *14*, 1183481. [CrossRef]
98. Mir, R.; Jallu, S.; Singh, T.P. The shikimate pathway: Review of amino acid sequence, function and three-dimensional structures of the enzymes. *Crit. Rev. Microbiol.* **2015**, *41*, 172–189. [CrossRef]
99. Maeda, H.; Dudareva, N. The shikimate pathway and aromatic amino acid biosynthesis in plants. *Annu. Rev. Plant Biol.* **2012**, *63*, 73–105. [CrossRef]
100. Pandey, R.P.; Parajuli, P.; Koffas, M.A.G.; Sohng, J.K. Microbial production of natural and non-natural flavonoids: Pathway engineering, directed evolution and systems/synthetic biology. *Biotechnol. Adv.* **2016**, *34*, 634–662. [CrossRef]
101. Wang, Q.; Li, Z.; Ren, X.X. Optimization of the water-extraction method of total flavonoids from *Epimedium sagittatum* Maxin. *Tianjin Agric. Sci.* **2012**, *18*, 3.
102. Huang, R.H.; Zhou, Y.C.; Han, W.; Deng, X. Study on water extraction process of *Herba epimedii* with microwave technology. *Zhongguo Zhong Yao Za Zhi.* **2005**, *30*, 107–110.
103. Kazemi, M.; Khodaiyan, F.; Hosseini, S.S. Eggplant peel as a high potential source of high methylated pectin: Ultrasonic extraction optimization and characterization. *LWT* **2019**, *105*, 182–189. [CrossRef]
104. Spinei, M.; Oroian, M. Microwave-assisted extraction of pectin from grape pomace. *Sci. Rep.* **2022**, *12*, 12722. [CrossRef] [PubMed]
105. Yang, X.H.; Li, L.; Xue, Y.B.; Zhou, X.X.; Tang, J.H. Flavonoids from *Epimedium pubescens*: Extraction and mechanism, antioxidant capacity and effects on CAT and GSH-Px of *Drosophila melanogaster*. *PeerJ* **2020**, *8*, e8361. [CrossRef] [PubMed]
106. Karbuz, P.; Tugrul, N. Microwave and ultrasound assisted extraction of pectin from various fruits peel. *J. Food Sci. Technol.* **2021**, *58*, 641–650. [CrossRef] [PubMed]
107. Zhang, R.; Su, D.; Hou, F.; Liu, L.; Huang, F.; Dong, L.; Deng, Y.; Zhang, Y.; Wei, Z.; Zhang, M. Optimized ultra-high-pressure-assisted extraction of procyanidins from lychee pericarp improves the antioxidant activity of extracts. *Biosci. Biotechnol. Biochem.* **2017**, *81*, 1576–1585. [CrossRef]
108. Xu, Y.; Li, Z.; Wang, Y. Fourteen microsatellite loci for the Chinese medicinal plant *Epimedium sagittatum* and cross-species application in other medicinal species. *Mol. Ecol. Resour.* **2008**, *8*, 640–642. [CrossRef]
109. Mihaljević, S.; Vršek, I. In vitro shoot regeneration from immature seeds of *Epimedium alpinum* induced by thidiazuron and CPPU. *Sci. Hortic.* **2009**, *120*, 406–410. [CrossRef]
110. Wang, P.; Li, C.; Li, X.; Huang, W.; Wang, Y.; Wang, J.; Zhang, Y.; Yang, X.; Yan, X.; Wang, Y.; et al. Complete biosynthesis of the potential medicine icaritin by engineered *Saccharomyces cerevisiae* and *Escherichia coli*. *Sci. Bull.* **2021**, *66*, 1906–1916. [CrossRef]
111. Yin, T.; Li, Y.F.; Xu, N.; Meng, F.Y. Current situation and thinking of artificial cultivation of *Epimedium*. *Lishizhen Med. Mater. Medica Res.* **2020**, *31*, 1468–1471.
112. Zhang, S.; Lu, C.; Cao, S.; Li, Q.; Wu, G.; Zhao, L. Efficient production of icariin and baohuoside I from *Epimedium Folium* flavonoids by fungal α-L-rhamnosidase hydrolysing regioselectively the terminal rhamnose of epimedin C. *Biotechnol. Biofuels Bioprod.* **2023**, *16*, 107. [CrossRef]

113. Shen, Y.; Wang, M.; Zhou, J.; Chen, Y.; Xu, L.; Wu, M.; Xia, G.; Tam, J.P.; Yu, J.; Teng, X.; et al. Eco-efficient biphasic enzymatic hydrolysis for the green production of rare baohuoside I. *Enzyme Microb. Technol.* **2019**, *131*, 109431. [CrossRef]
114. Xie, J.; Zhao, J.; Zhang, N.; Xu, H.; Yang, J.; Ye, J.; Jiang, J. Efficient production of isoquercitin, icariin and icariside II by a novel thermostable α-l-rhamnosidase PodoRha from *Paenibacillus odorifer* with high α-1,6-/α-1,2-glycoside specificity. *Enzyme Microb. Technol.* **2022**, *158*, 110039. [CrossRef]
115. Li, Q.; Wu, T.; Zhao, L.; Pei, J.; Wang, Z.; Xiao, W. Highly efficient biotransformation of astragaloside IV to cycloastragenol by sugar-stimulated β-glucosidase and β-xylosidase from *Dictyoglomus thermophilum*. *J. Microbiol. Biotechnol.* **2019**, *29*, 1882–1893. [CrossRef]
116. Jin, X.; Zhang, Z.; Sun, E.; Li, S.; Jia, X. Statistically designed enzymatic hydrolysis of an icariin/β-cyclodextrin inclusion complex optimized for production of icaritin. *Acta Pharm. Sin. B* **2012**, *2*, 83–89. [CrossRef]
117. Liu, C.; Li, R.; Peng, J.; Qu, D.; Huang, M.; Chen, Y. Enhanced hydrolysis and antitumor efficacy of *Epimedium* flavonoids mediated by immobilized snailase on silica. *Process Biochem.* **2019**, *86*, 80–88. [CrossRef]
118. Lou, H.; Liu, X.; Liu, S.; Chen, Q. Purification and characterization of a novel α-L-rhamnosidase from *Papiliotrema laurentii* ZJU-L07 and its application in production of icariin from epimedin C. *J. Fungi* **2022**, *8*, 644. [CrossRef]
119. Lyu, Y.; Zeng, W.; Du, G.; Chen, J.; Zhou, J. Efficient bioconversion of epimedin C to icariin by a glycosidase from *Aspergillus nidulans*. *Bioresour. Technol.* **2019**, *289*, 121612. [CrossRef]
120. Wu, T.; Pei, J.; Ge, L.; Wang, Z.; Ding, G.; Xiao, W.; Zhao, L. Characterization of a α-l-rhamnosidase from *Bacteroides thetaiotaomicron* with high catalytic efficiency of epimedin C. *Bioorg. Chem.* **2018**, *81*, 461–467. [CrossRef]
121. Cheng, L.; Zhang, H.; Cui, H.; Cheng, J.; Wang, W.; Wei, B.; Liu, F.; Liang, H.; Shen, X.; Yuan, Q. A novel α-L-Rhamnosidase renders efficient and clean production of icaritin. *J. Clean. Prod.* **2022**, *341*, 130903. [CrossRef]
122. Li, Q.; Ge, L.; Zheng, D.; Zhang, X.; Zhao, L. Screening and characterization of a GH78 α-l-rhamnosidase from *Aspergillus terreus* and its application in the bioconversion of icariin to icaritin with recombinant β-glucosidase. *Enzyme Microb. Technol.* **2022**, *153*, 109940. [CrossRef] [PubMed]
123. Xie, J.; Zhao, D.; Zhao, L.; Pei, J.; Xiao, W.; Ding, G.; Wang, Z. Overexpression and characterization of a Ca(2+) activated thermostable β-glucosidase with high ginsenoside Rb1 to ginsenoside 20(S)-Rg3 bioconversion productivity. *J. Ind. Microbiol. Biotechnol.* **2015**, *42*, 839–850. [CrossRef] [PubMed]
124. Papageorgiou, A.C.; Chen, J.; Li, D. Crystal structure and biological implications of a glycoside hydrolase family 55 β-1,3-glucanase from *Chaetomium thermophilum*. *Biochim. Biophys. Acta Proteins Proteom.* **2017**, *1865*, 1030–1038. [CrossRef] [PubMed]
125. Cheng, T.; Yang, J.; Zhang, T.; Yang, Y.S.; Ding, Y. Optimized biotransformation of icariin into icariside II by β-glucosidase from *Trichoderma viride* using central composite design method. *Biomed. Res. Int.* **2016**, *2016*, 5936947. [CrossRef] [PubMed]
126. Yang, Q.; Wang, L.; Zhang, L.; Xiao, H. Baohuoside I production through enzyme hydrolysis and parameter optimization by using response surface and subset selection. *J. Mol. Catal. B. Enzym.* **2013**, *90*, 132–138. [CrossRef]
127. Cui, L.; Zhang, Z.; Sun, E.; Jia, X.; Qian, Q. Effect of β-cyclodextrin complexation on solubility and enzymatic hydrolysis rate of icariin. *J. Nat. Sci. Biol. Med.* **2013**, *4*, 201–206. [PubMed]
128. Su, J.; Wu, T.; Cao, S.; Pei, J.; Zhao, L. Screening and characterization of a β-xylosidase from *Bifidobacterium breve* K-110 and its application in the biotransformation of the total flavonoids of epimedium to icariin with α-l-rhamnosidase. *Bioorg. Chem.* **2023**, *132*, 106364. [CrossRef]
129. Tong, X.; Qi, Z.; Zheng, D.; Pei, J.; Li, Q.; Zhao, L. High-level expression of a novel multifunctional GH3 family β-xylosidase/α-arabinosidase/β-glucosidase from *Dictyoglomus turgidum* in *Escherichia coli*. *Bioorg. Chem.* **2021**, *111*, 104906. [CrossRef]
130. Lu, Y.; Luo, Q.; Jia, X.; Tam, J.P.; Yang, H.; Shen, Y.; Li, X. Multidisciplinary strategies to enhance therapeutic effects of flavonoids from *Epimedii Folium*: Integration of herbal medicine, enzyme engineering, and nanotechnology. *J. Pharm. Anal.* **2023**, *13*, 239–254. [CrossRef]
131. Mohamad, N.R.; Marzuki, N.H.; Buang, N.A.; Huyop, F.; Wahab, R.A. An overview of technologies for immobilization of enzymes and surface analysis techniques for immobilized enzymes. *Biotechnol. Biotechnol. Equip.* **2015**, *29*, 205–220. [CrossRef]
132. De Carvalho, C.C. Enzymatic and whole cell catalysis: Finding new strategies for old processes. *Biotechnol. Adv.* **2011**, *29*, 75–83. [CrossRef] [PubMed]
133. Wachtmeister, J.; Rother, D. Recent advances in whole cell biocatalysis techniques bridging from investigative to industrial scale. *Curr. Opin. Biotechnol.* **2016**, *42*, 169–177. [CrossRef] [PubMed]
134. Gao, Q.; Wang, L.; Zhang, M.; Wei, Y.; Lin, W. Recent advances on feasible strategies for monoterpenoid production in *Saccharomyces cerevisiae*. *Front. Bioeng. Biotechnol.* **2020**, *8*, 609800. [CrossRef] [PubMed]
135. Guan, R.; Wang, M.; Guan, Z.; Jin, C.Y.; Lin, W.; Ji, X.J.; Wei, Y. Metabolic engineering for glycyrrhetinic acid production in *Saccharomyces cerevisiae*. *Front. Bioeng. Biotechnol.* **2020**, *8*, 588255. [CrossRef]
136. Ro, D.K.; Paradise, E.M.; Ouellet, M.; Fisher, K.J.; Newman, K.L.; Ndungu, J.M.; Ho, K.A.; Eachus, R.A.; Ham, T.S.; Kirby, J.; et al. Production of the antimalarial drug precursor artemisinic acid in engineered yeast. *Nature* **2006**, *440*, 940–943. [CrossRef]
137. Galanie, S.; Thodey, K.; Trenchard, I.J.; Filsinger Interrante, M.; Smolke, C.D. Complete biosynthesis of opioids in yeast. *Science* **2015**, *349*, 1095–1100. [CrossRef]
138. Chen, X.J.; Tang, Z.H.; Li, X.W.; Xie, C.X.; Lu, J.J.; Wang, Y.T. Chemical constituents, quality control, and bioactivity of *Epimedii folium* (Yinyanghuo). *Am. J. Chin. Med.* **2015**, *43*, 783–834. [CrossRef]

139. Szabó, R.; Rácz, C.P.; Dulf, F.V. Bioavailability improvement strategies for icariin and its derivates: A review. *Int. J. Mol. Sci.* **2022**, *23*, 7519. [CrossRef]
140. Jiang, W.; Ding, K.; Yue, R.; Lei, M. Therapeutic effects of icariin and icariside II on diabetes mellitus and its complications. *Crit. Rev. Food Sci. Nutr.* **2023**, 1–26. [CrossRef]
141. Tan, H.L.; Chan, K.G.; Pusparajah, P.; Saokaew, S.; Duangjai, A.; Lee, L.-H.; Goh, B.-H. Anti-cancer properties of the naturally occurring aphrodisiacs: Icariin and its derivatives. *Front. Pharmacol.* **2016**, *7*, 191. [CrossRef]
142. Yang, X.; Lang, S.; Li, S.; Jiang, C.; Han, J. Preparation of icariside I and icariside II, an exploration of their protective mechanism against cyclophosphamide-induced bone marrow suppression in mice, and their regulatory effects on immune function. *Pharmazie* **2022**, *77*, 32–37. [PubMed]
143. Zhang, D.; Zhao, N.; Wan, C.; Du, J.; Lin, J.; Wang, H. Icariin and icariside II reciprocally stimulate osteogenesis and inhibit adipogenesis of multipotential stromal cells through ERK signaling. *Evid. Based Complement. Altern. Med.* **2021**, *2021*, 8069930. [CrossRef] [PubMed]
144. Son, N.; Shi, L.; Li, Y.; Wang, Q.A. Total synthesis of icaritin via microwave-assistance Claisen rearrangement. *Lett. Org. Chem.* **2014**, *11*, 677–681.
145. Li, L.; Wang, L.; Fan, W.; Jiang, Y.; Zhang, C.; Li, J.; Peng, W.; Wu, C. The application of fermentation technology in traditional Chinese medicine: A review. *Am. J. Chin. Med.* **2020**, *48*, 899–921. [CrossRef] [PubMed]
146. Zhang, X.; Miao, Q.; Pan, C.; Yin, J.; Wang, L.; Qu, L.; Yin, Y.; Wei, Y. Research advances in probiotic fermentation of Chinese herbal medicines. *iMeta* **2023**, *2*, e93. [CrossRef]
147. Xie, Y.; Ya, H.; Zhang, X.; Fan, Y.; Zhao, B.; Wang, R.; Yang, Z. Biotransformation of flavonoid glycosides in *Epimedium* by *Lactobacillus plantarum*. *China Brew.* **2022**, *41*, 103–109.

molecules

Review

From Plant to Yeast—Advances in Biosynthesis of Artemisinin

Le Zhao [1,2], Yunhao Zhu [1,2], Haoyu Jia [1], Yongguang Han [1], Xiaoke Zheng [1,2], Min Wang [3,4,*] and Weisheng Feng [1,2,*]

1 School of Pharmacy, Henan University of Chinese Medicine, Zhengzhou 450046, China
2 Co-Construction Collaborative Innovation Center for Chinese Medicine and Respiratory Diseases by Henan and Education Ministry of P. R. China, Henan University of Chinese Medicine, Zhengzhou 450046, China
3 College of Chemistry and Materials Engineering, Beijing Technology and Business University, Beijing 100048, China
4 Beijing Key Laboratory of Plant Research and Development, Beijing Technology and Business University, Beijing 100048, China
* Correspondence: wangmin_216@163.com (M.W.); fwsh@hactcm.edu.cn (W.F.); Tel.: +86-134-2629-2115 (M.W.); +86-371-60190296 (W.F.)

Abstract: Malaria is a life-threatening disease. Artemisinin-based combination therapy (ACT) is the preferred choice for malaria treatment recommended by the World Health Organization. At present, the main source of artemisinin is extracted from *Artemisia annua*; however, the artemisinin content in *A. annua* is only 0.1–1%, which cannot meet global demand. Meanwhile, the chemical synthesis of artemisinin has disadvantages such as complicated steps, high cost and low yield. Therefore, the application of the synthetic biology approach to produce artemisinin in vivo has magnificent prospects. In this review, the biosynthesis pathway of artemisinin was summarized. Then we discussed the advances in the heterologous biosynthesis of artemisinin using microorganisms (*Escherichia coli* and *Saccharomyces cerevisiae*) as chassis cells. With yeast as the cell factory, the production of artemisinin was transferred from plant to yeast. Through the optimization of the fermentation process, the yield of artemisinic acid reached 25 g/L, thereby producing the semi-synthesis of artemisinin. Moreover, we reviewed the genetic engineering in *A. annua* to improve the artemisinin content, which included overexpressing artemisinin biosynthesis pathway genes, blocking key genes in competitive pathways, and regulating the expression of transcription factors related to artemisinin biosynthesis. Finally, the research progress of artemisinin production in other plants (*Nicotiana*, *Physcomitrella*, etc.) was discussed. The current advances in artemisinin biosynthesis may help lay the foundation for the remarkable up-regulation of artemisinin production in *A. annua* through gene editing or molecular design breeding in the future.

Keywords: *Artemisia annua*; artemisinin; synthetic biology; genetic engineering; transcription factor

Citation: Zhao, L.; Zhu, Y.; Jia, H.; Han, Y.; Zheng, X.; Wang, M.; Feng, W. From Plant to Yeast—Advances in Biosynthesis of Artemisinin. *Molecules* **2022**, *27*, 6888. https://doi.org/10.3390/molecules27206888

Academic Editors: Yongjun Wei, Lingbo Qu and Xiao-Jun Ji

Received: 28 September 2022
Accepted: 12 October 2022
Published: 14 October 2022

1. Introduction

Plants fix carbon through photosynthesis and use solar energy to convert water and CO_2 into organic matter to produce primary metabolites, which are used as precursors to synthesize a series of secondary metabolites through secondary metabolism. These secondary products are not only important for plant adaptation and response to the environment, but are also the source of many clinical drugs [1]. According to the chemical structure and biosynthesis pathway, plant secondary metabolites can be divided into three major groups: terpenoids (54%), alkaloids (27%), phenolics (18%) and others (1%) [2]. Terpenoids are the largest number of plant secondary products, possessing diversified structures and functions. Based on the number of 5-carbon isoprene units, terpenoids are classified into hemiterpenoids (5C), monoterpenoids (10C), sesquiterpenoids (15C), diterpenoids (20C), sesterterpenoids (25C), triterpenoids (30C), tetraterpenoids (40C), and polyterpenoids (>40C) [3]. Paclitaxel (taxol), a diterpenoid anticancer drug isolated from

Taxus brevifolia, has been approved for the clinical treatment of breast cancer, non-small cell lung cancer, ovarian cancer and AIDS-related Kaposi sarcoma (used as second-line therapy) [4]. Many plant polyphenolics metabolites, such as flavonoids, phenolic acids, lignans and tannins, are powerful antioxidants. Curcumin, a polyphenol compound that is produced by *Curcuma longa*, has antioxidant, anti-inflammatory and anticancer properties [5]. Among other types of plant secondary metabolites, an example is melatonin, which is derived from tryptophan and was initially isolated from animals before being discovered in plants. Melatonin plays an important role in plant growth and development as well as the response to biotic and abiotic stress; in addition, it has strong antioxidant activity [6].

Artemisinin is a plant secondary metabolite of sesquiterpenoids, which was first isolated from the medicinal plant *Artemisia annua* L. by Tu Youyou in 1972 [7]. Artemisinin-based combination therapy (ACT) is currently the preferred treatment for malaria recommended by the World Health Organization (WHO). Malaria is a life-threatening disease caused by the *Plasmodium parasite*, which is transmitted to humans through the bite of infected female *Anopheles* mosquitoes. So far, the global malaria epidemic remains severe. According to the latest *World malaria report 2021*, malaria infection cases in 2020 were found in 85 countries, infecting more than 241 million people and causing 627,000 deaths, mainly in Africa, Southeast Asia and the Eastern Mediterranean. Compared with 2019, there were 14 million more malaria infections and 47,000 more deaths than in 2020 due to the disruption of medical services during the COVID-19 pandemic [8]. In 2015, Tu Youyou was awarded the Nobel Prize in Physiology and Medicine for her discovery of a new method to treat malaria with artemisinin. Recent studies have found that artemisinin and its derivatives have anti-cancer, antibacterial and immunomodulatory properties—in addition to the efficacy of treating malaria [9]. Moreover, artemisinin can also be used for the treatment of diabetes [10] and pulmonary tuberculosis [11], which have broad application prospects.

The content of artemisinin in *A. annua* is very low, being only 0.1–1% DW (dry weight) of the plant. The process of extracting artemisinin from *A. annua* is complicated and costly, which cannot meet the global demand for artemisinin [12]. Although the chemical synthesis of artemisinin was successful as early as 1983, it has not been put into commercial production due to the disadvantages of having multiple steps, a high cost and low yield [13]. Therefore, how to obtain artemisinin efficiently, economically and in large quantities has become an international research hotspot. With the development of molecular biology, the application of synthetic biology to produce active components of medicinal plants has broad application prospects [14]. In this review, the artemisinin biosynthesis pathway was summarized, then reconstituting the biosynthesis pathway of artemisinin using microorganisms as chassis cells was discussed, and finally, the research progress in the production of artemisinin from different plants was reviewed, which laid the foundation for the mass, rapid and low-cost acquisition of artemisinin.

2. The Artemisinin Biosynthesis Pathway

In order to generate artemisinin in microorganisms and plants with the synthetic biology method, the biosynthesis pathway of artemisinin needs to be dissected. Artemisinin was derived from the plant terpenoid biosynthesis pathway by using IPP (isopentenyl diphosphate) and its isomer, DMAPP (dimethylallyl diphosphate), as raw materials, which were produced by the MVA (mevalonate) pathway in the cytoplasm and the MEP (methylerythritol phosphate) pathway in the plastid. Derivatives of the MVA pathway were precursors to sesquiterpenes, triterpenes and polyterpenoids, while the MEP pathway was generally involved in the biosynthesis of monoterpenes, diterpenes and tetraterpenes [15]. Two molecules of IPP and one molecule of DMAPP were catalyzed by FPPS (farnesyl diphosphate synthase) to form 15-carbon FPP (farnesyl diphosphate), which entered the downstream metabolic pathway of artemisinin biosynthesis (Figure 1).

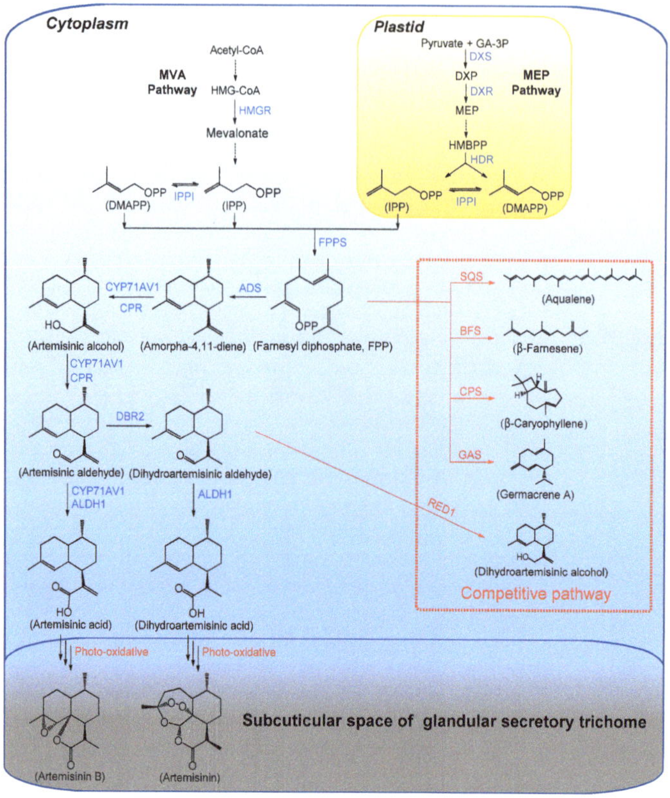

Figure 1. Artemisinin biosynthesis pathway and competitive metabolic pathway in *A. annua*. The biosynthesis of artemisinin is derived from the MVA and MEP pathways. FPP is the important intermediate at the branch point of artemisinin biosynthesis, which can not only be converted into amorpha-4,11-diene and artemisinin, but also into other terpenoids through competitive metabolic pathways, such as squalene, β-farnesene, β-caryophyllene, germacrene A, etc. The competitive pathways are marked with red boxes. HMG-CoA, 3-hydroxy-3-methylglutaryl coenzyme A; GA-3P, glyceraldehyde 3-phosphate; DXP, 1-deoxy-D-xylulose 5-phosphate; MEP, 2-C-methyl-D-erythritol-4-phosphate; HMBPP, 4-hydroxy-3-methyl-but-2-enyl diphosphate; IPP, isopentenyl diphosphate; DMAPP, dimethylallyl diphosphate; HMGR, 3-hydroxy-3-methylglutaryl coenzyme A reductase; DXS, 1-deoxy-D-xylulose 5-phosphate synthase; DXR, 1-deoxy-D-xylulose 5-phosphate reductoisomerase; HDR, 4-hydroxy-3-methyl-but-2-enyl diphosphate reductase; IPPI, isopentenyl diphosphate isomerase; FPPS, farnesyl diphosphate synthase; ADS, amorpha-4,11-diene synthase; CYP71AV1, cytochrome P450 monooxygenase; CPR, cytochrome P450 reductase; ALDH1, aldehyde dehydrogenase 1; DBR2, artemisinic aldehyde Δ11(13) reductase; SQS, squalene synthase; BFS, β-farnesene synthase; CPS, β-caryophyllene synthase; GAS, germacrene A synthase; RED1, dihydroartemisinic aldehyde reductase.

ADS (Amorpha-4,11-diene synthase) catalyzed the cyclization of FPP to form amorphine-4,11-diene, the precursor of artemisinin [16]. Under the coordination synergistic effect of CYP71AV1 (cytochrome P450 monooxygenase) and CPR (cytochrome P450 reductase), amorpha-4,11-diene was further oxidized to artemisinic alcohol and artemisinic aldehyde [17,18]. Artemisinic aldehyde was catalyzed by CYP71AV1 and ALDH1 (aldehyde dehydrogenase 1) to generate artemisinic acid [19]. Starting from artemisinic aldehyde, an additional branch emerged. Artemisinic aldehyde was catalyzed by DBR2 (artemisinic aldehyde Δ11(13) reductase) to generate DHAAA (dihydroartemisinic aldehyde), and

DHAAA was converted to DHAA (dihydroartemisinic acid) by ALDH1 [20]. Artemisinic acid and DHAA were thought to be the final compounds formed by the enzyme-catalyzed reactions in this pathway (Figure 1). In the subcuticular space of glandular-secreting trichomes (GSTs) on the surface of *A. annua* leaves, artemisinic acid was converted to artemisinin B and DHAA to artemisinin through photo-oxidation, which are a series of non-enzymatic reactions [21,22].

At present, artemisinin has been produced by the synthetic biology approach, mostly through the semi-synthetic route. The precursors of artemisinin, such as amorpha-4,11-diene, artemisinic acid and DHAA, were prepared in microorganisms by metabolic engineering, and then artemisinin was generated by chemical synthesis; however, some studies suggested that there may be a specifically expressed and highly active peroxidase in the GSTs of *A. annua*, which can catalyze the conversion of DHAA to artemisinin [23]. If such a peroxidase existed, the total biosynthesis of artemisinin would be carried out in microorganisms, rather than semi-synthesis.

3. Metabolic Engineering in Microorganisms

With the advancement of sequencing technology, more and more medicinal plant genomes have been sequenced, and the biosynthetic pathways of many natural products have been elucidated [24]. On this basis, the biosynthetic pathway of natural products can be reconstructed in microorganisms, and natural products and their precursors are able to be obtained rapidly and in large quantities through biological fermentation, thereby laying a foundation for the commercial application of natural plant products. As an effective antimalarial drug, the development of artemisinin was restricted by the cost of extracting compounds from natural *A. annua* and the complexity of chemical synthesis; however, for those who are most affected by malaria (such as African areas), classical plant breeding may not be enough to reduce the production cost of artemisinin to an affordable price. Therefore, using microorganisms (*Escherichia coli* and *Saccharomyces cerevisiae*) as chassis cells, the heterologous biosynthesis of artemisinin or its precursor through metabolic engineering may provide an alternative source of artemisinin supply. It took Keasling's laboratory ten years to achieve the semi-synthesis of artemisinin, thereby transferring the production of artemisinin from plants to microorganisms.

In 2003, Martin et al. transformed the codon-optimized *ADS* gene into *E. coli*, simultaneously co-expressed the *E. coli* SOE4 operon (encoding the rate-limiting enzyme genes *dxs*, *ippHp* and *ispA* of the *E. coli* MEP pathway), and heterologously expressed the MVA pathway genes (pMevT and pMBIS plasmids) from *S. cerevisiae*. This was the first time that amorpha-4,11-diene, the precursor of artemisinin, was synthesized in *E. coli* at a concentration of 24 mg/L [25] (Table 1). After optimization of the fermentation conditions, the yield of amorpha-4,11-diene increased 20-fold to 0.5 g/L [26]. Tsuruta et al. continued to optimize the genes of the yeast MVA pathway for heterologous expression in *E. coli* and replaced the yeast *HMGS* and *tHMGR* genes on the pMevT plasmid with more active equivalent genes from *Staphylococcus aureus*. Combined with the restricted supply of carbon and nitrogen, the amorpha-4,11-diene production in *E. coli* reached 27.4 g/L [27]. On the basis of the heterologous expression of the yeast MVA pathway genes (pAM92 plasmid), the *ADS*, *AMO* (amorpha-4,11-diene oxidase) and *CPR* genes of *A. annua* were simultaneously expressed to produce artemisinic acid in *E. coli* with the highest yield being 105 mg/L [28].

Table 1. Metabolic engineering to enhance the production of amorpha-4,11-diene and artemisinic acid in *E. coli* or yeast.

Enzyme Name	Expression Mode	Gene Source	Host	Products	Reference
pMevT (atoB [1], HMGS, tHMGR) pMBIS (EGR12, EGR8, MVD1, idi [2], ispA [3]) pSOE4 (dxs, ippHP [4], ispA) ADS	Overexpression Overexpression Overexpression	Yeast *E. coli* *A. annua*	*E. coli*	Amorpha-4,11-diene 24 mg/L	[25]
pAM45 (atoB1, HMGS, tHMGR, MK, PMK, PMD, idi [2], ispA [3],) pAM94 (ADS [5], MK)	Overexpression	Yeast	*E. coli*	Amorpha-4,11-diene 293 mg/L	[29]
pAM52 (atoB [1], mvaS, mvaA) [6] pMBIS (EGR12, EGR8, MVD1, idi [2], ispA [3]) ADS	Overexpression Overexpression	Yeast *A. annua*	*E. coli*	Amorpha-4,11-diene 27.4 g/L	[27]
pAM92 (atoB [1], HMGS, tHMGR, EGR12, EGR8, EGR19, idi [2], ispA [3], ADS [5]) AMO, CPR	Overexpression Overexpression	Yeast *A. annua*	*E. coli*	Artemisinic acid 105 mg/L	[28]
MVA pathway enzymes (EGR13, tHMGR, EGR12, EGR8, EGR20) EGR9 ADS, CYP71AV1, CPR	Overexpression Down-regulation Overexpression	Yeast Yeast *A. annua*	Yeast	Artemisinic acid 100 mg/L	[18]
All enzymes of MVA pathway (EGR10, EGR13, tHMG1, EGR12, EGR8, EGR19, IDI1, EGR20) EGR9 ADS, CYP71AV1, CPR	Overexpression Down-regulation Overexpression	Yeast Yeast *A. annua*	Yeast (CEN.PK2)	Amorpha-4,11-diene 40 g/L	[30]
Every gene of MVA pathway (EGR10, EGR13, tHMG1, EGR12, EGR8, EGR19, IDI1, EGR20) EGR9 ADS, CYP71AV1, CPR1, CYB5, ADH1, ALDH1	Overexpression Down-regulation Overexpression	Yeast Yeast *A. annua*	Yeast	Artemisinic acid 25 g/L	[31]

[1] atoB, acetoacetyl-CoA thiolase from *E. coli*. [2] idi, IPP isomerase from *E. coli*. [3] ispA, FPP synthase from *E. coli*. [4] ippHp, IPP isomerase from *Haematococcus pluvialis*. [5] ADS from *A. annua*. [6] pAM52 was the pMevT derivative, but HMGS and tHMGR were replaced by mvaS and mvaA from *S. aureus*.

In the downstream biosynthetic pathway of artemisinin, the conversion from amorphine-4,11-diene to artemisinic acid and DHAA required the synergistic catalysis of CYP71AV1 and CPR, both of which were associated with the cell's endomembrane system; however, *E. coli* lacked the CYP450 gene and the endomembrane system of eukaryotic cells, resulting in the affected expression of *CYP71AV1* and *CPR* genes. In 2006, Ro et al. modified the FPP biosynthesis pathway in *S. cerevisiae* to prompt FPP production, which down-regulated the expression of *EGR9* and inhibited the conversion of FPP to squalene, thus allowing more FPP to flow into the artemisinic acid biosynthesis pathway. Meanwhile, the *ADS*, *CYP71AV1* and *CPR* genes of *A. annua* were transformed into yeast strain EPY224 (Table 1). Finally, the biosynthesis of artemisinic acid was achieved in yeast with a yield of 100 mg/L [18]. Moreover, the biosynthesized artemisinic acid was able to be secreted outside of the engineered yeast, facilitating the purification of artemisinic acid from the yeast fermentation broth, which further prompted the industrialization process of artemisinin.

Westfall et al. adopted the yeast strain, CEN.PK2, to overexpress all the genes of the MVA pathway while down-regulating the *EGR9* gene so that the production of amorpha-4,11-diene reached 40 g/L [30]. In 2013, Paddon et al. transformed the *ADS*, *CYP71AV1*, *CPR1*, *CYB5*, *ADH1* and *ALDH1* genes of *A. annua* into yeast after codon optimization to reduce the accumulation of artemisinic aldehyde and improved the fermentation process. As a result, the yield of artemisinic acid reached 25 g/L (Table 1), which met the demand of industrial production. Then, artemisinic acid was extracted from the fermentation broth, and after four steps of chemical synthesis in vitro, artemisinin was finally acquired with an overall yield of 40–45% [31].

In summary, the use of synthetic biology methods to produce artemisinic acid in microorganisms mainly made the following improvements: 1. Codon optimization. The key genes of artemisinic acid biosynthesis pathway were codon-optimized and transformed into *E. coli* and yeast. 2. Reconstruction of metabolic pathways and selection of enzymes with high catalytic activity. Enzymes of the yeast MVA pathway were replaced with enzymes from *S. aureus*. 3. Choice of suitable chassis cells. The CYP450 oxidase and reductase of *A. annua* were expressed in eukaryotic yeast. 4. Control of the flow of metabolites. In yeast cells, the down-regulation of the *ERG9* gene (encoding squalene synthase) inhibited the conversion of FPP to squalene. Meanwhile, the genes of the MVA pathway were overexpressed under strong promoters, thereby allowing more FPP to flow into the artemisinic acid biosynthesis pathway.

4. Production of Artemisinin in Genetically Engineered Plants

4.1. Genetic Engineering in A. annua

The artemisinin content in *A. annua* can be improved by overexpressing artemisinin biosynthetic genes, blocking key genes in competitive pathways of artemisinin biosynthesis, or regulating the expression of transcription factors involved in artemisinin biosynthesis.

4.1.1. Overexpression of Key Genes in Artemisinin Biosynthesis

In the past two decades, key genes in the artemisinin biosynthesis pathway have been isolated and characterized (Figure 1). The content of artemisinin in *A. annua* was adjusted by regulating the expression of genes involved in the MVA and MEP pathways, or by overexpressing key genes in the downstream biosynthesis pathway of artemisinin (Table 2).

According to the genetic map and genomic data of *A. annua*, *HMGR* and *DXR* genes are closely related to artemisinin content [32,33]. HMGR and DXR are the rate-limiting enzymes of the MVA and MEP pathway, which provided the substrates IPP and DMAPP for the biosynthesis of artemisinin. By applying competitive inhibitors (fosmidomycin and mevinolin) to suppress the activities of DXR and HMGR, respectively, it was found that the production of artemisinin decreased by 14.2% and 80.4%, indicating that the MVA pathway was the main carbon donor for artemisinin biosynthesis [34]. The *HMGR* gene from *Catharanthus roseus* was overexpressed in *A. annua* through *Agrobacterium*-mediated transformation, and the artemisinin content of transgenic plants reached 0.60 mg/g DW, while artemisinin in non-transgenic lines was only 0.37 mg/g DW [35]. Overexpression of the *DXR* gene from *C. roseus* in *A. annua* prompted a 2.33-fold increase in the yield of artemisinin to 1.21 mg/g DW [36].

Overexpression of the *IPPI1* and *HDR1* genes in *A. annua* were able to increase the accumulation of artemisinin and artemisinin B, while inhibition of the expression of the *HDR1* gene reduced the artemisinin content [37,38]. FPP was the precursor of secondary metabolites such as sesquiterpenes (including artemisinin), triterpenes, coenzyme Q, plastoquinone, etc., so increasing the production of FPP can promote the biosynthesis of artemisinin. The *FPPS* gene of *A. annua* was first isolated by Matsushita in 1996 [39], and overexpression of the *FPPS* gene increased the yield of artemisinin by 1.38–3.36 times [40–42].

ADS catalyzed the conversion of FPP to amorpha-4,11-diene, the first key enzyme in the artemisinin biosynthesis pathway, and the *ADS* gene was specifically expressed in the GST cells of *A. annua* [43]. The haplotype genome comparison of *A. annua* revealed that the copy number of the *ADS* gene was highly correlated with artemisinin yield [44]. Overexpression of the *ADS* gene significantly increased not only artemisinin content, but also the content of artemisinic acid and DHAA [45]. DBR2 was another key enzyme in the artemisinin biosynthesis pathway, which was also specifically expressed in the GSTs of *A. annua*. DBR2 catalyzed artemisinic aldehyde to form dihydroartemisinic aldehyde, which further generated artemisinin. If the activity of DBR2 decreased, more artemisinic aldehyde was converted into artemisinic acid, resulting in an increased accumulation of artemisinin B [20]. The content of artemisinin in DBR2-overexpressed transgenic plants was 1.50–2.14 mg/g DW, which was 1.59–2.26 times that of the control [46].

The overexpression of a single key gene involved in the artemisinin biosynthesis can increase the artemisinin content, but the increase was not significant. Therefore, simultaneous overexpression of two or more key genes was able to remarkably improve the content of artemisinin [33]. When *HMGR* and *ADS* genes were overexpressed in *A. annua*, the artemisinin content was increased by 8.65-fold to 1.73 mg/g DW [47]. The *ADS* and *FPPS* genes were fused together and overexpressed under the regulation of the CaMV35S or CYP71AV1 promoter. The CaMV35S promoter significantly up-regulated the expression of the *ADS* and *FPPS* genes, but the accumulation of artemisinin did not increase significantly; however, under the control of the CYP71AV1 promoter, the *ADS-FPPS* fusion gene was specifically expressed in GSTs and significantly enhanced the artemisinin concentration to 26 mg/g DW [48]. Co-expression of the *CYP71AV1* and *CPR* genes can improve the content of artemisinin by 1.38–2.68 times, reaching 0.98–2.44 mg/g DW [36,49]; when the three genes, *FPPS*, *CYP71AV1* and *CPR*, were co-expressed, the artemisinin content was increased by 3.6 times, reaching 2.98 mg/g FW [50]; when the four genes, *ADS*, *CYP71AV1*, *CPR* and *ALDH1*, were co-expressed, the artemisinin content reached 27 mg/g DW, which was 3.4 times that of the control [51].

Therefore, the use of transgenic methods to regulate the expression of key genes in the biosynthesis pathway of artemisinin, and the overexpression of one or more key genes, can remarkably improve the content of artemisinin in transgenic plants (Table 2), which was of great significance for obtaining new varieties of *A. annua* with a high artemisinin content.

Table 2. Genetic engineering to increase artemisinin production in *A. annua*.

Gene Name	Expression Mode	Artemisinin (mg/g DW)		Fold Change	Reference
		Control	Transgenic		
		MVA or MEP pathway key genes			
CrHMGR [1]	Overexpression	0.32	0.39	1.22	[52]
		0.37	0.60	1.62	[35]
CrDXR [1]	Overexpression	0.52	1.21	2.33	[36]
IPPI1	Overexpression	0.75	2.5	3.33	[38]
HDR1	Overexpression	0.05	0.09	1.8	[37]
HDR1	Down-regulation	0.05	0.03	0.6	
GaFPPS [2]	Overexpression	3	10.08	3.36	[42]
FPPS	Overexpression	6.5	0.9	1.38	[40]
		5.3	13.3	2.5	[41]
		Artemisinin biosynthesis pathway key genes			
ADS	Overexpression	0.6	1.2	2	[45]
DBR2	Overexpression	0.94	2.14	2.26	[46]
		8	22.6	2.83	[53]
ALDH1	Overexpression	8	25.6	3.2	[54]
HMGR, *FPPS*	Overexpression	5	9	1.8	[55]
HMGR, *ADS*	Overexpression	0.2	1.73	8.65	[47]
ADS-FPPS fusion	Overexpression	10	26	2.6	[48]
CYP71AV1, *CPR*	Overexpression	0.91	2.44	2.68	[36]
		0.71 (mg/g FW)	0.98 (mg/g FW)	1.38	[49]
FPPS, *CYP71AV1*, *CPR*	Overexpression	0.83 (mg/g FW)	2.98 (mg/g FW)	3.6	[50]
ADS, *CYP71AV1*, *CPR*	Overexpression	6.4	15.1	2.4	[56]
ADS, *CYP71AV1*, *CPR*, *ALDH1*	Overexpression	8	27	3.4	[51]

Table 2. *Cont.*

Gene Name	Expression Mode	Artemisinin (mg/g DW)		Fold Change	Reference
		Control	Transgenic		
		Competitive pathway key genes			
SQS	Down-regulation	10	31.4	3.14	[57]
CPS	Down-regulation	2.32	3.6	1.55	[58]
CPS	Down-regulation	6.26	11.08	1.77	[59]
BFS	Down-regulation	6.26	11.08	1.77	
GAS	Down-regulation	6.26	12.71	2.03	
SQS	Down-regulation	6.26	10.70	1.71	

[1] *CrHMGR* and *CrDXR* genes were from *C. roseus*. [2] *GaFPPS* gene was from *Gossypium arboreum*.

4.1.2. Suppression of Competitive Metabolic Pathways

FPP has different metabolic flows in plants. It can not only be converted into amorpha-4,11-diene and enter the artemisinin biosynthesis pathway, but can also be converted into terpenoids such as β-farnesene, β-caryophyllene, germacrene A, squalene and *epi*-cedrol, which enter the biosynthesis pathway of other secondary metabolites (Figure 1). Therefore, by suppressing the flow of FPP into other competing pathways, more FPP can enter the artemisinin biosynthesis pathway and increase the yield of artemisinin [60,61] (Table 2).

SQS (squalene synthase) was the first key enzyme that catalyzed the sterol biosynthesis pathway. Using the hairpin RNA-mediated RNAi method to inhibit the expression of the *SQS* gene, the content of sterol was reduced by 37–58%, and the content of artemisinin was significantly increased, reaching 31.4 mg/g DW—3.14 times that of the control [57].

CPS (β-caryophyllene synthase) catalyzed the conversion of FPP to β-caryophyllene [62]. When the antisense RNA approach was adopted to down-regulate the expression of the *CPS* gene, the expression levels of the genes related to the artemisinin biosynthesis pathway were significantly up-regulated in transgenic plants. The results showed that the content of β-caryophyllene decreased by 40–62.7%, while the content of artemisinin increased by 54.9% [58].

Through RNA interference, the key enzyme genes, *CPS*, *BFS*, *GAS* and *SQS*, involved in the competitive pathway of artemisinin biosynthesis were gene-silenced. The results showed that the expression levels of key genes related to the artemisinin biosynthesis pathway (*ADS*, *CYP71AV1*, *DBR2*, *ALDH1*) were up-regulated, and the contents of artemisinin and DHAA were significantly increased. Compared with non-transgenic plants, the content of artemisinin in anti-CPS, anti-BFS, anti-GAS and anti-SQS transgenic plants rose by 77%, 77%, 103% and 71%, respectively [59].

In the biosynthesis pathway of artemisinin, RED1 (dihydroartemisinic aldehyde reductase) competed with ALDH1 to bind dihydroartemisinic aldehyde so that dihydroartemisinic aldehyde was converted into dihydroartemisinic alcohol, which means that artemisinin cannot be produced (Figure 1). If the expression level of the *RED1* gene was reduced, more dihydroartemisinic aldehyde can enter the artemisinin biosynthesis pathway [63]. Ranjbar et al. analyzed the content of artemisinin in eight *Artemisia* species and the changes in the expression levels of key genes in the biosynthesis pathway of artemisinin. The results indicated that the content of artemisinin in *A. annua* was the highest, followed by *Artemisia absinthium*. This may be because the expression levels of the *ADS* and *DBR2* genes in *A. annua* were higher than those in the other seven *Artemisia* species, while in *A. absinthium*, the expression of the *ALDH1* gene was down-regulated and the expression of the *RED1* gene was up-regulated, which reduced the artemisinin content in *A. absinthium* [64].

4.1.3. Regulation of Transcription Factors Expression

In the secondary metabolism of plants, transcription factors (TFs) can regulate the expression of a series of genes in metabolic pathways, and overexpression or suppression of these TFs is able to effectively regulate the accumulation of plant secondary metabolites [60,61]. A variety of TFs have been identified to be related to the biosynthesis of artemisinin, including WRKY, bHLH, AP2/ERF, bZIP, MYB, etc. (Table 3).

WRKY TF Family

WRKY TFs are one of the largest TF families in plants, with a conserved WRKYGQK sequence at the N-terminus and a typical zinc finger protein structure at the C-terminus [65]. WRKY TFs participate in the regulation of plant growth, development, senescence and response to environmental stress [66]. It has been reported that AaWRKY1, AaWRKY4, AaWRKY9, AaWRKY17, AaGSW1 and AaGSW2 are involved in the regulation of artemisinin biosynthesis (Table 3).

AaWRKY1 was the first cloned TF of *A. annua* [67]. AaWRKY1 activates the expression of the *ADS*, *CYP71AV1* and *DBR2* genes, and AaWRKY1 takes part in the jasmonic acid (JA) signaling pathway [68]. Overexpression of the AaWRKY1 gene increased the content of artemisinin by 1.3–2.0 times [60]. In AaWRKY4-overexpressed plants, the expression levels of the *ADS*, *CYP71AV1*, *DBR2* and *ALDH1* genes were significantly improved, and the production of artemisinin was increased by 35–50% [69]. AaWRKY9 was specifically expressed in the GSTs of *A. annua*, which responded to both light and JA signals, and positively regulated the biosynthesis of artemisinin. AaWRKY9 can bind to the promoters of the *AaDBR2* and *AaGSW1* genes to up-regulate the expression of artemisinin biosynthesis pathway genes, and the artemisinin production was increased by 1.6–2.2 times in AaWRKY9-overexpressed plants [70]. AaWRKY17 was also a TF that positively regulated artemisinin biosynthesis. Moreover, AaWRKY17 can activate the expression of two defense marker genes, *PR5* (Pathogenesis-Related 5) and *NHL10* (NDR1/HIN1-LIKE 10), and improve the tolerance of *A. annua* to *Pseudomonas syringae*, so AaWRKY17 could be used in the transgenic breeding of *A. annua* to improve artemisinin content and resistance to pathogens [71].

AaGSW1 (GLANDULAR TRICHOME-SPECIFIC WRKY 1) is a GSTs specific WRKY TF, which is regulated by AaMYC2 in the JA signaling pathway and AabZIP1 in the ABA signaling pathway. Overexpression of AaGSW1 activated the transcript levels of the *ADS*, *CYP71AV1*, *DBR2*, *ALDH1* and *ORA* genes, and improved the content of artemisinin by 55–100% [72]. A recent study reported that AaGSW1 can bind to the promoters of TF AaTCP15 (teosinte branched1/cycloidea/proliferating 15) and AaORA (octadecanoid-derivative responsive AP2-domain protein), forming the AaGSW1-AaTCP15/AaORA module network that regulates artemisinin biosynthesis through JA and ABA signaling transduction [73]. When AaGSW2 was overexpressed in *A. annua*, the density of GSTs was significantly increased in the overexpressed plants, and the artemisinin content went up 2-fold compared with the control [74].

bHLH TF Family

bHLH TFs are a class of TFs containing the basic helix-loop-helix (bHLH) domain that recognizes the E-box motif (CANNTG) of the promoter region [75]. According to the genome data, the bHLH TFs in *A. annua* have nearly 200 members, which makes it one of the largest TF families in *A. annua* [33]. So far, it has been reported that AabHLH1, AaMYC2 and AabHLH112 positively regulate the biosynthesis of artemisinin, while AabHLH2 and AabHLH3 are the negative regulators in artemisinin accumulation (Table 3).

AabHLH1 can bind to the promoters of the *ADS* and *CYP71AV1* genes and positively regulate artemisinin biosynthesis. In AabHLH1-overexpressed plants, the expression levels of the *ADS*, *CYP71AV1*, *DBR2* and *HMGR* genes are significantly up-regulated [76], and the artemisinin content is increased 1.3-fold [77]. Moreover, the expression of AabHLH1 is induced by JA and AabHLH1 interacted with all nine AaJAZ proteins, so AabHLH1 is also involved in the regulation of JA-induced artemisinin biosynthesis [77].

AaMYC2 belongs to the bHLH TF superfamily and is the core factor of the JA signaling pathway, which can bind to the G-box motif of the *CYP71AV1* and *DBR2* gene promoters. Overexpression of AaMYC2 remarkably improved the transcription levels of the *CYP71AV1* and *DBR2* genes, and the content of artemisinin increased by 23.55% compared with the control, while in the RNAi plants of AaMYC2, the artemisinin content decreased to 54.81% of the control [78].

As a low-temperature-inducible TF, AabHLH112 positively regulates artemisinin biosynthesis through AaERF1. Yeast one-hybrid results showed that AabHLH112 cannot bind to the promoters of artemisinin biosynthesis-related genes, but could bind to the promoter of the *AaERF1* gene [79]. AaERF1 is a member of the AP2/ERF TF family and can up-regulate the expression of artemisinin biosynthesis genes [80]. Overexpression of AabHLH112 significantly activated the transcript of the *AaERF1* gene, which further promoted the expression of artemisinin biosynthesis genes. In AabHLH112-overexpressed plants, the artemisinin content reached 14.35 mg/g DW, which increased by 70.42% compared to the control (8.42 mg/g DW) [79].

AabHLH2 and AabHLH3 are MYC-type bHLH TFs, which suppress the expression of the *ADS* and *CYP71AV1* genes by antagonizing AaMYC2 and participate in the negative regulation of artemisinin biosynthesis. Compared with the control, the artemisinin accumulation decreased by 27–66% and 21–61% in the overexpressed plants of AabHLH2 and AabHLH3, respectively. In contrast, the artemisinin content was increased by 42–87% and 35–60% in AabHLH2 and AabHLH3 RNAi plants, respectively [81]. As negative regulators, AabHLH2 and AabHLH3 may be good targets for prompting artemisinin production in *A. annua* by gene editing.

Table 3. Transcription factors involved in artemisinin biosynthesis pathway.

Name	Function	Expression Mode	Fold Change of Artemisinin Content	Reference
	WRKY TF family			
AaWRKY1	AaWRKY1 activates the expression of *ADS*, *CYP71AV1*, *DBR2* genes.	Overexpression	1.30–2.00	[68]
AaWRKY4	AaWRKY4 prompts the expression of *ADS*, *CYP71AV1*, *DBR2*, *ALDH1* genes.	Overexpression	1.35–1.50	[69]
AaWRKY9	AaWRKY9 positively regulates the expression of *AaDBR2* and *AaGSW1* genes.	Overexpression Down-regulation	1.60–2.20 0.55–0.65	[70]
AaWRKY17	AaWRKY17 activates the transcription of *ADS*, *PR5* and *NHL10* genes and enhances the resistance of *A. annua* to the pathogen *P. syringae*.	Overexpression Down-regulation	1.50–1.87 0.79–0.85	[71]
AaGSW1	AaGSW1 is a GST-specific TF, and positively regulates the expression of *CYP71AV1* and *AaORA* genes.	Overexpression	1.55–2.00	[72]
AaGSW2	As a GST-specific TF, AaGSW2 positively regulates the initiation of GST.	Overexpression Down-regulation	2.00 0.54	[74]
	bHLH TF family			
AabHLH1	AabHLH1 activates transcription of *ADS* and *CYP71AV1* genes and responds to JA induction.	Overexpression Down-regulation	1.30 0.60	[77]
AaMYC2	AaMYC2 prompts the expression of *CYP71AV1* and *DBR2* genes and responses to JA induction.	Overexpression Down-regulation	1.24 0.55	[78]
AabHLH112	AabHLH112 promotes artemisinin biosynthesis by activating TF AaERF1.	Overexpression	1.70	[79]
AabHLH2 AabHLH3	AabHLH2 and AabHLH3 suppress the expression of *ADS* and *CYP71AV1* genes by antagonizing AaMYC2, and negatively regulate artemisinin biosynthesis.	Overexpression Down-regulation Overexpression Down-regulation	0.34–0.73 1.42–1.87 0.39–0.79 1.35–1.60	[81]
	AP2/ERF TF family			
AaERF1 AaERF2	AaERF1 and AaERF2 activate the transcription of *ADS* and *CYP71AV1* genes and respond to JA signaling.	Overexpression Overexpression	1.19–1.67 1.24–1.51	[80]
AaORA	AaORA is a trichome-specific TF, which promotes the expression of *ADS*, *CYP71AV1*, *DBR2* and *AaERF1* genes, and improves the resistance of *A. annua* to the pathogen *B. cinerea*.	Overexpression Down-regulation	1.40–1.53 0.64–0.52	[82]
AaTAR1	AaTAR1 activates the transcription of *ADS* and *CYP71AV1* genes and regulates trichome development.	Overexpression Down-regulation	1.22–1.38 0.36–0.61	[83]
	bZIP TF family			
AabZIP1	AabZIP1 activates the transcription of *ADS* and *CYP71AV1* genes through ABA signaling, indirectly promotes the expression of *AaDBR2* and *AaALDH1* genes through AaMYC2 and improves the drought resistance of *A. annua*.	Overexpression	1.40–1.60	[84,85]
AabZIP9	AabZIP9 activates the expression of *ADS* gene and positively regulates artemisinin biosynthesis.	Overexpression	1.23–1.67	[86]
AaHY5	AaHY5 positively regulates light-induced artemisinin biosynthesis through interacting with TF AaGSW1.	Overexpression Down-regulation	2.00 0.50	[87]
AaTGA6	AaTGA6 takes part in artemisinin biosynthesis regulated by SA signaling and promotes artemisinin biosynthesis by activating the TF AaERF1.	Overexpression Down-regulation	1.90–2.20 0.40–0.80	[88]

Table 3. *Cont.*

Name	Function	Expression Mode	Fold Change of Artemisinin Content	Reference
MYB TF family				
AaMYB1	AaMYB1 participates in the biosynthesis of artemisinin and GA and regulates the trichome development.	Overexpression	2.00	[89]
AaMIXTA1	AaMIXTA1 is a GST-specific TF, which regulates GST formation and cuticle biosynthesis.	Overexpression	2.00	[90]
		Down-regulation	0.75	
AaTAR2	AaTAR2 positively regulates GST development, as well as artemisinin and flavonoid biosynthesis.	Overexpression	~1.38–1.54 [1]	[91]
		Down-regulation	~0.57–0.84 [1]	
AaMYB16 (AaMIXTA-like 2)	AaMYB5 and AaMYB16 show the antagonism and regulate the development of GST by competitively binding to the *AaHD1* promoter to form AaHD1-AaMYB5 or AaHD1-AaMYB16 complexes.	Overexpression	1.34–1.56	[92]
		Down-regulation	0.73–0.81	
AaMYB5		Overexpression	0.73–0.85	
		Down-regulation	1.45–1.84	
AaMYB17	AaMYB17 is a GST-specific TF and positively regulates artemisinin biosynthesis.	Overexpression	1.88	[93]
		Down-regulation	0.75	
AaMYB15	As a GST-specific TF, AaMYB15 negatively regulates the biosynthesis of artemisinin by inhibiting the expression of *AaORA* gene.	Overexpression	~0.60–0.74 [1]	[94]
		Down-regulation	~1.52 [1]	
AaTLR1	AaTLR1 and AaTLR2 interact with AaWOX1 to form the AaTLR1-AaWOX1-AaTLR2 complex to negatively regulate GST formation and artemisinin biosynthesis.	Overexpression	0.51–0.88	[95]
		Down-regulation	1.32–1.84	
AaTLR2		Overexpression	0.57–0.81	
Other TF family				
AaZFP1	As a C2H2-type TF, AsZFP1 activates the transcription of *AaIPPI1* gene and prompts artemisinin biosynthesis.	Overexpression	1.60	[96]
AaSPL9	AaSPL9 positively regulates glandular trichomes initiation and artemisinin biosynthesis by activating TF AaHD1.	Overexpression	1.33–1.60	[97]
AaSEP4	AaSEP4, as a MADS TF, promotes artemisinin biosynthesis by directly activating TF AsGSW1.	Overexpression	1.19–1.72	[98]

[1] The change fold was calculated according to the data in the references.

AP2/ERF TF Family

The APELATA 2/ethylene response factor (AP2/ERF) TFs are one of the most important TF families in plants, and are involved in plant responses to biotic and abiotic stresses, as well as regulating various plant developmental and secondary metabolism processes [99]. AP2/ERF TF family members all contain the AP2 domain consisting of about 60 amino acid residues. According to the number of AP2 domains and whether they contain other domains, the AP2/ERF TF family is divided into AP2, ERF, DREB, RAV and Soloist 5 subfamilies [100,101].

AaERF1 and AaERF2 (Table 3), which respond to JA signaling, activate the transcription of the *ADS* and *CYP71AV1* genes by binding to the CBF2 and RAA motifs of these two gene promoters. Compared with the wild type, overexpression of AaERF1 prompted an increase in yield of artemisinin by 19–67% and overexpression of AaERF2 by 24–51% [80].

AaORA (Table 3) is a trichome specific AP2/ERF TF, which is expressed in GSTs and non-glandular T-shaped trichomes (TSTs). In AaORA-overexpressed plants, the expression levels of the *ADS*, *CYP71AV1*, *DBR2* and *AaERF1* genes are all significantly up-regulated, and the contents of artemisinin and DHAA are increased by 40–53% and 22–35%, respectively [82].

In addition, overexpression of both AaERF1 and AaORA activated the transcription of defense marker genes, *PDF1.2* (PLANT DEFENSIN1.2) and *B-CHI* (BASIC CHITINASE), thereby enhancing the resistance of *A. annua* to the pathogenic fungus *Botrytis cinerea* [82,102]. A recent study reported that AaORA formed a transcriptional complex with the TF AaTCP14, which prompted the expression of *DBR2* and *ALDH1* genes, leading to an increased accumulation of artemisinin, and the AaORA-AaTCP14 complex was induced and activated by JA [103].

AaTAR1 (TRICHOME AND ARTEMISININ REGULATOR 1) (Table 3), as an AP2/ERF TF, directly binds to CBF2 and RAA motifs, activates the expression of the *ADS* and *CYP71AV1* genes, and participates in the positive regulation of glandular trichome development and artemisinin and cuticular wax biosynthesis. The contents of artemisinin and DHAA were enhanced by 22–38% and 69–130% in AaTAR1-overexpressed plants, respectively [83].

bZIP TF Family

bZIP (basic leucine zipper) TFs contain two specific motifs, one that is located at the C-terminal, which is composed of basic amino acids for binding to the target DNA sequence, and the other that is located at the N-terminal with the leucine zipper motif, which is required for the dimerization of bZIP TFs [104]. bZIP TFs participate in the regulation of plant growth and development to cope with various environmental stresses [105].

There are 75 bZIP family members in the *A. thaliana* genome, which are divided into 10 groups based on sequence similarity and conserved motifs [106]. The group A AtbZIP proteins contain ABRE (abscisic acid-responsive element) *cis*-elements, which are involved in ABA signal transduction [107]. ABA-dependent bZIP TFs can bind conserved DNA sequences with ACGT core *cis*-elements, such as ABRE (ACGTGG/TC), G-box (CACGTG), etc. [108]. AabZIP1, AabZIP9, AaHY5 and AaTGA6 are reported to be positive regulators of artemisinin biosynthesis (Table 3).

ABA treatment can prompt the artemisinin accumulation in *A. annua* [109]. AabZIP1 is a member of the group A bZIP family, and its expression is induced by abiotic stresses such as ABA, drought and high salt. AabZIP1 can bind to the ABRE *cis*-element of the *ADS* and *CYP71AV1* gene promoters through the N-terminal C1 domain and activate the expression of the *ADS* and *CYP71AV1* genes. When AabZIP1 is overexpressed, the artemisinin content in transgenic plants showed a 1.5-fold increase [84].

A recent study reported that AabZIP1 can also directly bind to the promoter of the *AaMYC2* gene and up-regulate the expression of AaMYC2, while AaMYC2 bound to the promoters of the *AaDBR2* and *AaALDH1* genes, which indicated that AabZIP1 indirectly activated the expression of the *AaDBR2* and *AaALDH1* genes through AaMYC2 and further improved artemisinin biosynthesis [85]. In addition, AabZIP1 can directly activate the transcription of cuticle wax biosynthesis genes *AaCER1* and *AaCYP86A1* and enhance the drought resistance of *A. annua* [85]. The results showed that the AabZIP1–AaMYC2 transcriptional module was a cross-talk between the ABA and JA signaling pathways in artemisinin biosynthesis, which provided a useful candidate gene for the genetic breeding of *A. annua* with a high artemisinin content and strong drought resistance.

As a member of the group C bZIP family in *A. annua*, AabZIP9 can bind to the "ACGT" *cis*-element of the *ADS* and *CYP71AV1* gene promoters to up-regulate the expression of the *ADS* gene. Overexpression of AabZIP9 prompted the content of artemisinin and DHAA to increase by 23.2–67.1% and 34.5–92.8%, indicating that AabZIP9 was a positive regulator of artemisinin biosynthesis [86].

AaHY5, a member of the group H of bZIP TFs, regulated light-induced artemisinin biosynthesis by interacting with AaGSW1 in the WRKY family. When AaHY5 was overexpressed, the contents of artemisinin and DHAA were up-regulated approximately 2-fold [87].

SA (salicylic acid) treatment prompted the accumulation of artemisinin [110], and AaTGA6, a member of group D bZIP TFs, was involved in the artemisinin biosynthesis regulated by the SA signal [88]. As an important regulator of the SA signaling pathway, AaNPR1 up-regulated the expression of *AaTGA6*. AaTGA6 directly bound to the "TGACG" element of the *AaERF1* gene promoter and activated the expression of key genes in artemisinin biosynthesis (*ADS*, *CYP71AV1*, *DBR2*, *ALDH1*). The artemisinin content was increased by 90–120% in AaTGA6-overexpressed plants, while AaTGA3 inhibited the binding of AaTGA6 to the *AaERF1* gene promoter [88].

MYB TF Family

The MYB TFs contain a highly conserved DNA-binding domain: the MYB domain, which consists of about 52 amino acid sequence repeats (Rs), each of which form three α-helices [111]. According to the number of adjacent repeats, the MYB TFs can be divided into four classes: R2R3-MYB, 1R-MYB, 3R-MYB and 4R-MYB [111]. R2R3-MYB TFs mainly participate in cell morphogenesis [112], phytohormone signaling response [113], secondary metabolism regulation [114] and response to environmental stress [115]. The reported MYB TFs of *A. annua* all belong to the R2R3-MYB class. AaMYB1, AaMIXTA1, AaTAR2, AaMYB16 (AaMIXTA-like 2) and AaMYB17 positively regulate artemisinin biosynthesis, while AaMYB5, AaMYB15, AaTLR1 and AaTLR2 are the negative regulators in artemisinin accumulation (Table 3).

AaMYB1 promoted the glandular trichome initiation and artemisinin accumulation. Overexpression of AaMYB1 improved the density of glandular trichome, up-regulated the key genes of artemisinin biosynthesis, and significantly increased the content of artemisinin. Furthermore, both AaMYB1 and its orthologue, AtMYB61, participated in the regulation of trichome and root development, stomatal aperture and gibberellin biosynthesis [89]. AaMIXTA1 is predominantly expressed in the basal cells of GSTs and takes part in GSTs formation and cuticle biosynthesis. Overexpression of AaMIXTA1 remarkably improved the accumulation of artemisinin [90]. AaTAR2 is also involved in the initiation and development of GSTs and regulated the secondary metabolism of terpenoids and flavonoids in *A. annua*. Overexpression of AaTAR2 significantly up-regulated the expression of key genes in artemisinin and flavonoid biosynthesis pathways, such as *HMGR*, *DXS*, *CYP71A1*, *DBR2* and *PAL*, *C4H*, *CHS*, *F3H*, and *DFR*. Furthermore, the HD-ZIP (homeodomain-leucine zipper) TFs, AaHD1 and AaHD8, prompted the transcription of *AaTAR2* by binding to the L1 box (TAAAGATA) of the *AaTAR2* promoter [91]. AaMYB17 was specifically expressed in GSTs of shoot tips. In AaMYB17-overexpressed *A. annua* plants, the density of GSTs was enhanced by 1.3–1.6 times, and the content of artemisinin was increased from 8 mg/g to 15 mg/g. In AaMYB17 RNAi plants, the GSTs density and artemisinin content were significantly reduced [93].

AaMYB15 is also a GSTs-specific TF, but unlike the previous R2R3-MYB TFs, AaMYB15 participates in the negative regulation of artemisinin accumulation. Overexpression of AaMYB15 resulted in the significant suppression of the expression of key genes for artemisinin biosynthesis, such as *ADS*, *CYP71AV1*, *DBR2* and *ALDH1*, and ultimately reduced the artemisinin content. The opposite results were observed in antisense-AaMYB15 *A. annua* plants. Yeast one-hybrid results indicated that AaMYB15 could not bind to the promoters of these key genes, but rather directly bound to the promoters of *AaORA*. AaORA is an AP2/ERF TF that positively regulates artemisinin biosynthesis, and AaMYB15 inhibited artemisinin biosynthesis by suppressing the expression of the *AaORA* gene [94].

AaMYB5 and AaMYB16 are two antagonistic MYB TFs of *A. annua*, which regulate the development of GSTs. Overexpression of AaMYB5 inhibited the initiation of GSTs, resulting in the decline in artemisinin content, while AaMYB16 had the opposite effect; however, none of them could independently regulate the formation of GSTs, but played a regulatory role by competitively binding to the *AaHD1* gene promoter to form AaHD1–AaMYB5 or AaHD1–AaMYB16 complexes [92]. AaHD1 was an HD-ZIP TF that promoted the initiation of GSTs by directly activating the transcription of *AaGSW2*. In addition, JA was also associated with the AaHD1–AaMYB5/AaMYB16 regulatory network, which regulates the transcription of glandular trichome development-related genes through the AaJAZ8 protein [92].

A recent study reported two MYB TFs that negatively regulated GSTs development and artemisinin production: AaTLR1 and AaTLR2. Overexpression of AaTLR1 and AaTLR2 decreased the artemisinin content by 11.5–49.4% and 19–43%, respectively, while knockdown of the *AaTLR1* and *AaTLR2* genes resulted in the opposite effect. Yeast three-hybrid results showed that AaTLR1 and AaTLR2 interacted with AaWOX1 (WUSCHEL homeobox 1) to form the AaTLR1–AaWOX1–AaTLR2 complex, which negatively regulated the initiation of GSTs [95].

Other TF Families

In addition to the TFs discussed above, there are other TFs that also participate in the biosynthesis of artemisinin, such as AaZFP1, AaSPL9 and AaSEP4, all of which are positive regulators.

AaZFP1 (zinc finger protein 1), as a C2H2-type TF, positively regulates the transcription of the *AaIPPI1* gene by directly binding to the promoter of *AaIPPI1*. In transient AaZFP1 overexpressed *A. annua* plants, the artemisinin content was increased from 2.39 mg/g to 3.72 mg/g, which was 1.6 times the control [96]. AaSPL9 (SQUAMOSA promoter-binding protein like 9) can directly bind to the promoter of the *AaHD1* gene and activates the expression of the *AaHD1* gene. When AaSPL9 was overexpressed in *A. annua*, the density of glandular trichome increased by 45–60%, and the content of artemisinin went up by

33–60%—from 11.5 mg/g DW to 18.5 mg/g DW [97]. As a member of the MADS-box TF family, AaSEP4 is predominantly expressed in GSTs and activates the expression of TF AsGSW1 by directly binding to the CArG motif of the *AaGSW1* promoter. Overexpression of AaSEP4 remarkably up-regulated the transcription levels of the *AaGSW1*, *ADS*, *CYP71AV1*, *DBR2* and *ALDH1* genes, and compared with the wild type, the content of artemisinin increased by 19–72% in AaSEP4-overexpressed plants [98].

Tables 2 and 3 summarize the current work of genetic engineering in *A. annua* to improve the generation of artemisinin. In general, the three strategies of "overexpression", "suppression" and "global regulation" are adopted to achieve up-regulation of artemisinin accumulation. These strategies include overexpressing key enzymes that are upstream and downstream of the artemisinin biosynthesis pathway, blocking key genes in competitive metabolic pathways, and overexpressing or knocking-down TFs to globally regulate artemisinin biosynthesis. Based on the currently reported TFs of *A. annua*, the MYB TFs may act as key regulators, as they have the most binding sites in the promoters of artemisinin biosynthesis genes [116].

4.2. Genetic Engineering in Nicotiana Species

With *Nicotiana* as bioreactors, artemisinin had been successfully produced, but the yield of artemisinin was relatively low [21]. Tobacco had the advantages of large biomass and rapid growth, making it a suitable alternative plant for *A. annua* to study the heterologous biosynthesis of artemisinin [117] (Table 4).

When the *ADS* gene was expressed in *Nicotiana tabacum*, amorpha-4,11-diene, a precursor of artemisinin, can be produced in leaves at concentrations of 0.2–1.7 ng/g FW [118]. Through *Agrobacterium*-mediated transformation, five genes, including *FPPS*, *ADS*, *CYP71AV1*, *DBR2* and *ALDH1*, were transformed into *N. tabacum*, and about 4 μg/g FW of amorpha-4,11-diene was accumulated in transgenic tobacco leaves [119]. Five genes (*HMGR*, *ADS/mtADS*, *CYP71AV1*, *CPR* and *DBR2*) derived from the MVA pathway and the artemisinin biosynthesis pathway were constructed into the same vector and overexpressed in *N. tabacum*. The highest concentration of artemisinin was 6.8 μg/g DW in transgenic plants [120].

Van Herpen overexpressed the *HMGR*, *FPPS*, *ADS* and *CYP71AV1* genes of *A. annua* in *Nicotiana benthamiana* in order to produce the precursor of artemisinin—artemisinic acid in transgenic plants; however, artemisinic acid was not detected in the leaves, but the glycosylation product of artemisinic acid, artemisinic acid-12-β-diglucoside, was detected at a concentration of 39.5 μg/g FW [121]. This may be because when the key genes of artemisinin biosynthesis were heterologously expressed in *N. benthamiana*, glycosylated artemisinin precursors were mainly produced—with little free artemisinic acid or DHAA [122,123]. AaLTP3 (lipid transfer protein 3) and AaPDR2 (pleiotropic drug resistance 2) from *A. annua* prompted the accumulation of artemisinic acid and DHAA in leaf apoplasts of *N. benthamiana* and prevented the reflux of artemisinic acid and DHAA from the apoplast back into the cytoplasm [123].

Table 4. Comparison of artemisinin (or related products) biosynthesis in heterologous plants.

Gene Name	Plant Species	Expression System	Product Yield	Reference
ADS	*N. tabacum*	Nuclear	Amorpha-4,11-diene 1.7 ng/g FW	[118]
FPPS, ADS, CYP71AV1, DBR2, ALDH1	*N. tabacum*	Nuclear	Amorpha-4,11-diene 4 μg/g FW	[119]
HMGR, ADS/mtADS, CYP71AV1, CPR, DBR2	*N. tabacum*	Nuclear	Artemisinin 6.8 μg/g DW	[120]
HMGR, FPPS, ADS, CYP71AV1	*N. benthamiana*	Nuclear	Artemisinic acid-12-β-glucoside 39.5 μg/g FW	[121]

Table 4. *Cont.*

Gene Name	Plant Species	Expression System	Product Yield	Reference
AACT [1], *HMGS* [1], *HMGR* [1], *MVK* [1], *PMK* [1], *PMD* [1], *IDI* [2], *FPPS* [2], *ADS*, *CYP71AV1*, *CPR*, *aad*A [3]	*N. tabacum*	Chloroplast	Artemisinic acid 100 μg/g FW	[124]
FPPS, *ADS*, *CYP71AV1*, *CPR*, *CYPB5*, *ADH1*, *ALDH1*, *DBR2*	*N. tabacum*	Chloroplast and nuclear	Artemisinic acid 120 μg/g FW	[125]
AACT [1], *HMGS* [1], *HMGR* [1], *MVK* [1], *PMK* [1], *PMD* [1], *IDI* [2], *FPPS* [2], *ADS*, *CYP71AV1*, *CPR*, *DBR2*	*N. tabacum*	Chloroplast and nuclear	Artemisinin 0.8 mg/g DW	[126]
ADS, *CYP71AV1*, *ADH1*, *DBR2*, *ALDH1*	*P. patens*	Nuclear	Artemisinin 0.21 mg/g DW	[127]
ADS, *DBR2*, *ALDH1*	*P. patens*	Nuclear	Artemisinin 0.04 mg/g DW	[128]
HMGR, *ADS*, *CYP71AV1*, *CPR*, *DBR2*	*C. morifolium*	Nuclear	Artemisinin (Detect by GC-MS)	[129]

[1] MVA pathway genes from yeast. [2] *IDI* and *FPPS* genes from *E. coli*. [3] *aad*A, an antibiotic selection marker gene.

To address the issue of glycosylation in tobacco, researchers attempted to target key genes of the artemisinin biosynthesis to different cellular compartments, such as chloroplasts, mitochondria and the nucleus. A new transformation method, COSTREL (combinatorial super transformation of transplastomic recipient lines), was adopted to transform the complete pathway of artemisinic acid biosynthesis into chloroplasts of *N. tabacum*, and the content of the artemisinic acid in transgenic plants reached 120 μg/g FW [125]. In 2014, Kumar's laboratory transformed 12 genes of the MVA pathway and artemisinic acid biosynthesis pathway into chloroplasts of *N. tabacum* through a biolistic approach, but the yield of artemisinic acid was low (100 μg/g FW) in transgenic tobaccos and the growth of transgenic tobaccos was stunted. These results suggested that it is necessary to disperse the genes of the artemisinin biosynthetic pathway to different cellular compartments [124]. In 2016, Kumar's laboratory separately transformed these two biosynthesis pathways into chloroplast and nucleus genomes of *N. tabacum*, respectively, which did not interfere with tobacco growth. First, six genes (*AACT, HMGS, HMGRt, MVK, PMK* and *PMD*) derived from the yeast MVA pathway were introduced into the chloroplast genome by a biolistic approach to increase the yield of IPP. Then, six key genes related to the artemisinin biosynthesis pathway (*IDI, FPPS, ADS, CYP71AV1, CPR* and *DBR2*) were transformed into the nuclear genome by the *Agrobacterium*-mediated method. Finally, DBR2, CPR and CYP71AV1 were transported into the chloroplast through the targeting of chloroplast transit peptides, and the artemisinin content in transgenic tobaccos reached 0.8 mg/g DW [126].

Although various methods have been used to improve the production of artemisinin in tobacco, due to the complexity of the heterologous expression and regulation of artemisinin biosynthesis genes, as well as the high level of glycosylation catalyzed by tobacco endogenous glycosyltransferase, the yield of artemisinin was still very low, but it is of great significance to study the heterologous biosynthesis of artemisinin in tobacco.

4.3. Genetic Engineering in Physcomitrella patens and Chrysanthemum morifolium

For more than two decades, *Physcomitrella patens* has become a model organism in plant biology, biotechnology and synthetic biology [130]. The *P. patens* genome was sequenced as early as 2008 [131], and the chromosome-level genomic data were released in 2018 [132]. In addition, *P. patens* had the advantages of a high homologous recombination rate, short growth cycle and large-scale culture, making it a green cell factory for metabolic engineering [133]. Different kinds of biopharmaceuticals, such as human complement factor H (FH), epidermal growth factor (EGF), hepatocyte growth factor (HGF), etc., have been successfully produced in *P. patens* [134]. The *TXS* (taxadiene synthase) gene from

T. brevifolia was stably expressed in *P. patens*, and the content of taxadiene (the precursor of the anticancer drug paclitaxel) in the transgenic *P. patens* reached 0.05% FW [135].

In 2017, Ikram transformed five key genes of the artemisinin biosynthesis pathway (*ADS, CYP71AV1, ADH1, DBR2* and *ALDH1*) into *P. patens* through homologous recombination. A high initial yield of 0.21 mg/g DW artemisinin was detected in transgenic *P. patens* after only three days of culture [127]. In 2019, Ikram heterologously expressed these five artemisinin biosynthesis genes in *P. patens* with different combinations, and the results showed that both combinations of three different genes, ADS–CYP71AV1–ADH1 and ADS–DBR2–ALDH1, can produce artemisinin, indicating that there may be endogenous enzymes in *P. patens* that can complement the biosynthesis pathway of artemisinin. In ADS–DBR2–ALDH1 transgenic lines, 0.04 mg/g DW of artemisinin was accumulated, and artemisinin B was detected at 1.74 μg/g FW in the liquid medium [128] (Table 4).

As a moss without vascular tissue, *P. patens* has less glycosylase compared with higher plants. Using *P. patens* as a chassis to produce artemisinin, it was less likely to endogenously modify the intermediate metabolites of artemisinin; however, the yield of artemisinin in *P. patens* was very low, so new molecular tools needed to be developed for *P. patens* to improve artemisinin accumulation.

Both *Chrysanthemum morifolium* Ramat and *A. annua* belong to the *Compositae* family and are characterized by a high content of sesquiterpenes and their precursors [136]. Firsov et al. transformed five artemisinin pathway genes, including *HMGR, ADS, CYP71AV1, CPR* and *DBR2*, into *Chrysanthemum* by *Agrobacterium*-mediated transformation; artemisinin was detected in transgenic lines by both GC-MS and TLC [129,137]. The results suggested that artemisinin biosynthesis genes can be expressed in transgenic *Chrysanthemum* to generate artemisinin (Table 4).

5. Challenges and Perspectives

Malaria still poses a threat to human health, and artemisinin is the most effective drug to treat malaria. The precursor of artemisinin, artemisinic acid, was generated in yeast through metabolic engineering, which produced the semi-synthesis of artemisinin and reached the level of industrial application. This was a huge advance in the research process of artemisinin biosynthesis, thereby transferring artemisinin production from plant to yeast. With inexpensive carbon sources as substrates, artemisinic acid was synthesized by fermentation of engineering yeast, but subsequent extraction and chemical synthesis are required before artemisinic acid can be converted into artemisinin.

To date, the main source of artemisinin is extracted from *A. annua* planted in fields, but the small amount of artemisinin in *A. annua* can hardly meet the needs of the pharmaceutical market. The heterologous expression of artemisinin biosynthesis genes in *Nicotiana* and *Physcomitrella* can stably generate artemisinin, but the yield of artemisinin is very low. Therefore, we should focus on *A. annua* itself to improve the yield of artemisinin through new biotechnologies, such as gene editing or molecular design breeding, etc.

The biosynthesis of artemisinin occurs from plant to yeast and then back to plant—e.g., *A. annua, Nicotiana, P. patens*, etc. In terms of artemisinin biosynthesis, it remains to be studied whether it is more advantageous to produce artemisinin by fermentation in microorganisms, or to extract artemisinin from *A. annua* cultivated in fields. Artemisinin is usually generated in the leaves of *A. annua* and *Nicotiana* by field planting. On the contrary, yeast and *P. patens* are commonly used for fermentation, and algae can even be fermented by light. Through high-cost fermentation, the desired product can be harvested under controlled conditions with shorter production cycles. On the other hand, the cost of planting in fields is lower, but the production cycle is longer. The following are several questions and challenges worth considering:

1. The final step of artemisinin biosynthesis was generally considered to be a photo-oxidation reaction. Can the artemisinic acid produced by the fermentation of engineered yeast be used for photochemical reactions in a large-scale photoreactor to increase the production of artemisinin?

2. The biosynthesis of artemisinin was mainly carried out in the GSTs on the surface of leaves. In the future, artemisinin will be produced in all tissues of *A. annua* leaves by new gene editing biotechnology and will not be limited to GSTs.

3. The molecular modification and directed evolution of key enzymes in artemisinin biosynthesis are performed by protein engineering methods to improve the activities of these key enzymes and improve the content of artemisinin.

4. *Chlamydomonas reinhardtii* is a model organism for the study of photosynthesis, known as "green yeast". The genetic background of *C. reinhardtii* is clear, and its genome was released in 2007. Moreover, the chloroplast and nuclear transformation methods of *C. reinhardtii* have been established. At present, there is no report on the biosynthesis of artemisinin in *C. reinhardtii*. Is it possible to transform the genes of artemisinin biosynthesis into *C. reinhardtii* through codon optimization and generate artemisinin by photo-fermentation?

5. Various TFs have been reported to regulate artemisinin biosynthesis. Are these TFs acting independently or cooperatively, and which TF is the most important?

6. Conclusions

Based on the analysis of the *A. annua* genome, the biosynthesis pathway of artemisinin and its regulatory mechanism were elucidated, and the genes of key enzymes as well as TFs involved in artemisinin biosynthesis were isolated and identified. With microorganisms or plants as the chassis, the biosynthesis pathway of artemisinin was reconstructed to increase the content of artemisinin or its precursors in order to obtain artemisinin in an efficient and low-cost manner. Such research strategies provided a paradigm for the use of synthetic biology methods to produce natural products of medicinal plants. Many natural products have important physiological activities and are important sources for the development of new drugs. We are confident that by reconstructing and modulating the biosynthesis pathway through the synthetic biology approach, the efficient biosynthesis of rare and important natural products can be achieved, thereby addressing both the quality and quantity issues. In summary, there are broad application prospects for the promotion of high-value natural products from laboratory to industrialization.

Author Contributions: Conceptualization, L.Z., Y.Z. and H.J.; resources, Y.Z. and H.J.; data curation, L.Z., H.J. and Y.H.; writing—original draft preparation, L.Z.; writing—review and editing, M.W.; supervision, X.Z. and W.F.; funding acquisition, L.Z., X.Z. and W.F. All authors have read and agreed to the published version of the manuscript.

Funding: This work was supported by National Key Research and Development Project (2017YFC1702800, 2019YFC1708802), Major Science and Technology Project in Henan Province (171100310500), Henan Province High-Level Personnel Special Support "ZhongYuan One Thousand People Plan" (ZYQR201810080), Training Program for Young Teachers in Colleges and Universities of Henan Province (2021GGJS086).

Institutional Review Board Statement: Not applicable.

Informed Consent Statement: Not applicable.

Data Availability Statement: The data presented in this study are available on request from the corresponding author.

Acknowledgments: We thank Kai Tan for assisting in the preparation of this manuscript.

Conflicts of Interest: The authors declare no conflict of interest.

References

1. Yang, L.; Yang, C.; Li, C.; Zhao, Q.; Liu, L.; Fang, X.; Chen, X.Y. Recent advances in biosynthesis of bioactive compounds in traditional Chinese medicinal plants. *Sci. Bull.* **2016**, *61*, 3–17. [CrossRef]
2. Zwenger, S.; Basu, C. Plant terpenoids: Applications and future potentials. *Microbiol. Mol. Biol. Rev.* **2008**, *3*, 1–7. [CrossRef]
3. Nagegowda, D.A.; Gupta, P. Advances in biosynthesis, regulation, and metabolic engineering of plant specialized terpenoids. *Plant Sci.* **2020**, *294*, 110457. [CrossRef]

4. Alqahtani, F.Y.; Aleanizy, F.S.; El Tahir, E.; Alkahtani, H.M.; AlQuadeib, B.T. Chapter Three—Paclitaxel. In *Profiles of Drug Substances, Excipients and Related Methodology*; Academic Press: Cambridge, MA, USA, 2019; Volume 44, pp. 205–238.
5. Hewlings, S.J.; Kalman, D.S. Curcumin: A review of its effects on human health. *Foods* **2017**, *6*, 92. [CrossRef] [PubMed]
6. Khan, M.; Ali, S.; Manghwar, H.; Saqib, S.; Ullah, F.; Ayaz, A.; Zaman, W. Melatonin function and crosstalk with other phytohormones under normal and stressful conditions. *Genes* **2022**, *13*, 1699. [CrossRef]
7. Tu, Y. The discovery of artemisinin (qinghaosu) and gifts from Chinese medicine. *Nat. Med.* **2011**, *17*, 1217–1220. [CrossRef] [PubMed]
8. World Health Organization. World Malaria Report 2021. Available online: https://www.who.int/teams/global-malaria-programme/reports/world-malaria-report-2021 (accessed on 6 December 2021).
9. Wong, Y.K.; Xu, C.; Kalesh, K.A.; He, Y.; Lin, Q.; Wong, W.S.F.; Shen, H.M.; Wang, J. Artemisinin as an anticancer drug: Recent advances in target profiling and mechanisms of action. *Med. Res. Rev.* **2017**, *37*, 1492–1517. [CrossRef] [PubMed]
10. Li, J.; Casteels, T.; Frogne, T.; Ingvorsen, C.; Honoré, C.; Courtney, M.; Huber, K.V.M.; Schmitner, N.; Kimmel, R.A.; Romanov, R.A.; et al. Artemisinins target $GABA_A$ receptor signaling and impair α cell identity. *Cell* **2017**, *168*, 86–100.e115. [CrossRef]
11. Zheng, H.; Colvin, C.J.; Johnson, B.K.; Kirchhoff, P.D.; Wilson, M.; Jorgensen-Muga, K.; Larsen, S.D.; Abramovitch, R.B. Inhibitors of *Mycobacterium tuberculosis* DosRST signaling and persistence. *Nat. Chem. Biol.* **2017**, *13*, 218–225. [CrossRef]
12. Wallaart, T.E.; Pras, N.; Beekman, A.C.; Quax, W. Seasonal variation of artemisinin and its biosynthetic precursors in plants of *Artemisia annua* of different geographical origin: Proof for the existence of chemotypes. *Planta Med.* **2000**, *66*, 57–62. [CrossRef]
13. Schmid, G.; Hofheinz, W. Total synthesis of qinghaosu. *J. Am. Chem. Soc.* **1983**, *105*, 624–625. [CrossRef]
14. Zhu, X.; Liu, X.; Liu, T.; Wang, Y.; Ahmed, N.; Li, Z.; Jiang, H. Synthetic biology of plant natural products: From pathway elucidation to engineered biosynthesis in plant cells. *Plant Commun.* **2021**, *2*, 100229. [CrossRef] [PubMed]
15. Vranová, E.; Coman, D.; Gruissem, W. Network analysis of the MVA and MEP pathways for isoprenoid synthesis. *Annu. Rev. Plant Biol.* **2013**, *64*, 665–700. [CrossRef] [PubMed]
16. Mercke, P.; Bengtsson, M.; Bouwmeester, H.J.; Posthumus, M.A.; Brodelius, P.E. Molecular cloning, expression, and characterization of amorpha-4,11-diene synthase, a key enzyme of artemisinin biosynthesis in *Artemisia annua* L. *Arch. Biochem. Biophys.* **2000**, *381*, 173–180. [CrossRef] [PubMed]
17. Teoh, K.H.; Polichuk, D.R.; Reed, D.W.; Nowak, G.; Covello, P.S. *Artemisia annua* L. (Asteraceae) trichome-specific cDNAs reveal CYP71AV1, a cytochrome P450 with a key role in the biosynthesis of the antimalarial sesquiterpene lactone artemisinin. *FEBS Lett.* **2006**, *580*, 1411–1416. [CrossRef] [PubMed]
18. Ro, D.K.; Paradise, E.M.; Ouellet, M.; Fisher, K.J.; Newman, K.L.; Ndungu, J.M.; Ho, K.A.; Eachus, R.A.; Ham, T.S.; Kirby, J.; et al. Production of the antimalarial drug precursor artemisinic acid in engineered yeast. *Nature* **2006**, *440*, 940–943. [CrossRef] [PubMed]
19. Teoh, K.H.; Polichuk, D.R.; Reed, D.W.; Covello, P.S. Molecular cloning of an aldehyde dehydrogenase implicated in artemisinin biosynthesis in *Artemisia annua*. *Botany* **2009**, *87*, 635–642. [CrossRef]
20. Zhang, Y.; Teoh, K.H.; Reed, D.W.; Maes, L.; Goossens, A.; Olson, D.J.H.; Ross, A.R.S.; Covello, P.S. The molecular cloning of artemisinic aldehyde Δ11(13) reductase and its role in glandular trichome-dependent biosynthesis of artemisinin in *Artemisia annua*. *J. Biol. Chem.* **2008**, *283*, 21501–21508. [CrossRef] [PubMed]
21. Ikram, N.K.B.K.; Simonsen, H.T. A review of biotechnological artemisinin production in plants. *Front. Plant Sci.* **2017**, *8*, 1966. [CrossRef] [PubMed]
22. Al-Khayri, J.M.; Sudheer, W.N.; Lakshmaiah, V.V.; Mukherjee, E.; Nizam, A.; Thiruvengadam, M.; Nagella, P.; Alessa, F.M.; Al-Mssallem, M.Q.; Rezk, A.A.; et al. Biotechnological approaches for production of artemisinin, an anti-malarial drug from *Artemisia annua* L. *Molecules* **2022**, *27*, 3040. [CrossRef]
23. Bryant, L.; Flatley, B.; Patole, C.; Brown, G.D.; Cramer, R. Proteomic analysis of *Artemisia annua*—Towards elucidating the biosynthetic pathways of the antimalarial pro-drug artemisinin. *BMC Plant Biol.* **2015**, *15*, 175. [CrossRef] [PubMed]
24. Sun, Y.; Shang, L.; Zhu, Q.H.; Fan, L.; Guo, L. Twenty years of plant genome sequencing: Achievements and challenges. *Trends Plant Sci.* **2022**, *27*, 391–401. [CrossRef] [PubMed]
25. Martin, V.J.J.; Pitera, D.J.; Withers, S.T.; Newman, J.D.; Keasling, J.D. Engineering a mevalonate pathway in *Escherichia coli* for production of terpenoids. *Nat. Biotechnol.* **2003**, *21*, 796–802. [CrossRef] [PubMed]
26. Newman, J.D.; Marshall, J.; Chang, M.; Nowroozi, F.; Paradise, E.; Pitera, D.; Newman, K.L.; Keasling, J.D. High-level production of amorpha-4,11-diene in a two-phase partitioning bioreactor of metabolically engineered *Escherichia coli*. *Biotechnol. Bioeng.* **2006**, *95*, 684–691. [CrossRef] [PubMed]
27. Tsuruta, H.; Paddon, C.J.; Eng, D.; Lenihan, J.R.; Horning, T.; Anthony, L.C.; Regentin, R.; Keasling, J.D.; Renninger, N.S.; Newman, J.D. High-level production of amorpha-4,11-diene, a precursor of the antimalarial agent artemisinin, in *Escherichia coli*. *PLoS ONE* **2009**, *4*, e4489. [CrossRef] [PubMed]
28. Chang, M.C.Y.; Eachus, R.A.; Trieu, W.; Ro, D.K.; Keasling, J.D. Engineering *Escherichia coli* for production of functionalized terpenoids using plant P450s. *Nat. Chem. Biol.* **2007**, *3*, 274–277. [CrossRef]
29. Anthony, J.R.; Anthony, L.C.; Nowroozi, F.; Kwon, G.; Newman, J.D.; Keasling, J.D. Optimization of the mevalonate-based isoprenoid biosynthetic pathway in *Escherichia coli* for production of the anti-malarial drug precursor amorpha-4,11-diene. *Metab. Eng.* **2009**, *11*, 13–19. [CrossRef]

30. Westfall, P.J.; Pitera, D.J.; Lenihan, J.R.; Eng, D.; Woolard, F.X.; Regentin, R.; Horning, T.; Tsuruta, H.; Melis, D.J.; Owens, A.; et al. Production of amorphadiene in yeast, and its conversion to dihydroartemisinic acid, precursor to the antimalarial agent artemisinin. *Proc. Natl. Acad. Sci. USA* **2012**, *109*, E111–E118. [CrossRef]
31. Paddon, C.J.; Westfall, P.J.; Pitera, D.J.; Benjamin, K.; Fisher, K.; McPhee, D.; Leavell, M.D.; Tai, A.; Main, A.; Eng, D.; et al. High-level semi-synthetic production of the potent antimalarial artemisinin. *Nature* **2013**, *496*, 528–532. [CrossRef]
32. Graham, I.A.; Besser, K.; Blumer, S.; Branigan, C.A.; Czechowski, T.; Elias, L.; Guterman, I.; Harvey, D.; Isaac, P.G.; Khan, A.M.; et al. The genetic map of *Artemisia annua* L. identifies loci affecting yield of the antimalarial drug artemisinin. *Science* **2010**, *327*, 328–331. [CrossRef]
33. Shen, Q.; Zhang, L.; Liao, Z.; Wang, S.; Yan, T.; Shi, P.; Liu, M.; Fu, X.; Pan, Q.; Wang, Y.; et al. The genome of *Artemisia annua* provides insight into the evolution of asteraceae family and artemisinin biosynthesis. *Mol. Plant* **2018**, *11*, 776–788. [CrossRef] [PubMed]
34. Ram, M.; Khan, M.A.; Jha, P.; Khan, S.; Kiran, U.; Ahmad, M.M.; Javed, S.; Abdin, M.Z. HMG-CoA reductase limits artemisinin biosynthesis and accumulation in *Artemisia annua* L. plants. *Acta Physiol. Plant.* **2010**, *32*, 859–866. [CrossRef]
35. Nafis, T.; Akmal, M.; Ram, M.; Alam, P.; Ahlawat, S.; Mohd, A.; Abdin, M.Z. Enhancement of artemisinin content by constitutive expression of the HMG-CoA reductase gene in high-yielding strain of *Artemisia annua* L. *Plant Biotechnol. Rep.* **2011**, *5*, 53–60. [CrossRef]
36. Xiang, L.; Zeng, L.; Yuan, Y.; Chen, M.; Wang, F.; Liu, X.; Zeng, L.; Lan, X.; Liao, Z. Enhancement of artemisinin biosynthesis by overexpressing *dxr*, *cyp71av1* and *cpr* in the plants of *Artemisia annua* L. *Plant Omics* **2012**, *5*, 503–507.
37. Ma, D.; Li, G.; Zhu, Y.; Xie, D. Overexpression and suppression of *Artemisia annua* 4-Hydroxy-3-methylbut-2-enyl diphosphate reductase 1 gene (*AaHDR1*) differentially regulate artemisinin and terpenoid biosynthesis. *Front. Plant Sci.* **2017**, *8*, 77. [CrossRef] [PubMed]
38. Ma, D.; Li, G.; Alejos-Gonzalez, F.; Zhu, Y.; Xue, Z.; Wang, A.; Zhang, H.; Li, X.; Ye, H.; Wang, H.; et al. Overexpression of a type-I isopentenyl pyrophosphate isomerase of *Artemisia annua* in the cytosol leads to high arteannuin B production and artemisinin increase. *Plant J.* **2017**, *91*, 466–479. [CrossRef] [PubMed]
39. Matsushita, Y.; Kang, W.; Charlwood, B.V. Cloning and analysis of a cDNA encoding farnesyl diphosphate synthase from *Artemisia annua*. *Gene* **1996**, *172*, 207–209. [CrossRef]
40. Han, J.; Liu, B.; Ye, H.; Wang, H.; Li, Z.; Li, G. Effects of overexpression of the endogenous farnesyl diphosphate synthase on the artemisinin content in *Artemisia annua* L. *J. Integr. Plant Biol.* **2006**, *48*, 482–487. [CrossRef]
41. Banyai, W.; Kirdmanee, C.; Mii, M.; Supaibulwatana, K. Overexpression of farnesyl pyrophosphate synthase (*FPS*) gene affected artemisinin content and growth of *Artemisia annua* L. *Plant Cell Tissue Organ Cult.* **2010**, *103*, 255–265. [CrossRef]
42. Chen, D.; Ye, H.; Li, G. Expression of a chimeric farnesyl diphosphate synthase gene in *Artemisia annua* L. transgenic plants via *Agrobacterium tumefaciens*-mediated transformation. *Plant Sci.* **2000**, *155*, 179–185. [CrossRef]
43. Wang, H.; Olofsson, L.; Lundgren, A.; Brodelius, P. Trichome-specific expression of amorpha-4,11-diene synthase, a key enzyme of artemisinin biosynthesis in *Artemisia annua* L., as reported by a promoter-GUS fusion. *Am. J. Plant Sci.* **2011**, *2*, 619–628. [CrossRef]
44. Liao, B.; Shen, X.; Xiang, L.; Guo, S.; Chen, S.; Meng, Y.; Liang, Y.; Ding, D.; Bai, J.; Zhang, D.; et al. Allele-aware chromosome-level genome assembly of *Artemisia annua* reveals the correlation between *ADS* expansion and artemisinin yield. *Mol. Plant* **2022**, *15*, 1310–1328. [CrossRef]
45. Ma, C.; Wang, H.; Lu, X.; Wang, H.; Xu, G.; Liu, B. Terpenoid metabolic profiling analysis of transgenic *Artemisia annua* L. by comprehensive two-dimensional gas chromatography time-of-flight mass spectrometry. *Metabolomics* **2009**, *5*, 497–506. [CrossRef]
46. Yuan, Y.; Liu, W.; Zhang, Q.; Xiang, L.; Liu, X.; Chen, M.; Lin, Z.; Wang, Q.; Liao, Z. Overexpression of artemisinic aldehyde Δ11 (13) reductase gene–enhanced artemisinin and its relative metabolite biosynthesis in transgenic *Artemisia annua* L. *Biotechnol. Appl. Biochem.* **2015**, *62*, 17–23. [CrossRef]
47. Alam, P.; Abdin, M.Z. Over-expression of HMG-CoA reductase and amorpha-4,11-diene synthase genes in *Artemisia annua* L. and its influence on artemisinin content. *Plant Cell Rep.* **2011**, *30*, 1919–1928. [CrossRef] [PubMed]
48. Han, J.; Wang, H.; Kanagarajan, S.; Hao, M.; Lundgren, A.; Brodelius, P.E. Promoting artemisinin biosynthesis in *Artemisia annua* plants by substrate channeling. *Mol. Plant* **2016**, *9*, 946–948. [CrossRef] [PubMed]
49. Shen, Q.; Chen, Y.; Wang, T.; Wu, S.; Xu, L.; Zhang, L.; Zhang, F.; Jiang, W.; Wang, G.; Tang, K. Overexpression of the cytochrome P450 monooxygenase (*cyp71av1*) and cytochrome P450 reductase (*cpr*) genes increased artemisinin content in *Artemisia annua* (Asteraceae). *Genet. Mol. Res.* **2012**, *11*, 3298–3309. [CrossRef] [PubMed]
50. Chen, Y.; Shen, Q.; Wang, Y.; Wang, T.; Wu, S.; Zhang, L.; Lu, X.; Zhang, F.; Jiang, W.; Qiu, B.; et al. The stacked over-expression of *FPS*, *CYP71AV1* and *CPR* genes leads to the increase of artemisinin level in *Artemisia annua* L. *Plant Biotechnol. Rep.* **2013**, *7*, 287–295. [CrossRef]
51. Shi, P.; Fu, X.; Liu, M.; Shen, Q.; Jiang, W.; Li, L.; Sun, X.; Tang, K. Promotion of artemisinin content in *Artemisia annua* by overexpression of multiple artemisinin biosynthetic pathway genes. *Plant Cell Tissue Organ Cult.* **2017**, *129*, 251–259. [CrossRef]
52. Aquil, S.; Husaini, A.M.; Abdin, M.Z.; Rather, G.M. Overexpression of the HMG-CoA reductase gene leads to enhanced artemisinin biosynthesis in transgenic *Artemisia annua* plants. *Planta Med.* **2009**, *75*, 1453–1458. [CrossRef]
53. Tang, K.; Shen, Q.; Chen, Y.; Wang, T.; Wu, S.; Lu, X. Overexpression DBR2 Gene Increased Artemisinin Content in *Artemisia annua* L. China Patent Application CN201210014227, 8 July 2015.

54. Tang, K.; Chen, Y.; Shen, Q.; Wang, T.; Wu, S.; Wang, G. Overexpression ALDH1 Gene Increased Artemisinin Content in *Artemisia annua* L. China Patent Application CN201210014242, 22 August 2012.
55. Wang, Y.; Jing, F.; Yu, S.; Chen, Y.; Wang, T.; Liu, P.; Wang, G.; Sun, X.; Tang, K. Co-overexpression of the *HMGR* and *FPS* genes enhances artemisinin content in *Artemisia annua* L. *J. Med. Plants Res.* **2011**, *5*, 3396–3403.
56. Lu, X.; Shen, Q.; Zhang, L.; Zhang, F.; Jiang, W.; Lv, Z.; Yan, T.; Fu, X.; Wang, G.; Tang, K. Promotion of artemisinin biosynthesis in transgenic *Artemisia annua* by overexpressing *ADS*, *CYP71AV1* and *CPR* genes. *Ind. Crops Prod.* **2013**, *49*, 380–385. [CrossRef]
57. Zhang, L.; Jing, F.; Li, F.; Li, M.; Wang, Y.; Wang, G.; Sun, X.; Tang, K. Development of transgenic *Artemisia annua* (Chinese wormwood) plants with an enhanced content of artemisinin, an effective anti-malarial drug, by hairpin-RNA-mediated gene silencing. *Biotechnol. Appl. Biochem.* **2009**, *52*, 199–207. [CrossRef] [PubMed]
58. Chen, J.; Fang, H.; Ji, Y.; Pu, G.; Guo, Y.; Huang, L.; Du, Z.; Liu, B.; Ye, H.; Li, G.; et al. Artemisinin biosynthesis enhancement in transgenic *Artemisia annua* plants by downregulation of the β-caryophyllene synthase gene. *Planta Med.* **2011**, *77*, 1759–1765. [CrossRef] [PubMed]
59. Lv, Z.; Zhang, F.; Pan, Q.; Fu, X.; Jiang, W.; Shen, Q.; Yan, T.; Shi, P.; Lu, X.; Sun, X.; et al. Branch pathway blocking in *Artemisia annua* is a useful method for obtaining high yield artemisinin. *Plant Cell Physiol.* **2016**, *57*, 588–602. [CrossRef] [PubMed]
60. Tang, K.; Shen, Q.; Yan, T.; Fu, X. Transgenic approach to increase artemisinin content in *Artemisia annua* L. *Plant Cell Rep.* **2014**, *33*, 605–615. [CrossRef] [PubMed]
61. Wani, K.I.; Choudhary, S.; Zehra, A.; Naeem, M.; Weathers, P.; Aftab, T. Enhancing artemisinin content in and delivery from *Artemisia annua*: A review of alternative, classical, and transgenic approaches. *Planta* **2021**, *254*, 29. [CrossRef] [PubMed]
62. Cai, Y.; Jia, J.W.; Crock, J.; Lin, Z.X.; Chen, X.Y.; Croteau, R. A cDNA clone for β-caryophyllene synthase from *Artemisia annua*. *Phytochemistry* **2002**, *61*, 523–529. [CrossRef]
63. Rydén, A.M.; Ruyter-Spira, C.; Quax, W.J.; Osada, H.; Muranaka, T.; Kayser, O.; Bouwmeester, H. The molecular cloning of dihydroartemisinic aldehyde reductase and its implication in artemisinin biosynthesis in *Artemisia annua*. *Planta Med.* **2010**, *76*, 1778–1783. [CrossRef] [PubMed]
64. Ranjbar, M.; Naghavi, M.R.; Alizadeh, H.; Soltanloo, H. Expression of artemisinin biosynthesis genes in eight *Artemisia* species at three developmental stages. *Ind. Crops Prod.* **2015**, *76*, 836–843. [CrossRef]
65. Rushton, P.J.; Somssich, I.E.; Ringler, P.; Shen, Q.J. WRKY transcription factors. *Trends Plant Sci.* **2010**, *15*, 247–258. [CrossRef] [PubMed]
66. Chen, X.; Li, C.; Wang, H.; Guo, Z. WRKY transcription factors: Evolution, binding, and action. *Phytopathol. Res.* **2019**, *1*, 13. [CrossRef]
67. Ma, D.; Pu, G.; Lei, C.; Ma, L.; Wang, H.; Guo, Y.; Chen, J.; Du, Z.; Wang, H.; Li, G.; et al. Isolation and characterization of AaWRKY1, an *Artemisia annua* transcription factor that regulates the amorpha-4,11-diene synthase gene, a key gene of artemisinin biosynthesis. *Plant Cell Physiol.* **2009**, *50*, 2146–2161. [CrossRef]
68. Jiang, W.; Fu, X.; Pan, Q.; Tang, Y.; Shen, Q.; Lv, Z.; Yan, T.; Shi, P.; Li, L.; Zhang, L.; et al. Overexpression of *AaWRKY1* leads to an enhanced content of artemisinin in *Artemisia annua*. *BioMed Res. Int.* **2016**, *2016*, 7314971. [CrossRef] [PubMed]
69. Huang, H.; Xing, S.; Tang, K.; Jiang, W. AaWRKY4 upregulates artemisinin content through boosting the expressions of key enzymes in artemisinin biosynthetic pathway. *Plant Cell Tissue Organ Cult.* **2021**, *146*, 97–105. [CrossRef]
70. Fu, X.; Peng, B.; Hassani, D.; Xie, L.; Liu, H.; Li, Y.; Chen, T.; Liu, P.; Tang, Y.; Li, L.; et al. AaWRKY9 contributes to light- and jasmonate-mediated to regulate the biosynthesis of artemisinin in *Artemisia annua*. *New Phytol.* **2021**, *231*, 1858–1874. [CrossRef] [PubMed]
71. Chen, T.; Li, Y.; Xie, L.; Hao, X.; Liu, H.; Qin, W.; Wang, C.; Yan, X.; Wu-Zhang, K.; Yao, X.; et al. AaWRKY17, a positive regulator of artemisinin biosynthesis, is involved in resistance to *Pseudomonas syringae* in *Artemisia annua*. *Hortic. Res.* **2021**, *8*, 217. [CrossRef] [PubMed]
72. Chen, M.; Yan, T.; Shen, Q.; Lu, X.; Pan, Q.; Huang, Y.; Tang, Y.; Fu, X.; Liu, M.; Jiang, W.; et al. GLANDULAR TRICHOME-SPECIFIC WRKY 1 promotes artemisinin biosynthesis in *Artemisia annua*. *New Phytol.* **2017**, *214*, 304–316. [CrossRef] [PubMed]
73. Ma, Y.N.; Xu, D.B.; Yan, X.; Wu, Z.K.; Kayani, S.I.; Shen, Q.; Fu, X.Q.; Xie, L.H.; Hao, X.L.; Hassani, D.; et al. Jasmonate- and abscisic acid-activated AaGSW1-AaTCP15/AaORA transcriptional cascade promotes artemisinin biosynthesis in *Artemisia annua*. *Plant Biotechnol. J.* **2021**, *19*, 1412–1428. [CrossRef] [PubMed]
74. Xie, L.; Yan, T.; Li, L.; Chen, M.; Ma, Y.; Hao, X.; Fu, X.; Shen, Q.; Huang, Y.; Qin, W.; et al. The WRKY transcription factor AaGSW2 promotes glandular trichome initiation in *Artemisia annua*. *J. Exp. Bot.* **2021**, *72*, 1691–1701. [CrossRef] [PubMed]
75. Sun, X.; Wang, Y.; Sui, N. Transcriptional regulation of bHLH during plant response to stress. *Biochem. Biophys. Res. Commun.* **2018**, *503*, 397–401. [CrossRef] [PubMed]
76. Ji, Y.; Xiao, J.; Shen, Y.; Ma, D.; Li, Z.; Pu, G.; Li, X.; Huang, L.; Liu, B.; Ye, H.; et al. Cloning and characterization of AabHLH1, a bHLH transcription factor that positively regulates artemisinin biosynthesis in *Artemisia annua*. *Plant Cell Physiol.* **2014**, *55*, 1592–1604. [CrossRef]
77. Li, L.; Hao, X.; Liu, H.; Wang, W.; Fu, X.; Ma, Y.; Shen, Q.; Chen, M.; Tang, K. Jasmonic acid-responsive AabHLH1 positively regulates artemisinin biosynthesis in *Artemisia annua*. *Biotechnol. Appl. Biochem.* **2019**, *66*, 369–375. [CrossRef]
78. Shen, Q.; Lu, X.; Yan, T.; Fu, X.; Lv, Z.; Zhang, F.; Pan, Q.; Wang, G.; Sun, X.; Tang, K. The jasmonate-responsive AaMYC2 transcription factor positively regulates artemisinin biosynthesis in *Artemisia annua*. *New Phytol.* **2016**, *210*, 1269–1281. [CrossRef] [PubMed]

79. Xiang, L.; Jian, D.; Zhang, F.; Yang, C.; Bai, G.; Lan, X.; Chen, M.; Tang, K.; Liao, Z. The cold-induced transcription factor bHLH112 promotes artemisinin biosynthesis indirectly via ERF1 in *Artemisia annua*. *J. Exp. Bot.* **2019**, *70*, 4835–4848. [CrossRef]
80. Yu, Z.X.; Li, J.X.; Yang, C.Q.; Hu, W.L.; Wang, L.J.; Chen, X.Y. The jasmonate-responsive AP2/ERF transcription factors AaERF1 and AaERF2 positively regulate artemisinin biosynthesis in *Artemisia annua* L. *Mol. Plant* **2012**, *5*, 353–365. [CrossRef] [PubMed]
81. Shen, Q.; Huang, H.; Xie, L.; Hao, X.; Kayani, S.-I.; Liu, H.; Qin, W.; Chen, T.; Pan, Q.; Liu, P.; et al. Basic helix-loop-helix transcription factors AabHLH2 and AabHLH3 function antagonistically with AaMYC2 and are negative regulators in artemisinin biosynthesis. *Front. Plant Sci.* **2022**, *13*, 885622. [CrossRef] [PubMed]
82. Lu, X.; Zhang, L.; Zhang, F.; Jiang, W.; Shen, Q.; Zhang, L.; Lv, Z.; Wang, G.; Tang, K. *AaORA*, a trichome-specific AP2/ERF transcription factor of *Artemisia annua*, is a positive regulator in the artemisinin biosynthetic pathway and in disease resistance to *Botrytis cinerea*. *New Phytol.* **2013**, *198*, 1191–1202. [CrossRef] [PubMed]
83. Tan, H.; Xiao, L.; Gao, S.; Li, Q.; Chen, J.; Xiao, Y.; Ji, Q.; Chen, R.; Chen, W.; Zhang, L. *TRICHOME AND ARTEMISININ REGULATOR* 1 is required for trichome development and artemisinin biosynthesis in *Artemisia annua*. *Mol. Plant* **2015**, *8*, 1396–1411. [CrossRef] [PubMed]
84. Zhang, F.; Fu, X.; Lv, Z.; Lu, X.; Shen, Q.; Zhang, L.; Zhu, M.; Wang, G.; Sun, X.; Liao, Z.; et al. A basic leucine zipper transcription factor, AabZIP1, connects abscisic acid signaling with artemisinin biosynthesis in *Artemisia annua*. *Mol. Plant* **2015**, *8*, 163–175. [CrossRef] [PubMed]
85. Shu, G.; Tang, Y.; Yuan, M.; Wei, N.; Zhang, F.; Yang, C.; Lan, X.; Chen, M.; Tang, K.; Xiang, L.; et al. Molecular insights into AabZIP1-mediated regulation on artemisinin biosynthesis and drought tolerance in *Artemisia annua*. *Acta Pharm. Sin. B* **2022**, *12*, 1500–1513. [CrossRef] [PubMed]
86. Shen, Q.; Huang, H.; Zhao, Y.; Xie, L.; He, Q.; Zhong, Y.; Wang, Y.; Wang, Y.; Tang, K. The transcription factor AabZIP9 positively regulates the biosynthesis of artemisinin in *Artemisia annua*. *Front. Plant Sci.* **2019**, *10*, 1294. [CrossRef] [PubMed]
87. Hao, X.; Zhong, Y.; Nützmann, H.W.; Fu, X.; Yan, T.; Shen, Q.; Chen, M.; Ma, Y.; Zhao, J.; Osbourn, A.; et al. Light-induced artemisinin biosynthesis is regulated by the bZIP transcription factor AaHY5 in *Artemisia annua*. *Plant Cell Physiol.* **2019**, *60*, 1747–1760. [CrossRef] [PubMed]
88. Lv, Z.; Guo, Z.; Zhang, L.; Zhang, F.; Jiang, W.; Shen, Q.; Fu, X.; Yan, T.; Shi, P.; Hao, X.; et al. Interaction of bZIP transcription factor TGA6 with salicylic acid signaling modulates artemisinin biosynthesis in *Artemisia annua*. *J. Exp. Bot.* **2019**, *70*, 3969–3979. [CrossRef] [PubMed]
89. Matías-Hernández, L.; Jiang, W.; Yang, K.; Tang, K.; Brodelius, P.E.; Pelaz, S. AaMYB1 and its orthologue AtMYB61 affect terpene metabolism and trichome development in *Artemisia annua* and *Arabidopsis thaliana*. *Plant J.* **2017**, *90*, 520–534. [CrossRef]
90. Shi, P.; Fu, X.; Shen, Q.; Liu, M.; Pan, Q.; Tang, Y.; Jiang, W.; Lv, Z.; Yan, T.; Ma, Y.; et al. The roles of *AaMIXTA1* in regulating the initiation of glandular trichomes and cuticle biosynthesis in *Artemisia annua*. *New Phytol.* **2018**, *217*, 261–276. [CrossRef]
91. Zhou, Z.; Tan, H.; Li, Q.; Li, Q.; Wang, Y.; Bu, Q.; Li, Y.; Wu, Y.; Chen, W.; Zhang, L. *TRICHOME AND ARTEMISININ REGULATOR* 2 positively regulates trichome development and artemisinin biosynthesis in *Artemisia annua*. *New Phytol.* **2020**, *228*, 932–945. [CrossRef] [PubMed]
92. Xie, L.; Yan, T.; Li, L.; Chen, M.; Hassani, D.; Li, Y.; Qin, W.; Liu, H.; Chen, T.; Fu, X.; et al. An HD-ZIP-MYB complex regulates glandular secretory trichome initiation in *Artemisia annua*. *New Phytol.* **2021**, *231*, 2050–2064. [CrossRef]
93. Qin, W.; Xie, L.; Li, Y.; Liu, H.; Fu, X.; Chen, T.; Hassani, D.; Li, L.; Sun, X.; Tang, K. An R2R3-MYB transcription factor positively regulates the glandular secretory trichome initiation in *Artemisia annua* L. *Front. Plant Sci.* **2021**, *12*, 657156. [CrossRef] [PubMed]
94. Wu, Z.; Li, L.; Liu, H.; Yan, X.; Ma, Y.; Li, Y.; Chen, T.; Wang, C.; Xie, L.; Hao, X.; et al. AaMYB15, an R2R3-MYB TF in *Artemisia annua*, acts as a negative regulator of artemisinin biosynthesis. *Plant Sci.* **2021**, *308*, 110920. [CrossRef]
95. Lv, Z.; Li, J.; Qiu, S.; Qi, F.; Su, H.; Bu, Q.; Jiang, R.; Tang, K.; Zhang, L.; Chen, W. The transcription factors TLR1 and TLR2 negatively regulate trichome density and artemisinin levels in *Artemisia annua*. *J. Integr. Plant Biol.* **2022**, *64*, 1212–1228. [CrossRef] [PubMed]
96. Deng, Y.A.; Li, L.; Peng, Q.; Feng, L.F.; Yang, J.F.; Zhan, R.T.; Ma, D.M. Isolation and characterization of AaZFP1, a C2H2 zinc finger protein that regulates the *AaIPPI1* gene involved in artemisinin biosynthesis in *Artemisia annua*. *Planta* **2022**, *255*, 122. [CrossRef] [PubMed]
97. He, Y.; Fu, X.; Li, L.; Sun, X.; Tang, K.; Zhao, J. AaSPL9 affects glandular trichomes initiation by positively regulating expression of *AaHD1* in *Artemisia annua* L. *Plant Sci.* **2022**, *317*, 111172. [CrossRef] [PubMed]
98. Chen, T.T.; Yao, X.H.; Liu, H.; Li, Y.P.; Qin, W.; Yan, X.; Wang, X.Y.; Peng, B.W.; Zhang, Y.J.; Shao, J.; et al. MADS-box gene *AaSEP4* promotes artemisinin biosynthesis in *Artemisia annua*. *Front. Plant Sci.* **2022**, *13*, 982317. [CrossRef]
99. Agarwal, P.K.; Agarwal, P.; Reddy, M.K.; Sopory, S.K. Role of DREB transcription factors in abiotic and biotic stress tolerance in plants. *Plant Cell Rep.* **2006**, *25*, 1263–1274. [CrossRef]
100. Nakano, T.; Suzuki, K.; Fujimura, T.; Shinshi, H. Genome-wide analysis of the ERF gene family in *Arabidopsis* and rice. *Plant Physiol.* **2006**, *140*, 411–432. [CrossRef]
101. Song, X.; Li, Y.; Hou, X. Genome-wide analysis of the AP2/ERF transcription factor superfamily in Chinese cabbage (*Brassica rapa* ssp. pekinensis). *BMC Genom.* **2013**, *14*, 573. [CrossRef]
102. Lu, X.; Jiang, W.; Zhang, L.; Zhang, F.; Zhang, F.; Shen, Q.; Wang, G.; Tang, K. AaERF1 positively regulates the resistance to *Botrytis cinerea* in *Artemisia annua*. *PLoS ONE* **2013**, *8*, e57657. [CrossRef]

103. Ma, Y.N.; Xu, D.B.; Li, L.; Zhang, F.; Fu, X.Q.; Shen, Q.; Lyu, X.Y.; Wu, Z.K.; Pan, Q.F.; Shi, P.; et al. Jasmonate promotes artemisinin biosynthesis by activating the TCP14-ORA complex in *Artemisia annua*. *Sci. Adv.* **2018**, *4*, eaas9357. [CrossRef]
104. Wang, S.; Zhang, X.; Li, B.; Zhao, X.; Shen, Y.; Yuan, Z. Genome-wide identification and characterization of bZIP gene family and cloning of candidate genes for anthocyanin biosynthesis in pomegranate (*Punica granatum*). *BMC Plant Biol.* **2022**, *22*, 170. [CrossRef]
105. Yu, Y.; Qian, Y.; Jiang, M.; Xu, J.; Yang, J.; Zhang, T.; Gou, L.; Pi, E. Regulation mechanisms of plant basic leucine zippers to various abiotic stresses. *Front. Plant Sci.* **2020**, *11*, 1258. [CrossRef] [PubMed]
106. Jakoby, M.; Weisshaar, B.; Dröge-Laser, W.; Vicente-Carbajosa, J.; Tiedemann, J.; Kroj, T.; Parcy, F. bZIP transcription factors in *Arabidopsis*. *Trends Plant Sci.* **2002**, *7*, 106–111. [CrossRef]
107. Choi, H.I.; Hong, J.H.; Ha, J.O.; Kang, J.Y.; Kim, S.Y. ABFs, a family of ABA-responsive element binding factors. *J. Biol. Chem.* **2000**, *275*, 1723–1730. [CrossRef] [PubMed]
108. Foster, R.; Izawa, T.; Chua, N.H. Plant bZIP proteins gather at ACGT elements. *FASEB J.* **1994**, *8*, 192–200. [CrossRef]
109. Jing, F.; Zhang, L.; Li, M.; Tang, Y.; Wang, Y.; Wang, Y.; Wang, Q.; Pan, Q.; Wang, G.; Tang, K. Abscisic acid (ABA) treatment increases artemisinin content in *Artemisia annua* by enhancing the expression of genes in artemisinin biosynthetic pathway. *Biologia* **2009**, *64*, 319–323. [CrossRef]
110. Pu, G.B.; Ma, D.M.; Chen, J.L.; Ma, L.Q.; Wang, H.; Li, G.F.; Ye, H.C.; Liu, B.Y. Salicylic acid activates artemisinin biosynthesis in *Artemisia annua* L. *Plant Cell Rep.* **2009**, *28*, 1127–1135. [CrossRef]
111. Dubos, C.; Stracke, R.; Grotewold, E.; Weisshaar, B.; Martin, C.; Lepiniec, L. MYB transcription factors in *Arabidopsis*. *Trends Plant Sci.* **2010**, *15*, 573–581. [CrossRef]
112. Pesch, M.; Hülskamp, M. One, two, three ... models for trichome patterning in *Arabidopsis*? *Curr. Opin. Plant Biol.* **2009**, *12*, 587–592. [CrossRef]
113. Yuan, Y.; Xu, X.; Luo, Y.; Gong, Z.; Hu, X.; Wu, M.; Liu, Y.; Yan, F.; Zhang, X.; Zhang, W.; et al. R2R3 MYB-dependent auxin signalling regulates trichome formation, and increased trichome density confers spider mite tolerance on tomato. *Plant Biotechnol. J.* **2021**, *19*, 138–152. [CrossRef] [PubMed]
114. Dubos, C.; Le Gourrierec, J.; Baudry, A.; Huep, G.; Lanet, E.; Debeaujon, I.; Routaboul, J.M.; Alboresi, A.; Weisshaar, B.; Lepiniec, L. MYBL2 is a new regulator of flavonoid biosynthesis in *Arabidopsis thaliana*. *Plant J.* **2008**, *55*, 940–953. [CrossRef]
115. Zhang, Z.; Zhang, L.; Liu, Y.; Shang, X.; Fang, S. Identification and expression analysis of R2R3-MYB family genes associated with salt tolerance in *Cyclocarya paliurus*. *Int. J. Mol. Sci.* **2022**, *23*, 3429. [CrossRef]
116. Hassani, D.; Fu, X.; Shen, Q.; Khalid, M.; Rose, J.K.C.; Tang, K. Parallel transcriptional regulation of artemisinin and flavonoid biosynthesis. *Trends Plant Sci.* **2020**, *25*, 466–476. [CrossRef] [PubMed]
117. Molina-Hidalgo, F.J.; Vazquez-Vilar, M.; D'Andrea, L.; Demurtas, O.C.; Fraser, P.; Giuliano, G.; Bock, R.; Orzáez, D.; Goossens, A. Engineering metabolism in *Nicotiana* species: A promising future. *Trends Biotech.* **2021**, *39*, 901–913. [CrossRef] [PubMed]
118. Wallaart, T.E.; Bouwmeester, H.J.; Hille, J.; Poppinga, L.; Maijers, N.C.A. Amorpha-4,11-diene synthase: Cloning and functional expression of a key enzyme in the biosynthetic pathway of the novel antimalarial drug artemisinin. *Planta* **2001**, *212*, 460–465. [CrossRef] [PubMed]
119. Zhang, Y.; Nowak, G.; Reed, D.W.; Covello, P.S. The production of artemisinin precursors in tobacco. *Plant Biotechnol. J.* **2011**, *9*, 445–454. [CrossRef] [PubMed]
120. Farhi, M.; Marhevka, E.; Ben-Ari, J.; Algamas-Dimantov, A.; Liang, Z.; Zeevi, V.; Edelbaum, O.; Spitzer-Rimon, B.; Abeliovich, H.; Schwartz, B.; et al. Generation of the potent anti-malarial drug artemisinin in tobacco. *Nat. Biotechnol.* **2011**, *29*, 1072–1074. [CrossRef]
121. Van Herpen, T.W.J.M.; Cankar, K.; Nogueira, M.; Bosch, D.; Bouwmeester, H.J.; Beekwilder, J. *Nicotiana benthamiana* as a production platform for artemisinin precursors. *PLoS ONE* **2010**, *5*, e14222. [CrossRef] [PubMed]
122. Ting, H.M.; Wang, B.; Rydén, A.M.; Woittiez, L.; van Herpen, T.; Verstappen, F.W.A.; Ruyter-Spira, C.; Beekwilder, J.; Bouwmeester, H.J.; van der Krol, A. The metabolite chemotype of *Nicotiana benthamiana* transiently expressing artemisinin biosynthetic pathway genes is a function of *CYP71AV1* type and relative gene dosage. *New Phytol.* **2013**, *199*, 352–366. [CrossRef] [PubMed]
123. Wang, B.; Kashkooli, A.B.; Sallets, A.; Ting, H.M.; de Ruijter, N.C.A.; Olofsson, L.; Brodelius, P.; Pottier, M.; Boutry, M.; Bouwmeester, H.; et al. Transient production of artemisinin in *Nicotiana benthamiana* is boosted by a specific lipid transfer protein from *A. annua*. *Metab. Eng.* **2016**, *38*, 159–169. [CrossRef] [PubMed]
124. Saxena, B.; Subramaniyan, M.; Malhotra, K.; Bhavesh, N.S.; Potlakayala, S.D.; Kumar, S. Metabolic engineering of chloroplasts for artemisinic acid biosynthesis and impact on plant growth. *J. Biosci.* **2014**, *39*, 33–41. [CrossRef] [PubMed]
125. Fuentes, P.; Zhou, F.; Erban, A.; Karcher, D.; Kopka, J.; Bock, R. A new synthetic biology approach allows transfer of an entire metabolic pathway from a medicinal plant to a biomass crop. *eLife* **2016**, *5*, e13664. [CrossRef] [PubMed]
126. Malhotra, K.; Subramaniyan, M.; Rawat, K.; Kalamuddin, M.; Qureshi, M.I.; Malhotra, P.; Mohmmed, A.; Cornish, K.; Daniell, H.; Kumar, S. Compartmentalized metabolic engineering for artemisinin biosynthesis and effective malaria treatment by oral delivery of plant cells. *Mol. Plant* **2016**, *9*, 1464–1477. [CrossRef] [PubMed]
127. Khairul Ikram, N.K.B.; Beyraghdar Kashkooli, A.; Peramuna, A.V.; van der Krol, A.R.; Bouwmeester, H.; Simonsen, H.T. Stable production of the antimalarial drug artemisinin in the moss *Physcomitrella patens*. *Front. Bioeng. Biotech.* **2017**, *5*, 47. [CrossRef] [PubMed]

128. Ikram, N.K.; Kashkooli, A.B.; Peramuna, A.; Krol, A.R.v.d.; Bouwmeester, H.; Simonsen, H.T. Insights into heterologous biosynthesis of arteannuin B and artemisinin in *Physcomitrella patens*. *Molecules* **2019**, *24*, 3822. [CrossRef] [PubMed]
129. Firsov, A.; Pushin, A.; Motyleva, S.; Pigoleva, S.; Shaloiko, L.; Vainstein, A.; Dolgov, S. Heterologous biosynthesis of artemisinin in *Chrysanthemum morifolium* Ramat. *Separations* **2021**, *8*, 75. [CrossRef]
130. Reski, R.; Bae, H.; Simonsen, H.T. *Physcomitrella patens*, a versatile synthetic biology chassis. *Plant Cell Rep.* **2018**, *37*, 1409–1417. [CrossRef] [PubMed]
131. Rensing, S.A.; Lang, D.; Zimmer, A.D.; Terry, A.; Salamov, A.; Shapiro, H.; Nishiyama, T.; Perroud, P.-F.; Lindquist, E.A.; Kamisugi, Y.; et al. The *Physcomitrella* genome reveals evolutionary insights into the conquest of land by plants. *Science* **2008**, *319*, 64–69. [CrossRef] [PubMed]
132. Lang, D.; Ullrich, K.K.; Murat, F.; Fuchs, J.; Jenkins, J.; Haas, F.B.; Piednoel, M.; Gundlach, H.; Van Bel, M.; Meyberg, R.; et al. The *Physcomitrella patens* chromosome-scale assembly reveals moss genome structure and evolution. *Plant J.* **2018**, *93*, 515–533. [CrossRef] [PubMed]
133. King, B.C.; Vavitsas, K.; Ikram, N.K.B.K.; Schrøder, J.; Scharff, L.B.; Bassard, J.-É.; Hamberger, B.; Jensen, P.E.; Simonsen, H.T. In vivo assembly of DNA-fragments in the moss, *Physcomitrella patens*. *Sci. Rep.* **2016**, *6*, 25030. [CrossRef] [PubMed]
134. Reski, R.; Parsons, J.; Decker, E.L. Moss-made pharmaceuticals: From bench to bedside. *Plant Biotechnol. J.* **2015**, *13*, 1191–1198. [CrossRef] [PubMed]
135. Anterola, A.; Shanle, E.; Perroud, P.F.; Quatrano, R. Production of taxa-4(5),11(12)-diene by transgenic *Physcomitrella patens*. *Transgenic Res.* **2009**, *18*, 655–660. [CrossRef] [PubMed]
136. Yang, L.; Nuerbiye, A.; Cheng, P.; Wang, J.H.; Li, H. Analysis of floral volatile components and antioxidant activity of different varieties of *Chrysanthemum morifolium*. *Molecules* **2017**, *22*, 1790. [CrossRef] [PubMed]
137. Firsov, A.; Mitiouchkina, T.; Shaloiko, L.; Pushin, A.; Vainstein, A.; Dolgov, S. *Agrobacterium*-mediated transformation of *Chrysanthemum* with artemisinin biosynthesis pathway genes. *Plants* **2020**, *9*, 537. [CrossRef] [PubMed]

molecules

MDPI

Review

Research Progress on Active Ingredients and Product Development of *Lycium ruthenicum* Murray

Ming-Lu Xu [1], Yun-Feng He [1,2], Liang Xie [1], Ling-Bo Qu [3,*], Guang-Ri Xu [1,*] and Cheng-Xing Cui [1,*]

[1] School of Chemistry and Chemical Engineering, Institute of Computational Chemistry, Henan Institute of Science and Technology, Xinxiang 453003, China; mingluxu0827@outlook.com (M.-L.X.); yunfenghe0620@outlook.com (Y.-F.H.); xieliang1449129616@outlook.com (L.X.)
[2] School of Food Science, Henan Institute of Science and Technology, Xinxiang 453003, China
[3] School of Chemistry and Chemical Engineering, Zhengzhou University, Zhengzhou 450001, China
* Correspondence: qulingbo@zzu.edu.cn (L.-B.Q.); xugr70@163.com (G.-R.X.); chengxingcui@hist.edu.cn (C.-X.C.)

Abstract: *Lycium ruthenicum* Murray possesses significant applications in both food and medicine, including antioxidative, anti-tumor, anti-fatigue, anti-inflammatory, and various other effects. Consequently, there has been a surge in research endeavors dedicated to exploring its potential benefits, necessitating the organization and synthesis of these findings. This article systematically reviews the extraction and content determination methods of active substances such as polysaccharides, anthocyanins, flavonoids, and polyphenols in LRM in the past five years, as well as some active ingredient composition determination methods, biological activities, and product development. This review is divided into three main parts: extraction and determination methods, their bioactivity, and product development. Building upon prior research, we also delve into the economic and medicinal value of *Lycium ruthenicum* Murray, thereby contributing significantly to its further exploration and development. It is anticipated that this comprehensive review will serve as a valuable resource for advancing research on *Lycium ruthenicum* Murray.

Keywords: *Lycium ruthenicum* Murray; extraction and identification methods; bioactivities; applied research

Citation: Xu, M.-L.; He, Y.-F.; Xie, L.; Qu, L.-B.; Xu, G.-R.; Cui, C.-X. Research Progress on Active Ingredients and Product Development of *Lycium ruthenicum* Murray. *Molecules* **2024**, *29*, 2269. https://doi.org/10.3390/molecules29102269

Academic Editor: Arjun H. Banskota

Received: 21 March 2024
Revised: 5 May 2024
Accepted: 8 May 2024
Published: 11 May 2024

1. Introduction

Lycium ruthenicum Murray (LRM), a member of the Solanaceae family, is a spiny, branching shrub known for its distinctive white and purplish-black fruit, which results from inhibited anthocyanin synthesis within the plant [1]. LRM exhibits significantly higher phytochemical content compared to *Lycium barbarum* (LB), displaying superior antioxidant activity as well [2]. Its beneficial effects include antioxidative, anti-tumor, anti-fatigue, and anti-inflammatory properties, attributed mainly to its rich array of active ingredients such as polysaccharides, anthocyanins, flavonoids, polyphenols, amino acids, and minerals.

LRM's therapeutic potential is evidenced by its protective effects, including mitigating alcoholic liver and kidney injury in animal models [3,4]. Li et al. [5] identified anti-inflammatory effects in LRM fruit extracts, potentially mediated by 5-hydroxymethylfurfural. Chemical extraction methods are commonly employed to isolate these active constituents. Polysaccharides, for instance, can be extracted through various techniques including hot water leaching, acid or alkali extraction, and low eutectic solvent extraction. Similarly, polyphenols can be obtained via ultrasonic, microwave, enzyme-assisted, or supercritical liquid extraction methods, among others.

These extracted fractions hold promise for pharmaceutical applications, either in their native form or following modification with functional groups. LRM's potential as a therapeutic agent extends to conditions like Alzheimer's disease (AD), with compounds like

N-acetyl-L-tryptophan possibly targeting calmodulin and exerting anti-AD effects. Moreover, LRM's multitargeted approach in modulating neurological, immune, and signaling pathways suggests potential efficacy in mitigating AD pathology [6,7].

This review is divided into three main parts: extraction and determination methods, their bioactivity, and product development. The purpose of this study is to provide a comprehensive and systematic summary and classification of the nutritional composition, efficacy, and product development over the past five years, for academic research or practical reference.

2. Extraction and Determination Methods

The extraction and determination details of LRM active ingredients are shown in Tables S1 and S2 (Supplementary Information).

2.1. Extraction and Identification of LRPs

2.1.1. LRP Extraction

Polysaccharides can be extracted through various methods, encompassing chemical, biological, physical, or hybrid approaches. Among these, the chemical method stands out as the most prevalent, employing techniques such as hot water leaching. Biological extraction relies on enzymatic processes, while physical extraction utilizes instruments like ultrasonic and microwave devices, as well as supercritical CO_2 extraction.

In traditional methodologies, He et al. [8] utilized hot water leaching to extract *Lycium barbarum* polysaccharides (LRPs), yielding a novel arab galactose LRP1-S2. Li et al. [9] optimized this approach by refining the hot water extraction, ultrasonic-assisted extraction, and enzymatic extraction methods. Their findings indicate that enzymatic extraction yielded the highest LRP content, followed by ultrasonic and hot water methods. Enzymatic extraction favored arabinose, ultrasonic-assisted extraction favored glucose, and hot water extraction yielded trace amounts of sucrose. Notably, hot water extraction exhibited superior extraction of sugar components compared to enzymatic and ultrasonic methods.

Additionally, Bai H. et al. [10] extracted LRPs using the conventional water bath method, ultrasound method, microwave method, and ultrasound-microwave synergistic extraction method. The results showed that the ultrasound–microwave synergistic extraction method > conventional water bath extraction method > microwave extraction method > ultrasound extraction method. In summary, the ultrasound–microwave synergistic extraction method is an effective method for extracting LRPs.

Recent advancements have introduced new extraction techniques. Jin et al. [11] optimized LRP extraction using high-hydrostatic-pressure-assisted extraction (HHPE), investigating its effects on LRP extraction yield. Their findings revealed that the optimal HHPE conditions included a pressure of 380 MPa, an extraction time of 8 min, and a ratio of 25 mL/g of LRM to water. Under these conditions, the polysaccharide yield closely matched predictions, showcasing HHPE as a viable method for LRP extraction. Compared to traditional methods, HHPE proved to be efficient, saving both time and energy, and holds potential for enhancing the extraction efficiency of polysaccharides from various plant sources.

2.1.2. Determination of LRP Content and Composition

In addition to extraction methods, determining content and composition is crucial. Traditional content determination methods include the phenol–sulfuric acid, anthrone–sulfuric acid, and 3,5-dinitrosalicylic acid methods, with the phenol–sulfuric acid method being the most commonly used [12]. Current techniques for analyzing the monosaccharide composition of plant polysaccharides include capillary electrophoresis, gas chromatography, high-performance liquid chromatography, and mass spectrometry [13].

Emerging alongside traditional methods are more convenient approaches. Zhang et al. [14] demonstrated the use of 1H-Nuclear Magnetic Resonance Spectrum (NMR) for rapid and accurate quantitative analysis of LRP. The highest content of LRP in July was

0.22 mg/g. This method offers advantages such as short measurement time, a strong linear relationship, stability, repeatability, and accurate control without sample decomposition.

Zhang [15] used the PMP (1-phenyl-3-methyl-5-pyrazolone) derivation method and the sugar alcohol acetate method combined with HPLC to determine the monosaccharide composition in polysaccharides and found that LRP1-S2 contains mannose, rhamnose, glucuronic acid, galacturonic acid, glucose, galactose, and arabinose. LRP3-S1 contains rhamnose, galacturonic acid, galactose, arabinose, and xylose.

Li et al. [16] established a tandem mass spectrometry method for the simultaneous determination of sucrose, fructose, rhamnose, arabinose, glucose, cottonseed sugar, and mannose in LRM. This method is simple, sensitive, and suitable for LRP determination. Contrasting with traditional methods, which measure content or composition individually, the new method allows simultaneous measurement and offers the advantages of shorter measurement time, precise control, and simplicity. Its application to other plant polysaccharides could significantly save time.

2.2. Extraction and Identification of LRM Polyphenols

2.2.1. LRM Polyphenol Extraction

Polyphenols, including tannins and flavonoids such as proanthocyanins and anthocyanins, are prevalent in medicinal plants. Extraction methods for polyphenols include ultrasonic extraction, microwave extraction, enzyme-assisted extraction, supercritical liquid extraction, ion precipitation extraction, and membrane extraction.

Guo et al. [17] optimized the ultrasonic-assisted aqueous two-phase extraction of proanthocyanidins from LRM. Under optimal conditions—ethanol volume fraction of 61%, ammonium sulfate mass concentration of 0.24 g/mL, material-to-liquid ratio of 1:26, ultrasound time of 49 min, and ultrasound power of 220 W—the extraction rate of *Lycium ruthenicum* Murr. proanthocyanidins (LRMPs) reached 6.113%. Wu et al. [18] compared solvent extraction with ultrasonic-assisted extraction of polyphenols from LRM, revealing the latter as a simple, economical, time-saving, low-energy-consumption method with high extraction efficiency. Mi et al. [19] optimized the ultrasonic-assisted pectinase extraction of anthocyanins from LRM using orthogonal experiments, yielding 19.4 ± 0.87 mg/g under the conditions of a 45 °C extraction temperature, 30 min extraction time, 0.5% enzyme content, 360 W ultrasonic power, and pH = 1. Stability experiments demonstrated the preservation of LRM-extracted anthocyanins in low-temperature and high-acidity storage solutions.

2.2.2. LRM Anthocyanins Content Determination

Common methods for determining anthocyanin content include the pH differential method, liquid chromatography, and spectrophotometry, among others.

Liu et al. [20] employed the pH differential method to assess anthocyanin content extracted through various processing methods of LRM. The results indicated the lowest content with spray drying, followed by ultra-fine crushing, notably increasing after purification with macroporous resin and gel chromatography.

Wang et al. [21] utilized high-performance liquid chromatography (HPLC) and UV detection to analyze anthocyanins in LRM extracts. The total anthocyanin content was 2.0 mg Cy/g. Meanwhile, the structure of anthocyanins in LRM was separated and identified by high-performance liquid chromatography–electrospray ionization–mass spectrometry (HPLC-ESI-MS). Two types of anthocyanins have been successfully identified, namely Petunidin-3-*O*-rutinoside(glucosyl-p-coumaroyl)-5-*O*-glucoside and Petunidin-3-*O*-rutinoside (p-coumaroyl)-5-*O*-glucoside. Gan et al. [22] optimized anthocyanin determination methods by employing ultra-high-performance liquid chromatography–time-of-flight mass spectrometry (UHPLC/FTL) for rapid and precise identification. The average sample recovery rate was 95.45%, with a relative standard deviation of 2.62%. This approach enhances the quality control of LRM due to its superior repeatability and precision compared to conventional methods.

Innovatively, Zhang et al. [23] developed a deep learning approach using near-infrared hyperspectral imaging (NIR-HSI), enabling the simultaneous determination of total phenols, total flavonoids, and total anthocyanins in LRM. Muhammad et al. [24] further refined this technique, optimizing phenolic and flavonoid assays using portable NIR spectroscopy integrated with chemometrics. The resulting models, assessed via calibration and prediction correlation coefficients, root mean square error of prediction, and residual predictive deviation, demonstrated the efficacy of this simple, rapid, and cost-effective approach for quantifying phenols and flavonoids in LRM samples.

2.3. Extraction and Composition Determination of LRM Flavonoids

Flavonoids in LRM are predominantly found in leaves and fruit, offering a range of physiological and biochemical benefits such as antioxidant, anti-aging, antibacterial, and lipid-modulating effects. However, research into their specific composition and extraction methods lags significantly behind that of polysaccharides and anthocyanins.

Presently, traditional solvent extraction methods like alcohol and alkali extraction are widely employed, alongside emerging technologies such as supercritical fluid extraction, enzyme engineering, and physical field assistance.

Shu et al. [25] conducted flavonoid extraction from Chaidamu LRM using supercritical CO_2, followed by analysis using ultra-high-performance liquid chromatography–mass spectrometry and peak view chromatography workstation fitting. Through secondary mass spectrometry analysis, UV spectroscopy, and comparison of mass spectrometry data, they identified 13 flavonoids in the supercritical CO_2 extraction of LRM, including rutin and Quercetin-3-β-Glucuronide and hyperoside, also known as Quercetin-3-*O*-galactopyranoside, dihydroquercetin, dihydroisorhamnosum, dihydrokaempferol, quercetin, apigenin, hesperidin, alfalfa, kaempferol, isorhamnosum, and kaempferol. Notably, hesperidin and alfalfa were first discovered during the study of LRM components.

2.4. LRM Betaine Content Determination

Betaine, originally discovered in sugar beets, lends its name to an alkaloid compound. Natural products sharing a chemical connection with betaine are collectively referred to as the "betaine class". The structural formula of betaine analogs is depicted in Figure 1.

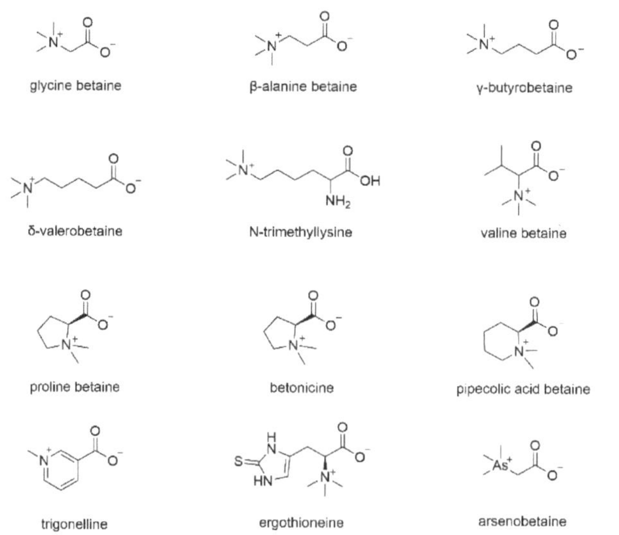

Figure 1. Chemical structure of betaine and its analogs.

Various methods are employed to determine betaine content, including non-aqueous titration, chemical assays, colorimetry, and gravimetry. Additionally, with technological advancements, HPLC has gained prominence in recent years for betaine quantification.

Currently, research on LRM betaine primarily centers on content detection, with limited exploration into extraction and utilization. Liu et al. [26] utilized reverse-phase (RP)-HPLC to determine the average betaine content in Qaidam LRM, finding it to be 1.34%. Geng et al. [27] employed ion chromatography to assess the betaine content in LRM from different regions, reporting levels as high as 19.58% in Minqin County, Gansu Province, and 9.44% in Golmud City, Qinghai Province.

2.5. Extraction and Content Determination of LRM Tannins

Various methods are employed for extracting tannins, including reflux extraction, cellulase hydrolysis, water temperature soaking, ultrasonic extraction, and supercritical CO_2 extraction, among others.

Commonly utilized techniques for determining tannin content include the leather powder method, potassium permanganate method, casein method, complexometric titration method, colorimetric method, and spectrophotometry.

Research on tannins in LRM remains limited. Meng et al. [28] utilized the homogenization method to extract tannins from Xinjiang LRM, determining the average tannin content in LRM through UV–visible spectrophotometry to be 2.47%.

2.6. Extraction, Composition, and Content Determination of LRM Fatty Acids

Various techniques are commonly employed for fatty acid extraction, including ultrasound-assisted extraction, Soxhlet extraction, supercritical fluid extraction, accelerated solvent extraction, microwave-assisted extraction, and solid-phase extraction.

Traditional methods for determining fatty acids primarily involve chromatography and chromatography–mass spectrometry. Rapid detection technologies such as NIR, NMR, and Raman Spectroscopy (RS), combined with chemometrics and hyperspectral imaging, are also utilized. Additionally, direct ionization techniques like Desorption Electrospray Ionization–Mass Spectrometry (DESI-MS), Extractive Electrospray Ionization Mass Spectrometry (EESI), and Internal Extractive Electrospray Ionization (iEESI)–MS, coupled with multivariate statistical methods, are gaining prominence.

In a study by Yossa Nzeuwa I.B. et al. [29], LRM was extracted with petroleum ether using a Soxhlet apparatus for 2 h. Analysis of the extracted oils via gas chromatography (GC)–MS revealed major fatty acid components including linoleic acid (59.38%), oleic acid (20.85%), palmitic acid (7.07%), linolenic acid (2.98%), and stearic acid (5.31%). Using inductively coupled plasma–atomic absorption spectrometry (ICP-AAS), the main mineral nutrients in LRM were identified as potassium, calcium, and magnesium, with traces of copper, iron, manganese, and zinc. Research on LRM fatty acids remains limited, and the mentioned measurement methods offer valuable insights for future investigations into LRM fatty acids.

2.7. Extraction and Determination of Phenylpropanoid Derivatives

Various extraction methods for phenylpropanoid include water extraction, organic solvent extraction, microwave-assisted extraction, and ultrasonic-assisted extraction, among others.

Methods for determining phenylpropanoid include ultraviolet spectrophotometry, second derivative spectroscopy, visible spectrophotometry, HPLC, RP-HPLC, solid-phase extraction microcolumn high-performance liquid chromatography (SPE-MHPLC), and others.

In a study by Zhao et al. [30], LRM was extracted using ethanol and condensed. Grades A–D were obtained through continuous elution of the concentrated solution via an HP-20 macroporous resin column. HPLC was employed to analyze the components, and similar compounds were further isolated using open silica CC, resulting in the identification of 26 compounds. Among these, 8 were identified as new phenylpropanoid derivatives (**1–7** and **15**), while the remaining 18 were known derivatives (**8–14**, **16–26**). Notably, compounds **16**, **17**, **24**, and **26** were isolated from *Lycium* for the first time (detailed information is shown in Figure 2).

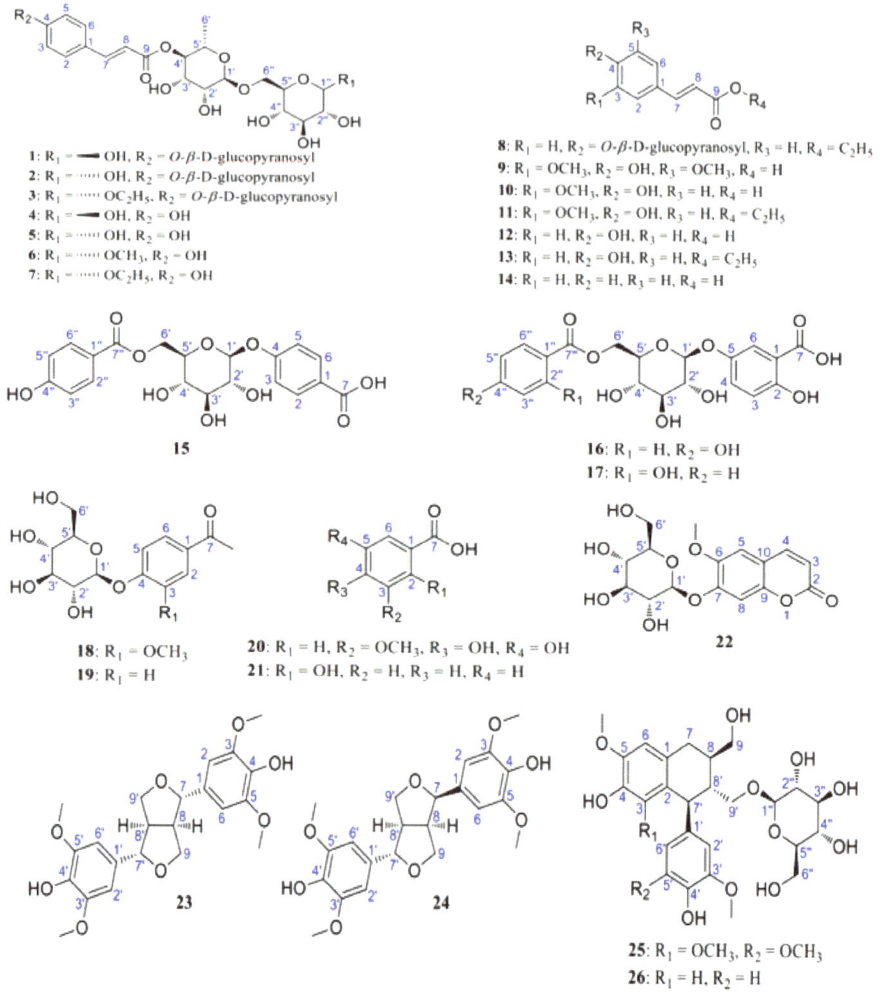

Figure 2. Chemical structures of compounds **1–26**. Reprinted with permission from reference [28], copyright 2022 Elsevier.

2.8. Extraction and Determination of Polyphenol Glycosides

Traditional extraction methods for polyphenolic glycosides typically involve impregnation and solvent extraction. However, with advancements in modern technology, researchers have enhanced and optimized these traditional methods, introducing newer techniques such as ultrasonic extraction and microwave-assisted extraction.

Determining polyphenolic glycosides necessitates a combination of various analytical methods, including chromatography, mass spectrometry, and Nuclear Magnetic Resonance.

In a study conducted by Hu et al. [31], polyphenolic glycosides were extracted using a solvent method and analyzed using high-resolution electrospray ionization–mass spectrometry (HRESIMS) and comprehensive NMR analysis. Through chemical hydrolysis, nine new polyphenol glycosides were isolated and identified, named lyciumserin

A–I (**1–9**), along with 16 known compounds (**10–25**). Remarkably, compounds **1** and **2** exhibited significant neuroprotective effects in a PC12 cell model of injury induced by 6-hydroxydopamine. The identified compounds (25) include a variety of glycosides such as 7-hydroxycoumarin-3-*O*-[6-*O*-(4-*O*-p-trans-coumaroyl)-α-L-rhamnopyranosyl]-β-D-glucopyranosyl-5-*O*-β-D-glucopyranoside (**1**), 7-hydroxycoumarin-3-*O*-[6-*O*-(4-*O*-(4-*O*-β-D-glucopyranosyl)-trans-p-coumaroyl)-α-L-rhamnopyranosyl]-β-D-glucopyranosyl-5-*O*-β-D-glucopyranoside (**2**), and others as listed in Figure 3.

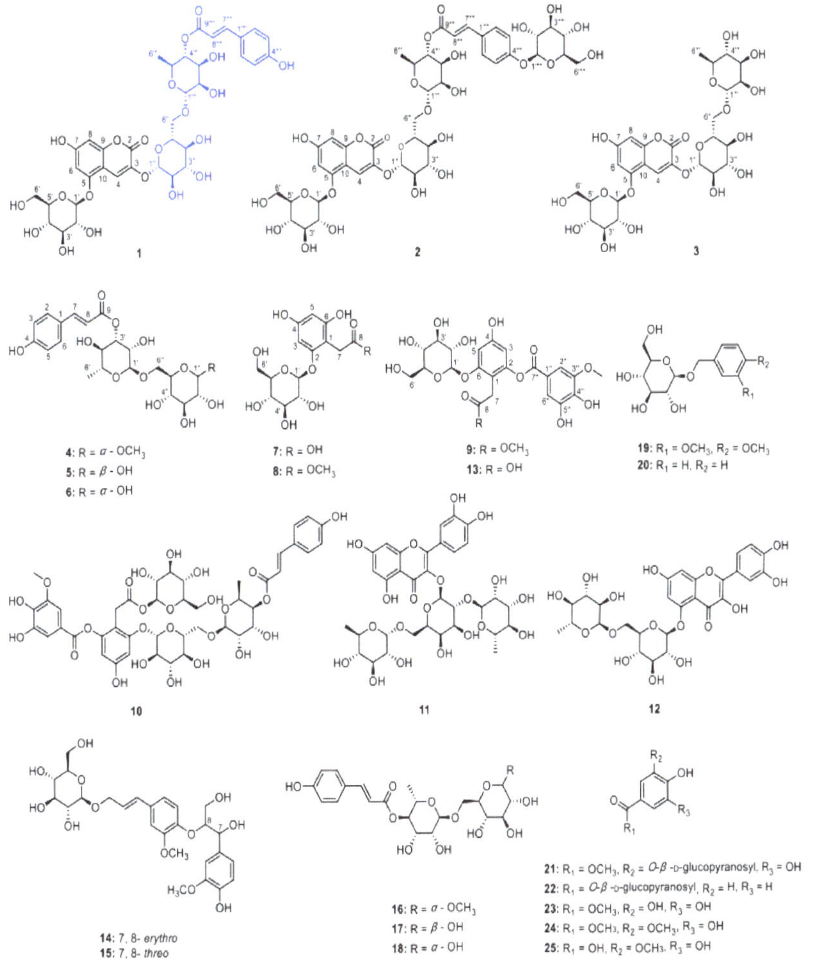

Figure 3. Structures of compounds **1–25** obtained from LRM. Reprinted with permission from reference [29], copyright 2022 American Chemical Society.

3. Bioactivities

LRM is rich in bioactive compounds, making it a promising candidate for developing various products (as summarized in Table S3 (Supplementary Information) and illustrated in Figure 4).

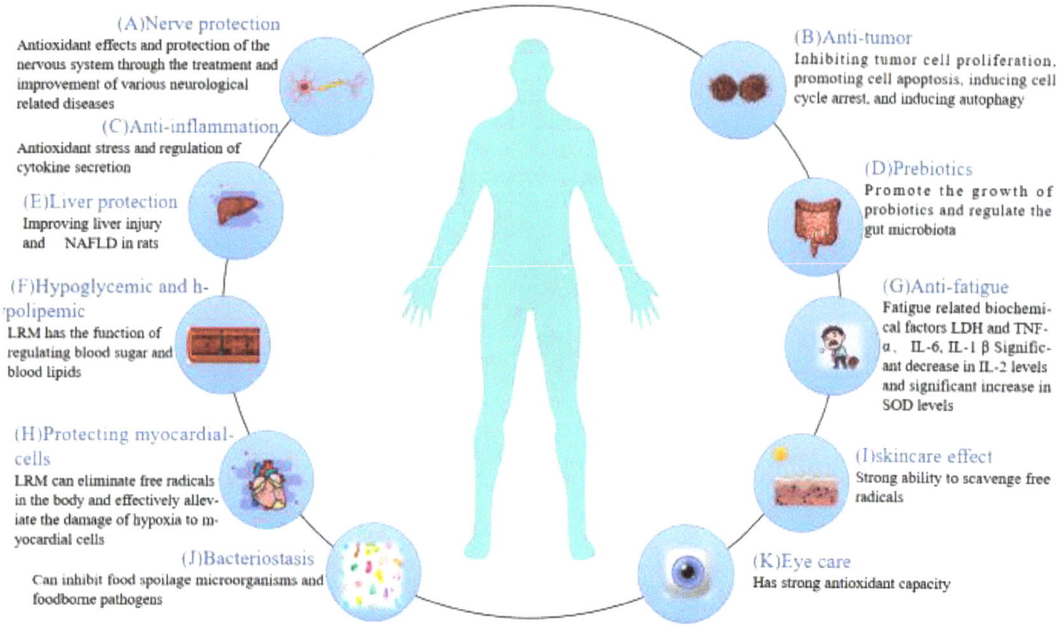

Figure 4. Biological activity of LRM extract. (**A**) Nerve protection; (**B**) anti-tumor; (**C**) anti-inflammation; (**D**) prebiotics; (**E**) liver protection; (**F**) hypoglycemic and hypolipemic; (**G**) anti-fatigue; (**H**) protecting myocardial cells; (**I**) skin care effect; (**J**) bacteriostasis; (**K**) eye care.

3.1. Anti-Tumor Activity

Polysaccharides and anthocyanins found in LRM exhibit potent anti-tumor properties. LRP1-S2, a novel polysaccharide extracted from LRM, can inhibit the proliferation of pancreatic cancer cells in vitro and in vivo, and it has no obvious toxicity to normal pancreatic HPDE 6-C7 cells and LO2 hepatocytes. It may be achieved by blocking P38 MAPK/NF-κB, and the GSK-3 β/β-Catenin signaling pathway induces the apoptosis of human pancreatic cancer BxPC-3 cells [8]. LRM anthocyanins effectively inhibit the proliferation and migration of human liver cancer HepG2 cells by upregulating proliferation factors (LATS1, LATS2, and MOB1) and autophagy factors (Beclin-1, LC3-II, AMPK, p-AMPK, and LC3-II), and downregulating the mRNA levels of the proliferation factor YAP, the protein expression of the autophagy factor p-mTOR, and the cell cycle factor CDK4 [32,33].

Qin X. et al. [34] found that LRP and LRM anthocyanidins alone (incubated for 48 h) failed to prevent tumor cell proliferation, but once applied together, significant anti-tumor activity was observed, resulting from a combination of LRP and LRM anthocyanidins, this inhibition of cancer cell proliferation occurs by blocking the G0-G1 cell cycle and inducing apoptosis of human colorectal cancer LoVo cells via a reactive oxygen species (ROS)-dependent pathway.

Wang et al. [35,36] found through their research that LRM anthocyanin Pt3G can inhibit the proliferation of prostate cancer DU-145, LNCaP, and PC-3 cells by inducing cell apoptosis. The mechanism of action may be related to the activation of the ROS/PTEN/PI3K/Akt/caspase pathway.

Additionally, LRM's total flavonoids induce HepG-2 liver cancer cell cycle arrest in the S phase, causing cell division and proliferation to stop. At the same time, apoptosis of HepG-2 liver cancer cells can be induced by upregulating Bax and Casp-3 and downregulating the expression level of Bcl-2 [37].

3.2. Prebiotic Effects

LRM polysaccharides serve as prebiotics, promoting the growth of *Lactobacillus, Bifidobacterium, Streptococcus thermophilus* G2, and *Lactobacillus paracasei* L9, and inhibiting the growth of *Enterobacterium* and *Enterococcus* [38–40]. They are fermented in the colon to produce short-chain fatty acids, which contribute to gut health and overall well-being. LRM anthocyanins have also been shown to modulate gut microbiota diversity and metabolic pathways, offering potential benefits for obesity and metabolic disorders [41]. Their mechanism of action may be related to increasing the diversity of the mouse gut microbiota, especially reducing the proportion of Firmicutes/Bacteroidota. On the other hand, it may be related to differences in gene expression in metabolic pathways or PPAR signaling pathways.

3.3. Anti-Fatigue Properties

LRM polysaccharides and tea polyphenols (TPs) exhibit anti-fatigue effects by reducing oxidative stress and inflammation [42,43]. Changes in fatigue-related biochemical factors lactic acid dehydrogenase (LDH), superoxide dismutase (SOD), tumor necrosis factor-α (TNF-α), interleukin-1β (IL-1β), interleukin-2 (IL-2), and interleukin-6 (IL-6) were measured in motor fatigue (EIF) mouse models treated with TPs and LRPs. The results showed that TP and LRP both showed significant anti-inflammatory effects and decreased oxidative stress. Compared to the control group, LDH, TNF-α, IL-6, IL-1β, and IL-2 levels were significantly reduced and SOD levels were significantly increased in the TP- or LRP-treated group. The results showed that TPs and LRPs were effective in the treatment of exercise fatigue. These compounds improve energy metabolism and mitigate fatigue-related biochemical changes, offering potential as natural remedies for combating fatigue.

3.4. Eye and Skin Care Benefits

LRM components, particularly polysaccharides and anthocyanins, show promise in eye and skin care. LRPs can enhance the antioxidant capacity of ARPE-19 cells, inhibit the apoptosis of ARPE-19 cells, and protect retinal pigment epithelial cells from oxidative stress. In addition, purified LRPs can enhance the antioxidant capacity of ARPE-19 cells without dose dependence. They can increase the activities of superoxide dismutase (SOD) and glutathione peroxidase (GSH Px) in ARPE-19 cells damaged by H_2O_2, reduce the content of malondialdehyde (MDA), and downregulate NLRP3, caspase-1, and IL-1β. The expression of proteins inhibits ARPE-19 cell apoptosis, thereby preventing age-related macular degeneration [44–47].

Additionally, LRM anthocyanins reduce the cell apoptosis rate through the death receptor pathway, decrease the expression of TNF-α and caspase-7, and increase the expression of survival proteins, thereby exerting a protective effect on human skin fibroblasts (HSFs) exposed to UVB [48,49]. These properties highlight the potential of LRM-derived compounds in promoting eye and skin health.

3.5. Neuroprotection

Neurological diseases often involve cellular damage caused by oxygen free radicals. Polysaccharides found in LRM exhibit antioxidant properties, protecting the nervous system from oxidative stress and apoptosis. LRP3, a component of LRM, has shown promising neuroprotective effects against oxidative stress and apoptosis in rat cortical neurons, suggesting its potential therapeutic role in neonatal hypoxic–ischemic encephalopathy (HIE) [50]. LRM anthocyanins significantly increased the autophagic flow of SH-SY 5Y cells exposed to oxygen–glucose deprivation (OGD), inhibited oxidative stress, reduced the inflammatory response, and decreased neuronal apoptosis. These effects enhance autophagy agonists and weaken autophagy inhibitors. Reducing OGD-induced damage by increasing autophagic flow may be a potential mechanism for anthocyanin protection or treatment of hypoxia and ischemia. Meanwhile, LRM anthocyanins (Pn3G5G) can alleviate cognitive impairment, reduce cortical and hippocampal neuronal loss, lower levels of age-related markers (advanced glycation end products (AGEs) and MDA), increase the ratio of oxidized/reduced glutathione, and inhibit hippocampal neuroinflammation. Pn3G5G can also regulate the biosynthesis of ABC transporters, protein digestion, and cofactors to prevent abnormal neuronal energy metabolism [51,52]. Anthocyanins present in LRM can penetrate the blood–brain barrier and protect nerve cells from damage. Studies indicate that LRM anthocyanins enhance autophagy and reduce oxidative stress, inflammation, and neuronal apoptosis, thereby protecting against hypoxia, ischemia, and age-related neuronal degeneration.

3.6. Anti-Inflammatory Effects

Inflammation is implicated in various diseases, including diabetes, Alzheimer's disease, and arthritis. LRM anthocyanidin can reduce MDA and increase antioxidant levels in H9c2 rat cardiomyocytes by increasing the activity of catalase (CAT), GSH-Px, SOD, and heme oxygenase-1 (HO-1). The energy metabolism of cardiomyocytes under hypoxia was enhanced by increasing Na^{+}-K^{+}-ATP and Ca^{2+}-ATP enzyme activity. By reducing the levels of the pro-inflammatory factors TNF-α and IL-6, increasing the levels of the anti-inflammatory factor IL-10, and inhibiting hypoxia-induced inflammation, H9c2 rat cardiomyocytes were protected from hypoxic damage [53]. Additionally, LRM anthocyanins have been found to alleviate acute gouty arthritis. LRM anthocyanidins inhibit abnormal proliferation and invasive synovial fibroblasts (SFs) in rheumatoid arthritis (RA) patients. In vitro assays confirmed that LRM anthocyanidins inhibit RASF growth through cell cycle arrest and cell invasion paralysis (c-Myc/p21/CDK2) and initiation of apoptosis (HIF-1α/CXCR4/ Bax/Bcl-2), and that LRM anthocyanidin selectively inhibits RASF proliferation without the immune-suppressive side effects normally associated with MTX (methotrexate) treatment of RA, suggesting their potential as natural remedies for autoimmune diseases [54,55].

3.7. Hepatoprotection

The liver plays a crucial role in metabolism, and its health is vital for overall well-being. LRPs, a component of LRM, have been shown to protect against liver injury induced by acrylamide poisoning in rats [56]. Moreover, proanthocyanidins from LRM have demonstrated hepatoprotective effects against non-alcoholic fatty liver disease (NAFLD) induced by a cholesterol-rich diet. LRM anthocyanins significantly reduced the accumulation of triglycerides in the liver, reduced inflammation, increased GSH Px activity, and decreased MDA levels. At the same time, it was found that LRM anthocyanins reduced ROS production and inflammation, increased fatty acid oxidation, and reduced fatty acid synthesis in the liver, thereby improving the development of NAFLD induced by a high Western diet. Therefore, oxidative stress and inflammation are also involved in the pathogenesis of NAFLD induced by a Western diet, and they are not solely associated with obesity. There is currently no effective drug or method to treat NAFLD. LRM contains water-soluble antioxidant anthocyanins, which can effectively improve NAFLD, suggesting its potential in preventing and treating NAFLD [57].

3.8. Bacteriostasis

Plant extracts rich in polyphenols exhibit natural antimicrobial properties, inhibiting various food spoilage microorganisms and foodborne pathogens. Anthocyanins derived from LRM have been shown to inhibit the growth of foodborne pathogens such as *Staphylococcus aureus* (*S. aureus*), *Escherichia coli* (*E. coli*), *Aspergillus*, and *Penicillium*. The antimicrobial effect varies among these pathogens, with *S. aureus* being the most susceptible, exhibiting a minimum antimicrobial concentration (MIC) of 3.125 mg/mL. Anthocyanins likely exert their antibacterial effects through mechanisms related to membrane permeability and integrity [58,59].

3.9. Cardioprotective Effects

Anthocyanins, water-soluble flavonoid pigments with potent antioxidant properties, have been found to mitigate myocardial cell damage induced by hypoxia. In vitro studies using H9c2 rat cardiomyocytes demonstrate that LRM anthocyanins promote cardiomyocyte proliferation and protect against hypoxic-induced cellular damage [60–62]. Investigation into the protective mechanisms revealed the involvement of circRNA-related regulatory networks, implicating novel circRNA-miRNA-mRNA interactions in cardiomyocyte protection. Novel_circ_0001555 in LRM anthocyanins may indirectly upregulate the expression of G protein signaling regulator 8 and dystrophin by targeting the sponge adsorption of rno-miR-30b-5p, thereby regulating the calcium signaling pathway, maintaining Ca^{2+} homeostasis, and ultimately protecting hypoxia-induced H9c2 rat cardiomyocytes [63].

3.10. Hypoglycemic and Hypolipemic Effects

LRM anthocyanins can promote the phosphorylation of IRS2 and Akt proteins, activate the P13K signaling pathway, promote downstream GSK3 β and FoxO1 phosphorylation, and play a role in inhibiting their activities, thus promoting glycogen synthesis, accelerating sugar consumption, and reducing gluconeogenesis to alleviate the insulin resistance of Hep-G2 cells, showing hypoglycemic and lipid-lowering effects [64]. Meanwhile, LRM anthocyanins bind to pancreatic lipase and inhibit its activity, thereby reducing the hydrolysis of dietary triglycerides and lipid uptake, which in turn plays a role in weight loss [65]. Additionally, the total flavonoid extract of LRM regulates lipid metabolism, improves antioxidant capacity, and significantly reduces the serum total cholesterol (TC), triglyceride (TG), and low-density lipoprotein (LDL-C) concentrations in the serum and liver tissue of hyperlipidemic Sprague Dawley (SD) rats, The concentration of high-density lipoprotein (HDL-C) in serum and liver tissue increased, indicating that the total flavonoids of LRM have the effect of regulating blood lipids [66].

3.11. Other Bioactivities

LRM exhibits various biological activities beyond those previously mentioned, including anti-influenza, anti-anxiety, and anti-aging properties, osteoporosis prevention, and gout relief. Kurskaya O. et al. [67] discovered that LRPs possess antiviral properties against influenza A/H3N2 virus, suggesting their potential as novel antiviral agents. Research by Luo J. [68] indicates that LRM anthocyanins alleviate withdrawal symptoms in nicotine-addicted mice, exhibiting anti-anxiety effects and reducing cravings. Xiong L. et al. [69] found that LRM and *Lycium barbarum* (LB) extracts contain phenols and flavonoids that promote longevity by enhancing physiological functions and stress tolerance in nematodes. Wang S.Q. et al. [70] identified a polysaccharide (LRP-S2A) from LRM that promotes osteoblast differentiation, potentially preventing osteoporosis. Li L. et al. [71] investigated the mechanism of LRM anthocyanins in treating gouty arthritis, identifying potential targets and pathways through tissue lipidomics, cyberpharmacology, and molecular docking. Guo S. et al. [72] demonstrated that LRM anthocyanins mitigate mitochondrial damage and autophagy induced by chromium (VI) in cells, suggesting protective effects against heavy metal toxicity.

4. LRM Product Development

The applications of LRM extract are shown in Table S4 (Supplementary Information).

4.1. LRM's Product Development in the Field of Packaging Technology

Smart tags and new packaging films are important components of modern packaging technology, providing additional value and functionality to products by integrating advanced materials and sensor technology. These innovative films, characterized by their biodegradability and renewability, provide potential solutions for environmental and health issues related to traditional plastic packaging. Smart tags can serve as indicators of the presence of specific substances in a product or its packaging, reaction signals between different substances, or concentrations of certain components to reflect the quality status of the product. Usually, these indicators are expressed through visual changes. An intelligently active packaging film incorporating LRM anthocyanins was developed. The findings suggest that anthocyanin-rich membranes should be refrigerated to delay degradation and maintain functionality [73].

To enhance the stability of LRM anthocyanins and mitigate oxidative degradation, a gelatin/polyvinyl oxide/LRM proanthocyanidin (GEL/PEO/LPC) fiber membrane was fabricated via electrospinning technology. Fiber membrane encapsulation effectively bolstered the photothermal stability of LRM proanthocyanidin, with higher encapsulation efficiency correlating with stronger antioxidant activity [74].

Xu F. et al. [75] engineered food packaging films by incorporating LRM polyphenols into a starch/polyvinyl alcohol mixture. The resulting film exhibited enhanced compactness and reduced light transmittance, water vapor transmittance, and oxygen transmittance. Moreover, the water contact angle, tensile strength, antioxidant, and antibacterial activities of the membrane were elevated. Notably, LRM extract induced significant color changes in the film across different buffer solutions and ammonia atmospheres, signifying promising prospects for food packaging applications.

LRM anthocyanins exhibit sensitivity to changes in meat food freshness, particularly reacting to ammonia gas, with reaction rates escalating alongside concentration increases. Notably, the color of the anthocyanin solution transitions from fuchsia to cyan to yellow. Additionally, anthocyanins undergo reactive discoloration in response to pH variations. Utilizing a straightforward approach, intelligent labels can be affixed to filter substrates to enable the swift, non-destructive testing of meat product freshness within packaging, thereby safeguarding consumer health [76].

In light of this, LRM anthocyanins were immobilized onto porous filter paper to fabricate LRM anthocyanin paper smart labels. These labels provided real-time and precise feedback on freshness, as demonstrated in subsequent fish preservation experiments. Furthermore, a novel pH-sensitive membrane gel incorporating Artemisia Sphaerocephala Krasch (ASKG), Soybean Protein Isolate (SPI), and LRM anthocyanins was developed. This membrane, prepared by impregnating solid ASKG and SPI substrates with anthocyanidin from LRM dissolved in an acidified alcohol solution, serves as an indicator for monitoring meat freshness during storage [77,78].

Subsequent research unveiled that derivatives from morning glory can chelate with metal ions such as Al^{3+} and Fe^{3+}, enhancing color stability across neutral and alkaline pH values. Purple potatoes and LRM, abundant in morning glory derivatives, emerge as promising natural colorant sources applicable over a wide pH range. Through metal chelation, these pigments offer vibrant purples, blues, and greens with heightened stability. They find utility in neutral to slightly alkaline pH food products like milk, milkshakes, ice cream, and creams. This innovation holds significance for the food industry, facilitating the adoption of natural colorants in line with clean labeling trends [79].

4.2. Product Development of Plant Growth Regulators

LRM significantly boosts antioxidant levels in the mycelial cells of *Agaricus bitorquis (Quél.) Sacc* (ABSC), amplifying biomass and polysaccharide production [80]. Wu S. et al. [81] found that using LRM anthocyanins as an inducer notably influenced the liquid fermentation of ABSC in large mushrooms, positively impacting ABSC extracellular polysaccharide (EPS) production and mycelial biomass during fermentation.

4.3. LRM in Product Development of Food and Health Drugs

In recent years, LRM has emerged as a notable agricultural product, capturing consumers' attention and gaining traction in the market due to its status as a nutritious small berry. Consequently, research on LRM processing technology is steadily advancing. However, the majority of LRM and its derivative products' development has yet to reach its full potential. Key LRM products include chewable tablets [82,83], composite beverages [84–87], yogurt [88], fruit vinegar [89], and more.

4.4. LRM Product Development in the Field of Materials Science

Microcapsule technology employs natural or synthetic polymer materials to encapsulate solids, gases, and other substances into small particles. This technique effectively shields the encapsulated substances from external factors, thereby preserving their activity and extending storage time. Shen M. [90] conducted experiments on LRM anthocyanins microcapsules and identified optimal preparation conditions: embedding time of 66 min, addition of chitosan (CS) at 0.2 mg/mL, and addition of hydroxypropyl-β-cyclodextrin (HP-β-CD) at 1.93 mg/mL. These conditions resulted in an encapsulation rate of 84.28 ± 0.34% and a tyrosinase inhibition rate of 50.75 ± 0.78%. Cell experiments demonstrated that microcapsules significantly enhanced anthocyanins' inhibition of tyrosinase activity and melanin content in B16 melanoma cells in mice.

4.5. LRM's Product Development in the Textile Industry

Wool, a vital raw material in the textile industry, boasts numerous advantages such as softness, smoothness, elasticity, moisture absorption, and warmth retention. Exploring the use of natural dyes in wool textiles can pave the way for new environmentally friendly products with health functions. Gu J. et al. [91] extracted LRM using an ethanol/water solution and deionized water. Subsequently, the wool fabric was dyed through immersion to produce multifunctional LRM-dyed wool fabric with vibrant colors. The influence of dyeing time on wool fabric dyeing properties was examined, and the fabric's antioxidant and antibacterial properties were assessed and compared. The results indicated that using an ethanol/water solvent yielded LRM extract with higher anthocyanin content, resulting in deeper surface penetration and stronger antioxidant and antibacterial properties in the dyed wool fabric compared to water alone.

5. Conclusions

This article systematically reviews the extraction and content determination methods of active substances such as polysaccharides, anthocyanins, flavonoids, and polyphenols in LRM over the past five years, as well as the determination methods of some active ingredient compositions, biological activities, and product development. These active ingredients have significant benefits for human health, such as anticancer, anti-fatigue, and neuroprotective effects. Through scientific research and product development, these nutrients can be transformed into various healthy foods, health products, drugs, or materials to meet market demand, providing reference values for the application of LRM. But in the past five years, research on LRM has mainly focused on polysaccharides and anthocyanins, while research on flavonoids, amino acids, minerals, and tannins is relatively scarce. Most of the research focuses on the extraction and determination of components, without further investigation into their activity and exploration of their mechanisms. In addition, LRM product development is still relatively shallow, only at the laboratory stage, which limits the exploration of its efficacy and results in the scarcity of commercial products.

However, there are several research issues that urgently need to be addressed, which are also key to LRM product development. Although LRM is considered to have multiple health benefits, current research on its pharmacological effects is relatively limited and lacks in-depth scientific research support. At the same time, research on the pharmacological effects of LRM has been limited to animal experiments, with limited progress in clinical applications. The variety of processed products of LRM is limited, and their market awareness is not high, especially the lack of comprehensive exploration of their auxiliary functions. Therefore, it is necessary to strengthen research on the pharmacological activity of LRM and provide a scientific basis for their application in the fields of medicine and health food. Simultaneously developing diversified processed products, as well as strengthening market promotion and brand building, will help to increase the added value and market share of products.

Supplementary Materials: The following supporting information can be downloaded at: https://www.mdpi.com/article/10.3390/molecules29102269/s1, Table S1: LRM active ingredient extraction; Table S2: LRM active ingredient determination; Table S3: Biological activity of LRM extract; Table S4: Product development of LRM extract.

Author Contributions: M.-L.X. collated documents, wrote the manuscript, and finalized the paper; Y.-F.H. and L.X. reviewed the literature, discussed the layout, finished the artworks (Figures and Tables), and finalized the paper; M.-L.X., Y.-F.H. and L.X. retrieved the relevant literature and discussed the layout. L.-B.Q., G.-R.X. and C.-X.C. designed this manuscript. All authors have read and agreed to the published version of the manuscript.

Funding: This research was funded by Horizontal project grant number 210010712007, Key Research Projects in Henan Province Universities number 23A230005, Henan Province Science and Technology Research Project number 232102110075.

Institutional Review Board Statement: Not applicable.

Informed Consent Statement: Not applicable.

Data Availability Statement: No new data were created.

Acknowledgments: We are grateful for the Funding for Internationalization Training High Level Talent in Henan Province.

Conflicts of Interest: The authors declare no conflict of interest.

References

1. Yang, J. *Functional Verification of Key Differential Genes in Black and White Fruit of Lycium ruthenicum Murr.*; Qinghai University: Xining, China, 2022.
2. Xiong, L.; Deng, N.; Zheng, B.; Li, T.; Liu, R.H. Goji berry (*Lycium* spp.) extracts exhibit antiproliferative activity via modulating cell cycle arrest, cell apoptosis, and the p53 signaling pathway. *Food Funct.* **2021**, *12*, 6513–6525. [CrossRef]
3. Hu, G.; Cao, J.; Zhou, H.; Zhang, J.; Tian, Y.; Song, Y.; Jiang, R. Protective effect of *Lycium ruthenicum* Murr. juice on alcoholic liver injury in rats. *Chin. J. Appl. Physiol.* **2022**, *38*, 241–246.
4. Sun, X.; Wang, H.; Cao, J.; Zhou, H.; Zhang, J.; Tian, Y.; Wang, X. Mechanism of regulation of Nrf2 signaling pathway related protein expression in *Lycium ruthenicum* Murr. juice to improve alcoholic kidney injury in rats. *J. Nutr.* **2021**, *43*, 485–491.
5. Li, J.; Su, R.; Liu, Z.; Dong, Q.; Wang, H.; Hu, N. Screening and mechanism of anti-inflammatory active ingredient in *Lycium ruthenicum* Murr. *Chin. Wildl. Resour.* **2023**, *42*, 9–15, 38.
6. Wang, M.; Xu, T.; Gao, L.; Huang, C.; Xu, P.; Gong, C.; Amakye, W.K.; Liao, L.; Yao, M.; Ren, J. *Lycium ruthenicum* Murr. Treatment attenuates APPswE/PS1ΔE9 mouse model-like mitochondrial dysfunction in Slc25a46 knockout mouse model. *Food Sci. Hum. Wellness* **2023**, *12*, 1618–1625. [CrossRef]
7. Luo, Z.; Yu, G.; Chen, X.; Liu, Y.; Zhou, Y.; Wang, G.; Shi, Y. Integrated phytochemical analysis based on UHPLC-LTQ-Orbitrap and network pharmacology approaches to explore the potential mechanism of *Lycium ruthenicum* Murr. For ameliorating Alzheimer's disease. *Food Funct.* **2020**, *11*, 1362–1372. [CrossRef]
8. He, F.; Zhang, S.; Li, Y.; Chen, X.; Du, Z.; Shao, C.; Ding, K. The structure elucidation of novel arabinogalactan LRP1-S2 against pancreatic cancer cells growth in vitro and in vivo. *Carbohydr. Polym.* **2021**, *267*, 118172. [CrossRef] [PubMed]
9. Li, J.; Wang, Y.; Luo, R.; Lu, S.; Xing, L. Effects of different extraction processes on sugar content of *Lycium ruthenicum* Murr. *Food Technol.* **2021**, *46*, 175–182.
10. Wang, H.; Bai, H.; Wang, J. Study on the extraction of polysaccharides from black fruit goji berries using ultrasound microwave synergistic extraction method. *Northwest Agric. J.* **2007**, *16*, 157–158.
11. Jin, X.; Huang, L.; Wang, H. Optimization of High Hydrostatic Pressure-Assisted Extraction of Polysaccharides from *Lycium ruthenicum* Murr. by Response Surface Methodology and Their Antioxidant Activity. *J. Biobased Mater. Bioenergy* **2022**, *16*, 715–720. [CrossRef]
12. Sun, L.; Du, Y.; Yang, X.; Luo, A. Research progress on chestnut polysaccharides. *Life Sci. Instrum.* **2020**, *18*, 33–39.
13. Guo, Y.; Zhang, L.; Cao, L.; Chen, J.; Liu, B.; Zhao, Q.; Zhao, B. Research progress on monosaccharide composition analysis techniques in plant polysaccharides. *Food Sci.* **2018**, *39*, 326–332.
14. Zhang, Z.; Guo, Y.; Jiang, K.; Ma, J.; Che, X. Quantitative magnetic resonance spectrometry for the determination of polysaccharides in *Lycium ruthenicum* Murr. *World TCM* **2022**, *17*, 349–352.
15. Zhang, S. *Structural Identification and Biological Activity Study of Polysaccharides from Hawthorn and Black Fruit Goji Berries*; Zunyi Medical University: Zunyi, China, 2019.

16. Li, J.; Xing, L.; Wang, Y.; Luo, R.; Lu, S. Analysis of Polysaccharides in *Lycium ruthenicum* Murr. Based on UPLC -MS/MS. *Process. Agric. Prod.* **2021**, *6*, 60–62.
17. Guo, J.; Wang, R.; Hao, N.; Wang, J. Optimization of ultrasonic assisted biaqueous extraction of proanthocyanidin from *Lycium ruthenicum* Murr. by response surface test. *J. Beijing United Univ.* **2020**, *34*, 85–92.
18. Wu, Y.; Zeng, W.; Wang, L.; Yang, N.; Hu, H. Comparison between solvent extraction and ultrasonic assisted extraction of polyphenols from *Lycium ruthenicum* Murr. *Food Indust. Technol.* **2021**, *42*, 108–114.
19. Mi, J.; Lu, L.; Luo, Q.; Zhang, Z.; Li, X.; Jin, B.; Ni, Z.; Yan, Y. Study on the optimization and stability of anthocyanins extraction from *Lycium ruthenicum* Murr. by ultrasonic assisted enzymatic method. *Food Technol.* **2020**, *45*, 187–191.
20. Liu, X.; Guan, Z.; Mi, J.; Yan, Y.; Lu, L.; Luo, Q.; Ran, L.; Jin, B.; Li, X. Effects of processing methods on anthocyanins content and pancreatic lipase activity of *Lycium ruthenicum* Murr. *Ningxia Agroforest. Technol.* **2021**, *62*, 61–66.
21. Wang, T.; Ma, Z.; Wu, J.; Wang, J.; Yu, Q.; Feng, Y. HPLC study of anthocyanins in *Lycium ruthenicum* Murr. *J. Analyt. Sci.* **2020**, *36*, 465–470.
22. Gan, X.; Wang, H.; Li, T.; Li, B. UPLC -Triple TOF/MS analysis of chemical constituents of *Lycium ruthenicum* Murr. and determination of total anthocyanins content. *Food Sci.* **2021**, *42*, 185–190.
23. Zhang, C.; Wu, W.; Zhou, L.; Cheng, H.; Ye, X.; He, Y. Developing deep learning-based regression approaches for determination of chemical compositions in dry black goji berries (*Lycium ruthenicum* Murr.) using near-infrared hyperspectral imaging. *Food Chem.* **2020**, *319*, 126536. [CrossRef] [PubMed]
24. Muhammad, A.; Zou, X.; Elrasheid, T.H.; Shi, J.; Zareef, M.; Allah, R.; Muhammad, B. Rapid Screening of Phenolic Compounds from Wild *Lycium ruthenicum* Murr. Using Portable near Infrared (NIR) Spectroscopy Coupled Multivariate Analysis. *Anal. Lett.* **2020**, *54*, 1–15.
25. Shu, T.; Ding, L.; Wang, X.; Jia, C.; Qiu, Y.; Ruan, H. Analysis of the Structure of Natural Flavonoids in the Supercritical CO_2 Extraction of *Lycium ruthenicum* Murr. by Ultra-High Performance Liquid Chromatography Mass Spectrometry. *Food Sci.* **2020**, *41*, 206–212.
26. Liu, Z.; Tao, Y.; Shao, Y.; Zhang, H. Determination of betaine in Qaidam Lycium barbarum and *Lycium ruthenicum* Murr. *Spectr. Lab.* **2012**, *29*, 694–697.
27. Geng, D.; Tan, L.; Xiao, Y.; Hu, F.; Dong, Q. Ion chromatography for determination of betaine in *Lycium ruthenicum* Murr. *Food Sci.* **2015**, *36*, 145–147.
28. Meng, Q.; Ma, G.; Bai, H. Optimization of determination method of tannin content in *Lycium ruthenicum* Murr. *J. Tarim Univ.* **2011**, *23*, 9–14.
29. Yossa Nzeuwa, I.B.; Xia, H.; Shi, Y.; Yang, C.; Shah, M.W.; Guo, B.; Wang, L.; Sun, G. Fatty acid and mineral contents of *Lycium ruthenicum* Murr. and antioxidant activity against isoproterenol-induced acute myocardial ischemia in mice. *Food Sci. Nutr.* **2020**, *8*, 1075–1081. [CrossRef]
30. Zhao, S.S.; Li, S.; Luo, Z.H.; Zhou, Z.Q.; Li, N.; Wang, Y.; Yao, X.S.; Gao, H. Bioactive phenylpropanoid derivatives from the fruits of *Lycium ruthenicum* Murr. *Bioorganic Chem.* **2021**, *116*, 105307. [CrossRef] [PubMed]
31. Hu, Y.K.; Bai, X.L.; Yuan, H.; Zhang, Y.; Emmanuel, A.; Liao, X. Polyphenolic Glycosides from the Fruits Extract of *Lycium ruthenicum* Murr. and Their Monoamine Oxidase B Inhibitory and Neuroprotective Activities. *J. Agric. Food Chem.* **2022**, *70*, 7968–7980. [CrossRef]
32. Liu, J.; Wang, M.; Zhang, L.; Xue, C.; Wu, Q.; Wu, H.; Wang, S.; Wu, H. Effect of anthocyanidin from *Lycium ruthenicum* Murr. on apoptosis and autophagy interaction in HepG2 cells in human hepatocellular carcinoma. *J. Qinghai Univ.* **2022**, *40*, 54–59.
33. Wang, M.; Li, Y.; Liu, J.; Zhang, L.; A, X.; Zhou, J.; Wu, H.; Chai, S. Effects of anthocyanidin on proliferation and autophagy of human hepatocellular carcinoma cells in *Lycium ruthenicum* Murr. of Qinghai-Tibet Plateau. *Nat. Prod. Res. Dev.* **2021**, *33*, 79–88.
34. Qin, X.; Wang, X.; Xu, K.; Yang, X.; Wang, Q.; Liu, C.; Wang, X.; Guo, X.; Sun, J.; Li, L.; et al. Synergistic antitumor effects of polysaccharides and anthocyanins from *Lycium ruthenicum* Murr. on human colorectal carcinoma LoVo cells and the molecular mechanism. *Food Sci. Nutr.* **2022**, *10*, 2956–2968. [CrossRef]
35. Li, S.; Li, Z.; Mi, J.; Lu, L.; Yan, Y.; Cao, Y.; Ran, L.; Wen, Y. Effects of *Lycium ruthenicum* Murr. anthocyanins Pt3G on proliferation and apoptosis of prostate cancer LNCaP and PC-3 cells. *Sci. Bull.* **2022**, *67*, 364–375.
36. Li, Z. *Study on the Mechanism of Anthocyanins Pt3G from Lycium ruthenicum Murr. Thunb on ROS/PTEN/PI3K/Akt Signaling Pathway in Prostate Cancer Cells*; Ningxia Medical University: Yinchuan, China, 2021.
37. Kawuli, G.; Ma, J.; Wulamu, A.; Waili, Y.; Aobulitalifu, A. Effects of total flavonoids from *Lycium ruthenicum* Murr. on HepG-2 cell proliferation and apoptosis. *J. Food Saf. Qual. Test.* **2021**, *12*, 7014–7020.
38. Hu, J. *Isolation Process, Antioxidant Activity and Probiotic Application of Polysaccharides from Mongolian Lycium ruthenicum Murr.*; Northwestern Agroforestry University: Yangling, China, 2021.
39. Xu, Y.; Zhu, Y.; Xue, Y.; Zhang, W.; Yang, X.; Zhang, Y.; Ma, R.; Wang, Y.; Wen, P. Extraction of Polysaccharides from *Lycium ruthenicum* Murr. and Its Effect on Growth Characteristics and Antioxidation of Lactobacillus paracasei L9 and Streptococcus thermophilus G2. *Food Ferment. Ind.* **2021**, *47*, 179–185.
40. Wang, L.; Lin, Z.; Pang, X.; Liu, J. Study on the regulation of gut microorganism in humanized mice by polysaccharides from *Lycium ruthenicum* Murr. *Anhui Agric. Sci.* **2020**, *48*, 178–180.
41. Liu, X. *Effects of Anthocyanins from Lycium ruthenicum Murr. on Intestinal Flora and Liver Transcriptome in Obese Mice Induced by High Fat Diet*; Ningxia Medical University: Yinchuan, China, 2021.

42. Dong, B.; Cai, Y.; Lv, L.; Zhu, Z.; Zhu, X. Research progress of multiple Chinese herbs in enhancing immunity and anti fatigue effects. *Tradit. Chin. Pat. Med. Simple Prep.* **2019**, *41*, 1119–1124.
43. Bi, Y.; Liu, X.; Liu, Y.; Wang, M.; Shan, Y.; Yin, Y.; Meng, X.; Sun, F.J.; Li, H.; Li, Z. Molecular and biochemical investigations of the anti-fatigue effects of tea polyphenols and fruit extracts of *Lycium ruthenicum* Murr. on mice with exercise-induced fatigue. *Front. Mol. Biosci.* **2023**, *10*, 1223411. [CrossRef]
44. Tan, Y.; Wulamu, A.; Waili, Y.; Kawuli, G. Protective effect of *Lycium ruthenicum* Murr. polysaccharide on H_2O_2-induced ARPE-19 cells. *Chin. J. Tradit. Chin. Med.* **2022**, *37*, 2289–2294.
45. Tan, Y. *Nutrient Composition Analysis of Lycium ruthenicum Murr. and Its Protective Effect of Polysaccharides on ARPE-19 Cells*; Xinjiang Medical University: Ürümqi, China, 2020.
46. Durai, P.C.; Thappa, D.M.; Kumari, R.; Malathi, M. Aging in elderly:chronological versus photoaging. *Indian J. Dermatol.* **2012**, *57*, 343–352.
47. Wei, L. *Study on the Effects of Qaidam Lycium ruthenicum Murr. Anthocyanidin on Senescence and p53 and p21 of Human Dermal Fibroblasts Cultured by UVB Radiation In Vitro*; Qinghai University: Xining, China, 2021.
48. Wang, L.; Wan, G.; Wang, G.; Zhang, M.; Li, N.; Zhang, Q.; Yan, H. Anthocyanins from *Lycium ruthenicum* Murr. in the Qaidam Basin Alleviates Ultraviolet-Induced Apoptosis of Human Skin Fibroblasts by Regulating the Death Receptor Pathway. *Clin. Cosmet. Investig. Dermatol.* **2022**, *15*, 2925–2932. [CrossRef] [PubMed]
49. Liu, R.; Chang, P.; Jiang, Z. Research progress on the neuroprotective effects of bioactive polysaccharides and their preventive and therapeutic effects on Alzheimer's disease. *Nat. Prod. Res. Dev.* **2014**, *26*, 2076–2081.
50. Deng, K.; Li, Y.; Xiao, M.; Wang, F.; Zhou, P.; Zhang, W.; Heep, A.; Li, X. *Lycium ruthenicum* Murr. polysaccharide protects cortical neurons against oxygen-glucose deprivation/reperfusion in neonatal hypoxic-ischemic encephalopathy. *Int. J. Biol. Macromol.* **2020**, *158*, 562–568. [CrossRef] [PubMed]
51. Cai, Y.; Li, X.; Pan, Z.; Zhu, Y.; Tuo, J.; Meng, Q.; Dai, G.; Yang, G.; Pan, Y. anthocyanins ameliorates hypoxia and ischemia induced inflammation and apoptosis by increasing autophagic flux in SH-SY5Y cells. *Eur. J. Pharmacol.* **2020**, *883*, 173360. [CrossRef] [PubMed]
52. Chen, S.; Hu, N.; Wang, H.; Li, G. The major anthocyanins of *Lycium ruthenicum* Murr. relieves cognitive deficits, oxidative stress, neuroinflammation, and hippocampal metabolome alterations in aging rats. *Funct. Foods* **2022**, *94*, 105104. [CrossRef]
53. Shen, T.; Wu, H.; Chen, X.; Diao, M.; Li, J.; Wang, S.; Wu, X.; Xing, Q. Effects of anthocyanidin from *Lycium ruthenicum* Murr. on antioxidant activity, inflammatory factors and energy metabolism of cardiomyocytes in H9c2 rats. *Feed Res.* **2023**, *46*, 61–65.
54. Yan, S.; Wang, S.; Chen, T.; Shen, C.; Zhao, J.; Ding, Y.; Kong, L.; Yang, L.; Li, Y. Effects of anthocyanidin extract from *Lycium ruthenicum* Murr. on acute gout arthritis in rats. *J. Food Sci. Technol.* **2022**, *40*, 105–111.
55. Xu, K.; Qin, X.; Zhang, Y.; Yang, M.; Zheng, H.; Li, Y.; Yang, X.; Xu, Q.; Li, Y.; Xu, P.; et al. *Lycium ruthenicum* Murr. anthocyanins inhibit hyperproliferation of synovial fibroblasts from rheumatoid patients and the mechanism study powered by network pharmacology. *Phytomedicine* **2023**, *118*, 154949. [CrossRef]
56. Wang, L.; Ye, W.; Wu, H.; Liu, J.; Pang, X.; Lin, Z. Protective effects of *Lycium ruthenicum* Murr. polysaccharides on liver injury in rats. *Food Ind. Technol.* **2020**, *41*, 287–290+296.
57. Lu, K.; Wang, J.; Yu, Y.; Wu, Y.; He, Z. *Lycium ruthenicum* Murr. alleviates nonalcoholic fatty liver in mice. *Food Sci. Nutr.* **2020**, *8*, 2588–2597. [CrossRef]
58. Dong, Y.; Yang, C.; Zhong, W.; Shu, Y.; Zhang, Y.; Yang, D. Antibacterial effect and mechanism of anthocyanins from *Lycium ruthenicum* Murr. *Front. Microbiol.* **2022**, *13*, 974602. [CrossRef] [PubMed]
59. Dong, Y. *Extraction of Anthocyanidin from Lycium ruthenicum Murr. and Its Antimicrobial and Antioxidant Mechanisms*; Jilin University: Changchun, China, 2022.
60. Li, J.; Yang, F.; Jian, G.; Dong, Q.; Wamg, H.; Hu, N. Research progress on anthocyanins components and pharmacological effects of *Lycium ruthenicum* Murr. *Huaxi Pharm. J.* **2022**, *37*, 331–336.
61. Xu, H. *A Study on the Induction of ROS Accumulation Mediated Leukocyte Phagocytosis and Apoptosis in HepG2 Cells by Lotus Seed Anthocyanins*; Jiangsu University: Zhenjiang, China, 2016.
62. Wang, M. *Effects of Anthocyanidin from Lycium ruthenicum Murr. on Hypoxic Adaptation of Cardiomyocytes in H9c2 Rats*; Qinghai University: Xining, China, 2022.
63. Li, J.; Wu, H.; Shu, M.; Shen, T.; Wu, X.Q.; Wang, S.; Li, W.P. *Lycium ruthenicum* Murr. anthocyanidin mediated the establishment of ceRNA regulatory networks in H9c2 rat cardiomyocytes induced by hypoxia. *J. Agric. Biotechnol.* **2023**, *31*, 1671–1683.
64. Zhang, J.; Mi, J.; Lu, L.; Luo, Q.; Yan, Y.; Ran, L.; Jin, B.; Cao, Y. Effect of anthocyanins extract from *Lycium ruthenicum* Murr. on pancreatic lipase activity. *Food Sci.* **2020**, *41*, 8–14.
65. Wang, Z. *Tracing the Origin of Lycium ruthenicum Murr. and Its Anthocyanins Hypoglycemia Mechanism*; Shaanxi Normal University: Xi'an, China, 2020.
66. Wulamu, A. *Isolation and Purification of Total Flavonoids from Lycium ruthenicum Murr. and Preliminary Study on its Lipid-Lowering and Anti-Liver Cancer Activities*; Xinjiang Medical University: Ürümqi, China, 2020.
67. Kurskaya, O.; Prokopeva, E.; Bi, H.; Sobolev, L.; Murashkina, T.; Shestopalov, A.; Wei, L.; Sharshov, K. Anti-Influenza Activity of Medicinal Material Extracts from Qinghai-Tibet Plateau. *Viruses* **2022**, *14*, 360. [CrossRef]
68. Luo, J. *Effect and Mechanism of Anthocyanins from Lycium ruthenicum Murr. on Anxiety and Craving after Nicotine Withdrawal in Mice*; Beijing University of Chinese Medicine: Beijing, China, 2020.

69. Xiong, L.; Deng, N.; Zheng, B.; Li, T.; Liu, R.H. HSF-1 and SIR-2.1 linked insulin-like signaling is involved in goji berry (*Lycium* spp.) extracts promoting lifespan extension of Caenorhabditis elegans. *Food Funct.* **2021**, *12*, 7851–7866. [CrossRef]
70. Wang, S.Q.; Liu, B.; Liu, S.; Xie, S.Z.; Pan, L.H.; Zha, X.Q.; Li, Q.M.; Luo, J.P. Structural features of an acidic polysaccharide with the potential of promoting osteoblast differentiation from *Lycium ruthenicum* Murr. *Nat. Prod. Res.* **2018**, *34*, 2249–2254. [CrossRef]
71. Li, L.; Jia, Q.; Zhang, H.; Yi, L.; Tang, Y.; Hu, P.; Wang, R.; Zhang, H. Tissue lipidomics, network pharmacology, and molecular docking to explore the therapeutic mechanism of anthocyanins from *Lycium ruthenicum* Murr. against gouty arthritis. *Food Funct.* **2023**, *14*, 7011–7023. [CrossRef]
72. Guo, S.; Qi, M.; Li, H.; Cui, Y.; Qi, C.; Cheng, G.; Lv, M.; Zheng, P.; Liu, J. The Protective Effect of *Lycium ruthenicum* Murr. anthocyanins in Cr (VI)-Induced Mitophagy in DF-1 Cells. *Life* **2022**, *12*, 1115. [CrossRef]
73. Qin, Y.; Yun, D.; Xu, F.; Li, C.; Chen, D.; Liu, J. Impact of storage conditions on the structure and functionality of starch/polyvinyl alcohol films containing *Lycium ruthenicum* anthocyanins. *Food Packag. Shelf Life* **2021**, *29*, 100693. [CrossRef]
74. Tan, Y. *Study on Extraction of Proanthocyanidin and Its Fibrillar Membrane Envelope in Lycium ruthenicum Murr.*; South China University of Technology: Guangzhou, China, 2021.
75. Xu, F.; Yan, Y.; Yong, H.; Yun, D.; Dan, C.; Liu, J. Comparison of the physical and functional properties of food packaging films containing starch and polyphenols from different varieties of wolfberry. *J. Food Meas. Charact.* **2022**, *16*, 4444–4456. [CrossRef]
76. Zhang, M.; Zeng, F.; Ye, Y.; Zheng, S.; Li, J.; Fei, P. Anthocyanins pH and ammonia sensitivity and antioxidant activity of *Lycium ruthenicum* Murr. *Food Ind. Technol.* **2021**, *42*, 22–27.
77. Chen, L.; Xiong, Z.; Jia, D.; Sun, X.; Huang, S.; Guo, T.; Zhan, H.; Liu, X. Intelligent labeling based on *Lycium ruthenicum* Murr. anthocyanidin: Freshness indicative performance and mechanism. *Mater. Guide* **2022**, *36*, 168–173.
78. Zhao, Y.; Gao, L.; Wang, J.; Xue, Z.; Zhang, M.; Ma, X.; Wang, G.; Lv, S. Preparation and Application of pH-Sensitive Film Containing anthocyanins Extracted from *Lycium ruthenicum* Murr. *Materials* **2023**, *16*, 3828. [CrossRef] [PubMed]
79. Tang, P.; Giusti, M.M. Metal Chelates of Petunidin Derivatives Exhibit Enhanced Color and Stability. *Foods* **2020**, *9*, 1426. [CrossRef] [PubMed]
80. Deng, Y.; Jiao, Y.; Yuan, Z. The antioxidant mechanism of exopolysaccharides' synthesis from *Agaricus bitorquis* (Quél.) Sacc. Chaidam under the influence of *Lycium ruthenicum* Murr. anthocyanins extract. *Macromol. Res.* **2023**, *31*, 1001–1017.
81. Wu, S.; Lu, H.Y.; Chen, Q.H.; Xie, H.C.; Jiao, Y.C. Anthocyanins extract from *Lycium ruthenicum* enhanced production of biomass and polysaccharides during submerged fermentation of *Agaricus bitorquis* (Quél.) Sacc. Chaidam. *Bioprocess Biosyst. Eng.* **2021**, *44*, 2303–2313. [CrossRef] [PubMed]
82. Yu, P.; Tian, K.; Wang, Q.; Zhang, J.; Zhao, L.; Pei, D. Extraction of proanthocyanidin and preparation of chewable tablets from *Lycium ruthenicum* Murr. *Food Ind.* **2020**, *41*, 152–157.
83. Zhou, W.; Lu, J.; Du, C.; Zhang, T. Study on Preparation Technology of *Lycium ruthenicum* Murr. Chewable Tablets. *Food Grease* **2021**, *34*, 82–86.
84. Heng, Y.; Zhou, Z.; Chen, C.; Song, Y.; Ku, J.; Ji, Z.; Mao, J. Study on the production technology of fermented beverage by Lactobacillus *Lycium ruthenicum* Murr. *Food Ferment. Ind.* **2023**, *49*, 114–121.
85. Jia, Q.; Zhu, L. Development and antioxidant properties of *Lycium ruthenicum* Murr. passion fruit milkdrink based on fuzzy mathematical evaluation. *Chin. Brew.* **2022**, *41*, 165–171.
86. Liu, Y.; Yin, J.; Lei, Y.; Dai, S. Study on the technology of compound beverage of Aronia melanocarpa (Michx.) Elliott and *Lycium ruthenicum* Murr. *Process. Agric. Prod.* **2020**, *504*, 48–51.
87. Ma, X.; Cao, L.; Xiao, M.; Sun, X.; Cui, M. Study on the Process and Stability of *Lycium ruthenicum* Murr. Functional Drinks. *Food Res. Dev.* **2020**, *41*, 80–85.
88. Jiang, B.; Xu, X.; Chen, Q.; Han, F.; Duan, H. Study on Optimization of *Lycium ruthenicum* Murr. Yogurt Technology by Response Surface Method. *J. Jamus Univ. (Nat. Sci.)* **2021**, *39*, 107–111.
89. Qi, H. *Optimization of Technology and Evaluation of Antioxidant Activity of Lycium ruthenicum Murr. Vinegar*; Stone River University: Stanford, CA, USA, 2022.
90. Shen, M. *Preparation and Activity Study of Anthocyanins Microcapsules of Lycium ruthenicum Murr.*; Shaanxi Normal University: Xi'an, China, 2020.
91. Gu, J.; Sun, X.; Fu, L.; Dong, Y. Extraction of *Lycium ruthenicum* Murr. pigment and dyeing of wool fabric. *Print. Dye.* **2023**, *49*, 37–41.

molecules

MDPI

Review

Bioactive Compounds Produced by Endophytic Bacteria and Their Plant Hosts—An Insight into the World of Chosen Herbaceous Ruderal Plants in Central Europe

Piotr Drożdżyński *, Natalia Rutkowska, Magdalena Rodziewicz and Olga Marchut-Mikołajczyk

Institute of Molecular and Industrial Biotechnology, Faculty of Biotechnology and Food Sciences, Lodz University of Technology, Stefanowskiego 2/22, 90-537 Lodz, Poland; natalia.rutkowska@dokt.p.lodz.pl (N.R.); rodziewicz.mag@gmail.com (M.R.); olga.marchut-mikolajczyk@p.lodz.pl (O.M.-M.)

* Correspondence: piotr.drozdzynski@p.lodz.pl; Tel.: +48-42-631-33-52

Abstract: The natural environment has been significantly impacted by human activity, urbanization, and industrialization, leading to changes in living organisms and their adaptation to harsh conditions. Species, including plants, adapt to these changes by creating mechanisms and modifications that allow them to survive in harsh environments. Also, endophytes, microorganisms that live inside plants, can support plant growth and defense mechanisms in these conditions by synthesizing antimicrobial secondary metabolites. What is more, endophytes produce bioactive metabolites, including alkaloids, amines, and peptides, which play a crucial role in the relationship between endophytes and their host organisms. Endophytes themselves benefit from this by creating a stable environment for their survival and development. The aim of this review is to gain insight into endophytic bioactive metabolites from chosen synanthropic ruderal plants. Industrial activities release pollutants like heavy metals, by-products, and waste, which challenge living organisms and require adaptation. Synanthropic plants, where endophytes are abundant, are particularly valuable for their bioactive compounds, which are used in agriculture and medicine. This review presents, among others, endophytes of herbaceous ruderal plants from central Europe—*Chelidonium majus* L., *Urtica dioica* L., *Plantago lanceolata* L., *Matricaria chamomilla* L., *Equisetum arvense* L., *Oenothera biennis* L., *Silybum marianum* L., and *Mentha piperita* L.

Keywords: synanthropic herbaceous plants; endophytic bacteria; biologically active compounds; biosynthetic gene clusters

Citation: Drożdżyński, P.; Rutkowska, N.; Rodziewicz, M.; Marchut-Mikołajczyk, O. Bioactive Compounds Produced by Endophytic Bacteria and Their Plant Hosts—An Insight into the World of Chosen Herbaceous Ruderal Plants in Central Europe. *Molecules* **2024**, *29*, 4456. https://doi.org/10.3390/molecules29184456

Academic Editors: Yongjun Wei, Xiao-Jun Ji and Lingbo Qu

Received: 18 July 2024
Revised: 9 September 2024
Accepted: 17 September 2024
Published: 19 September 2024

1. Introduction

Over the years, the natural environment has undergone many changes, often as the result of human activity, its broadly understood development, and progressive urbanization. Because of the shrinkage of natural water bodies, wetlands, and the gradual drying out of soils near cities, the transformation of the economy and thus industrialization has not only caused a dramatic change in the conditions prevailing there but has also made it impossible for some maladjusted species of living organisms to continue to exist. Synanthrophization refers to the whole of human impact, including direct and indirect acts that result in a specific modification to the environment and a need for adaptation by the organisms that inhabit it. Species, including plants, with different degrees of effectiveness respond to these needs by creating various types of mechanisms and modifications that allow them to develop in ruderal areas. Metabolism changes caused by changing environmental conditions are often associated with the production of biologically active compounds. By adapting to new living conditions, such plants can displace other species, which is why they are often called weeds, but some of them turn out to have healing potential. The functioning of plants in difficult conditions can also be facilitated by microorganisms inhabiting the inter-tissue spaces of plants, which not only do not harm the host but often produce many substances that allow plants to survive in unfavorable environmental conditions.

2. Endophytes and Their Sources

According to Wilson's 1995 definition [1], endophytes are fungi or bacteria which, for all or part of their life cycle, invade the tissues of living plants and cause unapparent and asymptomatic infections entirely within plant tissues but cause no symptoms of disease. These microbes are often found in intracellular and intercellular areas and vascular tissue and can colonize aerial parts or roots [2]. However, nowadays, it is clear that endophytic bacteria mostly derive from the rhizosphere environment [2]. The term "endophyte" has become a synonym for a "mutualist" or "commensalist" symbiont [2,3], and evolutionary endophytic bacteria are considered an intermediate group between saprophytic and pathogenic bacteria. Nevertheless, not all colonizers are harmless; some plant pests are often bacterial or fungal [4–6].

Based on current studies, medicinal and agricultural plants are commonly prioritized for the isolation of endophytes. Furthermore, fungal endophytes garner greater scrutiny compared to bacterial endophytes. Figure 1 illustrates a comparison of the endophytic hosts, revealing that medicinal plants, crops, and plants in specialized environments have been extensively investigated for the purpose of identifying antimicrobial endophytes [7].

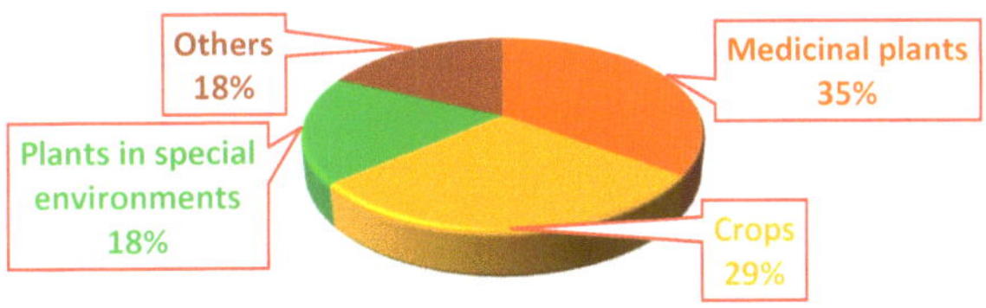

Figure 1. Examination of the ratio of biologically active strains from various sources that were evaluated for antibacterial activity based on research studies over the last decade [7,8].

Among the several division classifications suggested by Hardoim [9], endophytic microorganisms are classified as obligate, facultative, opportunistic, or "passenger", based on the distinct internal mechanisms used to colonize a plant, as depicted in Figure 2 [9–11].

Figure 2. Lifestyle-based categorization of endophytic bacterial species [12,13].

Opportunistic endophytes are microorganisms that invade plants to help them survive and thrive in harsh environments. The term "passenger endophyte" refers to microorganisms that inadvertently colonize a plant. Since obligatory endophytes are likely passed on through seeds instead of the rhizosphere, they cannot spread outside of plants. One the one hand, they depend on the plant's metabolic activity for their survival and spread inside the plant by vertical colonization or via the action of a vector; on the other hand, they are vital to the plant's ability to maintain its internal environment [4,12–14]. Facultative endophytes can live both within plants and outside of them; they are found living unrestrictedly in the soil, and when the chance presents itself, they will colonize inside plants through coordinated infection. The rhizosphere is the location where this process takes place.

A multitude of studies have conclusively shown that endophytic communities exhibit high levels of diversity, with nearly all of them being facultative. A number of researchers, such as Hallmann et al. (1997), Afzal et al. (2019), Santoyo et al. (2016), Surjit et al. (2014), and Rosenblueth and Martinez-Romero (2006), have presented data about the diverse range of endophytic communities [4,14–17]. Stochastic occurrences, such as open wounds along the root hairs, are what allow passive endophytes to colonize a plant. Passive endophytes do not actively strive to colonize the plant. They are referred to as associated bacteria since they are attached to the surface of the roots. In addition, bacteria that live in the rhizosphere (rhizobacteria) that are closely linked to the plant roots also occur. Endophytes that have a passive life strategy are less competitive than other endophytes because the cellular machinery that is necessary for plant colonization is absent from their cells [2,16,18]. Passive penetration may occur at the root tip, root hairs, or elongation zone level and at cracks such as those occurring at root emergence sites, but in general, many endophytes are equipped with cell-wall-degrading enzymes such as cellulases, which are probably needed for efficient plant colonization and spreading within plant tissues [19,20]. Structures such as stomata, hydathodes, lenticels, nectarines, and nectar glands, as well as tissue damage that is caused by the attack of insects or broken trichomes, are also entry routes for soil microbiota [21]. These colonizers are capable of establishing populations both intercellularly and intracellularly [22–24].

3. Endophytic Plant Colonization Mechanisms

The process of root colonization can be categorized into two primary phases. During the early phase, bacteria attach to the root cells and then reproduce, resulting in their widespread distribution throughout the whole surface. In this phase, there is a rapid increase in the number of microorganisms in habitats that have limited natural boundaries.

Within the rhizosphere, a specific cluster of microbes demonstrates a greater capacity to use root secretions as a carbon and energy substrate. This grants them a greater capacity to rival other bacterial species with enhanced effectiveness [24]. The colonization process begins with the identification of particular plant root secretions by bacteria. Certain bacteria demonstrate chemotaxis, a phenomenon in which they are drawn towards the origin of secretions due to their recognition of certain chemical compounds, such as phenolic acids or amino acids. The chemotactic response to certain components in root secretions is crucial for bacteria to establish colonization in plants [17]. Moreover, research has revealed that various plant tissues can harbor distinct endophyte communities with varying compositions [25].

The *Pseudomonas stutzeri* strains PPS96, PSR2, and PSR21, which showed the greatest ability to colonize wheat roots, were found to utilize p-hydroxyphenylacetic acid, bromosuccinic acid, benzoic acid, methyl pyruvate, N-acetyl-D-glucosamine, and D-trehalose as their sole carbon and energy sources. In contrast, *Pseudomonas stutzeri* P221, PPS16, PSR25, PSR67, PSR79, PSR81, PSR82, and PSR120 strains that had a restricted ability to colonize roots did not show this capability [26].

Some parts of plants on land can be colonized by different microbes. In addition, the surfaces of stems and leaves can potentially release exudates that attract bacteria.

During the process of phyllosphere colonization, bacterial strains first attach to the surface of the leaf and then spread out in a random fashion. Specific bacteria have the capacity to penetrate leaf tissue via natural openings, such as stomata and hydathodes, therefore influencing their local environment. Yaron and Römling (2014) found that bacterial strains in this specific region reproduce and create a fragile biofilm. However, a portion of these strains have the capability to infiltrate the leaf tissue and establish themselves as endophytes [27]. James et al. (2001) identified the stomata as the entry route for the colonization of the *Gluconobacter diazotrophicus* strain in sugarcane [28]. Regrettably, the presence of UV radiation, nutritional deficiency, and desiccation often hinders the colonization of leaf surfaces [29].

Passive mechanisms can also aid in the process of host invasion [9,30]. Soil-dwelling bacteria can unintentionally become endophytic by establishing themselves in natural wounds or when their roots are invaded by a parasite, among other ways. These bacteria are categorized as guest endophytes and commonly exist only in the root cortical tissue. Like passenger endophytes, opportunistic endophytes are also restricted to particular plant tissues, such as the root cortex. It is proposed that competent endophytes possess all the traits of opportunistic endophytes and demonstrate a significant level of adaptability to the plant environment. These organisms possess the capability to penetrate specific plant tissue, such as vascular tissue, and spread across the entire plant. They accomplish this by regulating plant metabolism, enabling them to coexist with the plant host even in significant quantities [30].

Horizontal transmission appears to be the main method of dispersion among endophytic species. Endophytes propagate through the production of spores, which are then dispersed by insects and wind. Endophytes can be horizontally transmitted, meaning they can spread from one infected plant to uncolonized ones. Endophytes that are transmitted by seeds and vertically inherited are commonly found as epiphytes, suggesting that endophytes may also establish colonies in the environments around host plants [31].

Endophytic bacteria can infect plants by vertical transmission, which refers to the transfer from the parent to successive generations. Specific endophytes that inhabit the seeds are crucial for guaranteeing their transfer to the progeny. This is also applicable to endophytes that travel together with seedlings of fruit trees, tubers, and rootstocks. The bacteria in question occupy a favorable habitat, and it is likely that other microorganisms in the rhizosphere may undergo slight changes in response to their presence [32].

When examining endophytes that come from the rhizosphere, it is crucial for them to successfully establish colonization in a new host by entering through the root or another portion of the plant.

3.1. *Physiological Mechanisms*

The primary mechanisms by which plants undergo bacterial colonization include the modulation of the root exudate composition and the activation of an innate immune response. The colonization of the plant is influenced by both random events and predictable bacterial factors. The process of colonizing plant roots entails the formation of microcolonies on the root surface. Opportunistic endophytes possess distinct characteristics that enable them to colonize roots, such as a chemotactic response; multiple studies [17,33] have demonstrated that endophytes exhibit a chemotactic response when exposed to root exudates from their host plants. Endophytes and other beneficial microorganisms actively acquire or distinguish the biomolecules that are abundant in root exudates. Exudates possess a prodigious supply of water and nutrients, which magnetize microorganisms chemotactically. A wide variety of plants facilitate endophytic contact with potential plant hosts via their flavonoid pigments, which function as chemoattractants [26,34]. The study conducted by Bacilio-Jimenez et al. (2003) revealed the presence of chemotactic mechanisms. The researchers observed that the rice root exudates were predominantly composed of glucose, arabinose, mannose, and galactose, as well as amino acids like histidine, proline, alanine, and glycine. Within the initial two weeks after the plants were planted, there was a significant rise in and variety of carbohydrates and amino acids [35].

The chemotactic response of endophytic isolates of *Corynebacterium flavescens* and *Bacillus pumilus* towards the components of rice root exudates exhibited a significantly higher magnitude, ranging from 3.9 to 5.1 times, compared to the tested isolate *Azospirillum brasiliense* [35]. The chemotaxis of the endophytic isolates mentioned was 2.2 to 2.8 times higher compared to the second strain, *Bacillus* sp. Potential endophytes display a major advantage by being able to use various parts of root secretions throughout the process of root colonization [26].

To preferentially attract the mutualistic microorganisms, especially the potential endophytes, plant roots release a variety of secondary metabolites, including sugars, amino acids, organic acids, flavonoids, phenolic compounds, and others [36]. The success of root colonization in rice and wheat by *Azorhizobium caulinodans* ORS571 and the endophytic *Serratia* sp. EDA2 was significantly enhanced in the presence of flavonoids, a category of metabolites that have also been implicated in the case of non-rhizobial endophytes [37,38].

What is more, the colonization of plant hosts depends on many factors and is affected by genotype, growth phase, physiological status, plant tissue type, taxon and strain type, agricultural practices, and environmental variables (temperature, water supply, and nutrients) [29].

3.2. Molecular Mechanisms

Molecular patterns also drive all stages of colonization. Several important genes found in bacterial genomes play a critical role in their ability to chemotax and migrate as well as adhere to the surface of roots and navigate throughout plants. The aforementioned discoveries were facilitated by the use of comparative genomics investigations and the examination of bacterial mutants, as extensively detailed by Pinski and coworkers [39]. In their 2002 work, de Weert and colleagues investigated the reaction of bacterial strains belonging to the species *Pseudomonas fluorescens* to chemotaxis while colonizing tomato roots. Wild strains and mutant strains were utilized, wherein mutations were introduced in the *cheA* gene, which is responsible for the flagella-mediated movement towards chemical attractants in tomatoes. The results of their experiments validated that the faulty *cheA* gene significantly reduced the competitive colonization of tomato root tips. All *cheA* mutants demonstrated a 10- to 1000-fold reduction in their ability to colonize the root tip of the vegetable compared to their wild counterpart after the seedlings received an inoculation with a 1:1 mixture [40]. In 2016, Rossi and colleagues investigated the *Azospirillum brasilense* strain and its capacity to adhere to maize roots. Cells of *A. brasilense* were rendered immotile as a consequence of alterations in the polar flagellum assembly caused by mutations in the *flmA* or *flmB* genes. Moreover, these alterations also impacted the competitive capacity of A. brasilense cells to attach to maize roots and generate lipopolysaccharides and exopolysaccharides [41]. Attachment and biofilm formation of new colonizers is the next step during colonization. Duque et al. employed a hybrid methodology using in silico algorithms and mutagenesis techniques to examine the genetic factors that influence the surface attachment of *Pseudomonas putida* strain KT2440. An analysis revealed that the genes encoding the synthesis of the major adhesion proteins LapA and LapF and flagellar proteins were highly conserved in all the sequenced strains of *P. putida*. Disabling these genes resulted in a reduced capacity to compete in colonizing maize roots [42].

In brief, the colonization of plant roots by bacteria is a progressive and multifaceted process that initiates in the rhizosphere with the migration of bacteria towards the root, followed by their adherence and subsequent development of a biofilm. Thus far, the majority of the genetic factors influencing bacterial colonization have been evaluated in experimental investigations involving a single strain, conducted in sterile environments. However, bacterial colonization can be influenced by both plant–bacteria and bacteria–bacteria interactions. Unfortunately, the tremendous variety and complexity of the root-associated microbiota hinder the characterization of the molecular pathways [43].

This reaction allows them to first colonize the rhizoplane and then infiltrate the interior tissues of the plant through fissures that occur at the points where lateral roots form and at the tips of the roots. The main cause of this phenomenon is the presence of methyl-accepting chemotaxis proteins (MCPs), which are sensors located in the bacterial membrane. These sensors detect chemicals in the surrounding environment and steer the bacteria towards attractive substances or away from repellents [35]. One study found that disabling the

Hsero_3720 gene, which encodes MCPs, in the mutant strain *Herbaspirillum seropedicae* SmR1 resulted in a 50% decrease in its capacity to attach to maize roots compared to a non-mutated strain [44].

4. Interaction between Plants and Endophytic Microorganisms

The relationship between an endophyte and a plant is considered symbiotic since both organisms benefit from this partnership. The co-evolutionary process results in the existence of an intimate relationship between the endophyte and its host, and this interaction is caused by changes at the cellular and molecular levels that affect the development of the plant, namely, improvements in plant growth and health through a variety of processes, including nitrogen fixation, the synthesis of plant hormones (phytohormones) and vitamins, and the enhancement of nutrient, mineral, and water uptake, as well as the induction of stress resistance [45]. Stress can be caused by abiotic factors such as osmotic stress as well as exposure to heavy metals and xenobiotic compounds. Endophytes can also accelerate digestion in plants and phosphorus solubilization, which provide resistance to biotic factors. Microbial interactions, which are advantageous to the host plant, are necessary for the development of potential endophyte colonies on the surface of the root. Through a mechanism referred to as cross-talk of signal molecules, an endophyte colonizes a host plant, which subsequently initiates communication with the invading endophyte. Following the initial colonization, certain endophytes have the ability to distribute themselves across various regions of the plant by utilizing the vascular tissues, a phenomenon referred to as systemic dissemination. Johnston-Monje and Raizada (2011) used endophytes that were labeled with green fluorescent protein (GFP) to demonstrate how these microorganisms move from seeds to the plant's roots and tissues. Additionally, they injected endophytes into stems and observed how they moved into the roots and rhizosphere. The results of this study imply that there may be a continuous flow of organisms throughout the root microbiome [46,47]. The relationship between plants and endophytic bacteria has a substantial impact on plant growth, health, development, yield, and soil quality. These interactions also give many benefits to the species involved and have numerous positive effects on the integrity and sustainability of agroecosystems. Then, they emphasize the relevance of the increase in growth, plant biomass, dry matter yield, or grain yield, thus contributing to a rise in agricultural income.

Endophytic microorganisms may outcompete invading pathogens through a variety of processes, including niche occupation through competition for space, resources, and physical niches of the rhizosphere or rhizoplane and endophytic tissues. Some beneficial endophytic bacteria can secrete antibiotics and lytic enzymes that enable the inhibition of various pathogens. Moreover, the artificial inoculation of endophytic microbes into plants can considerably minimize the attack of disease-causing factors such as fungi, bacteria, viruses, insects, and nematodes, thereby biologically controlling these pathogens [14,48].

These microorganisms may outcompete invading pathogens through a variety of processes, including niche occupation through competition for space, resources, and physical niches of the rhizosphere or rhizoplane and endophytic tissues. Some beneficial endophytic bacteria can secrete antibiotics and lytic enzymes that enable the inhibition of various pathogens [49,50].

5. Surface Sterilization of Plant Material

Endophyte isolation is a crucial step in the study of endophytic bacteria and fungi. Endophytic microorganisms, such as bacteria and fungi, are isolated in order to describe and explore their population dynamics and diversity. In addition, they are isolated in order to utilize their microbial inoculants to promote plant development and health and to obtain novel physiologically active secondary metabolites [51]. The crucial part of the isolation of endophytic microorganisms is the surface sterilization of the plant material. Table 1 lists the most frequent and widely recognized sterilizing methods. The success of surface sterilization is dependent on a variety of characteristics, including the sterilizing agent employed, the resistance of the plant tissue samples to the sterilizing agents, and the time and concentration of the sterilizing agent [52].

Table 1. The most frequent and widely recognized methods of plant material surface sterilization [52].

No.	Method of Plant Material Sterilization	Plant or Plants Part Example	Endophytes Isolated	Reference
		Single compound approach		
1	EtOH 70–99% (different times)	*Nepenthes species; Anemone nemorosa, Ranunculus ficaria, Vaccinium oxycoccus, Sambucus nigra*	*Acremonium* sp. *Libertella heveae*	[53–55]
2	NaOCl 2–30% (up to 30 min)	leaves stems and roots *Pistacia atlantica* L.; *Simmondsia chinensis*	*Total of 61 endophytes, with 10 belonging to genus Pseudomonas, Stenotrophomonas, Bacillus, Pantoea and Serratia; Bacillus* sp., *Methylobacterium aminovorans, Oceanobacillus kimchi, Rhodococcus pyridinivorans, and Streptomyces* sp.	[56,57]
3	0.1% $HgCl_2$ (1–10 min)	*Ziziphus nummularia*	*Microsporum gypseum, Aspergillus fumigatus, Aspergillus calidoustus, Penicillium viridicatum, Trichophyton tonsurans, and Penicillium marneffei*	[58]
4	Surfactants	Leaves, stems, and roots		[59]
5	Tween 20 0.01–10%	Leaves, stems, and roots	127 Endophytic fungi	[60]
6	Tween 80 0.1%	Fruits		[61]
7	Teepol 5%	Leaves, stems, and roots	228 Isolates representing at least 19 genera of actinobacteria	[62]
8	Triton X-100 0.05–0.1%	Leaves, stems, and roots	116 endophytic fungi	[63–65]
10	Hydrogen peroxide 3–90%	Leaves, stems, and roots		[59,66]
		Combined approach		
11	2% NaOCl + 0.1% Tween 20	Roots (tomato)	*Rhizobium, Bacillus, Microbacterium, Enterobacter species*	[67,68]
12	70% EtOH-3.125% NaOCl	*Psidium guajava and Ziziphus mauritiana*	7 Endophytic actinomycetes	[53]
13	75% EtOH (2 min) and 1.5%NaOCl (3 min)	*Brassica olerocea, B. rapa, and Raphanus sativus*	4178 Endophytic fungal isolates belonging to 51 different genera	[69]
14	70% EtOH (1 min), 2.5% NaOCl (3 min); 70% EtOH (1 min), 2.5% NaOCl (1 min), and 70% EtOH (30 s)	Roots and leaves of *Salvia abrotanoides*	56 Endophytes	[70]
15	75% EtOH (2 min), 2% NaOCl (3 min), and 75% EtOH (30 s)	*Gnetum gnemon*	*Staphylococcus warneri, Solibacillus isronensis, Bacillus megaterium, Caballeronia glebae; Bacillus licheniformis, Bacillus velezensis, and Bacillus atrophaeus*	[71]
16	75% EtOH (1 min), 1% $HgCl_2$ (4 min), and 75% EtOH (1 min)	*Lycoris radiata*	Total of 188 bacterial endophytes	[72]
17	70% EtOH (1 min), 2.5% NaOCl (2 min), and 70% EtOH (1 min)	Stems, leaves, and capsules *of Spiranthes spiralis, Serapiasvomeracea, and Neottia ovata*	*50 bacteria belonging to genera Sphingomonas, Microbacterium, Pantoea, Staphylococcus, Pseudomonas, Bacillus, and Streptomyces*	[73]

Table 1. *Cont.*

No.	Method of Plant Material Sterilization	Plant or Plants Part Example	Endophytes Isolated	Reference
18	75% EtOH (1 min), 1% NaOCl (3 min), and 75% EtOH (30 s)	*Artemisia nilagirica*	*Arthrobacter* sp., *Bacillus* sp., *Burkholderia* sp., *Pseudomonas* sp., *Psychrobacter* sp., *Serratia* sp., *Microbacterium* sp., *Enterobacter* sp., *Chromobacterium violaceum, and Kosakonia cowanii*	[74]
19	75% EtOH (1 min) and 5% NaOCl (8 min)	*Lycium ruthenicum*	109 Endophytic bacteria with 36 genera	[65,68]
20	Cycloheximide (100 mg/L) and nalidixic acid (20 mg/L) (1 min), 5% NaOCl 5 min), 2.5% $Na_2S_2O_3$ (10 min), and 10% $NaHCO_3$ (10 min)	*Glycine max*	70 Endophytic actinobacteria belonging to 14 genera	[75]
21	70% EtOH (3 min), 8% NaOCl (4–5 min), 2.5% $Na_2S_2O_3$ (10 min), and 70% EtOH (1 min)	*Camellia sinensis*	46 Actinobacteria belonging to families *Streptomycetaceae, Thermomonosporaceae, Camellia sinensis, Nocardioidaceae, Microbacteriaceae, Dermatophilaceae, Nocardiopsaceae, Nocardiaceae, Mycobacteriaceae, Dermacoccaceae, Micromonosporaceae*, and *Pseudonocardiaceae*	[76]
22	70% EtOH (1 min), 1.2% NaOCl (15 min), and 75% EtOH (30 s)	*Oryza sativa*	87 Endophytic bacteria	[77]
23	2.5% NaOCl (2 min), and 75% EtOH (30 s)	*Pinus cotorta*	*77 Endophytic diazotrophs Pseudomonas, Bacillus, Paenibacillus,* and *Rhizobium*	[78]
24	Surfactant-75% EtOH (10 min), 5.25% NaOCl (10 min), and 70% EtOH (2 min)	*Phoenix dactylifera* L.	14 Endophytic bacteria	[79]
25	5% NaOCl (4–10 min), 2.5% $Na_2S_2O_3$ (10 min), and 75% EtOH and 10% $NaHCO_3$ (10 min)	12 Forest trees	2174 Actinobacteria	[80]
26	0.1% Tween 20 and 70% EtOH (3 min), 0.4% NaOCl (1 min), and 70% EtOH (2 min)	7 Medicinal plants	13 *Streptomyces* sp., 2 *Microbacterium* sp., 1 *Leifsonia xyli*, 1 *Brevibacterium* sp., 1 *Actinomycete*	[81]
30	0.01% Tween 20 (1 min), 2.5% NaOCl (2.5 min), 2.5% $Na_2S_2O_3$ (5 min), 75% EtOH (2.5 min), and 10% $NaHCO_3$ (5 min)	*Dioscorea zingiberensis*	123 Strains of endophytic fungi, most abundant genera	[77]

In terms of isolating endophytic bacteria, the sterilizing agent is one of the most influential elements. One of the most prevalent sterilizing agents is ethyl alcohol (EtOH), an amphiphilic chemical that damages cell membranes, denatures proteins, and dissolves lipids inside the cell sap. As a result of ethanol application, the cell walls leak, and it dissolves important cell organelles, which eventually leads to the death of cells. The commonly used concentration oscillates between 60 and 90%. Higher concentrations are also used and are considered efficient against spores. Equally common to ethanol as a sterilizing agent is sodium hypochlorite, which has powerful oxidizing characteristics and a wide antibacterial spectrum [53,56,57,60,82]. Sodium hypochlorite in a solution forms hypochlorous acid, which can permeate microbial cell walls and membranes, causing protein denaturation and lipid breakdown inside the cell, leading to cell death. The concentrations used for surface sterilization typically range from 0.5% to 5% [83,84].

Occasionally, mercuric chloride is used for sterilization because it is efficient even at low concentrations. Mercury chloride, however, poses health risks to humans and results in the creation of toxic waste. As a consequence, there is a high potential for negative effects on the environment and human health when it is used [85]. It is unusual for a single sterilizing chemical to be sufficient for the surface sterilization of plant samples. In order to achieve successful sterilization results, it is best to use a combination of two or more chemicals. Common formulations contain 75% ethanol and 2% sodium hypochlorite [57,60,82]. This combination is useful in most types of isolation techniques. Some cases call for the use of surface-active agents to help with the process of sterilizing surfaces. Gupta et al. (2015) isolated endophytic bacteria using surface sterilization. Surface sterilization was carried out first by washing in 1% Savlon for 5 min. Then, the leaves and roots were treated with 0.1% mercuric chloride for surface sterilization [86]. Gohain et al. (2015) conducted a surface sterilization of plant samples (leaves, stems, and roots) from healthy plants in Gibbon. The samples were rinsed in 0.1% Tween 20 for 30 s, then surface-sterilized by immersing them sequentially in 70% ethanol for 2 min followed by 1% sodium hypochlorite for 3 min, 70% ethanol for 2 min, and then three washes with sterile distilled water for 5 min [87]. Qin et al. (2009) carried out a 'five-step' surface sterilization procedure consisting of a 3 min wash in sodium chlorate (5% $NaClO_3$), followed by a 10 min wash in sodium thiosulfate (2.5% $Na_2S_2O_3$), followed by a 3 min wash in 75% ethanol, followed by a 10 min wash in sodium bicarbonate (10% $NaHCO_3$), and a final wash in sterile water [80]. Coombs and Franco (2003) conducted a three-step surface sterilization procedure of plant roots as follows: a 60 s wash in 99% ethanol, followed by a 6 min wash in 3.125% NaOCl, another 30 s wash in 99% ethanol, and a final rinse in sterile reverse-osmosis-treated (RO) water [88].

There are also methods that avoid surface sterilization, such as vacuum or pressure extraction [14]. However, these techniques are not recommended for delicate, non-woody plant tissues, as they will collapse under vacuum. Gardner et al. (1982) and Bell et al. (1995) effectively isolated endophytic xylem-residing bacteria from the roots of a citrus tree in Florida and a grapevine tree in Nova Scotia [89,90]. Scholander et al. (1965) introduced a second method for extracting plant sap using a Scholander pressure bomb. This technique was utilized to successfully isolate endophytic bacteria from roots. The Scholander pressure bomb is less labor-intensive than trituration, which is an advantage [91]. However, limitations of this technique were encountered when working with immature, fleshy cucumber tissues [92].

6. Bioactive Metabolites from Endophytic Bacteria

Endophytes, as microorganisms inhabiting the interior of plant tissues, must adapt not only to the environmental conditions there but also to a variety of plant defense mechanisms. Their strategy has its foundation in the production of a wide range of bioactive compounds that support the growth of the host organism or defend it from adverse factors. The biochemistry of plants and microbes relies on many pathways encompassing numerous intermediate metabolites, all of which hold significant importance. Frequently, individual bioactive compounds exhibit many functions, resulting in a diverse range of relevance. These compounds engage in interactions that give rise to complex, multi-level connections, hence precluding the occurrence of singular outcomes for a given bioactive chemical [93,94]. The production of metabolites by endophytes serves several crucial functions, including supporting development in host tissues, supplying essential chemicals for survival, and promoting successful plant colonization, as shown in Figure 3 [45]. Endophytic bacteria and actinomycetes have a crucial function in the synthesis of bioactive chemicals (alkaloids, steroids, terpenoids, peptides, polyketones, flavonoids, quinols, phenols, and the natural insecticide azadirachtin).

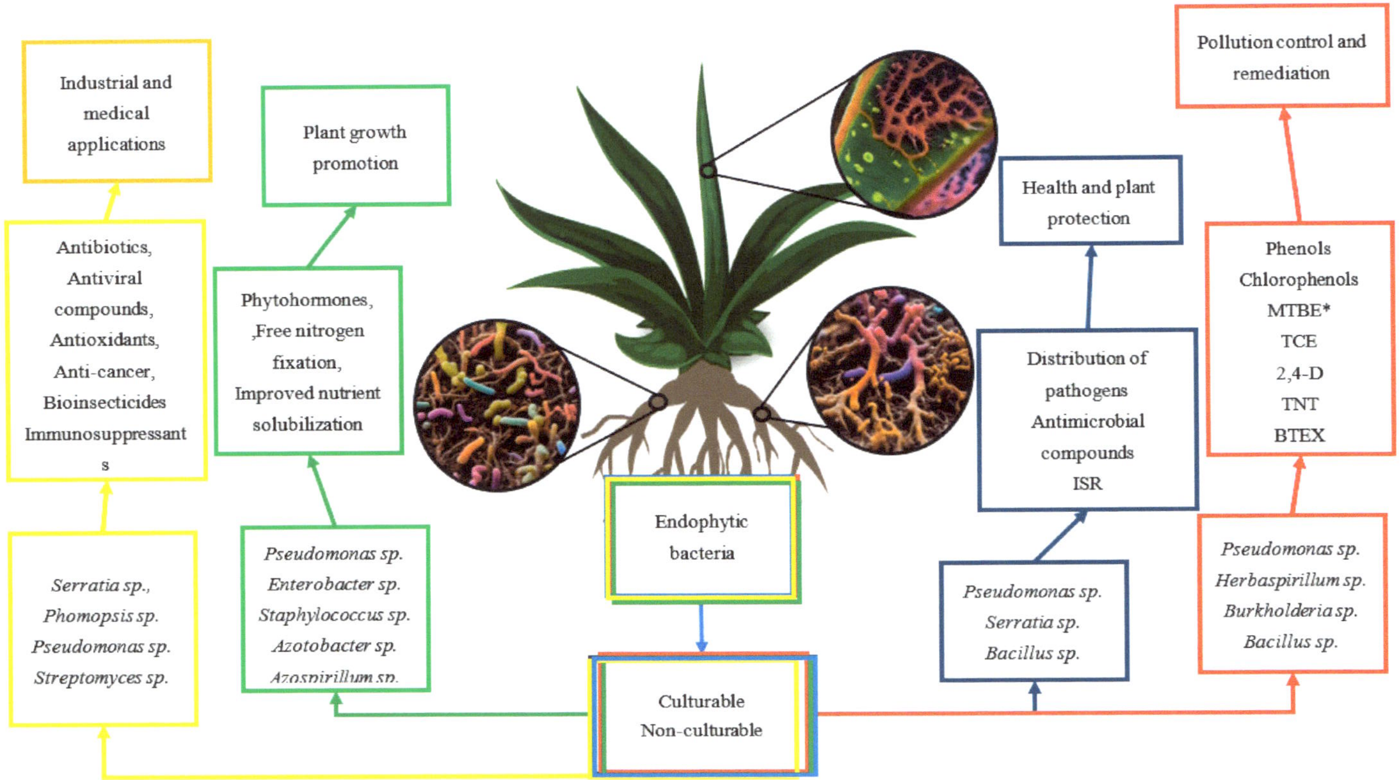

Figure 3. Compounds with biological activity produced by endophytic bacteria and their potential use in biotechnology [45,95–97]. * MTBE—tert-butyl methyl ether; TCE—1,1,1-trichloroethylene; 2,4-D—2,4-dichlorophenoxyacetic acid; TNT—2,4,6-trinitrotoluene; BTEX—benzene, toluene, ethylbenzene, and xylene isomers.

Table 2 gives a compilation of the bioactive chemicals synthesized by endophytic bacteria obtained from synanthropic herbaceous plants found throughout Europe.

Table 2. Bioactive compounds produced by endophytic bacteria isolated from synanthropic herbaceous plants occurring in Europe.

No	Host Plant	Endophytic Bacteria	Compound/-s	Activity	Reference
1	*Achillea fragrantissima*	*Streptomyces* sp., *Nocardioides* sp., *Kitasatosporia* sp., *Kibdelosporangium* sp.	chitinase, siderophores	enzymatical, antifungal	[98]
2	*Achillea millefolium (yarrow)*	*Bacillus safensis*	volatile metabolites (butanal, 3-methyl-, 2-heptanone, 6-methyl-5-methylene-, hydrogen azide, propene, 2-butene, 6-oxabicyclo3.1.0 hexane)	antifungal	[99]
3	*Alkanna tinctoria* (dyer's alkanet)	*Pseudomonas* sp., *Bacillus* sp.	IAA, ACC deaminase, siderophore; pectinase, ligninase	plant growth promotion; enzymatical	[100]
4	*Allium fistulosum* (spring onion)	*Streptomyces* sp. TP-AO569	fistupyrone	spore germination inhibition	[101]
5	*Allium tuberosum* (garlic chives)	*Streptomyces* sp. TP-A0595	6-prenylidole	antifungal	[102]
6	*Arctium lappa* (greater burdock)	*Bacillus* sp., *Pantoea* sp., *Microbacterium* sp., *Pseudomonas* sp.	IAA, ACC deaminase, siderophore, hydrolase	plant growth promotion	[103]
7	*Armoracia rusticana* (horseradish)	*Serratia ficaria*	siderophores, lipase, protease, chirinase	biocontrol, enzymatic, plant growth promotion	[104]
8	*Arnica montana* (mountain arnica)	*Streptomyces* sp.	glutarimide antibiotics (cycloheximide, actiphenol), diketopiperazines (cyclo-prolyl-valyl, cyclo-prolyl-isoleucyl, cyclo-prolyl-leucyl, cyclo-prolyl-phenylalanyl)	antibiotic, cytotoxic	[105]
9	*Artemisia vulgaris* (common mugworts)	*Bacillus pumilus*, *Bacillus safensis*	volatile metabolites (butanal, 3-methyl-, 2-heptanone, 6-methyl-5-methylene-, hydrogen azide, propene, 2-butene, 6-oxabicyclohexane)	antifungal	[99]
10	*Atropa belladonna* (deadly nightshade)	*Streptomyces* sp. AB100	piperazic-acid-containing peptides	antibiotic	[106]
11	*Calendula arvensis* (field marigold)	*Pseudomonas brassicacearum*	-	biocontrol, plant growth promotion	[107]
12	*Calendula officinalis* (medicinal calendula)	*Pseudomonas putida*	HCN, siderophores, lipase, protease, chitinase	biocontrol, enzymatic, plant growth promotion	[104]
13	*Matricaria chamomilla* L., *Calendula officinalis* L., and *Solanum distichum* Schumach. and Thonn	*Bacillus subtilis* subsp. *subtilis*, *Bacillus subtilis* subsp. *spizizenii*	-	antibacterial, antifungal, nematicidal	[108]

Table 2. *Cont.*

No	Host Plant	Endophytic Bacteria	Compound/-s	Activity	Reference
14	*Iris pseudacorus* L.	*Pseudomonas rhizosphaereae* FST5	IAA, ACC deaminase, HCN, chitinase, protease, glucanase, lipase	antifungal, plant growth promotion	[109]
15	*Chelidonium majus* (greater celandine)	*Bacillus velezensis*	IAA, siderophores, iturin B, iturin D, fengycin, surfactin, cellulase	biocontrol, plant growth promotion	[110]
16	*Chelidonium majus* (greater celandine)	*Bacillus pumilus* 2A	glycolipid	emulsyfing (biosurfactant)	[111]
17	*Chelidonium majus* (greater celandine)	*Bacillus amyloliquefaciens* 30B, *Erwinia persicinia* 2–5b	IAA	plant growth promotion	[112]
18	*Echinacea purpurea* (purple coneflower)	*Bacillus thuringensis*, *Bacillus amyloliquefaciens*	-	antifungal	[113]
19	*Echinacea purpurea* (purple coneflower)	*Rheinheimera* sp. EpRS3	lipases, phospholipase, protease	antibacterial, antimicrobial	[114]
20	*Elymus repens* (*couch grass*)	*Pseudomonas azotoformans*	IAA	plant growth promotion	[115]
21	*Equisetum arvense* (common horsetail)	*Psychrobacillus insolitus*, *Curtobacterium oceanosedimentum*	-	anticandidal	[116]
22	*Euphorbia helioscopia* (sun spurge)	*Comamonas koreensis*, *Stenotrophomonas maltophilia*, *Rhizobium* sp., *Brevundimonas* sp.	IAA-like compounds, siderophores	plant growth promotion	[117]
23	*Euphorbia helioscopia* (sun spurge)	*Bacillus cereus*, *Bacillus amyloliquefaciens*	-	biocontrol, plant growth promotion	[107]
24	*Euphorbia peplus* (petty spurge)	*Pseudomonas brassicacearum*	-	biocontrol, plant growth promotion	[107]
25	*Foeniculum vulgare* (fennel)	*Klebsiella pneumoniae*	siderophores, lipase, protease, chitinase	biocontrol, enzymatic, plant growth promotion	[104]
26	*Hypericum perforatum* (St. John's wort)	*Achromobacter* sp., *Erwinia persicina*, *Stenotrophomonas* sp.	IAA, HCN, cellulase, protease, beta-1,3-glucanase	biocontrol, plant growth promotion, antifungal	[118]
27	*Iris pseudacours* (pale yellow iris)	*Pseudomonas azotoformans*	HCN, siderophores, lipase, protease	biocontrol, enzymatic, plant growth promotion	[104]
28	*Matricaria chamomilla* L., *Calendula officinalis* L., and *Solanum distichum* Schumach. and Thonn	*Pseudomonas gessardii* HRT18	IAA, HCN, siderophores, lipase, protease, chitinase, glucanase	biocontrol, enzymatic, plant growth promotion	[109]
29	*Lavandula dentata* (*fringed lavender*)	*Pseudomonas* sp., *Bacillus* sp.	IAA, HCN, siderophores, lipase, protease, cellulase, pectinase	plant growth promotion	[11]
30	*Lotus corniculatus* (*bird's foot trefoil*)	*Serratia plymuthica*, *Pseudomonas mandelii*, *Tsukamurella pulmonis*	IAA, HCN, siderophores, cellulase	plant growth promotion	[119]
31	*Matricaria chamomilla* (*German chamomile*)	*Bacillus pumilus*	lipopeptides	antifungal	[99]

Table 2. *Cont.*

No	Host Plant	Endophytic Bacteria	Compound/-s	Activity	Reference
32	*Calendula officinalis*	*Bacillus mojavensis*, *Bacillus subtilis* subsp. *subtilis*, *Bacillus subtilis* subsp. *spizizenii*, *Bacillus endophyticus*, *Paenibacillus brasilensis*, *Paenibacillus polymyxa*, *Lysobacter enzymogenes*	-	antibacterial, antifungal, nematicidal	[108]
33	*Mentha longifolia* (wild mint)	*Bacillus* sp., *Bacillus aryabhattai*, *Bacillus pumilus*, *Bacillus megaterium*, *Bacillus toyonensis*	IAA, ACC deaminase, siderophores	plant growth promotion	[120]
34	*Mentha spicata* (spearmint)	*Bacillus anthracis*, *Bacillus toyonensis*	cellulase, xylanase, amylase, pectinase	enzymatical	[121]
35	*Polygonum cuspidatum*	*Bacillus safensis*	volatile metabolites (butanal, 3-methyl-, 2-heptanone, 6-methyl-5-methylene-, hydrogen azide, propene, 2-butene, 6-oxabicyclo3.1.0 hexane)	antifungal	[99]
36	*Oenothera biennis* (common evening primrose)	*Rhodococcus erythropolis*, *Rhizobium* sp.	IAA, HCN, cellulase	plant growth promotion	[119]
37	*Plantago lanceolata* (ribwort plantain)	*Pseudomonas brassicacearum*, *Bacillus methylotrophicus*	-	biocontrol, plant growth promotion	[107]
38	*Calendula officinalis*	*Bacillus halotolerans* Cal.l.30	surfactin, iturin, fengycin, bacillaene, bacillibactin, FAS-PKS, subtilosin A, bacilysin	antibacterial, antifungal, nematicidal	[122]
39	*Peperomia dindygulensis*	*Streptomyces* sp. YINM00001	cyclohexiamide, dinactin, warkmycin, anthramycin, alkylresorcinol, lanthipeptide, melanin, ectoine, geosmin	antibacterial, antifungal	[123]
40	*Origanum vulgare*	*Bacillus* sp.	paeninodin, terpene, paenilarvins	biocontrol	[124]
		Paenibacillus sp.	polymyxin, paenicidin A	antimicrobial	
41	*Mikania micrantha*	*Sphingomonas paucimobiliz*	Zeaxanthin	antioxidant	[105]
		Micrococcus yunnanensis	microansamycin, stenothricin	biocontrol	
42	*Mimosa pudica*	*Staphylococcus caprae*	Aureusimine, staphyloferrin A	Plant growth promotion	
		Neobacillus drentensis	Fengycin	biocontrol	
		Priestia megaterium	Surfactin, bacitracin, carotenoid	biocontrol	
43	*Millettia pachycarpa* Benth	*Paenibacillus peoriae* IBSD35	NRPSs, Fusaricidin synthetase, paenibacterin, gramicidin synthase	Biocontrol, antibiotic	[114]
44	*Panicum turgidum*	*Cellulosimicrobium* sp. JZ28	alkylresorcinol	Biocontrol	[125]
45	*Brassica napus* L.	*Pseudomonas fluorescens* BRZ63	alginate, LPS, siderophore, bacterioferrin, tryptophan, PQQ cofactor	biocontrol, enzymatic, plant growth promotion	[126]

Table 2. *Cont.*

No	Host Plant	Endophytic Bacteria	Compound/-s	Activity	Reference
46	*Suaeda fruticosa*, *Suaeda mollis*, *Mesembryanthmum nodiflorum*, *Arthrocnemum indicum*	*Bacillus albus* strains	petrobactin, bacitracin, thuricin	Biocontrol	[127]
47	*Bacopa monnieri*	*Bacillus* sp. LCF1	surfactin, iturin, fengycin, type I PKS	Biocontrol	[128]
48	*Arnica montana* L.	*Streptomyces* sp.	siderophore, ectoine, terpene, bacteriocin, butyrolactone, lantipeptide, melanin, NRPS, PKS, cyclohexiaminde	biocontrol, enzymatic, plant growth promotion	[128]
49	*Echinacea purpurea*	*Rheinheimera* sp. RS3	Resorcinol, lantipeptide, hserlactone, bacteriocin, NRPS	Antimycobacterial potential, biocontrol	

Note: LPS—lipopolysaccharide, NRPS—nonribosomal peptide synthetase, PKS—polyketide synthase.

The beneficial impact exerted by endophytes on the plant host results in plants refraining from actively eliminating them. Endophytic metabolites that have nitrogen in their chemical structures, such as alkaloids, amines, non-protein amino acids, peptides, glucosinolates, alkamides, and cyanogenic glycosides, play a significant role in the natural nitrogen cycle process and the mutual relationship between endophytes and their host organisms. Several of these factors are crucial for the survival of plants, such as amino acids, which are the fundamental components of plant proteins and enzymes, and tetrapyrrole structures, which are crucial components of chlorophyll molecules [34,129–131]. Moreover, a significant proportion of these molecules have medicinal properties and have been effectively employed in the synthesis of pharmaceuticals, including treatments for presently incurable conditions such as cancer [132]. However, the functions of metabolites produced by endophytic microorganisms vary and depend on the structure of a given compound. Fungal endophytes are a prominent category of endophytic microorganisms that have garnered significant attention due to their production of bioactive chemicals. This recognition stems from their frequent isolation from plant tissues that possess therapeutic properties. However, despite a limited understanding of this topic, an increasing body of scientific research suggests that more classes of endophytic microbes possess the ability to generate useful metabolites [133].

Synanthropic plants are a particularly intriguing source of endophytes that produce bioactive metabolites. These types of plants inhabit environments that are subject to intense anthropological pressure; as a result, they are exposed to numerous negative factors associated with the growth of civilization. The presence of endophytes appears to be invaluable for these plants. The bioactive compounds that have been extracted from synanthropic herbaceous plants encompass several groups, including phenols, alkaloids, benzopyrenes, glycosides, flavonoids, terpenoids, non-protein amino acids, cyanogenic glycosides, and numerous others [32,134–138]. Certain chemicals with distinct characteristics and the potential for microbial synthesis are isolated and employed in commercial applications, such as agriculture, for their natural pesticide properties as well as in medicinal contexts for the treatment of diverse human ailments [137].

Industrial activities are responsible for the production of a wide range of pollutants, including heavy metals, by-products, and waste. These pollutants impose significant challenges on living organisms, compelling them to adapt to adverse conditions. The growth and development of plants in challenging conditions primarily rely on the synthesis of compounds that serve to inhibit the entry of detrimental substances into cells or enhance the tolerance range of individuals, thereby ensuring that their presence does not hinder plant development. Endophytes, being microorganisms with the ability to synthesize diverse secondary metabolites, significantly enhance and occasionally even permit adaptation and persistence in challenging environments resulting from anthropogenic activities. Minimizing the detrimental impacts of non-biodegradable heavy metals is of paramount importance. Endophytic microorganisms have demonstrated their use and substantial contribution in enhancing the efficacy of the phytoremediation process [95,139,140]. The occurrence of stress factors has a direct effect on the endophytic microorganism population in plant tissues. In the case of synanthropic plants inhabiting areas polluted with heavy metals, an advantage in interspecific competition is gained by endophytes that can live in the presence of this pollution, while also demonstrating the capacity to effectively promote plant development in areas subject to intense anthropopressure [93,141]. This is illustrated by the action of endophytes of synanthropic plants in an iron-ion-polluted environment. This element effectively and rapidly inhibits growth and negatively affects leaves and roots, so limiting its availability to plants appears crucial for their proper development. Numerous bacterial genera like *Pseudomonas, Azotobacter, Bacillus, Rhizobium, and Enterobacter* are able to produce siderophores, compounds that have a direct role in modulating the bioavailability of iron ions, by chelating iron ions and binding them to complexes, hence facilitating plant development in environments characterized by elevated iron concentrations [142,143]. Among these, *Pseudomonas* strain GRP3, *Pseudomonas fluorescens* C7, *Streptomyces* spp., *Escherichia coli, Arthrobacter* spp., and *Enterobacter quasihormaechei* are endophytic representatives [144,145]. As with all secondary metabolites generated by endophytes, this is essential to their survival, as due to the synthesis of siderophores, endophytic microorganisms can absorb Fe ions and use them as one of the essential minerals for their survival. In addition, they diminish the bioavailability of this element for other microorganisms that are potentially deleterious to the plant and are undeniably endophyte competitors [146]. Siderophores predominantly bind iron ions, but it has been demonstrated that they can also bind ions of other metals, including copper, gallium, cadmium, aluminum, and lead [21,133,146]. In the case of *Polygonum acuminatum*, a plant often found in Japan, several endophytic fungi were identified and isolated during research that examined the composition of endophytic microflora in settings polluted with cadmium, lead, and copper, as well as in non-contaminated situations. Significantly, the study revealed that the abundance of fungi with the ability to produce siderophores was greater in the environment subjected to the stress factor compared to conditions that were considered optimal. Moreover, *Polygonum pubescens*, serving as an exemplar of the identical genus, exhibits the capacity to thrive in environments characterized by elevated levels of cadmium, lead, and iron. This finding suggests that the plant sustains the growth of endophytic microorganisms due to their assistance in managing heavy metal ions [147]. Within the realm of endophytic bacteria, the *Rahnella* strain exhibits remarkable resilience to three specific heavy metals, namely, cadmium carbonate ($CdCO_3$), lead carbonate ($PbCO_3$), and zinc phosphate ($Zn_3(PO_4)_2$). This strain, obtained from plant root tissues obtained from a polluted environment, not only demonstrates an exceptionally high tolerance to these heavy metals but also effectively solubilizes their respective compounds [148–150]. Additionally, the *Rahnella* strain exhibits the ability to produce siderophores, and plant phytohormones such as indole-3-acetic acid, and it can even dissolve inorganic phosphates. The investigation into the unique characteristics of this particular strain, along with its impact on enhancing the accumulation of heavy metals in plant tissues, specifically in the seedlings of rapeseed (*Brassica napus*), presents a significant area of interest for future studies in the realm of efficient phytoremediation of lead (Pb), cadmium (Cd), and zinc (Zn) ions [150–154].

The synthesis of phytohormones and the regulation of their synthesis by the plant host represent a prominent mechanism via which endophytes can effectively promote the growth of their host plant. Endophytic microorganisms strive to enhance the synthesis of auxins (e.g., indole-3-acetic acid (IAA)), gibberellins, or cytokinins, thus influencing the orientation of tissue development towards the root system, specifically by promoting its expansion [2]. This is accompanied by an increase in the absorptive surface, through which nutrients are absorbed and utilized not only by the plant but also by endophytes. Indole-3-acetic acid (IAA) is a phytohormone that is synthesized in significant quantities by a wide range of endophytic microbes, including both bacterial and fungal species. It is noteworthy that the action of IAA extends beyond growth stimulation [155]. The *Pseudomonas fluorescens* Sasm05 strain resides inside the tissues of *Sedum alfredii*. Hence, it exhibits the ability to accumulate zinc and cadmium from its surroundings. Additionally, this strain serves as a signaling molecule, facilitating communications between the bacterium and the host organism. Furthermore, the molecule is generated by the endophyte in response to elevated zinc concentrations in the surrounding environment. Additionally, it regulates the expression of genes that are important for the absorption and storage of this substance inside the plant [156].

From the perspective of the plant host, a significant concern is the protection against detrimental microbes, diseases, and predators, in which endophytes play a role. The host is protected against unfavorable biotic influences due to the synthesis of a wide range of secondary metabolites possessing antibacterial characteristics [74,157]. The advantages of this situation are reciprocal. By constraining or perhaps halting the proliferation of alternative bacteria, the likelihood of pathogenic germs infiltrating plant tissues, which frequently results in significant harm, is diminished. In relation to the endophytes themselves, this phenomenon offers two primary benefits. Firstly, it mitigates interspecies competition by creating a more stable and favorable environment for their survival and nutrient acquisition. Secondly, by counteracting the detrimental effects of pathogens on the host, endophytes create optimal conditions for their own growth and development. In the event of plant damage or a decline in its overall condition, the plant's tissues would initiate induced systematic resistance (ISR) mechanisms [158,159]. These mechanisms are designed to eliminate the threat and operate in a non-specific manner, which may potentially pose a risk to endophytes. This risk can be likened to the impact of antibiotics on probiotic microflora [67].

Genomics for Synthesis of Bioactive Compounds in Endophytic Bacteria

The complete elucidation of the biological roles of bacterial endophytes and their molecular interactions with plant hosts remains incompletely understood. Nonetheless, recent advancements in molecular biology techniques have significantly contributed to the advancement of knowledge in this field [4,24,39,160]. The involvement of endophytic genes and their expression has garnered significant attention in various areas, including the maintenance of endophytism, promotion of plant growth, biocontrol activity, and alleviation of different stresses [39,160]. Additionally, these genes have been implicated in the production of bioactive compounds [161–163]. The identification of these genes and a comprehensive understanding of their function, along with the examination of the genomic-level interactions between endophytes and their plant hosts, are imperative for effectively enhancing sustainable crop production. Furthermore, this knowledge is essential for the advancement of products that could potentially be utilized in the pharmaceutical and industrial sectors [137].

Advances in genomics and other "omics" techniques (metagenomics, metabolomics, transcriptomics, proteomics, etc.) have enabled a novel method for analyzing the role of gene expression in these previously mentioned traits [164]. A growing number of sequenced endophytic genomes has facilitated the exploration of several enigmatic aspects through comparative genomics investigations. The process of plant colonization by endophytic bacteria is widely recognized as a highly complex process [2]. However, recent findings have demonstrated that endophytic colonization has a significant impact on the genome of the host plant [165]. Additionally, it has been observed that genetic factors within the host plant play a crucial role in regulating the establishment of endophytes within root nodules [166]. Also, a study conducted by Monteiro et al. (2012) revealed that when closely related bacterial species were compared, they exhibited potential genes and other genetic factors associated with endophytic behavior during colonization rather than harmful activity [167]. Similarly, transcriptomics is used to investigate the endophytic microbial communities associated with other organisms and their potential role in gene expression regulation, as well as changes in gene expression patterns during exposure to potential pathogenic factors [168]. Proteomics provides information on the functional expression of the genome, primarily through mass spectrometry (MS) techniques, but it is insufficient and ineffective without genome studies [164,165].

Given the complex nature of the interactions between endophytes and their host organisms, as well as the diverse range of intermediary metabolites that are involved, analysis of the relationship should focus on the collective set of genes and their expression rather than individually [39,164]. Dudeja et al. (2021) and Pinski et al. (2019) provide comprehensive descriptions of the role of distinct genes in many aspects of endophytic bacteria–plant host interactions, including colonization, establishment of endophytism, and stress protection [39,160]. Despite the fact that previous studies have made significant progress in elucidating the genomic aspects of endophytes, it is evident that further investigation is necessary to fully comprehend this field of study.

The conventional strategy of searching for some putative properties of the given microbe strain and thus the bioactive compounds responsible for them includes cultivation- and bioassay-based screening, predicted on the basis of available literature data or knowledge about its isolation place, which may indicate its abilities (e.g., contaminated soil). Subsequently, this is followed by the purification, detection, and characterization of the compounds using chromatography in conjunction with mass spectrometry (MS) and/or nuclear magnetic resonance (NMR) techniques [169,170]. This basic method can be altered to induce gene expression and therefore produce more bioactive compounds by using the "one strain, many compounds" (OSMAC) approach (e.g., changing medium composition, cultivation conditions, co-cultivation with another strain, or adding epigenetic modifiers) [171,172]. Still, it is an untargeted method that does not give clear answers on which specific metabolites we can expect and does not exploit the strain's potential.

Genome mining presents an alternative approach to the conventional method of secondary metabolite discovery. This strategy holds the potential to uncover all genes responsible for encoding secondary metabolites, thereby reducing reliance on serendipitous discovery [135,161,173]. Within microorganisms, including endophytes of both fungal and bacterial origin, there exists a phenomenon wherein genes responsible for the synthesis of secondary metabolites are found in close proximity to one another within the genome. Examples of such gene clusters include nonribosomal peptide synthetases (NRPSs), ribosomally synthesized and post-translationally modified peptides (RiPPs), and polyketide synthetases (PKSs) [174–178]. One of the primary benefits of an organized operonic structure, which is extensively dispersed across the genome with approximately 30–50% of all genes being grouped in this manner, is the presence of a single promoter that governs the regulation of gene expression, leading to co-transcription [179].

The development of whole-genome sequencing (WGS) techniques facilitated the initial exploration of genome mining in *Streptomyces coelicolor* and *Streptomyces avermitilis*, two extensively studied species that were previously considered to have been fully utilized. These investigations unveiled the untapped capabilities of these organisms [173,180–183]. Subsequently, there has been substantial progress in bioinformatic methodologies and, therefore, genome mining, leading to the identification of numerous novel bioactive compounds derived from diverse endophytic bacteria [184,185]. This includes the isolation of compounds from herbaceous plants, such as those documented in Table 2. In a recent study conducted by Tsalgatidou et al. (2022), a wide range of genes encoding antimicrobial compounds were identified in *Bacillus halotolerans* from *Calendula officinalis*. These compounds included lipopeptides such as fengycin, surfacin, and mojavensin A, as well as siderophores such as bacillibactin [186]. This discovery builds upon previous research by Jasim et al. (2016), who also found various lipopeptide genes for surfactin, iturin, and fengycin in endophytic *Bacillus* sp. from *Bacopa* [187]. The process of genome mining encompasses more than just the discovery of genes and/or biosynthetic gene clusters (BGCs). It also involves the analysis of biosynthetic pathways for chemicals and the prediction of their functional and structural characteristics [125,126]. The approach relies on the utilization of synthetic biology and bioinformatic tools, including but not limited to antiSMASH [126,127], RODEO [128], and ThioFinder [188], among others. The ease of comparing collected data with database-presented biosynthetic gene clusters (BGCs) and identifying homologs has been facilitated by the accessibility of DNA sequences and other genomic data uploaded in public databases [125,189]. It has been observed that BGCs are predominantly inactive or exhibit low expression levels during standard laboratory cultivation, necessitating their activation for gene expression [161,190,191]. The various methods for awakening microbial cells from a dormant state encompass heterologous expression, ribosome engineering, co-cultivation, optimization of regulatory networks (such as employing more potent promoters), and the introduction of chemical elicitors. The activation mechanisms of silent (or cryptic) BGCs have been extensively examined in a comprehensive study by Rutledge [191]. Additionally, the utilization of mutasynthesis and/or genetic engineering, commonly referred to as combinatorial biosynthesis, may serve as a crucial measure in this process and offer insights into the pathways involved. By creating mutant microorganisms with inactivated or deleted biosynthetic gene clusters (BGCs) or specific individual genes, the production of certain compounds can be eliminated [125,192]. Ultimately, the process of purifying and isolating newly discovered bioactive compounds is executed, resulting in molecules that are now prepared for subsequent activity assessments and structural analysis.

One notable advantage of genome mining is its ease of application in laboratory settings, primarily due to its cost-effectiveness and lack of a need for specialized skills required by researchers. Additionally, genome mining offers the opportunity to explore novel compounds, as conventional methods often result in a high rate of rediscovery. However, it is important to acknowledge the limitations of this approach, including its ability to only identify previously synthesized BGCs or similar ones, as well as its inability to predict the biological activities of the identified compounds [173,180,181]. Encouragingly, there is a notable expansion in databases and significant advancements in synthetic biology, which have the potential to effectively address these challenges in coming years.

7. Ruderal Plants

For centuries, mankind has co-existed with synanthropic plants. They occur in locations where humans have either mistakenly or consciously destroyed the native plant cover, changing nature for their own benefit. Environmental factors in synanthropic settings necessitate specialized plant adaptations. These include reducing their growth cycle so that they can generate many generations in a single growing season, increasing the number of seeds produced, having a high ability for vegetative reproduction and competitiveness, and having broad adaptability, including adaptations to plant protection chemical compounds.

Synanthropic plants are divided into segetal and ruderal species. Segetal plants are typically weeds that grow alongside field and garden crops (from Latin *seges*—sowing, cultivated field) [193].

The term ruderal comes from the Latin word *rudus* and describes species adapted to disorders [194]. As a result of human activity, ruderal areas are constantly fed with nutrients and thus are characterized by the presence of highly productive flora and a variety of competitive relationships associated with dynamic, continuous, or occasional interference and hasty successive changes [146]. Ruderal plants include species from environments that are directly or indirectly influenced by humans and that can grow in disturbed artificial habitats, such as home surroundings, gardens, roadsides, railroads, etc. [10,195]. Additionally, they are frequently observed in semi-natural environments such as disturbed river margins and forest edges [193,196,197]. They flourish there despite human intervention. Because these ecosystems have a high concentration of mineral salts, the ruderal vegetation is mostly nitrophilous (nitrogen-loving) plants. Their communities are quite dynamic; their composition changes as they grow, and they occasionally initiate secondary succession [198].

Ruderal plants are resistant to competition as they easily adapt to new environments and rapidly form large quantities of seeds that can be dispersed even by car tires. That is why these plants are also known as pioneer plants. A common feature of most ruderal species is their ability to foster beneficial microorganisms around their roots, which, in turn, helps them cope better with environmental stress [194]. Moreover, their seeds require a small amount of nutrients for germination, their roots develop quickly, and they can form mycorrhizae [195].

7.1. *Ruderal Herbaceous Plants*

Ruderal plants are commonly referred to as pioneer plants because of their ability to easily adapt to new conditions and colonize new areas. The predominant species within this group include small annual herbaceous plants. These plants are a great source of biologically active compounds, and their use in medicine has been known for centuries [199].

The global market for plant-based products is estimated at USD 83 billion and continues to grow. According to Wangchuk, (2018), about 85% to 90% of the worldwide population relies on traditional medicine [200,201]. Nowadays, in less developed countries, more than 3.3 billion people utilize medicinal plants regularly. According to the World Health Organization (WHO), traditional medicinal plants are defined as natural resources that can be used in industrial processing for the treatment of diseases on a local or regional scale. It is estimated that between 750 thousand and 1 million different plant species grow in the world, where around 500 thousand have been identified and named. According to WHO, the number of plants used for the treatment is around 20 thousand. Traditional medicine and plant species are generally used by patients who have chronic medical conditions such as cancers (2%), liver diseases (21%), HIV (22%), asthma (24%), and rheumatologic disorders (26%) [202].

Herbaceous plants are an abundant source of biologically active compounds. Their usage is growing, as they serve as raw materials for the extraction of active ingredients that are later used in the synthesis of various drugs. Relevant laxatives, blood thinners, antibiotics, and antimalarial compounds are ingredients of many drugs, and they are mainly obtained from plants. Taxol, vincristine, and morphine isolated from foxglove, periwinkle or yew, and opium poppy, respectively, are examples of such compounds [203,204].

7.2. *Biotechnological Potential of Endophytic Bacteria from Ruderal Herbaceous Plants*

Ruderal plants experience a range of biotic and abiotic challenges throughout their lifespan because of their growing environment. These stresses include exposure to several xenobiotics, such as herbicides, pesticides, hydrocarbons, and heavy metals. Nevertheless, they manage to endure in such a hostile habitat due to the presence of microbes residing within their tissues [46].

The presence of endophytic microbes in plants provides several advantages, such as enhanced production and fitness, improved stress tolerance, and better resistance to diseases. Thus, their reliance on microbes that produce secondary metabolites is evidently crucial to their long-term survival strategy [10].

7.2.1. *Chelidonium majus* L.

Chelidonium majus L. is a ruderal plant indigenous to Europe, Asia, and South America. The plant, despite its toxicity, generates numerous bioactive compounds such as alkaloids that possess potent antibacterial, antiviral, and even anticancer properties [111,113,205]. Therefore, it is essential to isolate endophytes from the plant and examine the possibility of sharing the ability to produce the same biologically active compounds. Goryluk et al. (2009) isolated 34 endophytic bacteria from internal stem tissues of Chelidonium majus L. and investigated their antifungal properties against six fungal species, namely, *Alternaria alternata*, *Paecilomyces variotti*, *Aureobasidium pullulans*, *Byssochla-mysfulva*, *Chaetomium* sp., and *Exophiala mesophila*. The authors observed that eleven isolates demonstrated the suppression of fungal growth, with the exception of *B. fulva*. A single bacterial strain demonstrated broad-spectrum antifungal activity against all the fungi examined. The strain was categorized as *B. amyloliquefaciens* based on the API-20E, -50CHB tests, and the study of the 16S rDNA sequence [113]. Studies on endophytic microorganisms from *Chelidonium majus* L. were extended by Marchut-Mikolajczyk et al. (2018) [111]. The authors isolated 11 endophytic bacteria from the tissues of *Chelidonium majus L.* plants growing in a motorway neighborhood. The researchers investigated the capacity of isolated microorganisms to degrade hydrocarbons (namely, diesel oil and waste motor oil) and produce biosurfactants. Every strain that was examined exhibited degrading activity. Nevertheless, strain 2A exhibited the most pronounced degrading activity towards both diesel and waste engine oil. The strain also demonstrated the greatest biosurfactant production. The strain was categorized as *Bacillus pumilus* based on the study of the 16S rDNA sequencing. The authors assert that the biosurfactants created not only improve the breakdown of hydrophobic substances but also have the potential to stimulate plant development in a polluted environment [111]. While the data presented are novel and captivating, the utilization of the strain and biosurfactant can only occur once a comprehensive understanding of the underlying mechanism of the observed phenomenon is obtained. This knowledge is crucial in assessing the feasibility of applying biosurfactants to enhance plant growth, particularly in polluted regions.

7.2.2. *Urtica dioica* L.

Urtica dioica L. (stinging nettle) is a common plant found in northern Europe and much of Asia, primarily in rural areas. The beneficial impact of *Urtica dioica* on human well-being has been globally recognized since ancient times. It is utilized for the treatment of several ailments and conditions such as hemorrhoids, eczema, rheumatism, bronchitis, hyperthyroidism, and cancer. Importantly, it is noteworthy that this treatment does not have any adverse effects [206]. Thus, it is highly probable that endophytic microorganisms possessing unique capabilities reside within the internal tissues of stinging nettle [146].

Naoufal et al. (2018) successfully obtained 54 endophytic bacterial strains from *Urtica dioica* L. [207]. From this collection, the authors chose Gram-positive isolates belonging to the Bacilli genera that exhibited distinct morphologies and capabilities. The primary objective of this research endeavor was to identify the most effective phytopathogen inhibitors. The authors identified three isolates that demonstrated the most potent antagonistic activity against common phytopathogens (*Rhizoctonia solani, Fusarium oxysporum, Phytophthora parasitica,* and *Colletotrichum gloeosporioides*) based on biochemical analysis. The average value of phytopathogen growth inhibition was 73.21% when compared to the control sample, which lacked endophytic bacteria. Additionally, the authors documented that these species of endophytic bacteria might serve as a reliable reservoir of various enzymes and secondary metabolites [207].

Krimi et al. (2016) assessed the capacity of endophytic bacteria from four weeds (including stinging nettle) to inhibit the development of tomato plants and function as biocontrol agents against bacterial phytopathogens [107]. The seventy-three bacterial isolates isolated by the authors, each possessing eight distinct morphological profiles, exhibited antagonistic activity against a minimum of two pathogenic bacteria out of the seven that were tested [107]. Three isolates of *Bacillus amyloliquefaciens* OR2, *Bacillus pumilus* OS2, and *Bacillus methylothrophicus* OS4 were isolated from *Urtica dioica* L. The growth of *Agrobacterium vitis, Clavibacter michiganensis* subsp. *michiganesis,* and *Xanthomonas axonopodis* was significantly inhibited by 20–30.3 mm when exposed to the OR2 and OS2 strains. The growth of a greater variety of phytopathogens was significantly inhibited by the OS4 strain, including *Agrobacterium tumefaciens* (40 mm), A. vitis (34.6 mm), C. michiganensis subsp. michiganensis (40.3 mm), Pectobacerium spp. (33 mm), and *Ralstonia solanacearum* (41.3 mm). Moreover, Krimi et al. (2016) documented that endophytic strains derived from *U. dioica* L. exhibited significant growth-promoting activity on tomatoes, with the OS4 strain in particular significantly accelerating their development under in vitro and in vivo conditions. This acceleration was attributed to the endophytic strains' secretion of hormones including indole-3-acetic acid (IAA) and ethylene [107]. The potential of *U. dioica* endophytic mircoorganisms in biofertilizer and biopesticide formulation has been suggested by their capabilities [107]. Furthermore, Marchut-Mikolajczyk et al. (2023) successfully isolated nine strains of endophytic bacteria from common nettle [146]. Among these, three strains were chosen based on their capacity to biosynthesize polyphenols. The identified isolates comprised one strain of Bacillus mycoides and two strains of *Bacillus cereus* [146].

7.2.3. *Plantago lanceolata* L.

Plantago lanceolata L., also known as ribwort plantain, is a highly prevalent weed found in Europe, Asia, the Americas, and Australia, where it has been introduced as a non-native species. Traditionally, it was employed as a cure for wound healing in folk medicine. However, it also exhibits anti-inflammatory, analgesic, analeptic, antihistaminic, and other actions [208]. Plantains are often seen growing along roadsides and other polluted locations; thus, they might potentially be used as a bioindicator for the buildup of heavy metals in soil. However, our understanding of the bacteria residing within their tissues remains limited. Tello et al. (2014) reported the presence of the fungal endophyte species *Hygrocybe virginea* in the roots of *Plantago lanceolata* and provided evidence that it may be systemic endophyte transmitted with the plant seeds (vertically) [209]. In addition, Krimi et al. (2016) successfully isolated five endophytic bacterial strains from the roots. Among these strains, two were recognized as *Bacillus methylotropicus* and *Pseudomonas brassicacearum*, while the remaining strains belonged to the *Bacillus* spp. Genus [107]. These two strains have demonstrated the highest capacity as biocontrol agents against *Agrobacterium* spp. and *Pectobacterium* spp., as well as other isolates [107].

7.2.4. *Matricaria chamomilla* L.

Matricaria chamomilla L. (chamomile) is another widespread herbaceous plant indigenous to Europe and is broadly used in traditional medicine in the form of herbal tea, mostly for gastrointestinal issues. Essential oils derived from fresh or dried flower heads are currently used in a great number of cosmetics, perfumes, baked goods, and beverages. The potential benefits and bioactive properties of chamomile have been extensively studied over the years [208,210–212]. However, only a few endophytes have been isolated from its tissues to this day. Studies conducted by Köberl et al. (2013) identified endophytes from the inner root tissue of *Matricaria chamomilla* as *Paenibacillus polymexa* strain Mc5Re-14, and so far, this is the only endophytic bacteria from this plant that has been characterized and sequenced [115,213]. Its genome encodes many synthases (including a few polyketides), chitinase, extracellular glucanases, and genes responsible for auxin and spermidine production, which is reflected in its activity against phytopathogenic fungi (*Verticillium* sp., *Fusiarium culmorum*, *Rhizoctonia solani*) and the human pathogen *Escherichia coli*. In the work of Erjaee et al. (2019), chamomile was one of the plants that had the highest number of isolated bacterial endophytes compared to the studied medicinal plants. Two out of sixteen isolates exhibited antifungal activity for all five tested food spoilage fungi species [99]. Another interesting study conducted on *Marticaria chamomilla* by Schmidt et al. (2014) investigated the impact of various bacterial inoculants on the microbiome structure of the plant and the production of secondary metabolites. Pyrosequencing analysis of the 16S RNA gene libraries showed significant differences in the bacterial diversity between the treatments, and for two inoculating Gram-positive species—*B. subtilis* Co1–6 and the previously mentioned *P. polymyxa* Mc5Re-14—higher yields of apigenin-7-O-glucoside were also obtained in comparison to treatment with Gram-negative strains [214]. This indicates the importance of understanding the influence of bacterial inoculants and the risks of using them as biocontrol agents, which should be examined in-depth in further research.

7.2.5. *Equisetum arvense* L.

Equisetum arvense (field horsetail) is a plant common throughout Europe and the rest of the world, where it is found in a wide range of habitats, colonizing ruderal and segregated areas in abundance [208]. A characteristic feature of its tissues is a high content of silica and other compounds such as alkaloids, saponins, and phenolic compounds, which directly translates into the use of this plant as an herbal resource [211,212]. Infusions of the above-ground parts of the herb are a source of easily absorbed silica that has a particular effect on the circulatory and urinary systems. The herb itself also exhibits a diuretic effect, thus preventing excessive accumulation of this compound in the body. Horsetail is also used for things like diabetes and edema and for the general improvement of hair and nails [211].

E. arvense was analyzed for its endophytes, and the number of isolated bacteria showed how great their contribution to the colonization of this plant is. In their study, Das et al. (2017) obtained 103 bacterial strains, most of which were identified based on morphological characteristics and using 16S rDNA molecular techniques [116]. Ten bacteria showed promising anticandidial activity, and the bacteria of greatest importance appeared to be *Psychrobacillus insolitus*, generating the most extensive zones of inhibition for *Candida albicans* and *Curtobacterium oceanosedimentum* and effectively inhibiting the growth of *Candida glabrata*. Butanol metabolite extracts of these bacteria caused lysis of the cell membrane, resulting in fungal cell death [116]. Its antimicrobial activity has also been analyzed against common foodborne pathogens, specifically *Staphylococcus aureus* and *Escherichia coli* O157:H7. Three strains—*Streptomyces albolongus*, *Dermacoccus* sp., and *Mycobacterium* sp.—proved to be effective against S. aureus, while for the analyzed *E. coli* strain, the best growth inhibitors were *Streptomyces griseoaurantiacus* (EAL196) and *Paenibacillus* sp. (EAS116), whose results, however, were only moderate. The cell death of the pathogenic bacteria was most likely caused by the penetration of endophytic metabolites into their cells, which caused the disruption of various metabolic functions [215]. In their study, Woźniak et al. (2019) successfully isolated a total of twenty-three endophytic bacteria from six distinct plant species, one of which was *Equisetum arvense* L. Among these bacteria, they identified five unique endophytic strains [117]. The researchers investigated the process of phosphate solubilization, the synthesis of IAA-like compounds and siderophores, as well as nitrogen fixation. Additionally, they analyzed the phenotypic characteristics of the strains. The majority of the endophytic bacteria strains exhibited increased production of indole-3-acetic acid (IAA) at levels over 10 µg IAA/mL, with *Stenotrophomonas maltophilia* ES2 displaying one of the highest levels. Additionally, the strains were analyzed for their ability to produce siderophores. *Comamonas koreensis* ER1 was shown to be one of the most prolific producers of siderophores. The effectiveness of N2 binding by bacterial strains was also assessed. After being incubated in a bacterial culture medium for 72 h, the effectiveness of nitrogen fixation in *Rhizobium* sp. ES1 rose by a factor of 1.67. The findings indicated that the application of *Equisetum arvense* L. endophytes has the potential to enhance plant development [117]. The relatively high anticandidal and antimicrobial activity may be the start of new research to develop natural agents for both combating candidiasis and compounds for use in the food industry. *Equisetum arvense* L. still represents a reservoir of many potential secondary metabolites of endophytes and, in this regard, requires further research [116,117].

7.2.6. *Oenothera biennis* L.

Oenothera biennis (biennial evening primrose) is a native plant of South America and is common in Europe. The oil extracted from the seeds is used for medicinal purposes, as it lowers blood pressure and cholesterol and affects blood sugar levels positively. It serves as the best studied species among its family, and studies are underway to learn more about its chemical composition and biological activity, including studies of endophytes [208].

As it is a plant that readily grows in sandy, loose substrates and has a relatively high resistance to pollution (including hydrocarbon pollution), it is particularly keen to colonize industrial neighborhoods and railroad areas. This suggests a link to endophytes increasing tolerance to agents such as petroleum hydrocarbons. This aspect was investigated by Pawlik and others (2017), proving that the majority of endophytic bacteria colonizing evening primrose tissues were representatives of *Gammaproteobacteria* as well as *Alphaproteobacteria* and *Actinobacteria*. Species from the genera *Rhizobium*, *Rhodococcus*, and *Xanthomonas* were identified. Of the bacterial isolates, more than 90% were capable of utilizing diesel fuel as a carbon source, equally often kerosene oil, while some of them (about 30%, including all the *Rhodococcus* strains) also utilized n-hexadecane. As in other cases, the diversity of the endophytic flora in the stressed environments was relatively low [119].

Regarding the endophytic mechanisms promoting plant growth, many of the bacterial strains isolated from *O. biennis* tissues were capable of synthesizing IAA and hydrogen cyanide (*Stenotrophomonas* sp), and some were also siderophores (*Pseudomonas umsongensis*). Around half of the strains possessed the acdS gene in their genome, which enabled them to produce the enzyme ACC (1-aminocyclopropane-1-carboxylic acid) deaminase. This enzyme utilizes the precursor of ethylene to reduce its concentration in plants, thereby mitigating its negative impact on plant growth. Additionally, as *O. biennis* is a synanthropic plant, this ability to decrease ethylene levels also enhances its spread [119].

7.2.7. *Silybum marianum* L.

Silybum marianum L. (commonly known as milk thistle) is a highly demanded herb growing in central Europe; however, it originally grew in the Mediterranean basin, long known for its hepatoprotective effect on the liver. This is due to its extract—silymarin—which is a unique mixture of seven flavonolignans (silybin A and B, isosilybin A and B, silychristin, isosilychristin, and silydianin) and a flavonoid (taxifolin) [208,212]. Silymarin complexed with phosphatidylcholine and glycol conjugates for better absorption and delivery is safe for broad therapeutic use with minor side effects and no life-threatening adverse events [216–218]. An isolation of endophytic bacteria was conducted by Anwar et al. (2023). Anwar's team isolated a total of eleven endophytic bacteria from *S. marianum* plants that were harvested from locations contaminated with heavy metals. These bacteria were identified using 16s rRNA sequencing. Three isolates exhibited several characteristics, such as robust resistance to heavy metals, promotion of plant development, regulation of plant hormones, remediation of heavy metal toxicity, and antibacterial activity. The isolates SJLC and SJRB exhibited the highest production of indole-3-acetic acid (IAA). These isolates were recognized as *Bacillus* sp. and *Lysinibacillus* sp., respectively. The production of IAA led to an improvement in both root and shoot length. The SJLC isolate showed efficacy against four of the investigated infections. [147,219].

7.2.8. *Mentha piperita* L.

Mentha piperita (peppermint) is a naturally occurring hybrid, and although it is a plant native to the Mediterranean region, it is now settled and cultivated all over the globe. It is also abundant in European territories; in Poland, it is the most common species of mint. It is used in various industries, finding application in food and pharmaceuticals, among others. Peppermint essential oil has become the most widely used product of this type precisely because of its taste and a number of properties, such as antimicrobial activity. However, thanks to Juyal et al.'s (2017) research, we know that the substances found in the essential oil itself are not only responsible for this particular activity (e.g., menthol) but also the endophytes present in their tissues and the secondary metabolites they produce [220]. At this point, there is no evidence that *M. piperita* endophytes increase menthol levels as they synthesize it in *Picrorhiza kurroa* Royle ex Benth. The endophytic solvent fractions of the predominant strains among the bacterial endophytes inhabiting mainly the underground parts of *M. piperita* have been shown to have strong antimicrobial activity against selected strains of *Bacillus*, *Micrococcus*, and *Pseudomonas* [220]

Shokhiddinova and Normurodova conducted research on the production of enzymes by endophytes of medicinal plants, including *Mentha piperita* L. They found that all of the endophytic bacterial isolates (14 strains) extracted from the roots, stems, and leaves had affinity for 1% casein, which indicates protease production [221].

8. Future Perspectives

Synanthropic plants are widespread and can be found in almost every region of the globe. These organisms possess the capacity to withstand adverse environmental circumstances, including high salinity, drought, reduced light exposure, and the presence of toxic substances, such as heavy metals, hydrocarbons, or other pollutants, due to their ability to adapt to the escalating anthropogenic pressure they experience. These features and abilities are acquired via the process of adaptation. Consequently, this makes them a good place to find endophytic microorganisms that exhibit distinctive properties. Fungal endophytes have been described in several scientific articles as having the ability to produce physiologically active chemicals such as phytohormones, surfactants, immunosuppressants, anticancer agents, and antibiotics. Nevertheless, research on bacterial endophytes is still in its infancy, especially when it comes to synanthropic plants. These, in contrast to their non-anthropopressured plant relatives, may either generate more of the same physiologically active compounds or entirely new ones.

This suggests that these bacteria and the novel bioactive compounds they generate may have far-reaching effects in many fields, both now and in the future. These fields include the agriculture, pharmaceuticals, medicine, food, chemicals, tanning, and medical fields.

A comprehensive bioprospecting investigation of endophytic microorganisms from various ecological habitats, such as extreme environments (ruderal, salted, dry, hot, cold, devastated) and the marine environment, is essential for the identification and characterization of unique endophytes possessing specific attributes that could prove valuable in diverse applications within the agricultural, chemical, and medicinal industries. In the near future, there will likely be a shift in practice towards emphasizing the optimization of the interaction between plants and soil microorganisms and endophytes. The molecular pathways that govern the connection between plants and endophytes have not yet been discovered. They will offer a fresh avenue for the isolation and characterization of innovative compounds for human consumption as well as a revolutionary approach to enhancing agricultural productivity and environmental sustainability.

9. Conclusions

Synanthropic plants, which have successfully acclimated to harsh climatic conditions, serve as a valuable reservoir of endophytes. These microorganisms have the ability to generate distinct bioactive substances that could be extensively utilized in diverse sectors such as the agriculture, pharmaceutical, medicine, and the chemical industries. Nevertheless, there remains an immense number of plants yet to be investigated in the search for microorganisms possessing unique characteristics. The identification of novel endophytic species obtained from a plant thriving in harsh conditions is a chance to discover microbes, representing a promising opportunity for various biotechnological uses, namely, exhibiting strong metabolic activity against various pollutants, enhancing agricultural productivity, creating novel pharmaceuticals, and generating unique chemical substances. The continued investigation and advancement of genetic analysis techniques are essential in order to fully exploit this potential.

Author Contributions: P.D.: conceptualization, data curation, resources, writing—original draft preparation, writing—reviewing and editing, visualization, supervision, project administration. O.M.-M.: conceptualization, data curation, resources, writing—original draft, writing—reviewing and editing, supervision, project administration. N.R.: resources, writing—original draft preparation, writing—reviewing and editing. M.R.: writing—original draft preparation. All authors have read and agreed to the published version of the manuscript.

Funding: This research received no external funding.

Institutional Review Board Statement: Not applicable.

Informed Consent Statement: Not applicable.

Acknowledgments: This article was completed while the second author (N.R.) was a Doctoral Candidate in the Interdisciplinary Doctoral School at the Lodz University of Technology.

Conflicts of Interest: The authors declare no conflicts of interest.

References

1. Wilson, D. Endophyte: The Evolution of a Term, and Clarification of Its Use and Definition. *Oikos* **1995**, *73*, 274. [CrossRef]
2. Kandel, S.L.; Joubert, P.M.; Doty, S.L. Bacterial Endophyte Colonization and Distribution within Plants. *Microorganisms* **2017**, *5*, 77. [CrossRef]
3. Vandana, U.K.; Rajkumari, J.; Singha, L.P.; Satish, L.; Alavilli, H.; Sudheer, P.D.V.N.; Chauhan, S.; Ratnala, R.; Satturu, V.; Mazumder, P.B.; et al. The Endophytic Microbiome as a Hotspot of Synergistic Interactions, with Prospects of Plant Growth Promotion. *Biology* **2021**, *10*, 101. [CrossRef] [PubMed]
4. Afzal, I.; Shinwari, Z.K.; Sikandar, S.; Shahzad, S. Plant beneficial endophytic bacteria: Mechanisms, diversity, host range and genetic determinants. *Microbiol. Res.* **2019**, *221*, 36–49. [CrossRef]
5. Alvin, A.; Miller, K.I.; Neilan, B.A. Exploring the potential of endophytes from medicinal plants as sources of antimycobacterial compounds. *Microbiol. Res.* **2014**, *169*, 483–495. [CrossRef]
6. Backman, P.A.; Sikora, R.A. Endophytes: An emerging tool for biological control. *Biol. Control* **2008**, *46*, 1–3. [CrossRef]
7. Yu, H.; Zhang, L.; Li, L.; Zheng, C.; Guo, L.; Li, W.; Sun, P.; Qin, L. Recent developments and future prospects of antimicrobial metabolites produced by endophytes. *Microbiol. Res.* **2010**, *165*, 437–449. [CrossRef]
8. Thakur, D.; Sahani, K. In Vitro Antimicrobial Activity and Mic of the Extracellular Ethyl Acetate Crude Extract of Endophytic *Fungi fusarium* sp. Isolated From Tephrosia Purpurea Root. *Int. J. Pharm. Pharm. Sci.* **2019**, *11*, 48–53. [CrossRef]
9. Hardoim, P.R.; van Overbeek, L.S.; van Elsas, J.D. Properties of bacterial endophytes and their proposed role in plant growth. *Trends Microbiol.* **2008**, *16*, 463–471. [CrossRef]
10. Marchut-Mikolajczyk, O.; Drozdzynski, P. Endophytic Microorganisms from Synanthropic Plants. In *Bioremediation Science from Theory to Practice*; CRC Press: Boca Raton, FL, USA, 2021; pp. 318–329.
11. Pereira, S.I.A.; Monteiro, C.; Vega, A.L.; Castro, P.M.L. Endophytic culturable bacteria colonizing *Lavandula dentata* L. plants: Isolation, characterization and evaluation of their plant growth-promoting activities. *Ecol. Eng.* **2016**, *87*, 91–97. [CrossRef]
12. Kumar, A.; Singh, R.; Yadav, A.; Giri, D.D.; Singh, P.K.; Pandey, K.D. Isolation and characterization of bacterial endophytes of *Curcuma longa* L. *3 Biotech.* **2016**, *6*, 60. [CrossRef] [PubMed]
13. Nagda, V.; Gajbhiye, A.; Kumar, D. Isolation and Characterization of *Endophytic fungi* from Calotropis Procera for Their Antioxidant Activity. *Asian, J. Pharm. Clin. Res.* **2017**, *10*, 254–258. [CrossRef]
14. Hallmann, J.; Quadt-Hallmann, A.; Mahaffee, W.F.; Kloepper, J.W. Bacterial endophytes in agricultural crops. *Can. J. Microbiol.* **1997**, *43*, 895–914. [CrossRef]
15. Santoyo, G.; Moreno-Hagelsieb, G.; del Carmen Orozco-Mosqueda, M.; Glick, B.R. Plant growth-promoting bacterial endophytes. *Microbiol. Res.* **2016**, *183*, 92–99. [CrossRef] [PubMed]
16. Surjit, S.D.; Rupa, G. Beneficial properties, colonization, establishment and molecular diversity of endophytic bacteria in legumes and non legumes. *Afr. J. Microbiol. Res.* **2014**, *8*, 1562–1572. [CrossRef]
17. Rosenblueth, M.; Martínez-Romero, E. Bacterial endophytes and their interactions with hosts. *Mol. Plant-Microbe Interact.* **2006**, *19*, 827–837. [CrossRef]
18. Elbeltagy, A.; Nishioka, K.; Sato, T.; Suzuki, H.; Ye, B.; Hamada TIsawa, T.; Mitsui, H.; Minamisawa, K. Endophytic Colonization and in Planta Nitrogen Fixation by a Herbaspirillum sp. Isolated from Wild Rice Species. *Appl. Environ. Microbiol.* **2001**, *67*, 5285–5293. [CrossRef]
19. Afzal, M.; Khan, Q.M.; Sessitsch, A. Endophytic bacteria: Prospects and applications for the phytoremediation of organic pollutants. *Chemosphere* **2014**, *117*, 232–242. [CrossRef]
20. Compant, S.; Clément, C.; Sessitsch, A. Plant growth-promoting bacteria in the rhizo- and endosphere of plants: Their role, colonization, mechanisms involved and prospects for utilization. *Soil. Biol. Biochem.* **2010**, *42*, 669–678. [CrossRef]
21. Klama, J. Współżycie Endofitów Bakteryjnych z Roślinami (Artykuł Przeglądowy). *Acta Sci. Pol.* **2004**, *3*, 19–28.
22. Malfanova, N.; Lugtenberg, B.J.J.; Berg, G. Bacterial Endophytes: Who and Where, and What Are They Doing There? In *Molecular Microbial Ecology of the Rhizosphere*; John Wiley & Sons, Ltd.: Hoboken, NJ, USA, 2013.
23. Pisarksa, K.; Pietr, S. Bakterie endofityczne—Ich Pochodzenie i Interakcje z Roślinami. *Post. Mikrobiol.* **2014**, *53*, 141–151.
24. Santoyo, G.; Urtis-Flores, C.A.; Loeza-Lara, P.D.; Orozco-Mosqueda, M.D.C.; Glick, B.R. Rhizosphere Colonization Determinants by Plant Growth-Promoting Rhizobacteria (PGPR). *Biology* **2021**, *10*, 475. [CrossRef]
25. Compant, S.; Duffy, B.; Nowak, J.; Clément, C.; Barka, E.A. Use of plant growth-promoting bacteria for biocontrol of plant diseases: Principles, mechanisms of action, and future prospects. *Appl. Environ. Microbiol.* **2005**, *71*, 4951–4959. [CrossRef] [PubMed]
26. Oksinska, M.P.; Wright, S.A.I.; Pietr, S.J. Colonization of wheat seedlings (*Triticum aestivum* L.) by strains of Pseudomonas spp. with respect to their nutrient utilization profiles. *Eur. J. Soil. Biol.* **2011**, *47*, 364–373. [CrossRef]
27. Yaron, S.; Römling, U. Biofilm formation by enteric pathogens and its role in plant colonization and persistence. *Microb. Biotechnol.* **2014**, *7*, 496. [CrossRef]

28. James, E.K.; Olivares, F.L.; De Oliveira, A.L.M.; Reis, F.B.D.; Da Silva, L.G.; Reis, V.M. Further observations on the interaction between sugar cane and Gluconacetobacter diazotrophicus under laboratory and greenhouse conditions. *J. Exp. Bot.* **2001**, *52*, 747–760. [CrossRef] [PubMed]
29. Shahzad, R.; Khan, A.L.; Bilal, S.; Asaf, S.; Lee, I.J. What is there in seeds? Vertically transmitted endophytic resources for sustainable improvement in plant growth. *Front. Plant Sci.* **2018**, *9*, 24. [CrossRef]
30. Oukala, N.; Pastor, V.; Aissat, K. Bacterial Endophytes: The Hidden Actor in Plant Immune Responses against Biotic Stress. *Plants* **2021**, *10*, 1012. [CrossRef]
31. Frank, A.C.; Guzmán, J.P.S.; Shay, J.E. Transmission of Bacterial Endophytes. *Microorganisms* **2017**, *5*, 70. [CrossRef]
32. Gouda, S.; Das, G.; Sen, S.K.; Shin, H.S.; Patra, J.K. Endophytes: A treasure house of bioactive compounds of medicinal importance. *Front. Microbiol.* **2016**, *7*, 01538. [CrossRef]
33. Brader, G.; Compant, S.; Mitter, B.; Trognitz, F.; Sessitsch, A. Metabolic potential of endophytic bacteria. *Curr. Opin. Biotechnol.* **2014**, *27*, 30–37. [CrossRef] [PubMed]
34. Mushtaq, S.; Shafiq, M.; Tariq, M.R.; Sami, A.; Nawaz-ul-Rehman, M.S.; Bhatti, M.H.T.; Haider, M.S.; Sadiq, S.; Abbas, M.T.; Hussain, M.; et al. Interaction between bacterial endophytes and host plants. *Front. Plant Sci.* **2023**, *13*, 1092105. [CrossRef] [PubMed]
35. Bacilio-Jiménez, M.; Aguilar-Flores, S.; Ventura-Zapata, E.; Pérez-Campos, E.; Bouquelet, S.; Zenteno, E. Chemical characterization of root exudates from rice (Oryza sativa) and their effects on the chemotactic response of endophytic bacteria. *Plant Soil* **2003**, *249*, 271–277. [CrossRef]
36. Bernardi, D.I.; Chagas, F.O.D.; Monteiro, A.F.; Santos, G.F.D.; de Souza Berlinck, R.G. Secondary Metabolites of Endophytic Actinomycetes: Isolation, Synthesis, Biosynthesis, and Biological Activities. *Prog. Chem. Org. Nat. Prod.* **2019**, *108*, 207–296. [CrossRef]
37. Balachandar, D.; Sandhiya, G.S.; Sugitha, T.C.K.; Kumar, K. Flavonoids and growth hormones influence endophytic colonization and in planta nitrogen fixation by a diazotrophic Serratia sp. in rice. *World J. Microbiol. Biotechnol.* **2006**, *22*, 707–712. [CrossRef]
38. Lee, K.-B.; De Backer, P.; Aono, T.; Liu, C.-T.; Suzuki, S.; Suzuki, T.; Kaneko, T.; Yamada, M.; Tabata, S.; Kupfer, D.M.; et al. The genome of the versatile nitrogen fixer Azorhizobium caulinodans ORS571. *BMC Genom.* **2008**, *9*, 271. [CrossRef] [PubMed]
39. Pinski, A.; Betekhtin, A.; Hupert-Kocurek, K.; Mur, L.A.J.; Hasterok, R. Molecular Sciences Defining the Genetic Basis of Plant-Endophytic Bacteria Interactions. *Int. J. Mol. Sci.* **2019**, *20*, 1947. [CrossRef]
40. de Weert, S.; Vermeiren, H.; Mulders, I.H.M.; Kuiper, I.; Hendrickx, N.; Bloemberg, G.V.; Vanderleyden, J.; De Mot, R.; Lugtenberg, B.J.J. Flagella-driven chemotaxis towards exudate components is an important trait for tomato root colonization by Pseudomonas fluorescens. *Mol. Plant-Microbe Interact.* **2002**, *15*, 1173–1180. [CrossRef] [PubMed]
41. Rossi, F.A.; Medeot, D.B.; Liaudat, J.P.; Pistorio, M.; Jofré, E. In Azospirillum brasilense, mutations in flmA or flmB genes affect polar flagellum assembly, surface polysaccharides, and attachment to maize roots. *Microbiol. Res.* **2016**, *190*, 55–62. [CrossRef]
42. Duque, E.; de la Torre, J.; Bernal, P.; Molina-Henares, M.A.; Alaminos, M.; Espinosa-Urgel, M.; Roca, A.; Fernández, M.; de Bentzmann, S.; Ramos, J.-L. Identification of reciprocal adhesion genes in pathogenic and non-pathogenic Pseudomonas. *Environ. Microbiol.* **2013**, *15*, 36–48. [CrossRef]
43. Knights, H.E.; Jorrin, B.; Haskett, T.L.; Poole, P.S. Deciphering bacterial mechanisms of root colonization. *Environ. Microbiol. Rep.* **2021**, *13*, 428. [CrossRef] [PubMed]
44. Balsanelli, E.; Tadra-Sfeir, M.Z.; Faoro, H.; Pankievicz, V.C.S.; de Baura, V.A.; Pedrosa, F.O.; de Souza, E.M.; Dixon, R.; Monteiro, R.A. Molecular adaptations of Herbaspirillum seropedicae during colonization of the maize rhizosphere. *Environ. Microbiol.* **2016**, *18*, 2343–2356. [CrossRef] [PubMed]
45. Ryan, R.P.; Germaine, K.; Franks, A.; Ryan, D.J.; Dowling, D.N. Bacterial endophytes: Recent developments and applications. *FEMS Microbiol. Lett.* **2008**, *278*, 1–9. [CrossRef]
46. Hardoim, P.R.; van Overbeek, L.S.; Berg, G.; Pirttilä, A.M.; Compant, S.; Campisano, A.; Döring, M.; Sessitsch, A. The Hidden World within Plants: Ecological and Evolutionary Considerations for Defining Functioning of Microbial Endophytes. *Microbiol. Mol. Biol. Rev.* **2015**, *79*, 293–320. [CrossRef]
47. Johnston-Monje, D.; Raizada, M.N. Conservation and diversity of seed associated endophytes in Zea across boundaries of evolution, ethnography and ecology. *PLoS ONE* **2011**, *6*, 0020396. [CrossRef] [PubMed]
48. Fontana, D.C.; de Paula, S.; Torres, A.G.; de Souza, V.H.M.; Pascholati, S.F.; Schmidt, D.; Dourado Neto, D. Endophytic fungi: Biological control and induced resistance to phytopathogens and abiotic stresses. *Pathogens* **2021**, *10*, 570. [CrossRef]
49. Bashir, I.; War, A.F.; Rafiq, I.; Reshi, Z.A.; Rashid, I.; Shouche, Y.S. Phyllosphere microbiome: Diversity and functions. *Microbiol. Res.* **2021**, *254*, 126888. [CrossRef] [PubMed]
50. Li, J.; Wang, C.; Liang, W.; Liu, S. Rhizosphere Microbiome: The Emerging Barrier in Plant-Pathogen Interactions. *Front. Microbiol.* **2021**, *12*, 772420. [CrossRef]
51. Manghwar, H.; Zaman, W.; Kumari, M.; Swarupa, P.; Kesari, K.K.; Kumar, A. Microbial Inoculants as Plant Biostimulants: A Review on Risk Status. *Life* **2022**, *13*, 12. [CrossRef]
52. Sahu, P.K.; Tilgam, J.; Mishra, S.; Hamid, S.; Gupta, A.; Verma, S.K.; Kharwar, R.N. Surface sterilization for isolation of endophytes: Ensuring what (not) to grow. *J. Basic. Microbiol.* **2022**, *62*, 647–668. [CrossRef]

53. Yusof, H.; Saleh, N.N.; Sarmin, I.M. Optimization of the Isolation Method of Endophytic Actinomycetes from *Psidium guajava* (L.) and Ziziphus mauritiana/Hartini Yusof, Nurul Najwa Saleh and Nurul Izzah Mohd Sarmin. 2020. Available online: http://healthscopefsk.com/ (accessed on 11 April 2023).
54. Schulz, B.; Wanke, U.; Draeger, S.; Aust, H.J. Endophytes from herbaceous plants and shrubs: Effectiveness of surface sterilization methods. *Mycol. Res.* **1993**, *97*, 1447–1450. [CrossRef]
55. Bhore, S.J.; Komathi, V.; Kandasamy, K.I. Diversity of endophytic bacteria in medicinally important Nepenthes species. *J. Nat. Sci. Biol. Med.* **2013**, *4*, 431–434. [CrossRef] [PubMed]
56. Lacerda, L.T.; Gusmão, L.F.P.; Rodrigues, A. Diversity of endophytic fungi in Eucalyptus microcorys assessed by complementary isolation methods. *Mycol. Prog.* **2018**, *17*, 719–727. [CrossRef]
57. Perez-Rosales, E.; Alcaraz-Meléndez, L.; Puente, M.E.; Vázquez-Juárez, R.; Quiroz-Guzmán, E.; Zenteno-Savín, T.; Morales-Bojórquez, E. Isolation and characterization of endophytic bacteria associated with roots of jojoba (Simmondsia chinensis (Link) Schneid). *Curr. Sci.* **2017**, *112*, 396–401.
58. Compant, S.; Kaplan, H.; Sessitsch, A.; Nowak, J.; Barka, E.A.; Clément, C. Endophytic colonization of Vitis vinifera L. by Burkholderia phytofirmans strain PsJN: From the rhizosphere to inflorescence tissues. *FEMS Microbiol. Ecol.* **2008**, *63*, 84–93. [CrossRef]
59. Jahromi, R.; Mogharab, V.; Jahromi, H.; Avazpour, A. Synergistic effects of anionic surfactants on coronavirus (SARS-CoV-2) virucidal efficiency of sanitizing fluids to fight COVID-19. *Food Chem. Toxicol.* **2020**, *145*, 111702. [CrossRef]
60. Xiang, H.; Zhang, T.; Pang, X.; Wei, Y.; Liu, H.; Zhang, Y.; Ma, B.; Yu, L. Isolation of endophytic fungi from Dioscorea zingiberensis C. H. Wright and application for diosgenin production by solid-state fermentation. *Appl. Microbiol. Biotechnol.* **2018**, *102*, 5519–5532. [CrossRef]
61. Jaber, L.R.; Enkerli, J. Effect of seed treatment duration on growth and colonization of Vicia faba by endophytic Beauveria bassiana and Metarhizium brunneum. *Biol. Control* **2016**, *103*, 187–195. [CrossRef]
62. Rutala, W.A.; Weber, D.J. Disinfection, Sterilization, and Control of Hospital Waste. *Mand. Douglas Bennett's Princ. Pract. Infect. Dis.* **2015**, *2*, 3294. [CrossRef]
63. Munif, A.; Hallmann, J.; Sikora, R.A. Isolation of root endophytic bacteria from tomato and its biocontrol activity againts fungal diseases. *Microbiol. Indones.* **2012**, *6*, 2. [CrossRef]
64. Misaghi, I.J. Endophytic Bacteria in Symptom-Free Cotton Plants. *Phytopathology* **1990**, *80*, 808. [CrossRef]
65. Vega, F.E.; Pava-Ripoll, M.; Posada, F.; Buyer, J.S. Endophytic bacteria in *Coffea arabica* L. *J. Basic. Microbiol.* **2005**, *45*, 371–380. [CrossRef] [PubMed]
66. Mcdonnell, G.; Russell, A.D. Antiseptics and disinfectants: Activity, action, and resistance. *Clin. Microbiol. Rev.* **1999**, *12*, 147–179. [CrossRef] [PubMed]
67. Sahu, P.K.; Thomas, P.; Singh, S.; Gupta, A. Taxonomic and functional diversity of cultivable endophytes with respect to the fitness of cultivars against Ralstonia solanacearum. *J. Plant Dis. Prot.* **2020**, *127*, 667–676. [CrossRef]
68. Sahu, P.K.; Singh, S.; Gupta, A.; Singh, U.B.; Paul, S.; Paul, D.; Kuppusamy, P.; Singh, H.V.; Saxena, A.K. A simplified protocol for reversing phenotypic conversion of Ralstonia solanacearum during experimentation. *Int. J. Environ. Res. Public Health* **2020**, *17*, 4274. [CrossRef]
69. Chen, J.; Akutse, K.S.; Saqib, H.S.A.; Wu, X.; Yang, F.; Xia, X.; Wang, L.; Goettel, M.S.; You, M.; Gurr, G.M. Fungal Endophyte Communities of Crucifer Crops Are Seasonally Dynamic and Structured by Plant Identity, Plant Tissue and Environmental Factors. *Front. Microbiol.* **2020**, *11*, 1519. [CrossRef]
70. Teimoori-Boghsani, Y.; Ganjeali, A.; Cernava, T.; Müller, H.; Asili, J.; Berg, G. Endophytic Fungi of Native Salvia abrotanoides Plants Reveal High Taxonomic Diversity and Unique Profiles of Secondary Metabolites. *Front. Microbiol.* **2020**, *10*, 3013. [CrossRef]
71. Agarwal, H.; Dowarah, B.; Baruah, P.M.; Bordoloi, K.S.; Krishnatreya, D.B.; Agarwala, N. Endophytes from Gnetum gnemon L. can protect seedlings against the infection of phytopathogenic bacterium Ralstonia solanacearum as well as promote plant growth in tomato. *Microbiol. Res.* **2020**, *238*, 126503. [CrossRef]
72. Liu, Z.; Zhou, J.; Li, Y.; Wen, J.; Wang, R. Bacterial endophytes from Lycoris radiata promote the accumulation of Amaryllidaceae alkaloids. *Microbiol. Res.* **2020**, *239*, 126501. [CrossRef]
73. Alibrandi, P.; Lo Monaco, N.; Calevo, J.; Voyron, S.; Puglia, A.M.; Cardinale, M.; Perotto, S. Plant growth promoting potential of bacterial endophytes from three terrestrial mediterranean orchid species. *Plant Biosyst.* **2021**, *155*, 1153–1164. [CrossRef]
74. Ashitha, A.; Midhun, S.J.; Sunil, M.A.; Nithin, T.U.; Radhakrishnan, E.K.; Mathew, J. Bacterial endophytes from Artemisia nilagirica (Clarke) Pamp., with antibacterial efficacy against human pathogens. *Microb. Pathog.* **2019**, *135*, 103624. [CrossRef] [PubMed]
75. Liu, C.; Zhuang, X.; Yu, Z.; Wang, Z.; Wang, Y.; Guo, X.; Xiang, W.; Huang, S. Community Structures and Antifungal Activity of Root-Associated Endophytic Actinobacteria of Healthy and Diseased Soybean. *Microorganisms* **2019**, *7*, 243. [CrossRef] [PubMed]
76. Shan, W.; Zhou, Y.; Liu, H.; Yu, X. Endophytic actinomycetes from tea plants (Camellia sinensis): Isolation, abundance, antimicrobial, and plant-growth-promoting activities. *Biomed. Res. Int.* **2018**, *2018*, 1470305. [CrossRef] [PubMed]
77. Moronta-Barrios, F.; Gionechetti, F.; Pallavicini, A.; Marys, E.; Venturi, V. Bacterial Microbiota of Rice Roots: 16S-Based Taxonomic Profiling of Endophytic and Rhizospheric Diversity, Endophytes Isolation and Simplified Endophytic Community. *Microorganisms* **2018**, *6*, 14. [CrossRef] [PubMed]

78. Padda, K.P.; Puri, A.; Chanway, C.P. Isolation and identification of endophytic diazotrophs from lodgepole pine trees growing at unreclaimed gravel mining pits in central interior British Columbia, Canada. *Can. J. For. Res.* **2018**, *48*, 1601–1606. [CrossRef]
79. Yaish, M.W.; Antony, I.; Glick, B.R. Isolation and characterization of endophytic plant growth-promoting bacteria from date palm tree (Phoenix dactylifera L.) and their potential role in salinity tolerance. *Antonie Van Leeuwenhoek* **2015**, *107*, 1519–1532. [CrossRef]
80. Qin, S.; Li, J.; Chen, H.-H.; Zhao, G.-Z.; Zhu, W.-Y.; Jiang, C.-L.; Xu, L.-H.; Li, W.-J. Isolation, diversity, and antimicrobial activity of rare actinobacteria from medicinal plants of tropical rain forests in Xishuangbanna China. *Appl. Environ. Microbiol.* **2009**, *75*, 6176–6186. [CrossRef]
81. Passari, A.K.; Mishra, V.K.; Saikia, R.; Gupta, V.K.; Singh, B.P. Isolation, abundance and phylogenetic affiliation of endophytic actinomycetes associated with medicinal plants and screening for their in vitro antimicrobial biosynthetic potential. *Front. Microbiol.* **2015**, *6*, 273. [CrossRef]
82. Nawed, A.; Chandra, R. Endophytic Bacteria: Optimizaton of Isolation Procedure from Various Medicinal Plants and Their Preliminary Characterization. *Asian J. Pharm. Clin. Res.* **2015**, *8*, 233–238. Available online: https://journals.innovareacademics.in/index.php/ajpcr/article/view/6586/2724 (accessed on 23 May 2023).
83. Dukan, S.; Belkin, S.; Touati, D. Reactive oxygen species are partially involved in the bacteriocidal action of hypochlorous acid. *Arch. Biochem. Biophys.* **1999**, *367*, 311–316. [CrossRef]
84. Le Dantec, C.; Duguet, J.P.; Montiel, A.; Dumoutier, N.; Dubrou, S.; Vincent, V. Chlorine disinfection of atypical mycobacteria isolated from a water distribution system. *Appl. Environ. Microbiol.* **2002**, *68*, 1025–1032. [CrossRef] [PubMed]
85. ICSC 0979—MERCURIC CHLORIDE. Available online: https://www.inchem.org/documents/icsc/icsc/eics0979.htm (accessed on 6 August 2024).
86. Gupta, R.; Kale, P.S.; Rathi, M.L.; Jadhav, N. Isolation, characterization and identification of endophytic bacteria by 16S rRNA partial sequencing technique from roots and leaves of Prosopis cineraria plant. *Asian J. Plant Sci. Res.* **2015**, *5*, 36–43.
87. Gohain, A.; Gogoi, A.; Debnath, R.; Yadav, A.; Singh, B.P.; Gupta, V.K.; Sharma, R.; Saikia, R. Antimicrobial biosynthetic potential and genetic diversity of endophytic actinomycetes associated with medicinal plants. *FEMS Microbiol. Lett.* **2015**, *362*, 158. [CrossRef] [PubMed]
88. Coombs, J.T.; Franco, C.M.M. Isolation and Identification of Actinobacteria from Surface-Sterilized Wheat Roots. *Appl. Environ. Microbiol.* **2003**, *69*, 5603. [CrossRef] [PubMed]
89. Gardner, J.M.; Chandler, J.L.; Feldman, A.W. Growth promotion and inhibition by antibiotic-producing fluorescent pseudomonads on citrus roots. *Plant Soil* **1984**, *77*, 103–113. [CrossRef]
90. Bell, C.R.; Dickie, G.A.; Harvey, W.L.G.; Chan, J.W.Y.F. Endophytic bacteria in grapevine. *Can. J. Microbiol.* **2011**, *41*, 46–53. [CrossRef]
91. Scholander, P.F.; Hammel, H.T.; Bradstreet, E.D.; Hemmingsen, E.A. Sap Pressure in Vascular Plants: Negative hydrostatic pressure can be measured in plants. *Science* **1965**, *148*, 339–346. [CrossRef]
92. Achari, G.A.; Ramesh, R. Diversity, Biocontrol, and Plant Growth Promoting Abilities of Xylem Residing Bacteria from Solanaceous Crops. *Int. J. Microbiol.* **2014**, *2014*, 296521. [CrossRef]
93. Khare, E.; Mishra, J.; Arora, N.K. Multifaceted interactions between endophytes and plant: Developments and Prospects. *Front. Microbiol.* **2018**, *9*, 411604. [CrossRef]
94. Poveda, J.; Díaz-González, S.; Díaz-Urbano, M.; Velasco, P.; Sacristán, S. Fungal endophytes of Brassicaceae: Molecular interactions and crop benefits. *Front. Plant Sci.* **2022**, *13*, 932288. [CrossRef]
95. Gohain, A.; Gogoi, A.; Debnath, R.; Yadav, A.; Singh, B.P.; Gupta, V.K.; Sharma, R.; Saikia, R. Current Understanding of Endophytes: Their Relevance, Importance, and Industrial Potentials. *IOSR J. Biotechnol. Biochem.* **2017**, *3*, 43–59. [CrossRef]
96. de Oliveira, N.C. Endophytic bacteria with potential for bioremediation of petroleum hydrocarbons and derivatives. *Afr. J. Biotechnol.* **2012**, *11*, ajb10.2623. [CrossRef]
97. Wang, Y.; Dai, C.C. Endophytes: A potential resource for biosynthesis, biotransformation, and biodegradation. *Ann. Microbiol.* **2011**, *61*, 207–215. [CrossRef]
98. El-Shatoury, S.; El Kazzaz, W.; Elkraly, O. Antimicrobial Activities of Actinomycetes Inhabiting Achillea Fragrantissima (Family: Compositae). *Egypt J. Nat. Toxins* **2009**, *6*, 1–15. Available online: https://www.researchgate.net/publication/287492658 (accessed on 7 April 2024).
99. Erjaee, Z.; Shekarforoush, S.S.; Hosseinzadeh, S. Identification of endophytic bacteria in medicinal plants and their antifungal activities against food spoilage fungi. *J. Food Sci. Technol.* **2019**, *56*, 5262–5270. [CrossRef]
100. Rat, A.; Naranjo, H.D.; Krigas, N.; Grigoriadou, K.; Maloupa, E.; Varela Alonso, A.; Schneider, C.; Papageorgiou, V.P.; Assimopoulou, A.N.; Tsafantakis, N.; et al. Endophytic Bacteria From the Roots of the Medicinal Plant Alkanna tinctoria Tausch (Boraginaceae): Exploration of Plant Growth Promoting Properties and Potential Role in the Production of Plant Secondary Metabolites. *Front. Microbiol.* **2021**, *12*, 633488. [CrossRef]
101. Igarashi, Y. Screening of Novel Bioactive Compounds from Plant-Associated Actinomycetes. *Actinomycetologica* **2004**, *18*, 63–66. [CrossRef]
102. Sasaki, T.; Igarashi, Y.; Ogawa, M.; Furumai, T. Identification of 6-Prenylindole as an Antifungal Metabolite of Streptomyces sp. TP-A0595 and Synthesis and Bioactivity of 6-Substituted Indoles. *J. Antibiot.* **2002**, *55*, 1009–1012. [CrossRef]

103. Liu, J.-Q.; Chen, S.-M.; Zhang, C.-M.; Xu, M.-J.; Xing, K.; Li, C.-G.; Li, K.; Zhang, Y.-Q.; Qin, S. Abundant and diverse endophytic bacteria associated with medicinal plant Arctium lappa L. and their potential for host plant growth promoting. *Antonie Van Leeuwenhoek Int. J. General Mol. Microbiol.* **2022**, *115*, 1405–1420. [CrossRef]
104. Shurigin, V.; Egamberdieva, D.; Samadiy, S.; Mardonova, G.; Davranov, K. Endophytes from medicinal plants as biocontrol agents against fusarium caused diseases. *Mikrobiolohichnyi Zhurnal* **2020**, *82*, 41–52. [CrossRef]
105. Wardecki, T.; Brötz, E.; De Ford, C.; von Loewenich, F.D.; Rebets, Y.; Tokovenko, B.; Luzhetskyy, A.; Merfort, I. Endophytic Streptomyces in the traditional medicinal plant *Arnica montana* L.: Secondary metabolites and biological activity. *Antonie Van Leeuwenhoek Int. J. General Mol. Microbiol.* **2015**, *108*, 391–402. [CrossRef] [PubMed]
106. Bekiesch, P.; Oberhofer, M.; Sykora, C.; Urban, E.; Zotchev, S.B. Piperazic acid containing peptides produced by an endophytic Streptomyces sp. isolated from the medicinal plant Atropa belladonna. *Nat. Prod. Res.* **2021**, *35*, 1090–1096. [CrossRef] [PubMed]
107. Krimi, Z.; Alim, D.; Djellout, H.; Tafifet, L.; Mohamed-Mahmoud, F.; Raio, A. Bacterial endophytes of weeds are effective biocontrol agents of Agrobacterium spp., Pectobacterium spp., and promote growth of tomato plants. *Phytopathol. Mediterr.* **2016**, *55*, 184–196. [CrossRef]
108. Köberl, M.; Ramadan, E.M.; Adam, M.; Cardinale, M.; Hallmann, J.; Heuer, H.; Smalla, K.; Berg, G. Bacillus and streptomyces were selected as broad-spectrum antagonists against soilborne pathogens from arid areas in egypt. *FEMS Microbiol. Lett.* **2013**, *342*, 168–178. [CrossRef]
109. Shurigin, V.; Alimov, J.; Davranov, K.; Gulyamova, T.; Egamberdieva, D. The diversity of bacterial endophytes from *Iris pseudacorus* L. and their plant beneficial traits. *Curr. Res. Microb. Sci.* **2022**, *3*, 100133. [CrossRef]
110. Ait Kaki, A.; Kacem Chaouche, N.; Dehimat, L.; Milet, A.; Youcef-Ali, M.; Ongena, M.; Thonart, P. Biocontrol and Plant Growth Promotion Characterization of Bacillus Species Isolated from Calendula officinalis Rhizosphere. *Indian. J. Microbiol.* **2013**, *53*, 447–452. [CrossRef] [PubMed]
111. Marchut-Mikolajczyk, O.; Drozdzyński, P.; Pietrzyk, D.; Antczak, T. Biosurfactant production and hydrocarbon degradation activity of endophytic bacteria isolated from *Chelidonium majus* L. 06 Biological Sciences 0605 Microbiology 06 Biological Sciences 0607 Plant Biology 09 Engineering 0907 Environmental Engineering. *Microb Cell Fact* **2018**, *17*, 171. [CrossRef]
112. Goryluk-Salmonowicz, A.; Orzeszko-Rywka, A.; Piorek, M.; Rekosz-Burlaga, H.; Otlowska, A.; Gozdowski, D.; Blaszczyk, M. Plant growth promoting bacterial endophytes isolated from Polish herbal plants. *Acta Sci. Pol. Hortorum Cultus* **2018**, *17*, 101–110. [CrossRef]
113. Goryluk, A.; Rekosz-Burlaga, H.; Błaszczyk, M. Isolation and Characterization of Bacterial Endophytes of *Chelidonium majus* L. *Pol. J. Microbiol.* **2009**, *58*, 355–361.
114. Presta, L.; Bosi, E.; Fondi, M.; Maida, I.; Perrin, E.; Miceli, E.; Maggini, V.; Bogani, P.; Firenzuoli, F.; Di Pilato, V.; et al. Phenotypic and genomic characterization of the antimicrobial producer Rheinheimera sp. EpRS3 isolated from the medicinal plant Echinacea purpurea: Insights into its biotechnological relevance. *Res. Microbiol.* **2017**, *168*, 293–305. [CrossRef]
115. Goryluk-Salmonowicz, A.; Piórek, M.; Rekosz-Burlaga, H.; Studnicki, M.; Błaszczyk, M. Endophytic Detection in Selected European Herbal Plants. *Pol. J. Microbiol.* **2016**, *65*, 369–375. [CrossRef] [PubMed]
116. Das, G.; Patra, J.K.; Baek, K.H. Antibacterial Properties of Endophytic Bacteria Isolated from a Fern Species Equisetum arvense L. Against Foodborne Pathogenic Bacteria Staphylococcus aureus and Escherichia coli O157:H7. *Foodborne Pathog. Dis.* **2017**, *14*, 50–58. [CrossRef] [PubMed]
117. Woźniak, M.; Gałązka, A.; Tyśkiewicz, R.; Jaroszuk-ściseł, J. Endophytic bacteria potentially promote plant growth by synthesizing different metabolites and their phenotypic/physiological profiles in the biolog gen iii microplate^{TM} test. *Int. J. Mol. Sci.* **2019**, *20*, 5283. [CrossRef]
118. Egamberdieva, D.; Wirth, S.; Behrendt, U.; Ahmad, P.; Berg, G. Antimicrobial activity of medicinal plants correlates with the proportion of antagonistic endophytes. *Front. Microbiol.* **2017**, *8*, 00199. [CrossRef]
119. Pawlik, M.; Cania, B.; Thijs, S.; Vangronsveld, J.; Piotrowska-Seget, Z. Hydrocarbon degradation potential and plant growth-promoting activity of culturable endophytic bacteria of Lotus corniculatus and Oenothera biennis from a long-term polluted site. *Environ. Sci. Pollut. Res.* **2017**, *24*, 19640–19652. [CrossRef]
120. Alaylar, B. Isolation and Characterization of Culturable Endophytic Plant Growth Promoting Bacillus Species from *Mentha longifolia* L. *Turk. J. Agric. For.* **2021**, *46*, 73–82. [CrossRef]
121. Ting, A.S.Y.; Akinsanya, M.A.; Goh, J.K.; Lim, S.P. Antimicrobial and Enzymatic Activities of Endophytic Bacteria Isolated from *Mentha spicata* (MINT). *Malays. J. Microbiol.* **2015**, *11*, 102–108. Available online: https://www.researchgate.net/publication/282651721 (accessed on 7 April 2024).
122. Ngashangva, N.; Mukherjee, P.; Sharma, K.C.; Kalita, M.C.; Indira, S. Analysis of Antimicrobial Peptide Metabolome of Bacterial Endophyte Isolated From Traditionally Used Medicinal Plant Millettia pachycarpa Benth. *Front. Microbiol.* **2021**, *12*, 65689. [CrossRef] [PubMed]
123. Chlebek, D.; Pinski, A.; Żur, J.; Michalska, J.; Hupert-Kocurek, K. Genome Mining and Evaluation of the Biocontrol Potential of Pseudomonas fluorescens BRZ63, a New Endophyte of Oilseed Rape (*Brassica napus* L.) against Fungal Pathogens. *Int. J. Mol. Sci.* **2020**, *21*, 8740. [CrossRef]
124. Ben Mefteh, F.; Chenari Bouket, A.; Daoud, A.; Luptakova, L.; Alenezi, F.; Gharsallah, N.; Belbahri, L. Metagenomic Insights and Genomic Analysis of Phosphogypsum and Its Associated Plant Endophytic Microbiomes Reveals Valuable Actors for Waste Bioremediation. *Microorganisms* **2019**, *7*, 382. [CrossRef]

125. Albarano, L.; Esposito, R.; Ruocco, N.; Costantini, M. Genome mining as new challenge in natural products discovery. *Mar. Drugs* **2020**, *18*, 199. [CrossRef] [PubMed]
126. Blin, K.; Shaw, S.; Steinke, K.; Villebro, R.; Ziemert, N.; Lee, S.Y.; Medema, M.H.; Weber, T. AntiSMASH 5.0: Updates to the secondary metabolite genome mining pipeline. *Nucleic Acids Res.* **2019**, *47*, W81–W87. [CrossRef] [PubMed]
127. Medema, M.H.; Fischbach, M.A.; Gilbert, J.A.; Greenberg, E.P.; Loper, J.E.; Metcalf, W.W.; O'Connor, S.E.; Piel, J.; Raaijmakers, J.M.; Stachelhaus, T.; et al. AntiSMASH: Rapid identification, annotation and analysis of secondary metabolite biosynthesis gene clusters in bacterial and fungal genome sequences. *Nucleic Acids Res.* **2011**, *39* (Suppl. S2), gkr466. [CrossRef] [PubMed]
128. Tietz, J.I.; Schwalen, C.J.; Patel, P.S.; Maxson, T.; Blair, P.M.; Tai, H.-C.; Zakai, U.I.; Mitchell, D.A. A new genome-mining tool redefines the lasso peptide biosynthetic landscape. *Nat. Chem. Biol.* **2017**, *13*, 470–478. [CrossRef]
129. Castillo, U.F.; Strobel, G.A.; Mullenberg, K.; Condron, M.M.; Teplow, D.B.; Folgiano, V.; Gallo, M.; Ferracane, R.; Mannina, L.; Viel, S.; et al. Munumbicins E-4 and E-5: Novel broad-spectrum antibiotics from Streptomyces NRRL 3052. *FEMS Microbiol. Lett.* **2006**, *255*, 296–300. [CrossRef]
130. Li, J.; Zhao, G.-Z.; Huang, H.-Y.; Qin, S.; Zhu, W.-Y.; Zhao, L.-X.; Xu, L.-H.; Zhang, S.; Li, W.-J. Isolation and characterization of culturable endophytic actinobacteria associated with *Artemisia annua* L. *Antonie Van Leeuwenhoek* **2012**, *101*, 515–527. [CrossRef]
131. Taria, S.; Arora, A.; Alam, B.; Kumar, S.; Yadav, A.; Kumar, S.; Kumar, M.; Anuragi, H.; Kumar, R.; Meena, S.; et al. Introduction to Plant Nitrogen Metabolism. In *Advances in Plant Nitrogen Metabolism*; CRC Press: Boca Raton, FL, USA, 2022; pp. 1–18. [CrossRef]
132. Maela, M.P.; van der Walt, H.; Serepa-Dlamini, M.H. The Antibacterial, Antitumor Activities, and Bioactive Constituents' Identification of Alectra sessiliflora Bacterial Endophytes. *Front. Microbiol.* **2022**, *13*, 870821. [CrossRef]
133. Narayanan, Z.; Glick, B.R. Secondary Metabolites Produced by Plant Growth-Promoting Bacterial Endophytes. *Microorganisms* **2022**, *10*, 2008. [CrossRef]
134. Ancheeva, E.; Daletos, G.; Proksch, P. Bioactive Secondary Metabolites from Endophytic Fungi. *Curr. Med. Chem.* **2019**, *27*, 1836–1854. [CrossRef]
135. Bauman, K.D.; Butler, K.S.; Moore, B.S.; Chekan, J.R. Genome mining methods to discover bioactive natural products. *Nat. Prod. Rep.* **2021**, *38*, 2100–2129. [CrossRef]
136. Kaul, S.; Gupta, S.; Ahmed, M.; Dhar, M.K. Endophytic fungi from medicinal plants: A treasure hunt for bioactive metabolites. *Phytochem. Rev.* **2012**, *11*, 487–505. [CrossRef]
137. Rutkowska, N.; Drożdżyński, P.; Ryngajłło, M.; Marchut-Mikołajczyk, O. Plants as the Extended Phenotype of Endophytes—The Actual Source of Bioactive Compounds. *Int. J. Mol. Sci.* **2023**, *24*, 96. [CrossRef] [PubMed]
138. Strobel, G.A. Endophytes as Sources of Bioactive Products. 2003. Available online: www.elsevier.com/locate/micinf (accessed on 7 April 2024).
139. Ansari, A.A.; Gill, S.S.; Gill, R.; Lanza, G.R.; Newman, L. *Phytoremediation: Management of Environmental Contaminants*; Springer International Publishing: Berlin/Heidelberg, Germany, 2019; Volume 6. [CrossRef]
140. Kusari, S.; Spiteller, M. Are we ready for industrial production of bioactive plant secondary metabolites utilizing endophytes? *Nat. Prod. Rep.* **2011**, *28*, 1203–1207. [CrossRef] [PubMed]
141. Pratyusha, S. Phenolic Compounds in the Plant Development and Defense: An Overview. Available online: www.intechopen.com (accessed on 3 March 2024).
142. Rosconi, F.; Davyt, D.; Martínez, V.; Martínez, M.; Abin-Carriquiry, J.A.; Zane, H.; Butler, A.; de Souza, E.M.; Fabiano, E. Identification and structural characterization of serobactins, a suite of lipopeptide siderophores produced by the grass endophyte Herbaspirillum seropedicae. *Environ. Microbiol.* **2013**, *15*, 916–927. [CrossRef] [PubMed]
143. Cornelis, P. Iron uptake and metabolism in pseudomonads. *Appl. Microbiol. Biotechnol.* **2010**, *86*, 1637–1645. [CrossRef]
144. Misra, S.; Semwal, P.; Pandey, D.D.; Mishra, S.K.; Chauhan, P.S. Siderophore-Producing Spinacia Oleracea Bacterial Endophytes Enhance Nutrient Status and Vegetative Growth Under Iron-Deficit Conditions. *J. Plant Growth Regul.* **2024**, *43*, 1317–1330. [CrossRef]
145. Maheshwari, R.; Bhutani, N.; Suneja, P. Screening and characterization of siderophore producing endophytic bacteria from Cicer arietinum and Pisum sativum plants. *J. Appl. Biol. Biotechnol.* **2019**, *7*, 7–14. [CrossRef]
146. Marchut-Mikołajczyk, O.; Chlebicz, M.; Kawecka, M.; Michalak, A.; Prucnal, F.; Nielipinski, M.; Filipek, J.; Jankowska, M.; Perek, Z.; Drożdżyński, P.; et al. Endophytic bacteria isolated from *Urtica dioica* L.-preliminary screening for enzyme and polyphenols production. *Microb. Cell Fact.* **2023**, *22*, 169. [CrossRef]
147. Zahoor, M.; Irshad, M.; Rahman, H.; Qasim, M.; Afridi, S.G.; Qadir, M.; Hussain, A. Alleviation of heavy metal toxicity and phytostimulation of Brassica campestris L. by endophytic Mucor sp. MHR-7. *Ecotoxicol. Environ. Saf.* **2017**, *142*, 139–149. [CrossRef]
148. Cornu, J.Y.; Huguenot, D.; Jézéquel, K.; Lollier, M.; Lebeau, T. Bioremediation of copper-contaminated soils by bacteria. *World J. Microbiol. Biotechnol.* **2017**, *33*, 26. [CrossRef]
149. Do, H.; Che, C.; Zhao, Z.; Wang, Y.; Li, M.; Zhang, X.; Zhao, X. Extracellular polymeric substance from *Rahnella* sp. LRP3 converts available Cu into Cu5(PO4)2(OH)4 in soil through biomineralization process. *Environ. Pollut.* **2020**, *260*, 114051. [CrossRef] [PubMed]
150. Zhao, X.; Do, H.; Zhou, Y.; Li, Z.; Zhang, X.; Zhao, S.; Li, M.; Wu, D. Rahnella sp. LRP3 induces phosphate precipitation of Cu (II) and its role in copper-contaminated soil remediation. *J. Hazard. Mater.* **2019**, *368*, 133–140. [CrossRef] [PubMed]

151. He, H.; Ye, Z.; Yang, D.; Yan, J.; Xiao, L.; Zhong, T.; Yuan, M.; Cai, X.; Fang, Z.; Jing, Y. Characterization of endophytic *Rahnella* sp. JN6 from Polygonum pubescens and its potential in promoting growth and Cd, Pb, Zn uptake by *Brassica napus*. *Chemosphere* **2013**, *90*, 1960–1965. [CrossRef]
152. Jing, Y.X.; Yan, J.L.; He, H.D.; Yang, D.J.; Xiao, L.; Zhong, T.; Yuan, M.; Cai, X.; Li, S.B. Characterization of Bacteria in the Rhizosphere Soils of Polygonum Pubescens and Their Potential in Promoting Growth and Cd, Pb, Zn Uptake by *Brassica napus*. *Int. J. Phytoremediation* **2014**, *16*, 321–333. [CrossRef]
153. Tamariz-Angeles, C.; Huamán, G.D.; Palacios-Robles, E.; Olivera-Gonzales, P.; Castañeda-Barreto, A. Characterization of siderophore-producing microorganisms associated to plants from high-Andean heavy metal polluted soil from Callejón de Huaylas (Ancash, Perú). *Microbiol. Res.* **2021**, *250*, 126811. [CrossRef] [PubMed]
154. Ulloa-Muñoz, R.; Olivera-Gonzales, P.; Castañeda-Barreto, A.; Villena, G.K.; Tamariz-Angeles, C. Diversity of endophytic plant-growth microorganisms from Gentianella weberbaueri and Valeriana pycnantha, highland Peruvian medicinal plants. *Microbiol. Res.* **2020**, *233*, 126413. [CrossRef]
155. Majda, M.; Robert, S. The Role of Auxin in Cell Wall Expansion. *Int. J. Mol. Sci.* **2018**, *19*, 951. [CrossRef]
156. Chen, B.; Luo, S.; Wu, Y.; Ye, J.; Wang, Q.; Xu, X.; Pan, F.; Khan, K.Y.; Feng, Y.; Yang, X.; et al. The effects of the endophytic bacterium Pseudomonas fluorescens Sasm05 and IAA on the plant growth and cadmium uptake of Sedum alfredii hance. *Front. Microbiol.* **2017**, *8*, 2538. [CrossRef] [PubMed]
157. Tavarideh, F.; Pourahmad, F.; Nemati, M. Diversity and antibacterial activity of endophytic bacteria associated with medicinal plant, Scrophularia striata. *Vet. Res. Forum* **2022**, *13*, 409–415. [CrossRef]
158. Hammerschmidt, R. Induced disease resistance: How do induced plants stop pathogens? *Physiol. Mol. Plant Pathol.* **1999**, *55*, 77–84. [CrossRef]
159. Walters, D.; Walsh, D.; Newton, A.; Lyon, G. Induced resistance for plant disease control: Maximizing the efficacy of resistance elicitors. *Phytopathology* **2005**, *95*, 1368–1373. [CrossRef] [PubMed]
160. Dudeja, S.S.; Suneja-Madan, P.; Paul, M.; Maheswari, R.; Kothe, E. Bacterial endophytes: Molecular interactions with their hosts. *J. Basic. Microbiol.* **2021**, *61*, 475–505. [CrossRef] [PubMed]
161. Kenshole, E.; Herisse, M.; Michael, M.; Pidot, S.J. Natural product discovery through microbial genome mining. *Curr. Opin. Chem. Biol.* **2021**, *60*, 47–54. [CrossRef] [PubMed]
162. Cheffi, M.; Chenari Bouket, A.; Alenezi, F.N.; Luptakova, L.; Belka, M.; Vallat, A.; Rateb, M.E.; Tounsi, S.; Triki, M.A.; Belbahri, L. Olea europaea l. Root endophyte bacillus velezensis oee1 counteracts oomycete and fungal harmful pathogens and harbours a large repertoire of secreted and volatile metabolites and beneficial functional genes. *Microorganisms* **2019**, *7*, 314. [CrossRef] [PubMed]
163. Li, Z.; Wen, W.; Qin, M.; He, Y.; Xu, D.; Li, L. Biosynthetic Mechanisms of Secondary Metabolites Promoted by the Interaction Between Endophytes and Plant Hosts. *Front. Media* **2022**, *11*, 928967. [CrossRef]
164. Kaul, S.; Sharma, T.; Dhar, M.K. 'Omics' tools for better understanding the plant–endophyte interactions. *Front. Plant Sci.* **2016**, *7*, 955. [CrossRef]
165. Guo, D.-J.; Li, D.-P.; Singh, R.K.; Singh, P.; Sharma, A.; Verma, K.K.; Qin, Y.; Khan, Q.; Lu, Z.; Malviya, M.K.; et al. Differential Protein Expression Analysis of Two Sugarcane Varieties in Response to Diazotrophic Plant Growth-Promoting Endophyte Enterobacter roggenkampii ED5. *Front. Plant Sci.* **2021**, *12*, 727741. [CrossRef]
166. Zgadzaj, R.; James, E.K.; Kelly, S.; Kawaharada, Y.; de Jonge, N.; Jensen, D.B.; Madsen, L.H.; Radutoiu, S. A Legume Genetic Framework Controls Infection of Nodules by Symbiotic and Endophytic Bacteria. *PLoS Genet.* **2015**, *11*, e1005280. [CrossRef]
167. Monteiro, R.A.; Balsanelli, E.; Tuleski, T.; Faoro, H.; Cruz, L.M.; Wassem, R.; de Baura, V.A.; Tadra-Sfeir, M.Z.; Weiss, V.; DaRocha, W.D.; et al. Genomic comparison of the endophyte Herbaspirillum seropedicae SmR1 and the phytopathogen Herbaspirillum rubrisubalbicans M1 by suppressive subtractive hybridization and partial genome sequencing. *FEMS Microbiol. Ecol.* **2012**, *80*, 441–451. [CrossRef]
168. Tian, D.; Song, X.; Li, C.; Zhou, W.; Qin, L.; Wei, L.; Di, W.; Huang, S.; Li, B.; Huang, Q.; et al. Antifungal mechanism of Bacillus amyloliquefaciens strain GKT04 against Fusarium wilt revealed using genomic and transcriptomic analyses. *Microbiologyopen* **2021**, *10*, e1192. [CrossRef]
169. Luo, Y.; Cobb, R.E.; Zhao, H. Recent advances in natural product discovery. *Curr. Opin. Biotechnol.* **2014**, *30*, 230–237. [CrossRef] [PubMed]
170. Segers, K.; Declerck, S.; Mangelings, D.; Heyden, Y.V.; Van Eeckhaut, A. Analytical techniques for metabolomic studies: A review. *Bioanalysis* **2019**, *11*, 2297–2318. [CrossRef] [PubMed]
171. Armin, R.; Zühlke, S.; Mahnkopp-Dirks, F.; Winkelmann, T.; Kusari, S. Evaluation of Apple Root-Associated Endophytic Streptomyces pulveraceus Strain ES16 by an OSMAC-Assisted Metabolomics Approach. *Front. Sustain. Food Syst.* **2021**, *5*, 643225. [CrossRef]
172. Pan, R.; Bai, X.; Chen, J.; Zhang, H.; Wang, H. Exploring structural diversity of microbe secondary metabolites using OSMAC strategy: A literature review. *Front. Media* **2019**, *10*, 00294. [CrossRef] [PubMed]
173. Bachmann, B.O.; Van Lanen, S.G.; Baltz, R.H. Microbial genome mining for accelerated natural products discovery: Is a renaissance in the making? *J. Ind. Microbiol. Biotechnol.* **2014**, *41*, 175–184. [CrossRef]
174. Kudo, F.; Miyanaga, A.; Eguchi, T. Structural basis of the nonribosomal codes for nonproteinogenic amino acid selective adenylation enzymes in the biosynthesis of natural products. *J. Ind. Microbiol. Biotechnol.* **2019**, *46*, 515–536. [CrossRef]

175. Iacovelli, R.; Bovenberg, R.A.L.; Driessen, A.J.M. Nonribosomal peptide synthetases and their biotechnological potential in Penicillium rubens. *J. Ind. Microbiol. Biotechnol.* **2021**, *48*, 45. [CrossRef]
176. Wang, H.; Fewer, D.P.; Holm, L.; Rouhiainen, L.; Sivonen, K. Atlas of nonribosomal peptide and polyketide biosynthetic pathways reveals common occurrence of nonmodular enzymes. *Proc. Natl. Acad. Sci. USA* **2014**, *111*, 9259–9264. [CrossRef]
177. Le Govic, Y.; Papon, N.; Le Gal, S.; Bouchara, J.P.; Vandeputte, P. Non-ribosomal Peptide Synthetase Gene Clusters in the Human Pathogenic Fungus Scedosporium apiospermum. *Front. Microbiol.* **2019**, *10*, 476594. [CrossRef]
178. Gulick, A.M. Nonribosomal peptide synthetase biosynthetic clusters of ESKAPE pathogens. *Nat. Prod. Rep.* **2017**, *34*, 981–1009. [CrossRef]
179. Chu, H.Y.; Wegel, E.; Osbourn, A. From hormones to secondary metabolism: The emergence of metabolic gene clusters in plants. *Plant J.* **2011**, *66*, 66–79. [CrossRef] [PubMed]
180. Bentley, S.D.; Chater, K.F.; Cerdeño-Tárraga, A.-M.; Challis, G.L.; Thomson, N.R.; James, K.D.; Harris, D.E.; Quail, M.A.; Kieser, H.; Harper, D.; et al. Complete Genome Sequence of the Model Actinomycete *Streptomyces coelicolor* A3(2). *Nature* **2002**, *417*, 141–147. [CrossRef] [PubMed]
181. Challis, G.L.; Ravel, J. Coelichelin, a New Peptide Siderophore Encoded by the *Streptomyces coelicolor* Genome: Structure Prediction from the Sequence Of Its Non-Ribosomal Peptide Synthetase. *FEMS Microbiol. Lett.* **2000**, *187*, 111–114. Available online: www.fems-microbiology.org (accessed on 7 April 2024). [CrossRef] [PubMed]
182. Ikeda, H.; Shin-Ya, K.; Omura, S. Genome mining of the Streptomyces avermitilis genome and development of genome-minimized hosts for heterologous expression of biosynthetic gene clusters. *J. Ind. Microbiol. Biotechnol.* **2014**, *41*, 233–250. [CrossRef]
183. Malpartida, F.; Hopwood, D.A. Molecular cloning of the whole biosynthetic pathway of a Streptomyces antibiotic and its expression in a heterologous host. *Nature* **1984**, *309*, 462–464. [CrossRef]
184. Ali, M.A.; Lou, Y.; Hafeez, R.; Li, X.; Hossain, A.; Xie, T.; Lin, L.; Li, B.; Yin, Y.; Yan, J.; et al. Functional Analysis and Genome Mining Reveal High Potential of Biocontrol and Plant Growth Promotion in Nodule-Inhabiting Bacteria Within Paenibacillus polymyxa Complex. *Front. Microbiol.* **2021**, *11*, 618601. [CrossRef] [PubMed]
185. Kaewkla, O.; Franco, C.M.M. Genome mining and description of Streptomyces albidus sp. nov., an endophytic actinobacterium with antibacterial potential. *Antonie Van Leeuwenhoek Int. J. General Mol. Microbiol.* **2021**, *114*, 539–551. [CrossRef]
186. Tsalgatidou, P.C.; Thomloudi, E.-E.; Baira, E.; Papadimitriou, K.; Skagia, A.; Venieraki, A.; Katinakis, P. Integrated Genomic and Metabolomic Analysis Illuminates Key Secreted Metabolites Produced by the Novel Endophyte Bacillus halotolerans Cal.l.30 Involved in Diverse Biological Control Activities. *Microorganisms* **2022**, *10*, 399. [CrossRef]
187. Jasim, B.; Sreelakshmi, K.S.; Mathew, J.; Radhakrishnan, E.K. Surfactin, Iturin, and Fengycin Biosynthesis by Endophytic Bacillus sp. from Bacopa monnieri. *Microb. Ecol.* **2016**, *72*, 106–119. [CrossRef]
188. Li, J.; Qu, X.; He, X.; Duan, L.; Wu, G.; Bi, D.; Deng, Z.; Liu, W.; Ou, H.-Y. ThioFinder: A web-based tool for the identification of thiopeptide gene clusters in DNA sequences. *PLoS ONE* **2012**, *7*, e45878. [CrossRef]
189. Scheffler, R.J.; Colmer, S.; Tynan, H.; Demain, A.L.; Gullo, V.P. Antimicrobials, drug discovery, and genome mining. *Appl. Microbiol. Biotechnol.* **2013**, *97*, 969–978. [CrossRef] [PubMed]
190. Choi, S.S.; Katsuyama, Y.; Bai, L.; Deng, Z.; Ohnishi, Y.; Kim, E.S. Genome engineering for microbial natural product discovery. *Curr. Opin. Microbiol.* **2018**, *45*, 53–60. [CrossRef] [PubMed]
191. Rutledge, P.J.; Challis, G.L. Discovery of microbial natural products by activation of silent biosynthetic gene clusters. *Nat. Rev. Microbiol.* **2015**, *13*, 509–523. [CrossRef] [PubMed]
192. Olano, C.; Méndez, C.; Salas, J.A. Strategies for the design and discovery of novel antibiotics using genetic engineering and genome mining. In *Antimicrobial Compounds: Current Strategies and New Alternatives*; Springer: Berlin/Heidelberg, Germany, 2013; pp. 1–25. [CrossRef]
193. Nowak, A.; Nowak, S.; Nobis, M.; Nobis, A. A report on the conservation status of segetal weeds in Tajikistan. *Weed Res.* **2014**, *54*, 635–648. [CrossRef]
194. Siciński, J.T. Flora i roślinność (szata roślinna) synantropijna wsi oraz małych miasteczek. *Zesz. Wiej.* **2012**, *17*, 256–271.
195. Yalçinalp, E.; Meral, A. Ruderal Plants in Urban and Sub-Urban Walls and Roofs. *J. Agric. Fac. Ege Univ.* **2019**, *56*, 205–212. [CrossRef]
196. Fanfarillo, E. Segetal Plant Biodiversity in Italy: Floristic and Coenological Analyses at Different Spatial and Time Scales. Available online: https://www.researchgate.net/publication/339540030 (accessed on 14 December 2023).
197. Jackowiak, B. Man-Made Changes in the Flora and Vegetation of Poland: Current Review. *Diversity* **2023**, *15*, 618. [CrossRef]
198. Kelcey, J.G.; Müller, N. *Plants and Habitats of European Cities*; Springer: Berlin/Heidelberg, Germany, 2011; pp. 1–6851.
199. Hosseini, Z.; Zangari, G.; Carboni, M.; Caneva, G. Substrate preferences of ruderal plants in colonizing stone monuments of the pasargadae world heritage site, Iran. *Sustainability* **2021**, *13*, 9381. [CrossRef]
200. Wangchuk, P. Therapeutic Applications of Natural Products in Herbal Medicines, Biodiscovery Programs, and Biomedicine. *J. Biol. Act. Prod. Nat.* **2018**, *8*, 1426495. [CrossRef]
201. Nasim, N.; Sandeep, I.S.; Mohanty, S. Plant-derived natural products for drug discovery: Current approaches and prospects. *Nucl.* **2022**, *65*, 399. [CrossRef]
202. WHO Global Centre for Traditional Medicine. Available online: https://www.who.int/initiatives/who-global-centre-for-traditional-medicine (accessed on 11 April 2023).

203. Liu, K.; Ding, X.; Deng, B.; Chen, W. Isolation and characterization of endophytic taxol-producing fungi from Taxus chinensis. *J. Ind. Microbiol. Biotechnol.* **2009**, *36*, 1171–1177. [CrossRef] [PubMed]
204. Somjaipeng, S.; Medina, A.; Kwaśna, H.; Ortiz, J.O.; Magan, N. Isolation, identification, and ecology of growth and taxol production by an endophytic strain of Paraconiothyrium variabile from English yew trees (*Taxus baccata*). *Fungal Biol.* **2015**, *119*, 1022–1031. [CrossRef] [PubMed]
205. Drożdżyński, P. Biosurfaktanty Otrzymywane z Endofitycznych Bakterii Wyizolowanych z *Chelidonium majus* L.—Produkcja i Zastosowanie. 2021. Available online: https://politechnikalodzka.ssdip.bip.gov.pl/articles/pdf/277859 (accessed on 27 February 2024).
206. Zouari Bouassida, K.; Bardaa, S.; Khimiri, M.; Rebaii, T.; Tounsi, S.; Jlaiel, L.; Trigui, M. Exploring the *Urtica dioica* Leaves Hemostatic and Wound-Healing Potential. *Biomed. Res. Int.* **2017**, *2017*, 1047523. [CrossRef] [PubMed]
207. Naoufal, D.; Amine, H.; Ilham, B.; Khadija, O. Isolation and Biochemical Characterisation of Endophytic Bacillus spp from *Urtica dioica* and Study of Their Antagonistic Effect against Phytopathogens. *Annu. Res. Rev. Biol.* **2018**, *28*, 1–7. [CrossRef]
208. Chevallier, A. Encyclopedia of Herbal Medicine. Available online: https://www.dk.com/uk/book/9780241593370-encyclopedia-of-herbal-medicine-new-edition/ (accessed on 4 October 2023).
209. Tello, S.A.; Silva-Flores, P.; Agerer, R.; Halbwachs, H.; Beck, A.; Peršoh, D. Hygrocybe virginea is a systemic endophyte of Plantago lanceolata. *Mycol. Prog.* **2014**, *13*, 471–475. [CrossRef]
210. Bhukta, P.; Rath, D.; Pattnaik, G.; Kar, B.; Ranajit, S.K. A Comprehensive Review on Promising Phytopharmacological Applications of Chamomile Flower. *Asian J. Chem.* **2021**, *33*, 2864–2870. [CrossRef]
211. Presibella, M.M.; De Biaggi Villas-Bôas, L.; da Silva Belletti, K.M.; de Moraes Santos, C.A.; Weffort-Santos, A.M. Comparison of Chemical Constituents of *Chamomilla recutita* (L.) Rauschert Essential Oil and its Anti-Chemotactic Activity. *Braz. Arch. Biol. Technol.* **2006**, *49*, 717–724. [CrossRef]
212. Petronilho, S.; Maraschin, M.; Coimbra, M.A.; Rocha, S.M. In vitro and in vivo studies of natural products: A challenge for their valuation. The case study of chamomile (*Matricaria recutita* L.). *Ind. Crop. Prod.* **2012**, *40*, 1–12. [CrossRef]
213. Köberl, M.; Schmidt, R.; Ramadan, E.M.; Bauer, R.; Berg, G. The microbiome of medicinal plants: Diversity and importance for plant growth, quality, and health. *Front. Res. Found.* **2013**, *4*, 400. [CrossRef]
214. Schmidt, R.; Köberl, M.; Mostafa, A.; Ramadan, E.M.; Monschein, M.; Jensen, K.B.; Bauer, R.; Berg, G. Effects of bacterial inoculants on the indigenous microbiome and secondary metabolites of chamomile plants. *Front. Microbiol.* **2014**, *5*, 00064. [CrossRef]
215. Köberl, M.; White, R.A., III; Erschen, S.; Spanberger, N.; El-Arabi, T.F.; Jansson, J.K.; Berg, G. Complete Genome Sequence of Bacillus amyloliquefaciens Strain Co1-6, a Plant Growth-Promoting Rhizobacterium of Calendula officinalis. *Genome Announc.* **2015**, *3*, 862–877. [CrossRef] [PubMed]
216. Federico, A.; Dallio, M.; Loguercio, C. Silymarin/Silybin and chronic liver disease: A marriage of many years. *Molecules* **2017**, *22*, 191. [CrossRef]
217. Schmidt, A.G.H.H.-J. Silymarin as Supportive Treatment in Liver Diseases: A Narrative Review. *Adv. Ther.* **2020**, *37*, 1279–1301. [CrossRef]
218. Saller, R.; Brignoli, R.; Melzer, J.; Meier, R. An updated systematic review with meta-analysis for the clinical evidence of silymarin. *Forsch. Komplementärmedizin/Res. Complement. Med.* **2008**, *15*, 9–20. [CrossRef] [PubMed]
219. Anwar, Y.; Khan, S.U.; Ullah, I.; Hemeg, H.A.; Ashamrani, R.; Al-sulami, N.; Ghazi Alniami, E.; Hashem Alqethami, M.; Ullah, A. Molecular Identification of Endophytic Bacteria from Silybum marianum and Their Effect on *Brassica napus* Growth under Heavy Metal Stress. *Sustainability* **2023**, *15*, 3126. [CrossRef]
220. Juyal, P.; Shrivastava, V.; Mathur, A. Evaluation of Antimicrobial Potential of Endophytic Fractions of *Mentha piperita* against Pathogenic Microbes. *Int. J. Dev. Res.* **2017**, *7*, 14120–14124. Available online: http://www.journalijdr.com (accessed on 5 May 2024).
221. Shokhiddinova, M.N.; Normurodova, Q.T. Isolation of Endophytic Bacteria from Medicinal Plants and Screening for Their Enzyme Production Ability. *Appl. Microbiol. Open Access* **2024**, *9*, 1000281. [CrossRef]

molecules

MDPI

Article

Transcriptome and Metabolome Analyses Provide Insights into the Flavonoid Accumulation in Peels of *Citrus reticulata* 'Chachi'

Jianmu Su [1,†], Tianhua Peng [1,†], Mei Bai [1], Haiyi Bai [1], Huisi Li [1], Huimin Pan [1], Hanjun He [1], Huan Liu [2,*] and Hong Wu [1,*]

[1] Guangdong Laboratory for Lingnan Modern Agriculture, College of Life Sciences, South China Agricultural University, Guangzhou 510642, China
[2] State Key Laboratory of Agricultural Genomics, BGI-Shenzhen, Shenzhen 518083, China
* Correspondence: liuhuan@genomics.cn (H.L.); wh@scau.edu.cn (H.W.)
† These authors contributed equally to this work.

Abstract: The quality of Chinese medicinal materials depends on the content of bioactive components, which are affected by the environmental factors of different planting regions. In this research, integrated analysis of the transcriptome and metabolome of *C. reticulata* 'Chachi' was performed in two regions, and three orchards were included in the analysis. In total, only 192 compounds were found in fresh peels, and among 18 differentially accumulated flavonoid metabolites, 15 flavonoids were enriched in peels from the Xinhui planting region. In total, 1228 genes were up-regulated in peels from Xinhui, including the *CHS* and *GST* genes, which are involved in the salt stress response. Overall, based on the correlation analysis of flavonoid content and gene expression in peels of *C. reticulata* 'Chachi', we concluded that the authenticity of the GCRP from Xinhui may be closely related to the higher content of naringin and narirutin, and the increase in the content of these may be due to the highly saline environment of the Xinhui region.

Keywords: *Citrus reticulata* 'chachi'; transcriptome; metabolome; flavonoid

Citation: Su, J.; Peng, T.; Bai, M.; Bai, H.; Li, H.; Pan, H.; He, H.; Liu, H.; Wu, H. Transcriptome and Metabolome Analyses Provide Insights into the Flavonoid Accumulation in Peels of *Citrus reticulata* 'Chachi'. *Molecules* **2022**, *27*, 6476. https://doi.org/10.3390/molecules27196476

Academic Editor: Mohamed L. Ashour

Received: 4 September 2022
Accepted: 27 September 2022
Published: 1 October 2022

1. Introduction

Citri Reticulatae Pericarpium (CRP), the dried ripe pericarp of *Citrus reticulata* Blanco or its cultivars, is not only consumed as a dietary supplement in China and in other eastern countries, but is also among the most popular traditional medicinal herbs used in clinical practice for the treatment of indigestion and inflammatory syndromes of the respiratory tract [1,2]. As high-quality CRP raw materials, the pericarp of *C. reticulata* 'chachi' (PCRC) has been authenticated as a genuine regional drug [3,4]. *C. reticulata* 'chachi' has been cultivated in the Xinhui county of Guangdong province in China for hundreds of years. PCRC from Xinhui county has always been well regarded as the best national product in terms of geo-herbalism [5,6]. As a traditional Chinese herbal medicine, PCRC possesses various therapeutic activities, including strengthening the spleen and eliminating dampness and phlegm [7,8]. Recently, Liang et al. reported a possible effect of the abundant flavonoids in PCRC on the prevention and treatment of COVID-19 [9]. At present, annual market sales of PCRC from Xinhui have reached USD 145 million. Due to high market demand, *C. reticulata* 'chachi' has been widely introduced in Guangxi province with similar climate conditions to Guangdong. However, there are significant differences in the CRC from different regions in terms of the content of bioactive components and the pharmaceutical effect of the cultivars [10,11].

The synthesis and proper accumulation of secondary metabolites are strictly controlled in a spatial and temporal manner and are influenced by the changes in the abiotic and biotic environment [12,13]. Thus, environmental factors are crucial determinants of biosynthesis and fluctuations in plant secondary metabolites [12]. By comparing and analyzing the essential oil components of *Cinnamomum cassia* from different cultivation sites, Li et al. found that the composition of the essential oils from the *C. cassia* barks from the Guangdong and Guangxi provinces had marked differences [14].

Modern phytochemical studies have shown that the chemical constituents of PCRC are flavonoids, volatile oils, polysaccharides and alkaloids [15–17]. The flavonoid and bioactivity features of PCRC during storage showed that the flavonoid content changes dramatically [9,18]. The storage temperature and humidity also have an effect on the bioactive flavonoids and antioxidant activity in PCRC [19,20]. In recent years, some research has focused on the contents of bioactive components in cultivars from different planting areas. However, the original plants of used in those studies were also different [11,21]. Differential components may cause by cultivars and regions. However, it is unknown whether or how the flavonoids as the main bioactive components of PCRC are affected in different geographical areas when planting the same cultivar.

To compare the differences in the biosynthesis and accumulation of flavonoid content in PCRCs from the Guangdong and Guangxi planting regions, transcriptome and metabolome techniques were employed to identify genes related to their metabolic characteristics. Our study provided valuable data for the genes involved in the biosynthesis and accumulation of flavonoid content as well as quality evaluations and to maximize resource utilization in PCRC.

2. Results

2.1. Qualitative and Quantitative Analysis of Flavonoid Metabolites

The total flavonoid contents of the peels of *C. reticulata* 'Chachi' grown in three orchards were determined by spectrophotometry. The results showed that the total flavonoid contents are the difference distinguishing the three orchards. The samples from Maichong had the highest content, followed by the samples from Wubao and Daxin (Figure A1). Based on these results, the flavonoid metabolites of the citrus peels from three planting areas were investigated based on UPLC–ESI–MS/MS and databases. In the present study, 254 flavonoid metabolites were identified (Supplementary Table S1), including 110 flavonoids, 62 flavonols, 32 flavonoid C-glycosides, 21 dihydro-flavone, 10 isoflavones, 7 dihydro-flavonols, 5 tannins, 3 chalcones, 2 Proanthocyanidins, 1 flavanols, and 1 dihydro-isoflavone. Principle component analysis (PCA) uses several principal components to reveal the overall metabolic differences among the groups and the variability between intra-group samples. Two principal components, PC1 and PC2, were extracted and were 32.24% and 21.08%, respectively (Figure A2). In the PCA score plot, the Maichong planting region (MC), Wubao planting region (WB), and Daxin planting region (DX) were clearly separated, and the repeated samples were gathered together compactly.

Differential flavonoid metabolites were screened for each comparison group by combining the fold changes and variable importance in project (VIP) values of the OPLS-DA model. There were 48 significantly different flavonoid metabolites between MC and DX (29 down-regulated and 19 up-regulated) (Figure 1a), and 37 between WB and DX (28 down-regulated and 9 up-regulated) (Figure 1b). After taking an intersection of each comparison group in a Venn diagram (Figure A3), 18 common differential flavonoid metabolites were observed, 15 flavonoid metabolites, including isohemiphloin, naringin, and narirutin, were up-regulated in the MC and WB of the Xinhui genuine region, and 3 flavonoid metabolites were up-regulated in the DX of other regions (Figure 1c).

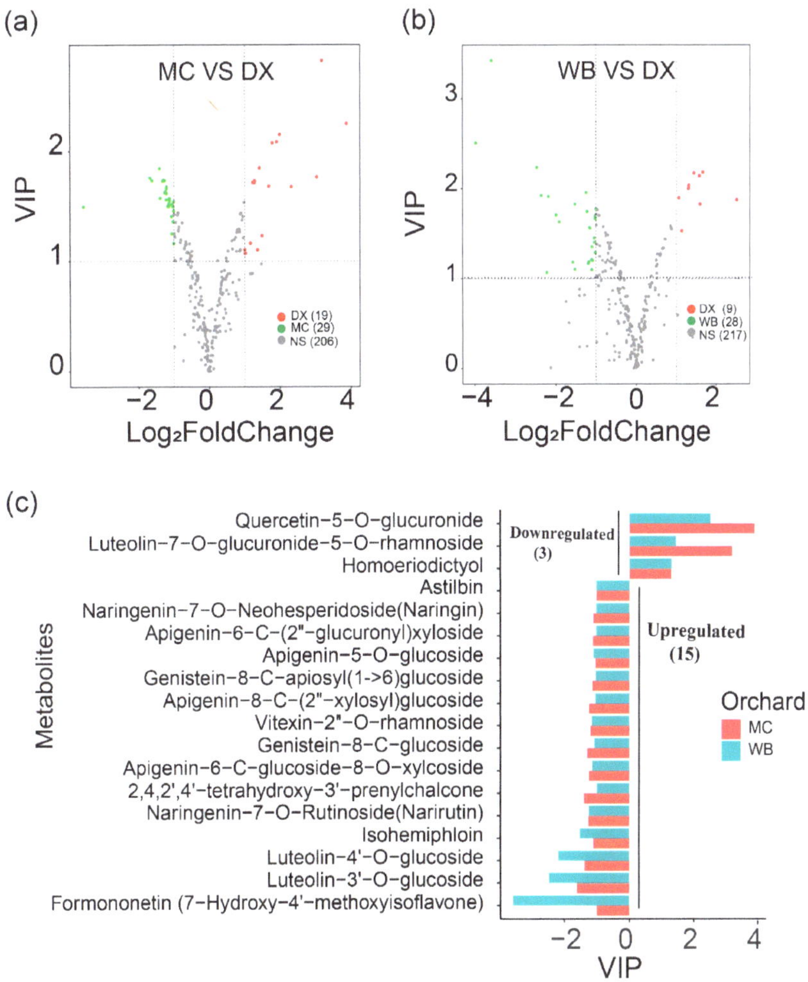

Figure 1. Differences in flavonoid content among three orchards. (**a**) Comparison of metabolite differences between Maichong (MC) orchard and Daxin (DX) orchard. The red dots represent the up-regulated metabolites in Daxin orchard, the green dots represent the up-regulated metabolites in Maichong orchard, the grey dots represent the metabolites with no significance, and the numbers in brackets are the metabolite numbers. VIP, variable importance in projection. (**b**) Comparison of metabolite differences between Wubao (WB) orchard and Daxin (DX) orchard. The abscissa is the log value of the fold change, and the ordinate is the VIP value. The metabolites with |log (fold change)| ≥ 1 indicate differential metabolites. The red dots represent the up-regulated metabolites in Daxin orchard, the green dots represent the up-regulated metabolites in Wubao orchard, the grey dots are the metabolites with no difference, and the numbers in brackets are the metabolite numbers. (**c**) The abscissa is VIP, and the ordinate is the metabolites. The red bar represents the up-regulated or down-regulated metabolites from Maichong, and blue represents the up-regulated or down-regulated metabolites from Wubao. VIP ≤ -1 indicates that the metabolite up-regulated by Maichong or Wubao, VIP ≥ 1 means the metabolite is down-regulated in Maichong or Wubao.

2.2. Differentially Expressed Genes between the Three Planting Regions

To investigate the molecular processes and genes involved in flavonoid biosynthesis in peels of *C. reticulata* 'chachi', we compared the RNA-seq data derived from citrus peels from the three planting regions. Genes with adjusted *p*-values ≤ 0.05 and absolute log2fold change ≥ 1 were considered as significantly differential expression genes (DEGs) and are shown in volcano plots (Figure 2a,b and Figure A4). Comparative transcriptome analysis revealed that the Maichong (MC) vs. Daxin (DX) comparison generated the greatest number of DEGs (2971), followed by the Wubao (WB) vs. Daxin (DX) comparison (2504), and the Maichong (MC) vs. Wubao (WB) comparison (95). More specifically, a total of 2971 genes were differentially expressed in the MC vs. DX comparison, including 1199 up-regulated genes and 1772 down-regulated genes. A total of 2504 DEGs were identified in the WB vs. DX comparison: 991 DEGs were up-regulated and 1513 DEGs were down-regulated. In the WB vs. MC comparison, there were only 96 DEGs, 51 of which were up-regulated and 45 were down-regulated (Figure A4).

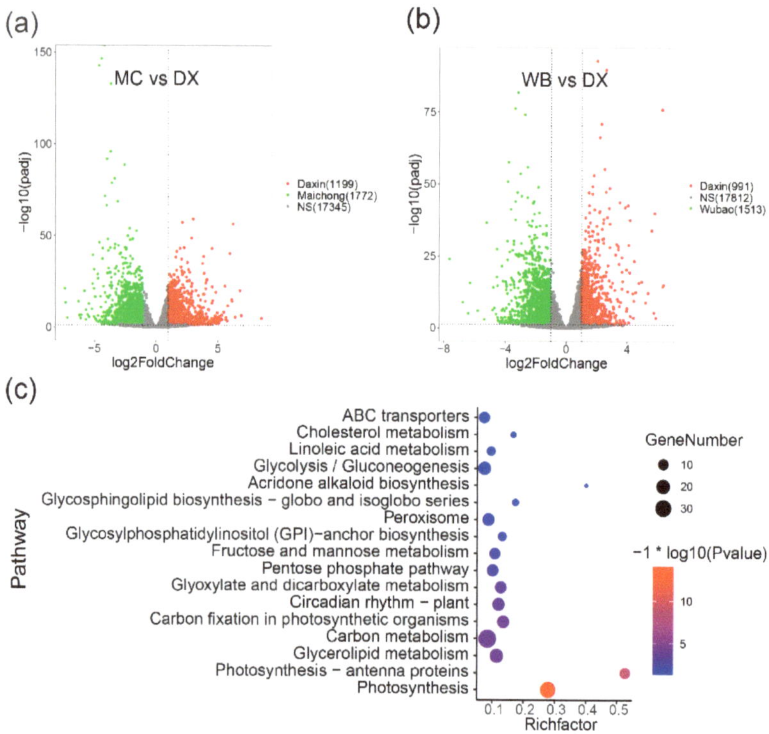

Figure 2. Volcano plots of DEGs in the pairwise comparisons and pathway enrichment. (**a**) Volcano plot of DEGs in the MC vs. DX comparison. Green dots represent genes up-regulated in Maichong. Red dots represent genes up-regulated in Daxin. Grey dots represent genes that show no significant difference between two orchards. (**b**) Volcano plot of DEGs in the WB vs. DX comparison. Green dots represent genes up-regulated in Wubao. Red dots represent genes up-regulated in Daxin. Grey dots represent genes that are not significantly different between two orchards. (**c**) Pathway enrichment of genes co-up-regulated in MC and WB. The pathway with *p* value ≤ 0.05 are shown. Node size represents the number of differential genes within the pathway and color represents the significance of the pathway. 'Gene number' is the number of genes enriched in a pathway. 'Rich factor' is the percentage of total DEGs in the given pathway term.

In order to find genes up-regulated in peels from the Xinhui genuine region, a intersect gene set was performed, and then the genes were used to perform pathway enrichment. The result showed that 1226 genes were co-up-regulated in MC and WB. The functional groups with a p value ≤ 0.05 are shown (Figure 2c), where the most enriched genes are those related to photosynthesis, photosynthesis-antenna protein, glycerolipid metabolism, carbon metabolism, carbon fixation in photosynthetic organisms, and circadian rhythm in plants. Among the 1226 differential genes, 77 genes were annotated to 28 transcription factors (TFs), which were clustered into two groups based on the Euclidean distance of gene expression (Figure A5). In the green group, we found that the expression of NAC was the highest, with an average FPKM of 377.8. This was followed by MYB, TUB, C2C2-CO-like, bHLH and Alfin-like, the FPKM values of which were 163.68, 110.73, 106.62, 75.84, and 68.65, respectively. The purple group was further divided into two subgroups, one group containing two TFs (C2CE-GATA, SBP), and the other group containing 19 TFs, such as GRAS, mTEFR, and HSF.

2.3. *WGCNA Identifies Candidate Genes Associated with Flavonoid content*

To further investigate the relationship between gene expression and metabolism, a WGCNA was performed. After clustering the samples, all of the samples were retained in the subsequent analysis (Figure 3a). Fifteen flavonoid content that were enriched in MC and WB are shown in the heatmap, and the sample clustering result showing gene expression exhibited a similar pattern with that of the flavonoid content. A soft threshold power of 9 ($\beta = 9$) was selected according to the preconditions of approximate scale-free topology (Figure A6). The analysis identified 31 distinct co-expression modules (labeled with diverse colors) which are shown in the dendrogram (Figure A7), with brown module (containing 1445 genes) being found to be positively associated with 12 flavonoids (Figure 3b), examples of which include genistein-8-C-glucoside (R = 0.87, p = 0.002), apigenin-5-O-glucoside (R = 0.86, p = 0.003), isohemiphloin (R = 0.91, $p = 1 \times 10^{-5}$, luteolin-4′-O-glucoside (R = 0.82, p = 0.007), luteolin-3′-O-glucoside (R = 0.83, p = 0.005), lpigenin-8-C-(2″-xylosyl) glucoside (R = 0.88, p = 0.002), genistein-8-C-apiosyl (1→6)glucoside (R = 0.89, p = 0.001), apigenin-6-C-glucoside-8-O-xylcoside (R = 0.86, p = 0.003), apigenin-6-C-(2″-glucuronyl) xyloside (R = 0.85, p = 0.004), vitexin-2″-O-rhamnoside (R=0.87, p = 0.003), and naringenin-7-O-rutinoside (narirutin) (R = 0.86, p = 0.003).

Considering that the brown module had a strong positive correlation with most flavonoids, we performed pathway enrichment analysis based on the genes in the module. In general, the genes in the brown module were enriched in glycolysis/Gluconeogenesis, mRNA surveillance pathway, ribosome, pyruvate metabolism, pantothenate and CoA biosynthesis (Figure 4a). Additionally, a total of 133 genes were discovered and were subsequently defined as hub genes, with a principles of adjacency threshold > 0.4. A co-expression network was constructed based on the adjacency between the hub genes in the brown module. We found that 24 genes intersected with the DEGs described above (Figure 4b). In the network, a dihydrolipoamide acetyltransferase (*DLAT*) and an unknown function gene Cs3g11540 showed a high correlation with other genes. The *DLAT* gene was found to be up-regulated in DX. Transcript factors such as MYB, C2C2-Dof, and Alfin-like were also found in the gene network. In the genes, two genes *CYP73A* and *CHS* encoding the enzymes of the phenylpropanoid biosynthesis pathway and the flavonoid biosynthesis pathway, respectively, were up-regulated in MC and WB (Figure 4c), implying that these two up-regulated genes may likely lead to higher of naringin accumulation.

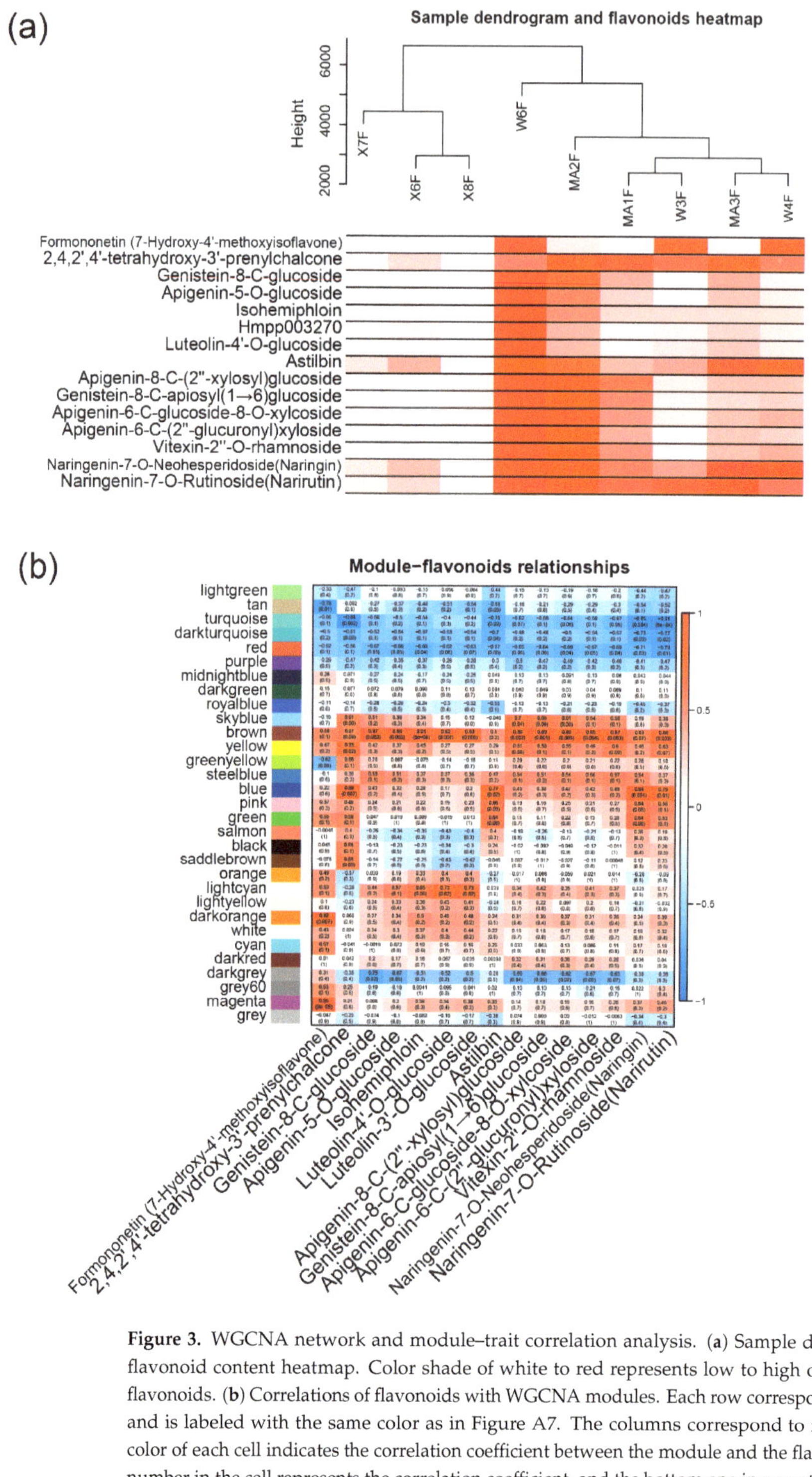

Figure 3. WGCNA network and module–trait correlation analysis. (**a**) Sample dendrogram and flavonoid content heatmap. Color shade of white to red represents low to high of the content of flavonoids. (**b**) Correlations of flavonoids with WGCNA modules. Each row corresponds to a module and is labeled with the same color as in Figure A7. The columns correspond to flavonoids. The color of each cell indicates the correlation coefficient between the module and the flavonoids (the top number in the cell represents the correlation coefficient, and the bottom one in parentheses represents the *p* value).

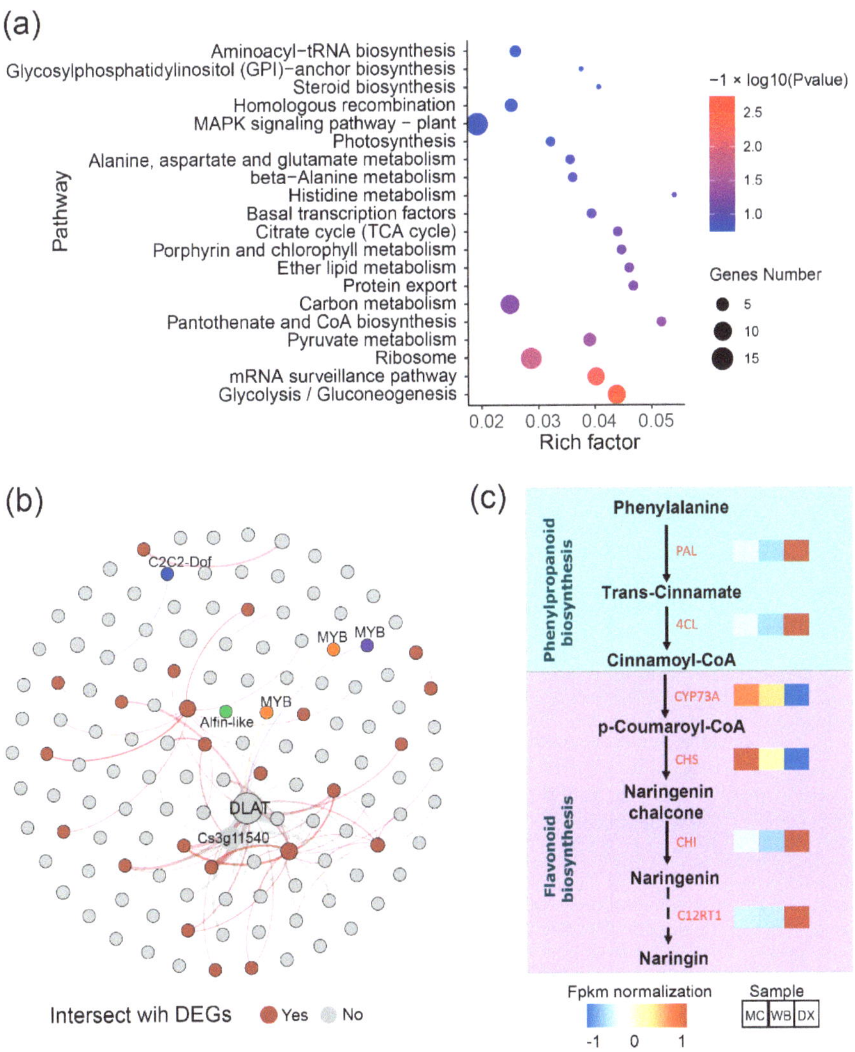

Figure 4. Co-occurrence network analysis based on the genes in module brown. (**a**) Pathway enrichment of genes in the brown module (p value $\leq$ 0.05 are shown). Node size represents the number of differential genes within the pathway and color represents the significance of the pathway. 'Gene number' is the number of genes enriched in a pathway. 'Rich factor' is the percentage of total DEGs in the given pathway term. (**b**) Co-occurrence network analysis based on hub genes with the principles of adjacency threshold > 0.4. Nodes that are red represent genes intersecting with DEGs between MC and WB vs. DX. Genes annotated to TFs are marked. Node size represents the degrees of the genes in the network. (**c**) Gene expression in the biosynthesis of naringin. FPKM was normalized and is shown in color. *PAL*, phenylalanine ammonia-lyase. *4CL*, 4-coumarate-CoA ligase. *CYP73A*, trans-cinnamate 4-monooxygenase. *CHS*, chalcone synthase. CHI, chalcone isomerase. *C12RT1*, flavanone 7-O-glucoside 2″-O-beta-L-rhamnosyltransferase.

The blue module (containing 3156 genes) was positively associated with 2,4,2′,4′-tetrahydroxy-3′-prenylchalcone (R = 0.88, p = 0.002), astilbin (R = 0.77, p = 0.02), naringenin-7-O-neohesperidoside (naringin) (R = 0.84, p = 0.004) and naringenin-7-O-rutinoside (narirutin) (R = 0.79, p = 0.01). Considering that naringin and its aglycone naringenin belong to this series of flavonoids and display strong anti-inflammatory and antioxidant activities. We focused on the correlations of naringin with gene modules and identified hub genes that were strongly correlated with naringin. Genes expressed in 80% of the samples and FPKM values > 1 were retained for WGCNA unsigned co-expression network analysis. In total, 108 hub genes were obtained by screening genes correlated with naringin ($|GTS| > 0.9$ and $|GMM| > 0.9$) in the blue module (Figure 5a). There were 23 hub genes that were enriched in DX and 85 hub genes that were enriched in MC and WB (Figure 5b). A co-expression network was constructed based on the adjacency > 0.4 between hub genes in the blue module. Twelve genes showed high correlation with other genes (Figure 5c), in which three genes, Cs8g20220 (protein glucosyltransferase), Cs2g14860 (unknown function) and Cs6g01210 (K01648-ATP citrate (pro-S)-lyase), were enriched in Daxin, and nine genes, Cs7g27960 (glutathione S-transferase), Cs6g02730 (ribD, diaminohydroxyphosphoribosylamino-pyrimidine deaminase), Cs8g10800 (unknown function), Cs8g01700 (unknown function), Cs7g31190 (light-harvesting complex II chlorophyll a/b binding protein 4), orange1.1t02492 (unknown function), Cs7g31690 (unknown function) and Cs7g03460 (ribulose-bisphosphate carboxylase/fructose-bisphosphate aldolase-lysine N-methyltransferase), were enriched in MC and WB (Figure 5b). Two MYB genes, three bHLH genes and a GRAS gene showed a high correlation with these twelve genes. Interestingly, among the 12 hub genes, we found a gene that encodes glutathione S-transferase, which is widely found in citrus. In total, there were 40 glutathione S-transferase genes that were found in our study (Figure A8), in which 23 genes were up-regulated in MC and WB, and 17 genes were up-regulated in DX.

In order to explore the abiotic stress factors that cause the differences in the flavonoid content in peels of *C. reticulata* 'Chachi' from different regions, the salinity content of the rhizosphere soil in which *C. reticulata* 'Chachi' is cultivated in the three regions was measured. The results show that the soil salinity of WB was higher than that of DX, and the differences between WB and DX were significant (Figure A9, t test p = 0.050). However, the differences between the soil salinity of WB and MC were not significant (t test p = 0.379). The soil salinity of MC was a little higher than that of DX, but the differences were not significant.

2.4. Salt Stress Plays an Important Role in the Accumulation of Flavonoids

To further verify whether salt stress may promote the biosynthesis of flavonoids in plants, a tobacco salt stress treatment was performed (Figure 6a). As a result, the fresh weight of the aerial part of tobacco under salt stress was significantly higher than that of the control (Figure 6b, t test $p = 7.12 \times 10^{-5}$), while the height of tobacco plant was not significant between control and treatment (Figure 6c, t test p = 0.381). Especially, the flavonoid contents in leaves of tobacco under salt stress were higher than in control (Figure 6d, t test p = 0.042). These results further verified our hypothesis that salt stress may increase the content of total flavonoids in plants.

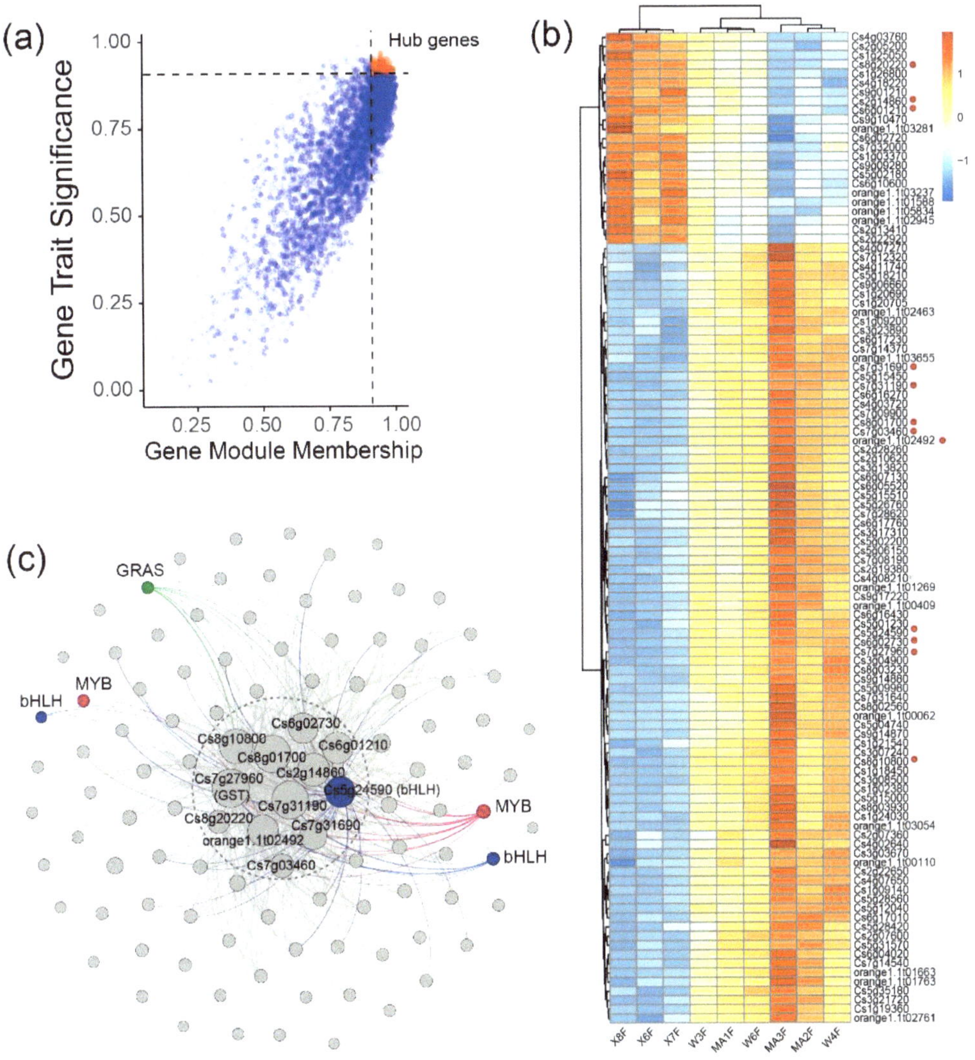

Figure 5. Identification of hub genes in blue modules. (**a**) The correlation of gene module membership and gene trait significance. Hub genes were obtained by screening genes which correlate with naringin (|GTS| > 0.9 and |GMM| > 0.9) in the blue module. Red dots represent hub genes. (**b**) FPKM values of 108 hub genes in the heatmap. FPKM was centered and scaled in the row. (**c**) Co-occurrence network analysis based on 108 hub genes with the principles of adjacency threshold > 0.4. Genes annotated to TFs were marked. Node size represents the degrees of the genes in the network.

Figure 6. Plant height, fresh weight and total flavonoid content of tobacco under salt stress. (**a**) The phenotypic changes in tobacco plant after 14 days of salt stress. (**b**) The fresh weight of the aerial part of tobacco. (**c**) The height of the tobacco. (**d**) Total flavonoid content of tobacco.

2.5. Verification of Gene Expression Patterns by qRT-PCR

In order to verify the accuracy of the RNA-seq results, we selected six genes that were involved in biosynthesis of naringin for qRT-PCR analysis. The results showed that the expression pattern of these six differential genes was similar to that of RNA-seq (Figure 7). The results indicated that the RNA-seq was reliable.

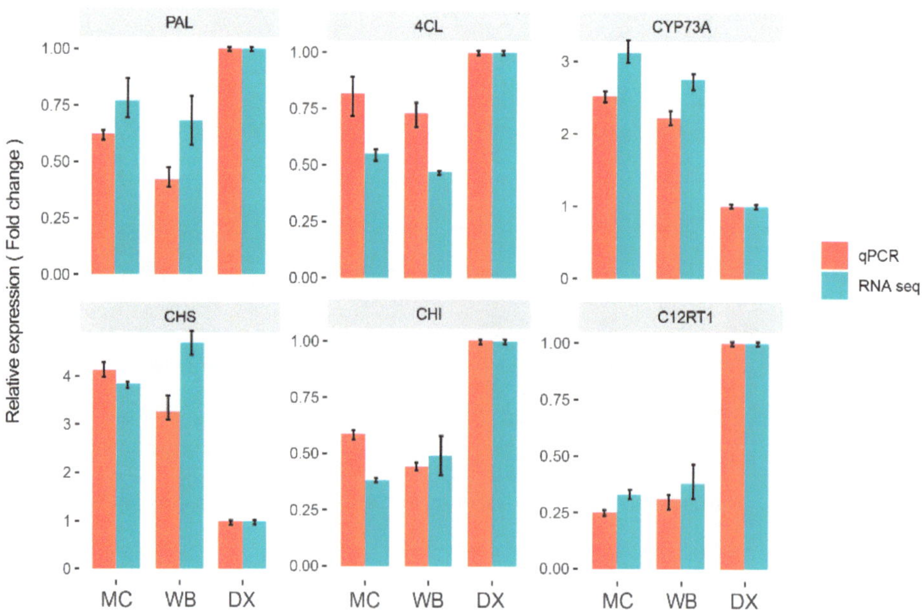

Figure 7. qRT-PCR validation profiles of six selected genes. The expression levels of genes in DX were used as the reference state, which was set to 1, and fold change values are shown here.

3. Discussion

PCRC produce flavonoids as secondary metabolites, which are extensively distributed and play vital roles in plant physiology, as well as exhibit high bioactivity for clinical applications [22,23]. In the present study, the differences in the transcripts and metabolites in the samples from three different orchards, MC, WB and DX were compared. UPLC–MS/MS and untargeted metabolomics analysis were employed to intensively investigate the composition of the flavonoids in PCRC grown in different planting regions. Up to 254 metabolites of flavonoids were detected in all samples, including 110 flavonoids, 62 flavonols, 32 flavonoid C-glycosides, 21 dihydroflavone, 10 isoflavones, 7 dihydroflavonol, 5 tannins, 3 chalcones, 2 proanthocyanidins, 1 flavanol, and 1 dihydroisoflavone. Based on the comparison of the metabolites from three orchards, 18 common differential flavonoid metabolites were observed, 15 flavonoid metabolites were from the Xinhui genuine region, and three flavonoid metabolites were up-regulated in DX. These results imply that peels from Xinhui contain a higher flavonoid content.

Transcriptome analysis results revealed 1226 differentially expressed genes in MC and WB. These differentially expressed genes were assigned to the photosynthesis, photosynthesis-antenna proteins, glycolipid metabolism, and carbon metabolism pathways. Flavonoid biosynthesis is light-dependent, and higher light intensities stimulate the synthesis of phenols and flavonoids to protect living plants [24,25]. Greater exposure to sunlight increases flavonoid contents and influences the color and nutritional quality of the fruit [26]. The pathway enrichment of differentially expressed genes suggests that the accumulation of flavonoids in citrus peels may be likely related to the amount of light received by the plant.

Using WGCNA analysis, we identified 31 distinct co-expression modules, and the brown module was found to be positively associated with 12 flavonoids (Figure 3b). The genes in the brown module were assigned to the glycolysis-gluconeogenesis mRNA surveillance pathway and to ribosome (Figure 4a). *DLAT* is an enzyme component of the multienzyme pyruvate dehydrogenase complex that is responsible for the pyruvate decarboxylation step that links glycolysis to the citric acid cycle, and it is closely related to flavonoid biosynthesis [27]. Additionally, the repression of dihydrolipoamide acetyltransferase (DLAT) led to improvements in the concentration of naringenin [28], which is a substrate of many flavonoids [29]. A co-expression network showed a high correlation between DLAT and Cs3g11540 (gene with an unknown function) with other genes. Transcription factors such as MYB, C2C2-Dof, and Alfin-like were also found in the hub gene network, suggesting that these TFs are an important regulator of flavonoid biosynthesis, and play a crucial role in flavonoid synthesis in citrus [30,31].

Chalcone synthase (*CHS*) catalyzes the first and most important step of in flavonoid biosynthesis by directing carbon flux from general phenylpropanoid metabolism to the flavonoid pathway [32]. The overexpression of EaCHS1 increases the production of downstream flavonoids and the expressions of related genes in the phenylpropanoid pathway [33,34]. *CHS* was discovered to be up-regulated in the MC and WB of Xinhui, Guangdong; this is thought to increase the accumulation of flavonoids in the PCRC from Xinhui. Out of 40 glutathione S-transferase (*GST*) genes, 23 genes were up-regulated in MC and WB, and 17 genes signaling were up-regulated in DX. It has been reported that *GST* is involved in *CHS* signaling in the plant cell and plays an important role in tolerance to salt stress [35–37]. The up-regulation of *CHS* found in MC and WB implies that the *GST* genes were in high correlation with *CHS*. Salt stress can increase the accumulation of antioxidant flavonoids in the leaves of capsicum and cyanobacteria [38,39]. In our result, we found that the soil salinity of WB and MC was higher than that of DX. The salt stress of tobacco showed that the total flavonoids of treatment were higher than control. These results suggest that the difference in flavonoid content in the peels of *C. reticulata* 'Chachi' between different regions may be caused by salt stress.

In summary, the peels from Xinhui contain a higher content of flavonoids, which may be caused by salt stress induced by the soil conditions of the Xinhui region. Firstly, salt stress may repress the *DLAT* expression level, which may lead to improvements in the concentration of naringenin. Some transcription factors such as MYB, C2C2-Dof, and Alfin-like likely play an important role during this process. Secondly, *GST* plays an important role in this salt stress process. Salt stress promotes the up-regulation of *CHS*, thereby further promoting the biosynthesis of naringenin.

4. Materials and Methods

4.1. Plant Materials

Two representative orchards (Maichong: 22°24′ N, 113°1′ E and Wubao: 22°28′ N, 112°57′ E) with prominent *C. reticulata* 'Chachi' production in Xinhui county (geo-authentic regional product), Guangdong province, and one orchard (Daxin: 23°17′ N, 117°27′ E) in Guangxi province were selected. For each orchard, the peels of three citrus trees were collected for transcriptome sequencing and untargeted metabolomic profiling. Peels were frozen in liquid nitrogen immediately and were stored at −80 °C until further use.

4.2. RNA Isolation and Transcriptome Sequencing

RNA from the peels was extracted using a Column Plant RNAout Kit according to the instructions (TIANDZ, Beijing, China). RNA extraction was quantified and assessed for integrity using NanoDrop (Thermo, Massachusetts, USA) and a 2100 Agilent Bioanalyzer (Agilent, California, USA) prior to subsequent experiments. To construct the BGI-based mRNA-seq library, 1 ug qualified RNA from each sample was used. Then, the library was sequenced on the BGISEQ platform using a 100 bp pair-ended strategy.

The raw reads were filtered by SOAPfilter [40], and an average of 4 Gb high-quality data were generated for each sample. Then, the resulting high-quality reads from each sample were mapped against the citrus sinensis reference genome [41] using hisat2 [42]. The expression level of each gene in each sample was calculated as the FPKM with StringTie2 [43]. The differential expression genes were identified using DEseq2 [44] based on the read's matrix, with an FDR p value correction value < 0.05 and absolute fold change value > 2. The differential genes were enriched into pathways using Fisher's exact test.

4.3. Identification of Co-Expression Modules and Hub Genes

A gene co-expression network was built using the WGCNA package (Version 4.0.2) [45] in R (Alcatel-Lucent Bell, Labs New Jersey, USA). Genes expressed in 80% of samples and with FPKM values > 1 were retained for WGCNA unsigned co-expression network analysis. Parameters were set up as power = 9, minModuleSize = 30, and MEDissThreshold = 0.25. A total of 31 different modules were derived. To construct co-expression genes and metabolites networks, genes in the modules correlated with naringin were used. Hub genes were obtained by screening genes correlate with metabolites (|GTS| > 0.9 and |GMM| > 0.9). The hub gene co-expression network and the resulting gene–gene interactions were used to visualize the subnetworks using Gephi (Version 0.92) [46].

4.4. Metabolic Profiling

The freeze-dried peels were crushed using a mixer mill (MM 400, Retsch, Haan, Germany) with a zirconia bead for 1.5 min at 30 Hz. Next, 100 mg powder was weighted and extracted overnight at 4 °C with 1.0 mL of 70% aqueous methanol. Following centrifugation at 10,000× g for 10 min, the extracts were absorbed and filtrated before LC–MS analysis.

The sample extracts were analyzed using an LC–ESI–MS/MS system (HPLC, Shim-pack UFLC SHIMADZU CBM30A system, www.shimadzu.com.cn/, accessed on 20 June 2021; MS, Applied Biosystems 4500 Q TRAP, www.appliedbiosystems.com.cn/, accessed on 20 June 2021). LIT and triple quadrupole (QQQ) scans were acquired on a triple quadrupole-linear ion trap mass spectrometer (Q TRAP), API 4500 Q TRAP LC/MS/MS System, equipped with an ESI Turbo Ion-Spray interface, operating in a positive ion mode.

4.5. Tobacco Salt Stress Treatment

Seeds of tobacco cultivars, Nicotiana tabacum, were planted into pots filled with soils. Tobacco plants were transplanted to a single pot after three weeks growing. Three days later, 0 mM/L (control, watering) and 50 mM/L NaCl (salt stress) were applied to each pot for 14 days. Each group has five independent plants as replicates. Total flavonoids were extracted via the approach descript above. The content of total flavonoids in the extract of tobacco leaves at the wavelength of 510 nm by UV and rutin as the standard.

4.6. Validation of the RNA Sequence Data Using qRT-PCR

The RNA was extracted as stated before. The RNA extraction was quantified and assessed for integrity using the NanoDrop (Thermo, Massachusetts, USA), and was reverse transcribed into cDNA using the PrimeScript RT reagent Kit (TaKaRa, Dalian, China). The expression levels of the six genes involved in the biosynthesis of naringin were measured by quantitative real-time PCR using SYBR Green Supermix on the CFX96 Real-Time PCR Detection System (BIO-RAD, California, USA) according to the manufacturer's protocol. Each sample had three biological replicates with three technical replicates for each biological replicate. The relative expression level was calculated by the equation ratio $2^{-\Delta\Delta Ct}$. The primers of these genes were designed using primer 6 software (PREMIER Biosoft International, San Francisco, USA) and primers can be seen in Table 1.

Table 1. The primer of six genes.

Gene	Former Primer	Reverse Primer
PAL	5′-ACGTTGAAGGGAGCCATTT-3′	5′-ACGTTGAAGGGAGCCATTT-3′
4CL	5′-TTCCGATCAAAGTTGCCCGA-3′	5′-CGGCAGCGAATTTACGTGAG-3′
CYP73A	5′-AGGGAGTGGAGTTCGGATCA-3′	5′-CGTGAAGAAAGGGACGGTCA-3′
CHS	5′-GGCCTCAACCCATCTGTCAA-3′	5′-AGCATACGACTAGAACGCGG-3′
CHI	5′-ATGCAAGTCCCAGTGCATCA-3′	5′-TTGGGCAACGGATTGCAAAG-3′
C12RT1	5′-GAGAGAGGTGGCCGACATTA-3′	5′-TCTTCGCACACTTGGATGCC-3′

4.7. Data Analysis

Unsupervised PCA (principal component analysis) was performed by the statistics function prcomp within R (www.r-project.org). The data were unit variance scaled before unsupervised PCA. Significantly regulated metabolites between groups were determined by VIP $\geq$ 1 and absolute log2FC (fold change) $\geq$ 1. VIP values were extracted from the OPLS-DA results, which also contained score plots and permutation plots, and were generated using the R package MetaboAnalystR [47]. The statistical analysis was carried out using Student's unpaired t-test to compare the total content of flavonoids of tobacco leaves. Effects are considered significant at $p < 0.05$.

5. Conclusions

We performed an integrated transcriptome and metabolome analysis on *C. reticulata* 'Chachi' peels collected from three orchards in the Guangdong and Guangxi provinces. A regulatory network of metabolites and transcripts involved in flavonoid biosynthesis and salt stress was revealed. We identified some candidate genes involved in flavonoids accumulation, including two MYB genes and three bHLH genes that are most likely involved in the regulation of flavonoids biosynthesis, and one glutathione S-transferase gene involved in salt stress that is probably promoted for flavonoids accumulations. Taken together, we concluded that PCRC from Xinhui region has high contents of naringin and narirutin, which may be due to the high-salt environment in Xinhui region. The findings provide an avenue for further investigation into the biosynthesis and genetic regulation of citrus flavonoid metabolism and may aid in the selection of germplasms with a richflavonoid content.

Supplementary Materials: The following supporting information can be downloaded at https://www.mdpi.com/article/10.3390/molecules27196476/s1: Supplementary Table S1. The metabolites detected in this study.

Author Contributions: Conceptualization, H.W.; data curation, J.S.; formal analysis, T.P.; investigation, J.S., T.P., M.B., H.B., H.L. (Huisi Li) and H.P.; methodology, J.S., T.P., H.B. and H.H.; project administration, H.L. (Huan Liu) and H.W.; resources, H.L. (Huan Liu), H.H. and H.W.; supervision, H.L. (Huan Liu); validation, M.B., H.B. and H.P.; visualization, J.S. and T.P.; writing—original draft, J.S.; writing—review and editing, H.W. All authors have read and agreed to the published version of the manuscript.

Funding: This research was funded by Key realm R&D Program of Guangdong Province (2020B020221001), the Laboratory of Lingnan Modern Agriculture Project (NZ2021024) and the Guangdong Provincial Special Fund for Modern Agriculture Industry Technology Innovation Teams (2019KJ125). The open competition program of top ten critical priorities of Agricultural Science and Technology Innovation for the 14th Five-Year Plan of Guangdong Province (2022SDZG07).

Institutional Review Board Statement: Not applicable.

Informed Consent Statement: Not applicable.

Data Availability Statement: The data presented in this study are available within this article and its Appendix A and Supplementary Materials.

Acknowledgments: We thank Sahu Sunil Kumar for revising the manuscript.

Conflicts of Interest: The authors declare no conflict of interest.

Appendix A

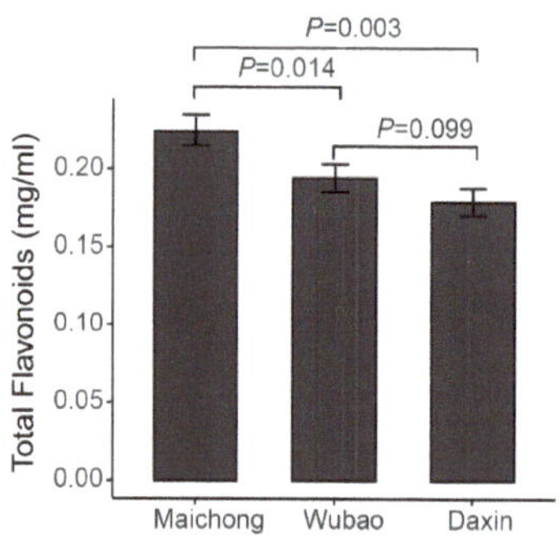

Figure A1. Total flavonoid content of *Citrus reticulata* 'Chachi'.

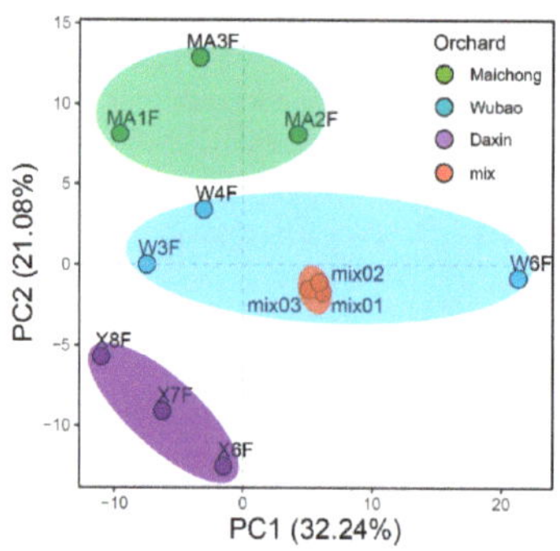

Figure A2. PCA of flavonoid metabolites of peels samples from three orchards. Horizontal ordinate shows PC1 value, and vertical coordinate shows PC2 value. Mix represents sample mix.

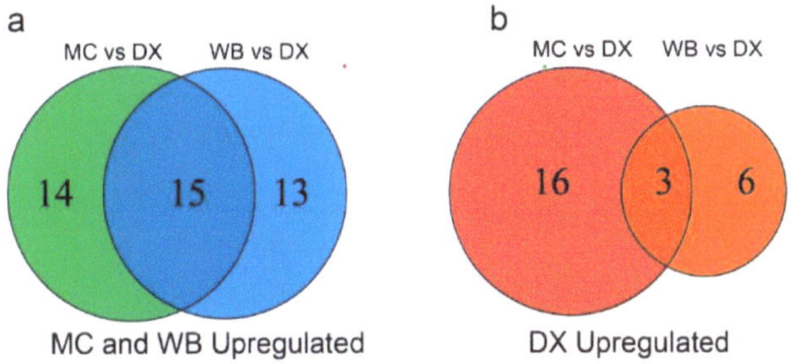

Figure A3. Numbers of up-regulated metabolites in MC, WB and DX. (**a**) The metabolites up-regulated by Maichong and Wubao. The circle on the left represents the number of metabolites up-regulated by Maichong, the circle on the right represents the number of metabolites up-regulated by Wubao, and the intersection is the number of metabolites up-regulated by the two orchards. (**b**) The number of metabolites up-regulated by Daxin. The circle on the left represents the number of metabolites up-regulated in Daxin when Maichong is compared to Daxin. The circle on the right represents the number of metabolites up-regulated in Daxin when Wubao is compared to Daxin. The intersection is the number of metabolites up-regulated in Daxin.

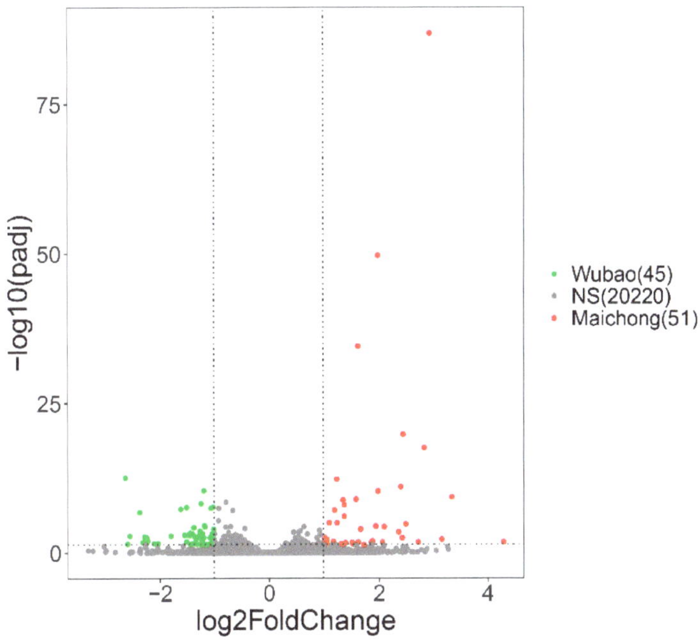

Figure A4. Volcano plot of DEGs in the comparisons of Wubao and Maichong. Green dots represent genes up-regulated in Wubao. Red dots represent genes up-regulated in Maichong. Grey dots represent genes that show no significant difference between two orchards.

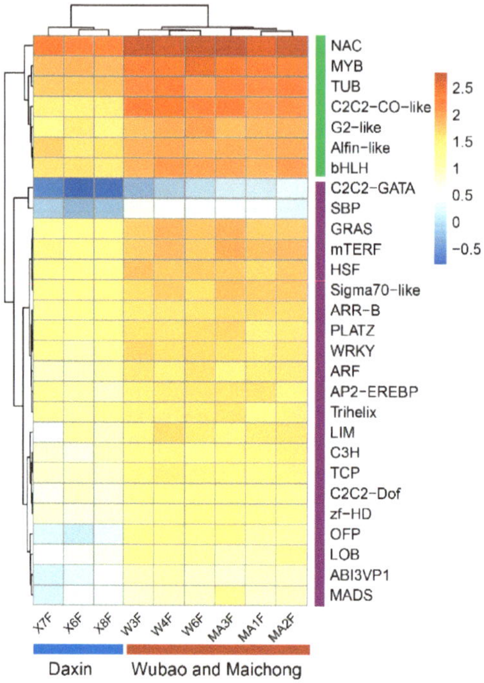

Figure A5. Heatmap base on the expression of TFs.

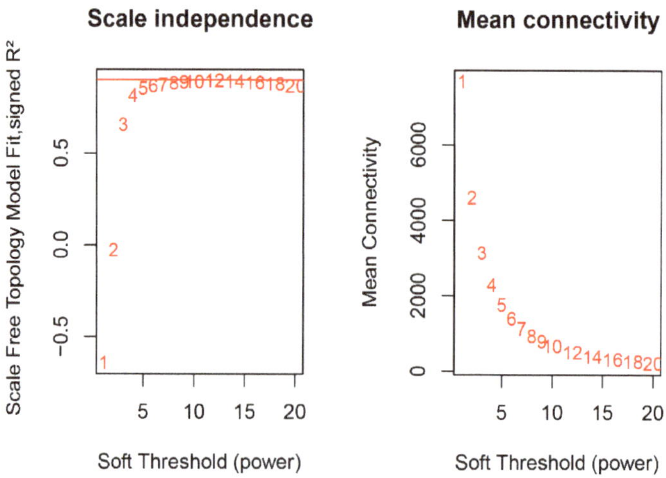

Figure A6. Approximate scale-free topology.

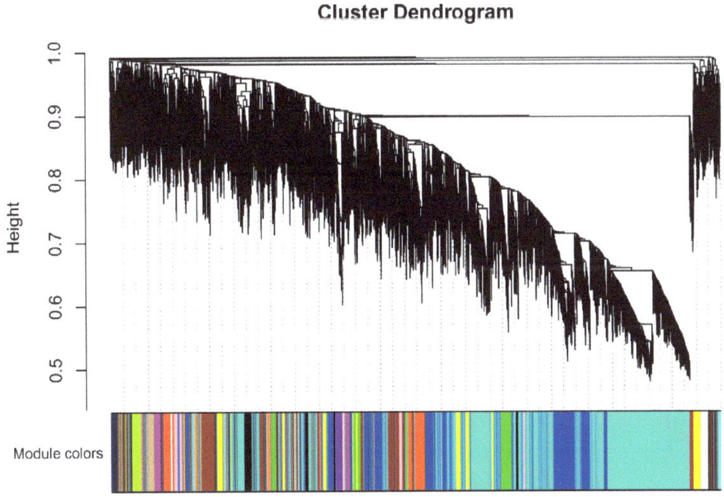

Figure A7. Co-expression modules cluster dendrogram.

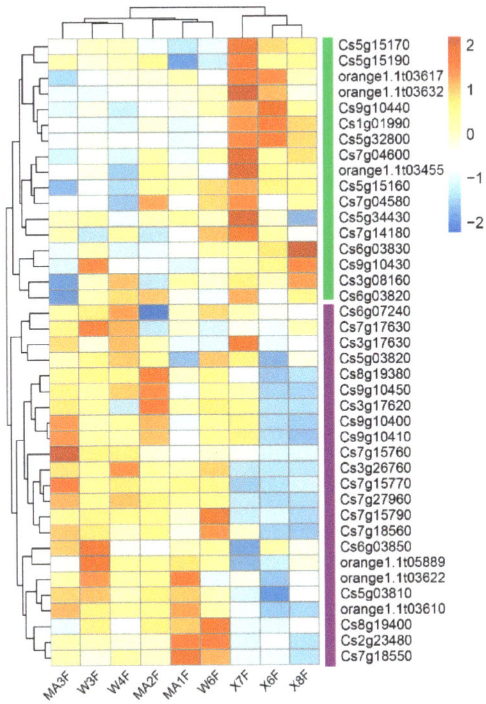

Figure A8. Heatmap base on genes expression of glutathione S-transferase genes.

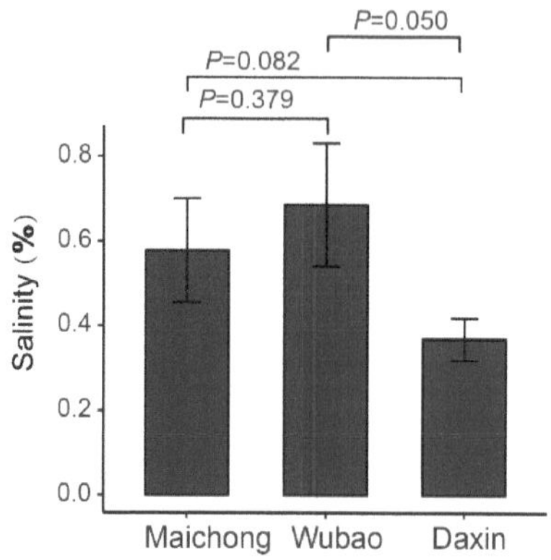

Figure A9. Soil salinity of three orchards.

References

1. Liu, E.H.; Zhao, P.; Duan, L.; Zheng, G.D.; Guo, L.; Yang, H.; Li, P. Simultaneous determination of six bioactive flavonoids in Citri Reticulatae Pericarpium by rapid resolution liquid chromatography coupled with triple quadrupole electrospray tandem mass spectrometry. *Food Chem.* **2013**, *141*, 3977–3983. [CrossRef] [PubMed]
2. National Pharmacopoeia Commission. *Pharmacopoeia of the People's Republic of China, One Sections*; China Medical Science Press: Beijing, China, 2020; pp. 197–200.
3. Zheng, Y.; Zeng, X.; Peng, W.; Wu, Z.; Su, W. Study on the Discrimination between Citri Reticulatae Pericarpium Varieties Based on HS-SPME-GC-MS Combined with Multivariate Statistical Analyses. *Molecules* **2018**, *23*, 1235. [CrossRef] [PubMed]
4. Wang, P.; Zhang, J.; Zhang, Y.; Su, H.; Qiu, X.; Gong, L.; Huang, J.; Bai, J.; Huang, Z.; Xu, W. Chemical and genetic discrimination of commercial Guangchenpi (*Citrus reticulata* 'Chachi') by using UPLC-QTOF-MS/MS based metabolomics and DNA barcoding approaches. *Rsc Adv.* **2019**, *9*, 23373–23381. [CrossRef] [PubMed]
5. Zheng, Y.Y.; Zeng, X.; Peng, W.; Wu, Z.; Su, W.W. Characterisation and classification of Citri Reticulatae Pericarpium varieties based on UHPLC-Q-TOF-MS/MS combined with multivariate statistical analyses. *Phytochem. Anal.* **2019**, *30*, 278–291. [CrossRef]
6. Yu, X.; Sun, S.; Guo, Y.; Liu, Y.; Yang, D.; Li, G.; Lü, S. Citri Reticulatae Pericarpium (Chenpi): Botany, ethnopharmacology, phytochemistry, and pharmacology of a frequently used traditional Chinese medicine. *J. Ethnopharmacol.* **2018**, *220*, 265–282. [CrossRef]
7. Qian, Z.M.; Huang, Q.; Feng, X.L.; Li, D.Q. Comparison of the chemical-antioxidant profiles of different parts of Citrus reticulata Blanco (Rutaceae) based on OLE-HPLC-DAD-MS/MS-ABTS assay. *J. Food Meas. Charact.* **2021**, *15*, 3873–3883. [CrossRef]
8. Ho, S.C.; Kuo, C.T. Hesperidin, nobiletin, and tangeretin are collectively responsible for the anti-neuroinflammatory capacity of tangerine peel (Citri reticulatae pericarpium). *Food Chem Toxicol.* **2014**, *71*, 176–182. [CrossRef]
9. Liang, S.; Wen, Z.; Tang, T.; Liu, Y.; Dang, F.; Xie, T.; Wu, H. Study on flavonoid and bioactivity features of the pericarp of Citri Reticulatae 'chachi' during storage. *Arab. J. Chem.* **2022**, *15*, 1–13. [CrossRef]
10. Liu, M.S.; Wang, J.; Deng, H.D.; Geng, L.L. Comparative study on main components and detection methods of Pericarpium Citri Reticulatae from different habitats. *IOP Conf. Ser. Earth Environ. Sci.* **2021**, *705*, 1–8. [CrossRef]
11. Luo, M.; Luo, H.; Hu, P.; Yang, Y.; Wu, B.; Zheng, G. Evaluation of chemical components in Citri Reticulatae Pericarpium of different cultivars collected from different regions by GC-MS and HPLC. *Food Sci. Nutr.* **2018**, *6*, 400–416. [CrossRef]
12. Li, Y.; Kong, D.; Fu, Y.; Sussman, M.R.; Wu, H. The effect of developmental and environmental factors on secondary metabolites in medicinal plants. *Plant Physiol. Bioch.* **2020**, *148*, 80–89. [CrossRef] [PubMed]
13. Verma, N.; Shukla, S. Impact of various factors responsible for fluctuation in plant secondary metabolites. *J. Appl. Res. Med. Aromat. Plants* **2015**, *2*, 105–113. [CrossRef]
14. Li, Y.Q.; Kong, D.X.; Huang, R.S.; Liang, H.L.; Xu, C.G.; Wu, H. Variations in essential oil yields and compositions of Cinnamomum cassia leaves at different developmental stages. *Ind. Crops Prod.* **2013**, *47*, 92–101. [CrossRef]
15. Luo, Y.; Zeng, W.; Huang, K.E.; Li, D.X.; Chen, W.; Yu, X.Q.; Ke, X.H. Discrimination of Citrus reticulata Blanco and *Citrus reticulata* 'Chachi' as well as the *Citrus reticulata* 'Chachi' within different storage years using ultra high performance liquid chromatography quadrupole/time-of-flight mass spectrometry based metabolomics approach. *J. Pharm. Biomed. Anal.* **2019**, *171*, 218–231.
16. He, X.; Huang, S.; Wu, M.; Wu, M.; Zhang, Y.; Ma, Z.; Liu, L.; Cao, H. Simultaneous quantitative analysis of ten bioactive flavonoids in Citri Reticulatae Pericarpium Viride (Qing Pi) by ultrahigh-performance liquid chromatography and high-resolution mass spectrometry combined with chemometric methods. *Phytochem. Anal.* **2021**, *32*, 1152–1161. [CrossRef]

17. Zheng, G.D.; Chao, Y.; Liu, M.; Yang, Y.; Zhang, D.; Wang, K.; Tao, Y.; Zhang, J.; Li, Y.; Wei, M. Evaluation of dynamic changes in the bioactive components in Citri Reticulatae Pericarpium (*Citrus reticulata* 'Chachi') under different harvesting and drying conditions. *J. Sci. Food Agric.* **2021**, *101*, 3280–3289. [CrossRef]
18. Wang, G.; Chen, S.; Gu, G.; Qiu, J.; Chen, Y.; Jia, Z.; Tang, H. Comparison of the Content of Flavonoids, Total Phenols, and Carotenoids and Antioxidant Activity in Guang Citri reticulatae Pericarpium during the Aging Time. *Pharmacogn. Mag.* **2020**, *16*, 375–381.
19. Fu, M.; An, K.; Xu, Y.; Chen, Y.; Wu, J.; Yu, Y.; Zou, B.; Xiao, G.; Ti, H. Effects of different temperature and humidity on bioactive flavonoids and antioxidant activity in Pericarpium Citri Reticulata (*Citrus reticulata* 'Chachi'). *LWT Food Sci. Technol.* **2018**, *93*, 167–173. [CrossRef]
20. Zhu, Y.; Zhu, Y.; Liu, X.; Qi, J. Change in nutrients of Citri Reticulatae 'Chachi' in relation to browning during aging. *J. Food Compos. Anal.* **2022**, *107*, 1–7. [CrossRef]
21. Zheng, G.; Chao, Y.; Luo, M.; Xie, B.; Zhang, D.; Hu, P.; Yang, X.; Yang, D.; Wei, M. Construction and Chemical Profile on "Activity Fingerprint" of Citri Reticulatae Pericarpium from Different Cultivars Based on HPLC-UV, LC/MS-IT-TOF, and Principal Component Analysis. *Evid.Based Complement. Altern. Med.* **2020**, *2020*, 4736152. [CrossRef]
22. Zeng, S.L.; Li, S.Z.; Xiao, P.T.; Cai, Y.Y.; Chu, C.; Chen, B.Z.; Li, P.; Li, J.; Liu, E.H. Citrus polymethoxyflavones attenuate metabolic syndrome by regulating gut microbiome and amino acid metabolism. *Sci. Adv.* **2020**, *6*, eaax6208. [CrossRef] [PubMed]
23. Lim, H.; Yeo, E.; Song, E.; Chang, Y.-H.; Han, B.-K.; Choi, H.-J.; Hwang, J. Bioconversion of Citrus unshiupeel extracts with cytolase suppresses adipogenic activity in 3T3-L1 cells. *Nutr. Res. Pract.* **2015**, *9*, 599–605. [CrossRef] [PubMed]
24. Liu, Y.; Fang, S.; Yang, W.; Shang, X.; Fu, X. Light quality affects flavonoid production and related gene expression in Cyclocarya paliurus. *J. Photochem. Photobiol. B* **2018**, *179*, 66–73. [CrossRef] [PubMed]
25. Genzeng, R.; Jianghui, C.; Xiaodong, X.; Yukun, G.; Shilin, W.; Jinhua, C.; Hao, X. Flavonoid Biosynthesis Pathway Participating in Salt Resistance in a Landrace Sweet Sorghum Revealed by RNA-Sequencing Comparison With Grain Sorghum. *J. Agric. Sci.* **2019**, *11*, 63–76. [CrossRef]
26. Zoratti, L.; Karppinen, K.; Luengo Escobar, A.; Häggman, H.; Jaakola, L. Light-controlled flavonoid biosynthesis in fruits. *Front. Plant Sci.* **2014**, *5*, 534. [CrossRef] [PubMed]
27. Wu, J.; Du, G.; Chen, J.; Zhou, J. Enhancing flavonoid production by systematically tuning the central metabolic pathways based on a CRISPR interference system in *Escherichia coli*. *Sci. Rep.* **2015**, *5*, 13477. [CrossRef]
28. Cress, B.F.; Toparlak, Ö.D.; Guleria, S.; Lebovich, M.; Stieglitz, J.T.; Englaender, J.A.; Jones, J.A.; Linhardt, R.J.; Koffas, M.A. CRISPathBrick: Modular Combinatorial Assembly of Type II-A CRISPR Arrays for dCas9-Mediated Multiplex Transcriptional Repression in *E. coli*. *ACS Synth. Biol.* **2015**, *4*, 987–1000. [CrossRef]
29. Frydman, A.; Weisshaus, O.; Bar-Peled, M.; Huhman, D.V.; Sumner, L.W.; Marin, F.R.; Lewinsohn, E.; Fluhr, R.; Gressel, J.; Eyal, Y. Citrus fruit bitter flavors: Isolation and functional characterization of the gene Cm1,2RhaT encoding a 1,2 rhamnosyltransferase, a key enzyme in the biosynthesis of the bitter flavonoids of citrus. *Plant J.* **2004**, *40*, 88–100. [CrossRef]
30. Dubos, C.; Le Gourrierec, J.; Baudry, A.; Huep, G.; Lanet, E.; Debeaujon, I.; Routaboul, J.-M.; Alboresi, A.; Weisshaar, B.; Lepiniec, L. MYBL2 is a new regulator of flavonoid biosynthesis in *Arabidopsis thaliana*. *Plant J.* **2008**, *55*, 940–953. [CrossRef]
31. Zhu, F.; Luo, T.; Liu, C.; Wang, Y.; Yang, H.; Yang, W.; Zheng, L.; Xiao, X.; Zhang, M.; Xu, R.; et al. An R2R3-MYB transcription factor represses the transformation of α- and β-branch carotenoids by negatively regulating expression of CrBCH2 and CrNCED5 in flavedo of Citrus reticulate. *New Phytol.* **2017**, *216*, 178–192. [CrossRef]
32. Wang, Y.; Li, J.; Xia, R. Expression of chalcone synthase and chalcone isomerase genes and accumulation of corresponding flavonoids during fruit maturation of Guoqing No. 4 satsuma mandarin (Citrus unshiu Marcow). *Sci. Hortic.* **2010**, *125*, 110–116. [CrossRef]
33. Chen, L.; Guo, H.; Lin, Y.; Cheng, H. Chalcone synthase EaCHS1 from Eupatorium adenophorum functions in salt stress tolerance in tobacco. *Plant Cell Rep.* **2015**, *34*, 885–894.
34. Lam, P.Y.; Wang, L.; Lui, A.C.W.; Liu, H.; Takeda-Kimura, Y.; Chen, M.X.; Zhu, F.Y.; Zhang, J.; Umezawa, T.; Tobimatsu, Y.; et al. Deficiency in flavonoid biosynthesis genes CHS, CHI, and CHIL alters rice flavonoid and lignin profiles. *Plant Physiol.* **2022**, *188*, 1993–2011. [CrossRef] [PubMed]
35. Xu, J.; Tian, Y.S.; Xing, X.J.; Peng, R.H.; Zhu, B.; Gao, J.J.; Yao, Q.H. Over-expression of AtGSTU19 provides tolerance to salt, drought and methyl viologen stresses in Arabidopsis. *Physiol. Plant* **2016**, *156*, 164–175. [CrossRef] [PubMed]
36. Petrussa, E.; Braidot, E.; Zancani, M.; Peresson, C.; Bertolini, A.; Patui, S.; Vianello, A. Plant Flavonoids—Biosynthesis, Transport and Involvement in Stress Responses. *Int. J. Mol. Sci.* **2013**, *14*, 14950–14973. [CrossRef]
37. Loyall, L.; Uchida, K.; Braun, S.; Furuya, M.; Frohnmeyer, H. Glutathione and a UV Light–Induced Glutathione S-Transferase Are Involved in Signaling to Chalcone Synthase in Cell Cultures. *Plant Cell* **2000**, *12*, 1939–1950.
38. Genzel, F.; Dicke, M.D.; Junker-Frohn, L.V.; Neuwohner, A.; Thiele, B.; Putz, A.; Usadel, B.; Wormit, A.; Wiese-Klinkenberg, A. Impact of Moderate Cold and Salt Stress on the Accumulation of Antioxidant Flavonoids in the Leaves of Two Capsicum Cultivars. *J. Agric. Food Chem.* **2021**, *69*, 6431–6443. [CrossRef]
39. Singh, D.P.; Prabha, R.; Meena, K.K.; Sharma, L.; Sharma, A.K. Induced Accumulation of Polyphenolics and Flavonoids in Cyanobacteria under Salt Stress Protects Organisms through Enhanced Antioxidant Activity. *Am. J. Plant Sci.* **2014**, *5*, 726–735. [CrossRef]

40. Luo, R.; Liu, B.; Xie, Y.; Li, Z.; Huang, W.; Yuan, J.; He, G.; Chen, Y.; Pan, Q.; Liu, Y.; et al. SOAPdenovo2: An empirically improved memory-efficient short-read *de novo* assembler. *GigaScience* **2012**, *1*, 1–18. [CrossRef]
41. Xu, Q.; Chen, L.L.; Ruan, X.; Chen, D.; Zhu, A.; Chen, C.; Bertrand, D.; Jiao, W.B.; Hao, B.H.; Lyon, M.P.; et al. The draft genome of sweet orange (*Citrus sinensis*). *Nat. Genet.* **2013**, *45*, 59–66. [CrossRef]
42. Kim, D.; Paggi, J.M.; Park, C.; Bennett, C.; Salzberg, S.L. Graph-based genome alignment and genotyping with HISAT2 and HISAT-genotype. *Nat. Biotechnol.* **2019**, *37*, 907–915. [CrossRef] [PubMed]
43. Kovaka, S.; Zimin, A.V.; Pertea, G.M.; Razaghi, R.; Salzberg, S.L.; Pertea, M. Transcriptome assembly from long-read RNA-seq alignments with StringTie2. *Genome Biol.* **2019**, *20*, 278–291. [CrossRef]
44. Love, M.I.; Huber, W.; Anders, S. Moderated estimation of fold change and dispersion for RNA-seq data with DESeq2. *Genome Biol.* **2014**, *15*, 550. [CrossRef] [PubMed]
45. Langfelder, P.; Horvath, S. WGCNA: An R package for weighted correlation network analysis. *BMC Bioinform.* **2008**, *9*, 1–13. [CrossRef] [PubMed]
46. Bastian, M.; Heymann, S.; Jacomy, M. Gephi: An open source software for exploring and manipulating networks. *Proc. Int. AAAI Conf. Web Soc. Media* **2009**, *3*, 361–362.
47. Pang, Z.; Zhou, G.; Ewald, J.; Chang, L.; Hacariz, O.; Basu, N.; Xia, J. Using MetaboAnalyst 5.0 for LC-HRMS spectra processing, multi-omics integration and covariate adjustment of global metabolomics data. *Nat. Protoc.* **2022**, *17*, 1735–1761. [CrossRef]

molecules

Article

Lanostane Triterpenoids and Ergostane Steroids from *Ganoderma luteomarginatum* and Their Cytotoxicity

Qingyun Ma [1,†], Shuangshuang Zhang [2,†], Li Yang [1], Qingyi Xie [1], Haofu Dai [1], Zhifang Yu [3,*] and Youxing Zhao [1,*]

1 Haikou Key Laboratory for Research and Utilization of Tropical Natural Products, Institute of Tropical Bioscience and Biotechnology, Chinese Academy of Tropical Agricultural Sciences & Hainan Key Laboratory for Protection and Utilization of Tropical Bioresources, Hainan Academy of Tropical Agricultural Resource, Haikou 571101, China
2 Jiangsu Food&Pharmaceutical Science College, Huaian 223003, China
3 College of Food Science and Technology, Nanjing Agricultural University, Nanjing 210095, China
* Correspondence: yuzhifang@njau.edu.cn (Z.Y.); zhaoyouxing@itbb.org.cn (Y.Z.)
† These authors contributed equally to this work.

Abstract: Macrofungus *Ganoderma luteomarginatum* is one of the main species of *Ganoderma* fungi distributed in Hainan province of China, the fruiting bodies of which have been widely used in folk as a healthy food to prevent tumors. To explore the potential cytotoxic constituents from *G. luteomarginatum*, the phytochemical investigation on the ethyl acetate soluble fraction of 95% ethanolic extract from the fruiting bodies of this fungus led to the isolation of twenty-six lanostane triterpenoids (**1–26**), including three undescribed ones (**1–3**), together with eight ergostane steroids (**27–34**). The structures of three new lanostane triterpenoids were elucidated as lanosta-7,9(11)-dien-3β-acetyloxy-24,25-diol (**1**), lanosta-7,9(11)-dien-3-oxo-24,26-diol-25-methoxy (**2**), and lanosta-8,20(22)-dien-3,11,23-trioxo -7β,15β-diol-26-oic acid methyl ester (**3**) by the analysis of 1D, 2D NMR, and HRESIMS spectroscopic data. All isolates were assayed for their cytotoxic activities using three human cancer cell lines (K562, BEL-7402, and SGC-7901) and seven lanostane triterpenoids (**1**, **2**, **7**, **13**, **18**, **22**, and **24**), and one ergostane steroid (**34**) showed definite cytotoxicity with IC_{50} values that ranged from 6.64 to 47.63 μg/mL. Among these cytotoxic lanostane triterpenoids, compounds **2** and **13** showed general cytotoxicity against three human cancer cell lines, while compounds **1** and **18** exhibited significant selective cytotoxicity against K562 cells with IC_{50} values of 8.59 and 8.82 μg/mL, respectively. Furthermore, the preliminary structure–cytotoxicity relationships was proposed.

Keywords: *Ganoderma luteomarginatum*; lanostane triterpenoids; ergostane steroids; cytotoxic activity

Citation: Ma, Q.; Zhang, S.; Yang, L.; Xie, Q.; Dai, H.; Yu, Z.; Zhao, Y. Lanostane Triterpenoids and Ergostane Steroids from *Ganoderma luteomarginatum* and Their Cytotoxicity. *Molecules* **2022**, *27*, 6989. https://doi.org/10.3390/molecules27206989

Academic Editor: Baoan Song

Received: 22 September 2022
Accepted: 14 October 2022
Published: 18 October 2022

1. Introduction

The genus *Ganoderma*, belonging to the family Ganodermataceae and known as "Lingzhi" in Chinese, has been widely used as traditional Chinese medicine and functional foods for health in China and Southeast Asia for thousands of years [1]. There are more than 100 species in this family growing on cut or rotten trees in China, and 78 wild ones were found in Hainan Province [2]. As the major genus in this family, *Ganoderma* is a prolific producer of novel natural products responsible for its health benefits, mainly containing polysaccharides with an immunostimulative effect and triterpenes with a cytotoxic action [3]. Two main species, *G. lucidum* and *G. sinensis*, are recorded in Pharmacopoeia of the People's Republic of China and used as an addition to conventional therapy in a clinical treatment of chronic bronchitis, bronchial asthma, leukopenia, coronary heart disease, arrhythmia, and acute infectious hepatitis. Recent research on chemical constituents of *Ganoderma* species showed that lanostane-type triterpenoids are the main characteristic natural products [4], and more than 400 lanostanoids have been isolated from the fungi of *Ganoderma*. These small molecule compounds have attracted considerable attention due to their extensive biological and pharmacological activities [5,6], including cytotoxic [7–9], hepatoprotective [10,11], anti-inflammatory [12–14], antidiabetic [15,16], neuroprotective [17], antiviral [18], antiaging [19], and antioxidant [20–22] effects. The genus *Ganoderma* is used as a healthy food and has been traditionally used for the prevention of numerous diseases or various pathological conditions, including complementary cancer therapy, especially a broad-spectrum application for the treatment of cancer.

Cancer has been considered as a huge threat to human health, and most governments are committed to diminishing this threat. The prevention and treatment of cancer becomes a key health goal. Finding antitumor drugs with high efficiency and low toxicity has become the urgent task, and countless researchers are dedicated to discovering bioactive ingredients from nature resources. *Ganoderma* is a promising anticancer immunotherapy agent owing to its low toxicology and efficacy as a combination therapy through the regulation of the immune system [23]. Polysaccharides and triterpenes from *Ganoderma* have been known to possess chemopreventive and antitumor activity. Many studies indicate that lanostane-type triterpenoids act as an inhibitor on different cancer cell lines, including the lung, liver, colon, pancreas, breast, skin, and prostate [6]. Among the reported active lanostanoids, the ganoderic acids are the main types of triterpene that play key roles in the biological activity. Lanostane-type triterpenoids can cause cell cycle arrest by the downregulation of cyclin D1 in the G1 phase of cell growth and inhibition of PKC activity in the G2 growth phase. Moreover, lanostane-type triterpenoids also prevent tumor metastasis by modulating MMPs and IL-8 and inhibit the excretion of inflammatory cytokines [24].

Ganoderma luteomarginatum, used as folk medicinal *Ganoderma* species, is a rare species mainly distributed in Hainan Province in China [25], where a pharmacodynamic molecular basis has been brought into focus in recent years [26,27]. In our ongoing endeavor to explore bioactive natural products, several species of *Ganoderma* have been studied, and a series of active compounds have been found [7,15,28–31]. The fruiting bodies of *G. luteomarginatum* have been widely used as a healthy food to prevent tumors. To explore the potential cytotoxic constituents from *G. luteomarginatum*, we performed a phytochemical investigation on this fungus, which resulted in the isolation of twenty-six lanostane triterpenoids (**1–26**) (Figure 1), including three undescribed ones: lanosta-7,9(11)-dien-3β-acetyloxy-24,25-diol (**1**), lanosta-7,9(11)-dien-3-oxo-24,26-diol-25- methoxy (**2**), and lanosta-8,20(22)-dien-3,11,23 -trioxo-7β,15β-diol- 26-oic acid methyl ester (**3**), together with eight ergostane steroids (**27–34**) (Figure 1). All these isolates were evaluated for their cytotoxic activity against three human cancer cell lines. Herein, we reported the isolation, structural elucidation, and cytotoxicity of the compounds isolated from the fruiting bodies of *G. luteomarginatum*.

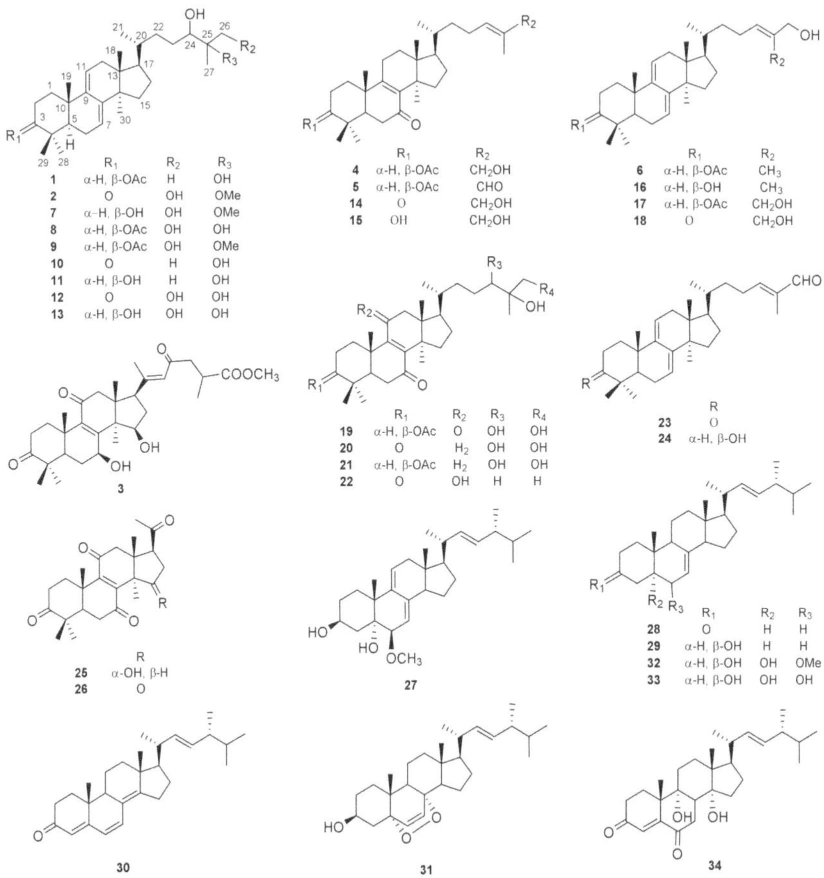

Figure 1. The structures of compounds **1–34**.

2. Results and Discussion

2.1. Structural Elucidation of Compounds

Compound **1** was obtained as white amorphous powder, and its molecular formula was determined to be $C_{32}H_{52}O_4$ on the basis of HRESIMS ion at *m/z* $[M + Na]^+$ 523.3754 (calcd. 523.3763 for $C_{32}H_{52}NaO_4^+$), indicating seven degrees of unsaturation. The IR spectrum revealed the presence of hydroxyl groups (3475 cm^{-1}), double bonds (1641 cm^{-1}), and ester carbonyl (1696 cm^{-1}). The ^{1}H NMR spectral data (Table 1) of **1** revealed the signals for nine methyls (δ_H 0.58; 0.89; 0.90; 0.93, d, J = 6.3 Hz; 0.97; 1.02; 1.18; 1.23; 2.07), one proton related to oxygenated carbon (δ_H 4.53, dd, J = 11.4, 4.6 Hz), and two olefinic protons (δ_H 5.47, t, J = 4.5 Hz; 5.34, d, J = 5.9 Hz). The ^{13}C NMR and DEPT (Table 1) spectra presented 32 carbon signals for nine methyls; eight methylenes; seven methines (two oxygenated and two olefinic); and eight non-protonated carbons (two olefinic, one oxygenated, and one ester carbonyl at δ_C 171.2). The above-mentioned NMR data were closely similar to those of lanosta-7,9(11)-dien-3b-acetyloxy-24,25,26-trihydroxy [29], suggesting that **1** had a lanostane skeleton and structurally similar to this compound. The only difference between them was that the methylol group (δ_C 67.8) at C-26 in lanosta-7,9(11)-dien-3b-acetyloxy-24,25,26-trihydroxy was replaced by one methyl (δ_C 25.7) in **1**, which was confirmed by HMBC correlations (Figure 2) from H$_3$-27 (δ_H 1.18) to C-26, C-24 (δ_C 79.7) and C-25 (δ_C 73.4). The attachment of acetate group to C-3 was proposed by the key HMBC correlation of H-3 (δ_H 4.53) with acetal carbonyl (δ_C 171.2). The other obvious HMBC correlations

(Figure 2) of **1** from H_3-28 (δ_H 0.90) and H_3-29 (δ_H 0.97) to C-3, C-4 (δ_C 37.9), and C-5 (δ_C 49.3); from H_3-18 (δ_H 0.55) to C-12 (δ_C 37.7), C-14 (δ_C 50.4), and C-17 (δ_C 51.1); from H_3-19 (δ_H 1.02) to C-1 (δ_C 35.5), C-5, and C-9 (δ_C 145.7); and from H_3-30 (δ_H 0.89) to C-8 (δ_C 142.8), C-13 (δ_C 43.8) and C-15 (δ_C 28.0) further assigned its planar structure of lanostanoid. The relative configuration of the tetracyclic core structure of **1** was determined to be the same as that of lanosta-7,9(11)-dien-3b-acetyloxy- 24,25,26-trihydroxy by comparison of their NMR and ROESY spectroscopic data (Figure 1), revealing ROESY correlations of H-18 with H-19 and H-20 (δ_H 1.39). The key ROESY correlation of H-3 (δ_H 4.53) with H_3-28 (δ_H 0.90) and H-5 (δ_H 1.18) suggested the assignment of their same orientation. Based on comprehensive analysis of 1D, 2D NMR, and HRESIMS spectrums (see Supplementary Materials), the structure of compound **1** was elucidated to be lanosta-7,9(11)-dien-3β -acetyloxy-24,25-diol.

Table 1. ^{1}H (500 MHz) and ^{13}C NMR (125 MHz) Data of Compounds **1–3** (in $CDCl_3$).

No.	1		2		3	
	δ_C	δ_H (J in Hz)	δ_C	δ_H (J in Hz)	δ_C	δ_H (J in Hz)
1	35.5	1.97 m 1.50 m	36.7	2.36 m 1.77 m	35.6	2.45 m 2.06 m
2	24.4	1.71 m	35.0	2.80 dt (14.7,5.7) 2.38 ddd (14.7, 4.5, 3.1)	34.4	2.85 m 2.36 m
3	81.0	4.53 dd (4.6, 11.4)	217.1		216.8	
4	37.9		47.6		46.9	
5	49.3	1.18 m	50.8	1.56 dd (3.7, 11.9)	49.3	1.67 dd (13.5, 2.2)
6	22.9	2.08 m	23.8	2.02 m 2.23 m	36.5	2.11 m 1.84 m
7	120.1	5.47 t (4.5)	120.0	5.52 d (6.8)	69.5	4.80 dd (7.1, 9.4)
8	142.8		143.0		157.0	
9	145.7		144.6		143.3	
10	37.3		37.3		38.3	
11	116.6	5.34 d (5.9)	117.4	5.40 d (6.2)	199.4	
12	37.7	2.23 d (17.6) 2.02 m	37.9	2.28 m 2.08 m	51.7	2.75 d (14.6) 2.37 d (14.6)
13	43.8		43.9		48.5	
14	50.4		50.4		57.1	
15	28.0	1.29 m	28.0	1.27 m	77.8	4.33 d (6.7)
16	31.6	1.60 m 1.36 m	31.6	1.67 m 1.40 m	35.3	2.35 m 1.42 m
17	51.1	1.58 m	51.1	1.61 m	54.5	2.76 m
18	15.8	0.55 s	15.9	0.61 s	19.6	1.10 s
19	23.3	1.02 s	22.2	1.21 s	19.7	1.36 s
20	36.7	1.39 m	36.7	1.44 m	156.8	
21	18.7	0.93 d (6.3)	18.7	0.94 d (6.5)	21.3	2.16 s
22	33.6	1.76 m	33.8	1.86 m	124.7	6.16 s
23	28.8	1.56 m	28.4	1.63 m	198.4	
24	79.7	3.31 dd (10.2, 2.1)	76.8	3.59 d (9.4)	47.8	2.94 m 2.57 m
25	73.4		78.5		35.0	2.96 m
26	25.7	1.23 s	64.4	3.75 d (12.0) 3.65 d (12.0)	176.7	
27	22.9	1.18 s	16.0	1.05 s	17.3	1.20 d (6.8)
28	28.2	0.90 s	25.4	1.10 s	27.0	1.13 s
29	17.1	0.97 s	22.6	1.14 s	21.0	1.12 s
30	25.7	0.89 s	25.6	0.89 s	26.0	1.24 s
OAc	171.2 21.5	2.07 s				
OMe			49.5	3.34 s	52.0	3.69 s

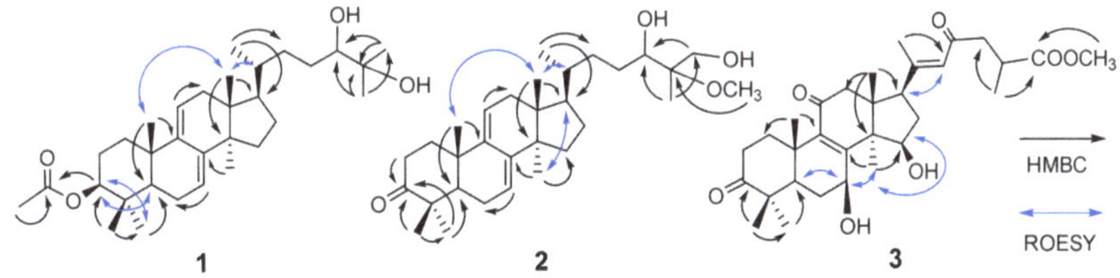

1 2 3

Figure 2. Key HMBC (H→C) and ROESY (↔) correlations of **1–3**.

Compound **2** had the molecular formula $C_{31}H_{50}O_4$, as determined by the HRESIMS ion peak at 509.3597 (calcd. 509.3607 for $C_{31}H_{50}NaO_4^+$). The ^{13}C NMR and DEPT (Table 1) spectra showed 31 carbon signals for eight methyls (one methoxy at δ_C 49.5); nine methylenes (one oxygenated at δ_C 64.4); six methines (one oxygenated at δ_C 76.8 and two olefinic); and eight non-protonated carbons (two olefinic, one oxygenated, and one ketone carbonyl at (δ_C 217.1). The NMR spectra of **2** resembled those of (24*S*,25*R*)-25 - methoxylanosta-7,9(11)-dien-3β,24,26-triol [32], except for the presence of a ketone carbonyl of C-3 (δ_C 217.1) in **2** replacing the hydroxylated methine in (24*S*,25*R*)-25 - methoxylanosta-7,9(11)-dien-3β,24,26-triol, which was corroborated by the HMBC correlations (Figure 2) from H_3-28 (δ_H 1.10), H_3-29 (δ_H 1.14), and H_2-2 (δ_H 2.38/2.80) to C-3 (δ_C 217.1). The attachment of methoxy to C-25 was assigned by the key HMBC correlation of the protons signal at δ_H 3.34 with C-25 (δ_C 78.5). The other clear HMBC correlations (Figure 2) of **1** from H_3-28 (δ_H 1.10), H_3-29 (δ_H 1.14), and H_3-19 (δ_H 1.21) to C-5 (δ_C 50.8) and from H_3-18 (δ_H 0.61) and H_3-30 (δ_H 0.89) to C-13 (δ_C 43.9) and C-14 (δ_C 50.4) further confirmed its planar structure of lanostanoid. The relative configuration of **2** was established as same as that of (24*S*, 25*R*)-25-methoxylanosta-7,9(11)-dien-3β,24,26-triol by the ROESY spectrum (Figure 2), revealing key ROESY correlations of H-30 with H-17 (δ_H 1.61) and of H-18 with H-19 and H-20 (δ_H 1.44), as well as their similar NMR data. On the basis of the above evidence, the structure of **2** was defined as lanosta-7,9(11)-dien-3-oxo-24,26-diol-25-methoxy.

The HRESIMS ion peak 529.3160 [M+H]$^+$) of compound **3** gave the molecular formula $C_{31}H_{44}O_7$, indicating ten degrees of unsaturation. The ^{13}C NMR and DEPT (Table 1) spectra showed 31 carbon signals for eight methyls (one methoxy at δ_C 52.0), six methylenes, six methines (two oxygenated, and one olefinic at δ_C 124.7) and eleven non-protonated carbons (three olefinic; three ketones at δ_C 216.8, 199.4, and 198.4; and one ester carbonyl δ_C 176.7). A detailed analysis of the 1D NMR data of **3** suggested that compound **3** possessed the same planar structure as methyl ganoderenate A [33] with a lanostane skeleton. The whole connectivity of compound **3** was also further demonstrated by 2D NMR data, including HMBC spectrum (Figure 2), exhibiting HMBC correlations from H_3-28 (δ_H 1.13), H_3-29 (δ_H 1.12), and H_3-19 (δ_H 1.36) to C-5 (δ_C 49.3) and from H_3-18 (δ_H 1.10) and H_3-30 (δ_H 1.24) to C-13 (δ_C 48.5) and C-14 (δ_C 57.1). The relative configuration of the tetracyclic core structure of **1** was determined to be the same as methyl ganoderenate A, except for chiral C-15 by analysis of its ROESY spectrum (Figure 2). The key ROESY correlations of H-7 (δ_H 4.80) with H_3-30 (δ_H 1.24) and H-5 (δ_H 1.67) proposed the β-orientation of hydroxy at C-7. The β-orientation of hydroxy at C-15 was assigned by the key ROESY correlation of H_3-30 (δ_H 1.24) with H-15 (δ_H 4.33). Moreover, the key ROESY correlation of H-22 (δ_H 6.16) with H-17 (δ_H 2.76) indicated that the geometry of double bond $\Delta^{20(22)}$ in **3** was *E* instead of *Z* in methyl ganoderenate A. Therefore, the structure of compound **3** was established as lanosta-8,20(22)-dien-3,11,23-trioxo-7β,15β- diol- 26-oic acid methyl ester.

The thirty-one known compounds, including twenty-one lanostane triterpenoids (**4–26**) and eight ergostane steroid (**27–34**), were identified as lanosta-8,24*E*-dien-7-oxo -3β- acetyloxy-26-ol (**4**) [26], lanosta-8,24*E*-dien-7-oxo-3β-acetyloxy-26-al (**5**) [26], lanosta -7,9(11),24-trien-3β-acetyloxy-26-ol (**6**) [26], (24*R*,25*S*)-lanosta-7,9(11)-dien- 3β,24,26-triol -25-methoxy (**7**) [16], lanosta-7,9(11)-dien-3β-acetyloxy-24,25,26-triol (**8**) [29], lanosta-7,9(11) -dien-3β-acetyloxy-24,26-dihydroxy-25-methoxy (**9**) [29], ganodermanondiol (**10**) [34], lucidumol B (**11**) [34], 26-hydroxy-ganodermanondiol (**12**) [35], ganoderiol A (**13**) [36], ganoderone A (**14**) [37], lucidadiol (**15**) [38], ganodermadiol (**16**) [39], lanosta-7,9(11),24*E*-trien-3β-acetyloxy-26,27-diol (**17**) [29], ganoderiol F (**18**) [40], lanosta-8-en-7,11-dioxo-3β-acetyloxy-24,25,26-triol (**19**) [29], ganoderiol D (**20**) [40], lanosta-8-en-7-oxo-3β-acetyloxy-24,25,26-trihydroxy (**21**) [34], lucidumol A (**22**) [41], lanosta-7,9(11),24*E*-trien-3-oxo-26-al (**23**) [42], lanosta-7,9(11),24-triene-3β-ol-26-al (**24**) [42], lucidone H (**25**) [29], lucidadone H (**26**) [43], ergosta-7,9(11),22*E*-triene-3β,5α-diol -6β-methoxy (**27**) [29], ergosta-7,22*E*-dien-3-one (**28**) [44], ergosta-7,22*E*-dien-3β-ol (**29**) [45], ergosta-4,6,8(14),22*E*-tetraen-3-one (**30**) [44], ergosterol peroxide (**31**) [46], ergosta-3β,5α-diol-7,22*E*-dien-6β-methoxy (**32**) [47], ergosta-7,22*E*-dien-3β,5α,6β-triol (**33**) [48], and calvasterol B (**34**) [49] by comparing their NMR data with those reported in the literature.

Ganoderma has been used as a healthy food and medicinal purposes for centuries particularly in China, Japan, and Korea. A great deal of work has been carried out on over thirty species of *Ganoderma*. Two types of natural products, lanostane triterpenoids (**1–26**) and ergostane steroids (**27–34**), were discovered from the fruiting bodies of *G. luteomarginatum* collected in Hainan Province, China. These two types of compounds are widely found in genus *Ganoderma* [1], and lanostanoids (*Ganoderma* triterpenoids) was the characteristic active metabolites in species of *Ganoderma*, which are a class of compounds with various chemical structures. Here, the isolated lanostane triterpenoids were divided into two groups according to the conjugated system at C-7, C-8, C-9, and C-11. The first group possessed the conjugated double bond $\Delta^{7,9(11)}$ as shown in **1**, **2**, **6–18**, **23**, and **24**. The second group had $(\Delta^8)\alpha,\beta$-unsaturated ketone at C-7 or C-9. The C-26 in lanostanoid structures is often oxidized to alcohols, aldehydes, and acids [5]. Among the structures of identified lanostanoids (**1–26**), ganoderiol derivative was the main type. In addition, three ganoderic aldehydes (**5**, **23**, **24**) and one ganoderic acid (**3**) were also found. The *β*-configuration of OH-15 in new ganoderic acid (**3**) from *G. luteomarginatum* was consistent with the previously discovered ganoderic acid derivatives from this fungus [27], which was opposite of that shown in the corresponding compounds from other *Ganoderma* species [7,12,50]. Moreover, norlanostanoids with 24 carbon atoms often occur in *Ganoderma*. In present study, two hexanorlanostanoids (**25** and **26**) were isolated. Some lanostane triterpenoids besides the three new ones in our study are structurally different from those previously reported lanostanoids [26,27] from *G. luteomarginatum* collected in Guangxi Province, China. This may be related to the different growth environment of this fungus or different growth period for collection, which needs further comparative analysis in our subsequent studies.

2.2. *Cytotoxic Activities of Compounds*

The cytotoxic activities of all the isolates were evaluated by MTT method toward three human cancer cell lines (K562, BEL-7402, and SGC-7901). The results were presented in Table 2. Of the compounds tested, seven lanostane triterpenoids (**1**, **2**, **7**, **13**, **18**, **22**, and **24**) showed definite cytotoxicity against K562 with IC_{50} values range from 6.64 to 17.38 μg/mL, among which compounds **1**, **13** and **18** showed the IC_{50} values of 8.59, 6.64, and 8.82 μg/mL, respectively. Compounds **2** and **13** also showed moderate cytotoxicity against two human cancer cell lines (BEL-7402 and SGC-7901). Moreover, compound **7** showed moderate cytotoxicity against human cancer cell lines BEL-7402 with IC_{50} value of 20.05 μg/mL. Compounds **1**, **18**, **22**, and **24** had no obvious cytotoxicity on BEL-7402 and SGC-7901 cell lines (IC_{50} > 50 μg/mL). Among these cytotoxic lanostanoids, compounds **2** and **13** showed general cytotoxicity against three human cancer cell lines, while compounds **1** and **18** exhibited significant selective cytotoxicity against K562 cell lines. One ergostane steroid (**34**) showed general cytotoxicity against three human cancer cell lines.

Table 2. Cytotoxic activities of compounds from *Ganoderma luteomarginatum* (IC_{50}, μg/mL).

Compounds	K562	BEL-7402	SGC-7901
1	8.59	>50	>50
2	16.05	24.27	33.38
7	11.69	20.05	>50
13	6.64	13.49	15.62
18	8.82	>50	>50
22	16.95	>50	>50
24	17.38	>50	>50
34	22.81	47.63	26.06
Paclitaxel	5.62	3.26	3.41

Lanostane-type triterpenoids of *Ganoderma* are considered to be the major pharmacologically active compounds that contribute to its antitumor efficacy. The lanostane-type triterpenoids were extensively evaluated for cytotoxic activities against a series of tumor cell lines [5] related to lung, liver, colon, pancreas, breast, skin, and prostate [6]. The lanostanoids with structural complexity and functional group variety may be specific to different cell lines and the structure-cytotoxicity relationships could be raised. Compounds **1**, **18**, **22**, and **24** exhibited selective cytotoxicity against K562 cell lines may be due to their unique structures. From the results of isolated lanostanoids against three human cancer cell lines (K562, BEL-7402, and SGC-7901), the conjugated double-bond $\Delta^{7,9(11)}$ system in tetracyclic skeleton (**1**, **2**, **7**, **13**, **18**, and **24**) seemed to be more important than (Δ^{8})α,β-unsaturated ketone system for potent cytotoxic activity. Comparing the cytotoxicity between **7**/**9**, **18**/**17**, and **24**/**5** with only difference at C-3, it suggested that acetylation may be the negative factor for cytotoxic activity. In addition, compound **13** exhibited significant cytotoxicity, while their keto-3 analog **12** was inactive, assumed that reduction of the keto-3 group to OH-3 in lanostane triterpenoids would improve the cytotoxicity against K562, BEL-7402, and SGC-7901 significantly. The above preliminary structure–cytotoxicity relationships provide an approach to understanding the structural requirements of lanostane-type triterpenoids.

3. Materials and Methods

3.1. General Experimental Procedures

The NMR spectra were recorded with a Bruker AV-500 spectrometer (Bruker, Bremen, Germany) with TMS as an internal standard. HRESIMS data were determined on a mass spectrometer API QSTAR Pulsar (Bruker, Bremen, Germany). Optical rotations were measured on a Rudolph Autopol III polarimeter (USA). UV spectra were obtained on a Shimadzu UV-2550 spectrometer. IR spectra were obtained on a Nicolet 380 FT-IR spectrometer with KBr pellets. Silica gel (60–80 and 200–300 mesh, Marine Chemical Industry Factory, Qingdao, China), Rp-C18 (20–45 mL; Fuji Silysia Chemical Ltd., Aichi, Japan), and Sephadex LH-20 (Merck, Germany) were used for column chromatography. Fractions were monitored by TLC and spots were visualized by heating after spraying with 5% H_2SO_4 in ethanol.

3.2. Fungal Material

Fruiting bodies of G. luteomarginatum were collected in Qiongzhong County, Hainan Province, China (June, 2012), and identified by Prof. Xing-Liang Wu of Hainan University. A voucher specimen (No. 2012HB01) is deposited at the Institute of Tropical Bioscience and Biotechnology, Chinese Academy of Tropical Agricultural Sciences.

3.3. Isolation and Characterization of Compounds

Dried and powdered fruiting bodies of G. luteomarginatum (2.5 kg) were extracted with EtOH-H_2O (10 L, 95:5, *v*/*v*) under reflux conditions three times at a duration of 4 h. The combined extracts were concentrated and suspended in H_2O, followed by successive partitioning with EtOAc and n-BuOH, respectively. The EtOAc extract (53.0 g) was separated by silica gel column chromatography (CC) under reduced pressure using a solvent gradient of petroleum ether (PE)-EtOAc (20:1→0:1, *v*/*v*) to afford six fractions (Fr1-Fr6). Fr2 (6.2 g) was subjected to silica gel CC under reduced pressure eluted with PE-EtOAc (3:1) to give 3 subfractions: 2a–2c. Subfraction 2a (925 mg) was repeatedly purified by silica gel CC eluted with PE-EtOAc (5:1) to obtain compounds **4** (7.3 mg), **5** (4.2 mg), **6** (5.2 mg), **14** (21.5 mg), **26** (28.7 mg), and **30** (16.4 mg). Subfraction 2b (816 mg) was separated by silica gel column to yield compounds **23** (7.7 mg), **24** (6.0 mg), and **28** (68.3 mg) using an eluent $CHCl_3$-MeOH (15:1), and compounds **25** (2.4 mg) and **29** (81.7mg) were separated from subfraction 2c (674 mg) using an eluent: $CHCl_3$-MeOH (12:1). Fr3 (9.6 g) was separated by Rp-18 CC with MeOH-H_2O (30:70→0:100) to give 3 subfractions: 3a–3c. Subfraction 3a (526 mg) was purified by silica gel CC eluted with PE-EtOAc (3:1) to obtain compounds **1** (5.5 mg), **2** (8.2 mg), **9** (8.6 mg), and **10** (9.0 mg). Subfraction 3b (603 mg) was separated by silica gel CC eluted with PE-EtOAc (2:1) to yield compounds **3** (4.6 mg), **15** (5.5 mg), **16** (5.1 mg), **22** (10.5 mg), and **31** (27.0 mg). Subfraction 3c (839 mg) was subjected to silica gel CC eluted with $CHCl_3$-MeOH (10:1) to yield compounds **17** (8.6 mg), **18** (9.8 mg), **27** (3.9 mg), **32** (3.8 mg), and **34** (5.1 mg). Fr4 (8.5 g) was treated by Rp-18 CC with MeOH-H_2O (30:70→0:100) to afford subfractions 4a–4d. Subfraction 4b (406 mg) was subjected to Sephadex LH-20 ($CHCl_3$/MeOH, 1:1), then by silica gel CC with PE-EtOAc (2:1) to yield compounds **7** (8.1 mg), **8** (7.6 mg), and **19** (7.4 mg). Compounds **11** (6.5 mg) and **12** (15.6 mg) was purified from subfraction 4b (365 mg) using an eluent $CHCl_3$-MeOH (8:1). Fr5 (9.0 g) was separated by Rp-18 CC with MeOH-H_2O (30:70→0:100) to afford subfractions 5a–5c. Subfraction 5b (582 mg) was subjected to silica gel CC with $CHCl_3$-EtOAc (2:1) to yield compounds **13** (7.8 mg) and **20** (10.2 mg). Subfraction 5c (264 mg) was separated by silica gel CC eluted with $CHCl_3$-MeOH (6:1) to obtain **21** (8.4 mg) and **33** (25.8 mg).

Lanosta-7,9(11)-dien-3β-acetyloxy-24,25-diol (**1**): White amorphous powder; $[\alpha]^{27}_D$ $-2.5°$ (c 0.02, MeOH); UV (MeOH) λ_{max} (logε) 252 (3.62), 242 (5.24), 238 (4.50), 210 (1.21); IR (KBr) ν_{max} cm^{-1} 3475, 2938, 1696, 1641, 1302, 1028; ^{1}H and ^{13}C NMR data; see Table 1; HRESIMS m/z $[M+Na]^+$ 523.3754 (calcd. 523.3763 for $C_{32}H_{52}NaO_4$).

Lanosta-7,9(11)-dien-3-oxo-24,26-diol-25-methoxy (**2**): White amorphous powder; $[\alpha]^{27}_D$ $+4.0°$ (c 0.02, MeOH); UV (MeOH) λ_{max} (log ε) 242 (5.60), 238 (5.03), 212 (2.32); IR (KBr) v_{max} cm^{-1} 3424, 2952, 1724, 1644, 1386, 1020; ^{1}H and ^{13}C NMR data; see Table 1; HRESIMS m/z $[M+Na]^+$ 509.3597 (calcd. 509.3607 for $C_{31}H_{50}NaO_4$).

Lanosta-8,20(22)-dien-3,11,23-trioxo-7β,15β-diol-26-oic acid methyl ester (**3**): White amorphous powder; $[\alpha]^{27}_D$ $-1.5°$ (c 0.02, MeOH); UV (MeOH) λ_{max} (log ε) 250 (3.54), 210 (1.37); IR (KBr) v_{max} cm^{-1} 3443, 2934, 1732, 1639, 1384, 1019; ^{1}H and ^{13}C NMR data; see Table 1; HRESIMS m/z 529.3160 $[M+H]^+$ (calcd. 529.3165 for $C_{31}H_{45}O_7$).

3.4. *Bioassay of Cytotoxic Activity*

All the compounds was assayed for their cytotoxic activity against three human tumor cell lines: K562 (leukemic cell line), BEL7402 (hepatoma cell line), and SGC7901 (gastric cancer cell line) using MTT methods reported previously [7]. Briefly, each tested compound was dissolved with DMSO at concentration of 10 mM and then diluted to the required concentrations with the medium. Cells were cultured in 96-well plates with initial density of 5000 cells/well 12 h before treatment and exposed to different concentrations (40, 8, 1.6, 0.32, and 0.064 μM, respectively) of compounds, with paclitaxel (Sigma, Livonia, MI, USA) as the positive control. After the culturing period, 20 μL of MTT (5 mg/mL) was added per well and incubated for 4 h at 37 °C. Finally, absorbance was measured at 570 nm using a microplate reader. Each assay was replicated three times. The effect of the compounds on cell viability was calculated and expressed as the IC_{50}.

4. Conclusions

Lanostane-type triterpenoids are the main characteristic natural products of the fungi of *Ganoderma*, which have extensive biological and pharmacological activities, especially possess cytotoxicity. The chemical investigation of the fruiting bodies of *G. luteomarginatum* led to the isolation of twenty-six lanostane triterpenoids (**1–26**), including three undescribed ones (**1–3**): lanosta-7,9(11)-dien-3β-acetyloxy-24,25-diol (**1**), lanosta-7,9(11)-dien-3-oxo -24,26-diol-25-methoxy (**2**), and lanosta-8,20(22)-dien-3,11,23-trioxo -7β,15β-diol-26-oic acid methyl ester (**3**), together with eight ergostane steroids (**27–34**). The cytotoxicity assay showed that seven lanostane triterpenoids (**1, 2, 7, 13, 18, 22,** and **24**) revealed definite cytotoxicity against K562, among which compounds **1, 13,** and **18** showed the IC_{50} values of 8.59, 6.64, and 8.82 μg/mL, respectively, indicating the prospect of an antitumor. Some preliminary structure–cytotoxicity relationships of these lanostane triterpenoids showed that the conjugated double-bond $\Delta^{7,9(11)}$ system in tetracyclic lanostane skeleton seemed to be more important than the $(\Delta^8)\alpha,\beta$-unsaturated ketone system for potential cytotoxic activity. The present study further enriched the understanding of the structural diversity of *G. luteomarginatum*, which also provides theoretical information for its subsequent anticancer drug development.

Supplementary Materials: The following supporting information about 1D, 2D NMR, and HRESIMS data can be downloaded at: https://www.mdpi.com/article/10.3390/molecules27206989/s1, Figure S1-1: ^{1}H NMR spectrum of compound **1**. Figure S1-2: ^{13}C NMR spectrum of compound **1**. Figure S1-3: HSQC spectrum of compound **1**. Figure S1-4: ^{1}H-^{1}H COSY spectrum of compound **1**. Figure S1-5: HMBC spectrum of compound **1**. Figure S1-6: ROESY spectrum of compound **1**. Figure S1-7: HRESIMS Data of compound **1**. Figure S2-1: ^{1}H NMR spectrum of compound **2**. Figure S2-2: ^{13}C NMR spectrum of compound **2**. Figure S2-3: HSQC spectrum of compound **2**. Figure S2-4: ^{1}H-^{1}H COSY spectrum of compound **2**. Figure S2-4: ^{1}H-^{1}H COSY spectrum of compound **2**. Figure S2-6: ROESY spectrum of compound **2**. Figure S2-7: HRESIMS Data of compound **2**. Figure S3-1: ^{1}H NMR spectrum of compound **3**. Figure S3-2: ^{13}C NMR spectrum of compound **3**. Figure S3-3 HSQC spectrum of compound **3**. Figure S3-4: ^{1}H-^{1}H COSY spectrum of compound **3**. Figure S3-5: HMBC spectrum of compound **3**. Figure S3-6: ROESY spectrum of compound **3**. Figure S3-7: HRESIMS Data of compound **3**.

Author Contributions: Q.M.: wrote the manuscript and isolated the compounds. S.Z.: data analysis and isolated the compounds. Q.X.: activity test. L.Y.: structural elucidation. H.D.: method adviser. Z.Y.: revise the manuscript. Y.Z.: theory and orientation adviser. All authors have read and agreed to the published version of the manuscript.

Funding: This work was financially supported by the National Natural Science Fund of China (81973568), China Agriculture Research System of MOF and MARA (CARS-21), Financial Fund of the Ministry of Agriculture and Rural Affairs, China (NFZX2021), and the Special Fund for Basic Scientific Research of Chinese Academy of Tropical Agricultural Sciences (1630052022030).

Institutional Review Board Statement: Not applicable.

Informed Consent Statement: Not applicable.

Data Availability Statement: Not applicable.

Acknowledgments: The authors are grateful to the members of the analytical group of the Institute of Tropical Bioscience and Biotechnology, Chinese Academy of Tropical Agricultural Sciences for the spectral measurements.

Conflicts of Interest: The authors declare no conflict of interest.

References

1. Paterson, R.R.M. *Ganoderma*-A therapeutic fungal biofactory. *Phytochemistry* **2006**, *67*, 1985–2001. [CrossRef] [PubMed]
2. Wu, X.L.; Guo, J.R.; Liao, Q.Z.; Xie, S.H.; Xiao, M. The resources and ecological distrbution of Ganodermataceae in Hainan island. *Mycosystema* **1998**, *17*, 122–129.
3. Kladar, N.V.; Gavarić, N.S.; Božin, B.N. *Ganoderma*: Insights into anticancer effects. *Eur. J. Cancer Prev.* **2016**, *25*, 462–471. [CrossRef]
4. Baby, S.; Johnson, A.J.; Govindan, B. Secondary metabolites from *Ganoderma*. *Phytochemistry* **2015**, *114*, 66–101. [CrossRef] [PubMed]
5. Xia, Q.; Zhang, H.; Sun, X.; Zhao, H.; Wu, L.; Zhu, D.; Yang, G.; Shao, Y.; Zhang, X.; Mao, X.; et al. A comprehensive review of the structure elucidation and biological activity of triterpenoids from *Ganoderma* spp. *Molecules* **2014**, *19*, 17478–17535. [CrossRef] [PubMed]

6. Cör, A.D.; Knez, Ž.; Knez, M.M. Antioxidant, antibacterial, antitumor, antifungal, antiviral, anti-inflammatory, and nevro-protective activity of *Ganoderma lucidum*: An overview. *Front. Pharmacol.* **2022**, *13*, 934982.
7. Huang, S.Z.; Ma, Q.Y.; Kong, F.D.; Guo, Z.K.; Cai, C.H.; Hu, L.L.; Zhou, L.M.; Wang, Q.; Dai, H.F.; Mei, W.L.; et al. Lanostane-type triterpenoids from the fruiting body of *Ganoderma calidophilum*. *Phytochemistry* **2017**, *143*, 104–110. [CrossRef] [PubMed]
8. Amen, Y.M.; Zhu, Q.; Tran, H.B.; Afifi, M.S.; Halim, A.F.; Ashour, A.; Mira, A.; Shimizu, K. Lucidumol C, a new cytotoxic lanostanoid triterpene from *Ganoderma lingzhi* against human cancer cells. *J. Nat. Med.* **2016**, *70*, 661–666. [CrossRef]
9. Zhang, J.J.; Wang, D.W.; Cai, D.; Lu, Q.; Cheng, Y.X. Meroterpenoids from *Ganoderma lucidum* mushrooms and their biological roles in insulin resistance and triple-negative breast cancer. *Front. Chem.* **2021**, *9*, 772740. [CrossRef]
10. Liu, L.Y.; Chen, H.; Liu, C.; Wang, H.Q.; Kang, J.; Li, Y.; Chen, R.Y. Triterpenoids of *Ganoderma theaecolum* and their hepatoprotective activities. *Fitoterapia* **2014**, *98*, 254–259. [CrossRef]
11. Zhang, X.; Gao, X.; Long, G.; Yang, Y.; Chen, G.; Hou, G.; Huo, X.; Jia, J.; Wang, A.; Hu, G. Lanostane-type triterpenoids from the mycelial mat of *Ganoderma lucidum* and their hepatoprotective activities. *Phytochemistry* **2022**, *198*, 113131. [CrossRef]
12. Wu, Y.L.; Han, F.; Luan, S.S.; Ai, R.; Zhang, P.; Li, H.; Chen, L.X. Triterpenoids from *Ganoderma lucidum* and their potential anti-inflammatory effects. *J. Agric. Food Chem.* **2019**, *67*, 5147–5158. [CrossRef] [PubMed]
13. Peng, X.R.; Wang, Q.; Wang, H.R.; Hu, K.; Xiong, W.Y.; Qiu, M.H. FPR2-based anti-inflammatory and anti-lipogenesis activities of novel meroterpenoid dimers from *Ganoderma*. *Bioorg. Chem.* **2021**, *116*, 105338. [CrossRef]
14. Kou, R.W.; Gao, Y.Q.; Xia, B.; Wang, J.Y.; Liu, X.N.; Tang, J.J.; Yin, X.; Gao, J.M. Ganoderterpene A, a new triterpenoid from *Ganoderma lucidum*, attenuates LPS-induced inflammation and apoptosis via suppressing MAPK and TLR-4/NF-κB pathways in BV-2 cells. *J. Agric. Food Chem.* **2021**, *69*, 12730–12740. [CrossRef] [PubMed]
15. Guo, J.C.; Yang, L.; Ma, Q.Y.; Ge, Y.Z.; Kong, F.D.; Zhou, L.M.; Zhang, F.; Xie, Q.Y.; Yu, Z.F.; Dai, H.F.; et al. Triterpenoids and meroterpenoids with α-glucosidase inhibitory activities from the fruiting bodies of *Ganoderma australe*. *Bioorg. Chem.* **2021**, *117*, 105448. [CrossRef] [PubMed]
16. Chen, X.Q.; Zhao, J.; Chen, L.X.; Wang, S.F.; Wang, Y.; Li, S.P. Lanostane triterpenes from the mushroom *Ganoderma resinaceum* and their inhibitory activities against α-glucosidase. *Phytochemistry* **2018**, *149*, 103–115. [CrossRef] [PubMed]
17. Lu, S.Y.; Peng, X.R.; Dong, J.R.; Yan, H.; Kong, Q.H.; Shi, Q.Q.; Li, D.S.; Zhou, L.; Li, Z.R.; Qiu, M.H. Aromatic constituents from *Ganoderma lucidum* and their neuroprotective and anti-inflammatory activities. *Fitoterapia* **2019**, *134*, 58–64. [CrossRef] [PubMed]
18. Cheng, P.G.; Teoh, T.C.; Rizman-Idid, M. Chemical compounds and computational prediction of their inhibitory effects on the HIV-1 gp120 receptor by lingzhi or reishi medicinal mushroom, *Ganoderma lucidum* (agaricomycetes), with antler-like morphology of fruiting bodies. *Int. J. Med. Mushrooms* **2021**, *23*, 63–77. [CrossRef]
19. Weng, Y.; Xiang, L.; Matsuura, A.; Zhang, Y.; Huang, Q.; Qi, J. Ganodermasides A and B, two novel anti-aging ergosterols from spores of a medicinal mushroom *Ganoderma lucidum* on yeast via UTH1 gene. *Bioorg. Med. Chem.* **2010**, *18*, 999–1002. [CrossRef]
20. Qiu, J.; Wang, X.; Song, C. Neuroprotective and antioxidant lanostanoid triterpenes from the fruiting bodies of *Ganoderma atrum*. *Fitoterapia* **2016**, *109*, 75–79. [CrossRef]
21. Zheng, S.; Zhang, W.; Liu, S. Optimization of ultrasonic-assisted extraction of polysaccharides and triterpenoids from the medicinal mushroom *Ganoderma lucidum* and evaluation of their in vitro antioxidant capacities. *PLoS ONE* **2020**, *15*, e0244749. [CrossRef] [PubMed]
22. Dong, Q.; He, D.; Ni, X.; Zhou, H.; Yang, H. Comparative study on phenolic compounds, triterpenoids, and antioxidant activity of *Ganoderma lucidum* affected by different drying methods. *Food Meas.* **2019**, *13*, 3198–3205. [CrossRef]
23. Cao, Y.; Xu, X.W.; Liu, S.J.; Huang, L.F.; Gu, J. *Ganoderma*: A cancer immunotherapy review. *Front. Pharmacol.* **2018**, *9*, 1217. [CrossRef] [PubMed]
24. Shahid, A.; Huang, M.; Liu, M.; Shamim, M.A.; Parsa, C.; Orlando, R.; Huang, Y. The medicinal mushroom *Ganoderma lucidum* attenuates UV-induced skin carcinogenesis and immunosuppression. *PLoS ONE* **2022**, *17*, e0265615. [CrossRef] [PubMed]
25. Wu, X.L.; Dai, Y.C. *Coloured Illustrations of Ganodermataceae of China*; Science Press: Beijing, China, 2005; pp. 89–91.
26. Su, H.G.; Zhou, Q.M.; Guo, L.; Huang, Y.J.; Peng, C.; Xiong, L. Lanostane triterpenoids from *Ganoderma luteomarginatum* and their cytotoxicity against four human cancer cell lines. *Phytochemistry* **2018**, *156*, 89–95. [CrossRef] [PubMed]
27. Li, X.C.; Liu, F.; Su, H.G.; Peng, C.; Zhou, Q.M.; Liu, J.; Huang, Y.J.; Guo, L.; Xiong, L. Twelve undescribed derivatives of ganoderic acid isolated from *Ganoderma luteomarginatum* and their cytotoxicity against three human cancer cell lines. *Phytochemistry* **2021**, *183*, 112617. [CrossRef] [PubMed]
28. Yang, L.; Kong, D.X.; Xiao, N.; Ma, Q.Y.; Xie, Q.Y.; Guo, J.C.; Deng, C.Y.; Ma, H.X.; Hua, Y.; Dai, H.F.; et al. Antidiabetic lanostane triterpenoids from the fruiting bodies of *Ganoderma weberianum*. *Bioorg. Chem.* **2022**, *127*, 106025. [CrossRef] [PubMed]
29. Zhang, S.S.; Ma, Q.Y.; Huang, S.Z.; Dai, H.F.; Guo, Z.K.; Yu, Z.F.; Zhao, Y.X. Lanostanoids with acetylcholinesterase inhibitory activity from the mushroom *Haddowia longipes*. *Phytochemistry* **2015**, *110*, 133–139. [CrossRef]
30. Huang, S.Z.; Cheng, B.H.; Ma, Q.Y.; Wang, Q.; Kong, F.D.; Dai, H.F.; Qiu, S.Q.; Zheng, P.Y.; Liu, Z.Q.; Zhao, Y.X. Anti-allergic prenylated hydroquinonesand akaloids from the fruiting body of *Ganoderma calidophilum*. *RSC Adv.* **2016**, *6*, 21139–21147. [CrossRef]
31. Guo, J.C.; Kong, F.D.; Ma, Q.Y.; Xie, Q.Y.; Zhang, R.S.; Dai, H.F.; Wu, Y.G.; Zhao, Y.X. Meroterpenoids with protein tyrosine phosphatase 1B inhibitory activities from the fruiting bodies of *Ganoderma ahmadii*. *Front. Chem.* **2020**, *8*, 279. [CrossRef] [PubMed]
32. Su, H.G.; Peng, X.R.; Shi, Q.Q.; Huang, Y.J.; Zhou, L.; Qiu, M.H. Lanostane triterpenoids with anti-inflammatory activities from *Ganoderma lucidum*. *Phytochemistry* **2020**, *173*, 112256. [CrossRef] [PubMed]

33. Chen, B.; Tian, J.; Zhang, J.; Wang, K.; Liu, L.; Yang, B.; Bao, L.; Liu, H. Triterpenes and meroterpenes from *Ganoderma lucidum* with inhibitory activity against HMGs reductase, aldose reductase and α-glucosidase. *Fitoterapia* **2017**, *120*, 6–16. [CrossRef] [PubMed]
34. Fujita, A.; Airsawa, M.; Sage, M.; Hayashi, T.; Morita, N. Two new terpenoids from *Ganoderma lucidum*. *J. Nat. Prod.* **1986**, *49*, 1122–1125. [CrossRef]
35. Sato, H.; Nishitoba, T.; Shirasu, S.; Oda, K.; Sakamura, S. Ganoderiol A and B, new triterpenoids from the fungus *Ganoderma lucidum* (Reishi). *Agric. Biol. Chem.* **1986**, *50*, 2887–2890. [CrossRef]
36. Kennedy, E.M.; P'Pool, S.J.; Jiang, J.; Sliva, D.; Minto, R.E. Semisynthesis and biological evaluation of ganodermanontriol and its stereoisomeric triols. *J. Nat. Prod.* **2011**, *74*, 2332–2337. [CrossRef] [PubMed]
37. Niedermeyer, T.H.J.; Lindequist, U.; Mentel, R.; Gordes, D.; Schmidt, E.; Thurow, K.; Lalk, M. Antiviral terpenoid constituents of *Ganoderma pfeifferi*. *J. Nat. Prod.* **2005**, *68*, 1728–1731. [CrossRef]
38. Gonzalez, A.G.; Leon, F.; Rivera, A.; Munoz, C.M.; Bermejo, J. Lanostanoid triterpenes from *Ganoderma lucidum*. *J. Nat. Prod.* **1999**, *62*, 1700–1701. [CrossRef]
39. Airsawa, M.; Fujita, A.; Sage, M.; Fukumura, H.; Hayashi, T.; Shimizu, M.; Morita, N. Three new lanostanoids from *Ganoderma lucidum*. *J. Nat. Prod.* **1986**, *49*, 621–625. [CrossRef]
40. Nishitoba, T.; Sato, H.; Oda, K.; Sakamura, S. Novel triterpenoids and a steroid from the fungus *Ganoderma lucidum*. *Agric. Biol. Chem.* **1988**, *52*, 211–216. [CrossRef]
41. Liu, S.Y.; Wang, Y.; He, R.R.; Qu, G.X.; Qiu, F. Chemical constituents of *Ganoderma lucidum* (Leys. ex Fr.) Karst. *J. Shenyang Pharm. Univ.* **2008**, *25*, 183–193.
42. Gonzalez, A.G.; Leon, F.; Rivera, A.; Munoz, C.M.; Bermejo, J. New lanostanoids from the fungus *Ganoderma concinna*. *J. Nat. Prod.* **2002**, *65*, 417–421. [CrossRef]
43. Nishitoba, T.; Sato, H.; Sakamura, S. Triterpenoids from the fungus *Ganoderma lucidum*. *Phytochemistry* **1987**, *26*, 1777–1784. [CrossRef]
44. Zhang, P.; Bao, H.Y.; Tolgor. Chemical constituents from sporophore of *Hericium coralloides* (I). *Chin. Trad. Herb. Drugs* **2012**, *43*, 2356–2360.
45. Keller, A.C.; Maillard, M.P.; Hostettmann, K. Antimicrobial steroids from the fungus *Fomitopsis pinicola*. *Phytochemistry* **1996**, *41*, 1041–1046. [CrossRef]
46. Wan, H.; Sun, R.Q.; Wu, D.J.; Guo, B.S. Three steriols from *Gyroporus castaneus*. *Nat. Prod. Res. Dev.* **1999**, *11*, 18–20.
47. Gao, H.; Hong, K.; Zhang, X.; Liu, H.W.; Wang, N.L.; Zhuang, L.; Yao, X.S. New steryl esters of fatty acids from the mangrove fungus *Aspergillus awamori*. *Helv. Chim. Acta* **2007**, *90*, 1165–1178. [CrossRef]
48. Piccialli, V.; Sica, D. Four new trihydroxylated sterols from the sponge *Spongionella gracilis*. *J. Nat. Prod.* **1987**, *50*, 915–920. [CrossRef]
49. Kawahara, N.; Sekita, S.; Satake, M. Two steroids from *Calvatia cyathiformis*. *Phytochemistry* **1995**, *38*, 947–950. [CrossRef]
50. Jiao, Y.; Xie, T.; Zou, L.H.; Wei, Q.; Qiu, L.; Chen, L.X. Lanostane triterpenoids from *Ganoderma curtisii* and their NO production inhibitory activities of LPS-induced microglia. *Bioorg. Med. Chem. Lett.* **2016**, *26*, 3556–3561. [CrossRef]

molecules

MDPI

Article

Antibacterial Activity of *Syzygium aromaticum* (Clove) Bud Oil and Its Interaction with Imipenem in Controlling Wound Infections in Rats Caused by Methicillin-Resistant *Staphylococcus aureus*

Abdulaziz Khaleef Alanazi [1,2], Mohammed Hussein Alqasmi [1], Mohammed Alrouji [1], Fahd A. Kuriri [1], Yasir Almuhanna [1,*], Babu Joseph [1] and Mohammed Asad [1]

1 Department of Clinical Laboratory Sciences, College of Applied Medical Sciences, Shaqra University, Shaqra 11961, Saudi Arabia
2 Medical Laboratory, Hafar Albatin Central Hospital, Hafar Albatin 39513, Saudi Arabia
* Correspondence: yalmuhanna@su.edu.sa; Tel.: +966-506253022

Abstract: Methicillin-resistant *Staphylococcus aureus* (MRSA) is one of the leading causes of infection worldwide. Clove oil's ability to inhibit the growth of MRSA was studied through in vitro and in vivo studies. The phytochemical components of clove oil were determined through gas chromatography-mass spectrometry (GC-MS) analysis. The antibacterial effects of clove oil and its interaction with imipenem were determined by studying *MIC*, MBC, and *FIC* indices in vitro. The in vivo wound-healing effect of the clove oil and infection control were determined using excision wound model rats. The GC-MS analysis of clove oil revealed the presence of 16 volatile compounds. Clove oil showed a good antibacterial effect in vitro but no interaction was observed with imipenem. Clove bud oil alone or in combination with imipenem healed wounds faster and reduced the microbial load in wounds. The findings of this study confirmed the antibacterial activity of clove oil in vitro and in vivo and demonstrated its interaction with imipenem.

Keywords: excision wound; epithelization; essential oils; wound healing

Citation: Alanazi, A.K.; Alqasmi, M.H.; Alrouji, M.; Kuriri, F.A.; Almuhanna, Y.; Joseph, B.; Asad, M. Antibacterial Activity of *Syzygium aromaticum* (Clove) Bud Oil and Its Interaction with Imipenem in Controlling Wound Infections in Rats Caused by Methicillin-Resistant *Staphylococcus aureus*. *Molecules* **2022**, 27, 8551. https://doi.org/10.3390/molecules27238551

Academic Editor: Yongjun Wei

Received: 23 October 2022
Accepted: 30 November 2022
Published: 5 December 2022

1. Introduction

Staphylococcus aureus is a commensal microbe that often causes wound infections [1]. It is a multidrug-resistant organism that colonizes skin, wounds, the nose, the perineum, and the throat, leading to infections in various organs [2]. Approximately one in ten hospital admissions worldwide results in hospital-acquired infections and of these 8% are caused by MRSA [3].

The hunt for novel antibacterial chemicals including herbal products is rapidly increasing in order to produce better medications for combating multidrug resistance organisms [4]. Medicinal products derived from aromatic plants are widely used in traditional medicines for controlling bacterial infections [5]. Many reports indicate that antibiotic-resistant strains can be controlled by using essential oils [6]. These oils are highly volatile, hydrophobic, and lipophilic with numerous biological and pharmacological properties [7].

Syzygium aromaticum, commonly called clove, (family-Myrtaceae), is indigenous to the islands of Indonesia. It is now cultivated worldwide as a flavoring agent, for medicinal purposes, and for use in perfumes [8]. It is widely used as a food preservative due to its antimicrobial properties [9]. Apart from its antimicrobial properties, antioxidant, analgesic, anesthetic, anti-inflammatory, and insecticidal activity have been reported [10]. It is also used for preventing degenerative diseases because of its antioxidant effects [11]. Clove bud oil is a common healing agent for wounds and burns in traditional medicine [12]. The antibacterial and wound infection control efficacy of clove bud oil and its interaction with imipenem were determined in the present study. A detailed phytochemical screening was also carried out by gas chromatography-mass spectrometry (GC-MS) to detect volatile components in the oil.

2. Results

2.1. Gas Chromatography-Mass Spectrometric Analysis

The clove oil was subjected to silylation for the derivatization of constituents to facilitate GC-MS analysis. The spectrum produced via gas chromatography is shown in Figure 1. The list of compounds detected is shown in Table 1. Eugenol, caryophyllene, and 2-(octadecyloxy)-ethanol were the major compounds identified by the GC-MS analysis. The calculated retention indices and their comparison with the retention indices from the literature, with references included, are given as supplementary data (Table S1).

Table 1. Major components identified by GCMS of *S. aromaticum* oil.

Peak	Name	Retention Time	Area (%)	RI $_{cal}$	RI $_{lit}$	Reference
1.	Cyclohexanone, 5-methyl-2-(1-methylethyl)-, trans-	7.431	0.13	1114	1124	[13]
2.	* Levomenthol	7.709	0.28	1123	1143.95	[14]
3.	* Phenol, 4-(2-propenyl)-	8.664	0.05	1151		–
4.	* Eugenol	10.097	72.54	1388	1378	[15]
5.	Caryophyllene	10.731	5.16	1405	1392.14	[14]
6.	Humulene (α- Caryophyllene)	11.097	1.55	1415	1428	[16]
7.	Bicyclosesquiphellandrene	11.219	0.06	1519	1521	[17]
8.	α-Farnesene	11.475	0.13	1525	1522	[18]
9.	Naphthalene, 1,2,3,5,6,8a-hexahydro-4,7-dimethyl-1-(1-methylethyl)-, (1S-cis)-	11.686	0.25	1531	1528	[19]
10.	Caryophyllene oxide	12.408	0.22	1538	1537	[20]
11.	* 1-Decanol, 2-hexyl	15.163	0.09	1492	1504	[21]
12.	Heptadecane	15.741	0.15	1719		–
13.	Hexatriacontyl pentafluoropropionate	17.308	0.50	1962		–
14.	Heptadecane, 2,6,10,15-tetramethyl-	17.563	0.59	2168		–
15.	1-Docosene	17.774	0.051	2179	2190	[22]
16.	* 2-(octadecyloxy)ethanol	18.285	1.00	1284		–

* Indicates compounds were derivatized by silylation.

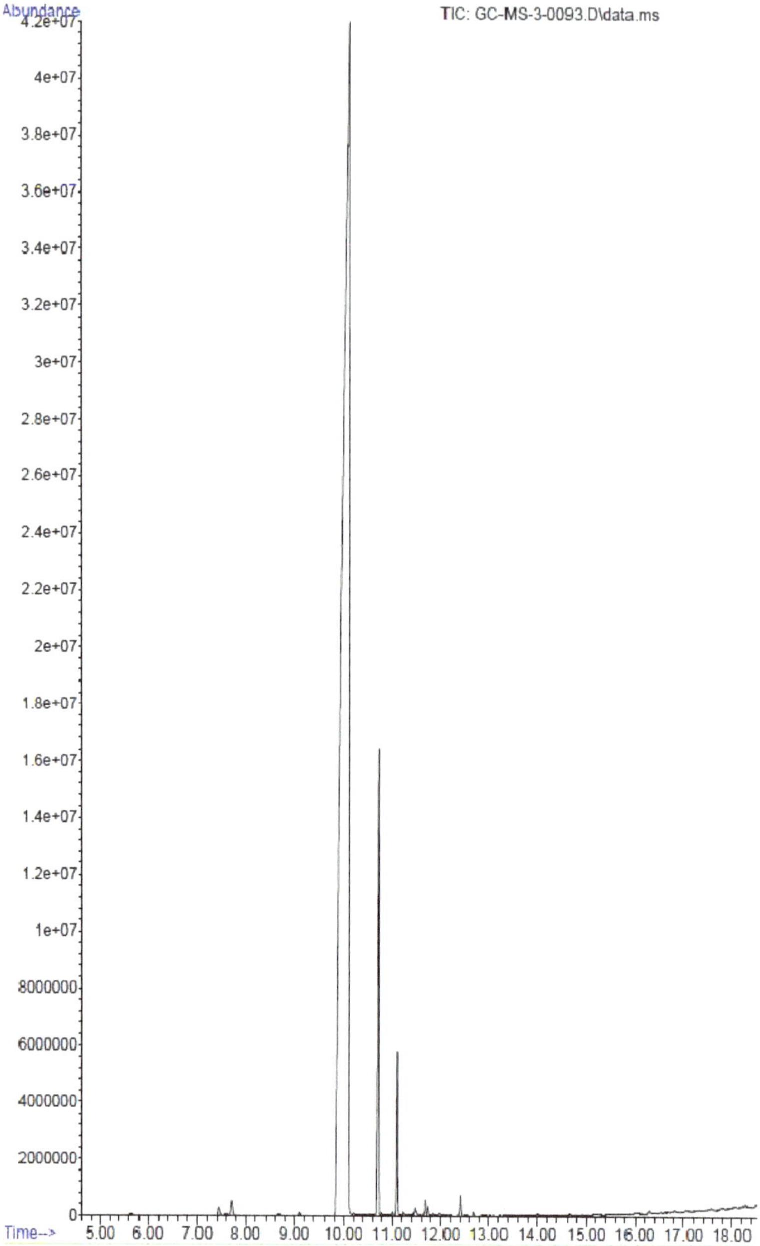

Figure 1. GC-MS spectra of clove oil.

2.2. Antibacterial Activity In Vitro

Clove oil showed good antibacterial effects at 20 µL/mL and 40 µL/mL against MRSA. The zone of inhibition was not affected when it was combined with imipenem (Table 2). Concentrations of less than 20 µL/mL showed a much lower inhibition zone.

Table 2. Antibacterial activity of clove oil and its combination with imipenem against MRSA.

Antibacterial Agent	Zone of Inhibition (mm)
Clove oil (20 µL/mL)	10
Clove oil (40 µL/mL)	13
Imipenem (4 µL/mL)	22
Imipenem (4 µL/mL) + Clove oil (40 µL/mL)	22

Values are replicates from three measurements.

The MRSA strain showed a resistant pattern toward most of the antibiotics, as shown in Table S2. It was sensitive to linezolid, vancomycin, and minocycline. An intermediate degree of sensitivity was observed for imipenem, and this was selected as a standard antibiotic for the study (Table S1).

The *MIC* of clove oil was found to be 2.5 µL/mL and a concentration of 5 µL/mL was found to be the MBC against MRSA. There were no colonies formed on the mannitol salt agar when a loop of culture was inoculated from the tubes containing 5 µL/mL clove oil and above (Table 3).

Table 3. *MIC* and MBC of clove oil against MRSA.

Clove Oil (µL/mL)	Visible Growth (*MIC*)	Growth on Mannitol Salt Agar (MBC)
20	−	−
10	−	−
5	−	−
2.5	−	+
1.25	++	+
0.62	++	+
0.31	++	++

+ indicates visible growth, ++ indicates more growth and – shows no observed growth.

The clove oil showed an antibacterial effect when tested separately; however, there were no antagonistic or synergetic effects when it was combined with imipenem (Table S2).

2.3. Wound-Healing and Skin Irritation Study

The period of epithelization of the wounds was determined on each day after the application of the ointment formulation. The inoculation of wounds with MRSA significantly reduced the epithelization period.

The wounds were healed, and epithelization was significantly quicker after different treatments in comparison to the control (Figure 2). Of all the treatments, the lowest period of epithelization was observed in the combination group followed by imipenem. The effect of 10% clove oil was comparable to that observed with imipenem. The lower concentration of clove oil (5%) was the least effective. However, there was no significant difference between the different concentrations of clove oil and imipenem, indicating a lack of dose-dependent response or interaction between imipenem and clove oil.

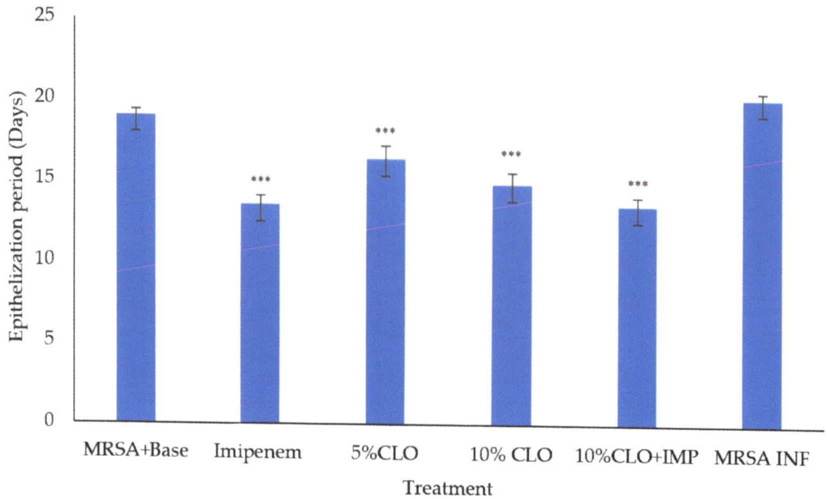

Figure 2. Epithelization period after different treatments. Values are mean±SEM for six animals; *** $p < 0.001$ in comparison to MRSA+base (CLO—clove oil, IMP—imipenem, and MRSA INF—MRSA infected).

The degree of wound contraction was measured at 4-day intervals for up to 20 days. The MRSA strain aggravated the wounds, leading to fluid exudation from the wounded tissues in all the animals with few deaths (two out of eight animals) in the control group. The measurements obtained from the six animals throughout the treatment period were ensured by adding more animals in the groups where mortality was observed. The clove oil (5 % *w*/*w*) did not show any significant change in wound healing on day 4, while it was less effective than clove oil (10% *w*/*w*) and imipenem in contracting the wounds on the other days of measurement (Figure 3). However, this difference was not significant. The clove oil (10% *w*/*w*) and imipenem improved the healing of wounds from the 4 day onwards. A similar result was observed in the combination group of clove oil 10% (*w*/*w*) and imipenem (Figure 4).

In the histological examination, a worn-out epidermis was seen in the skin sections of the control animals. Of all the treatments, the imipenem and clove oil combination showed better healing followed by clove oil (10% *w*/*w*) and clove oil (5% *w*/*w*). The epidermal layer was damaged in the control animals infected and not subjected to treatments, with a large bulk of inflammatory cells and fewer capillaries. Imipenem and clove oil (10% *w*/*w*) and their combination improved epidermal regeneration and angiogenesis (Figure 5).

The log CFU was significantly lower in all the treatment groups when compared to the control group with respect to all treatments. No significant difference was seen between the imipenem-treated group and the group that received a combination of imipenem and clove oil (10%). This indicates a lack of interaction between clove oil and imipenem (Table 4).

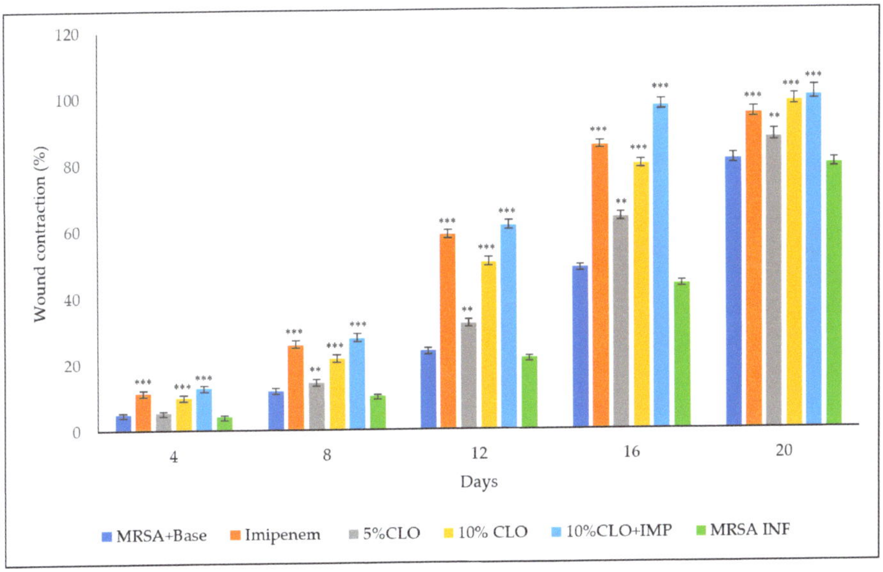

Figure 3. Wound contraction (%) after different treatments. Values are mean ± SEM for six animals; ** $p < 0.01$ and *** $p < 0.001$ in comparison to MRSA+base (CLO—clove oil. Imp—imipenem, and MRSA INF—MRSA-infected).

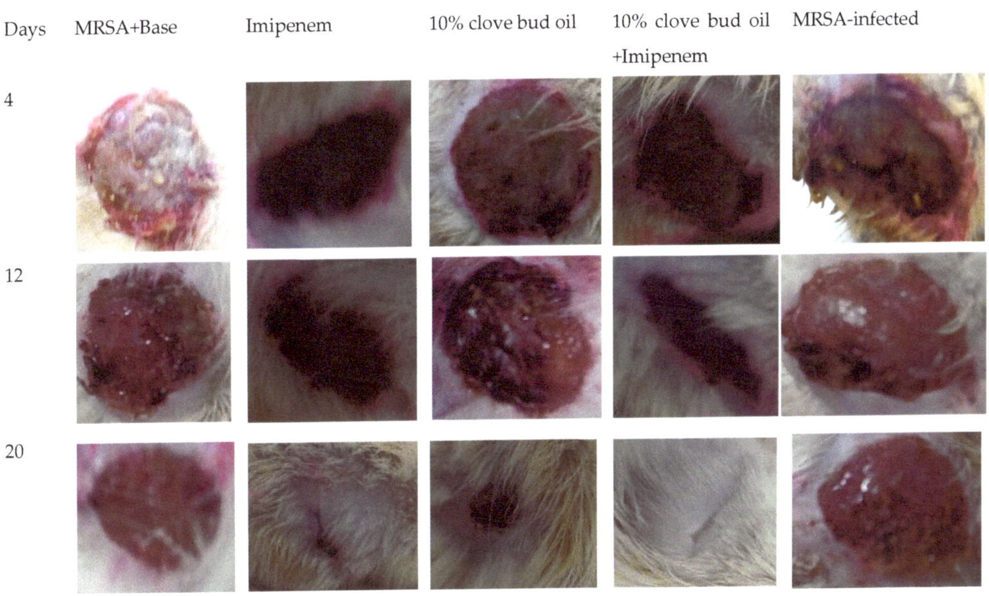

Figure 4. Images of wounds infected and treated with different concentrations of clove oil and imipenem.

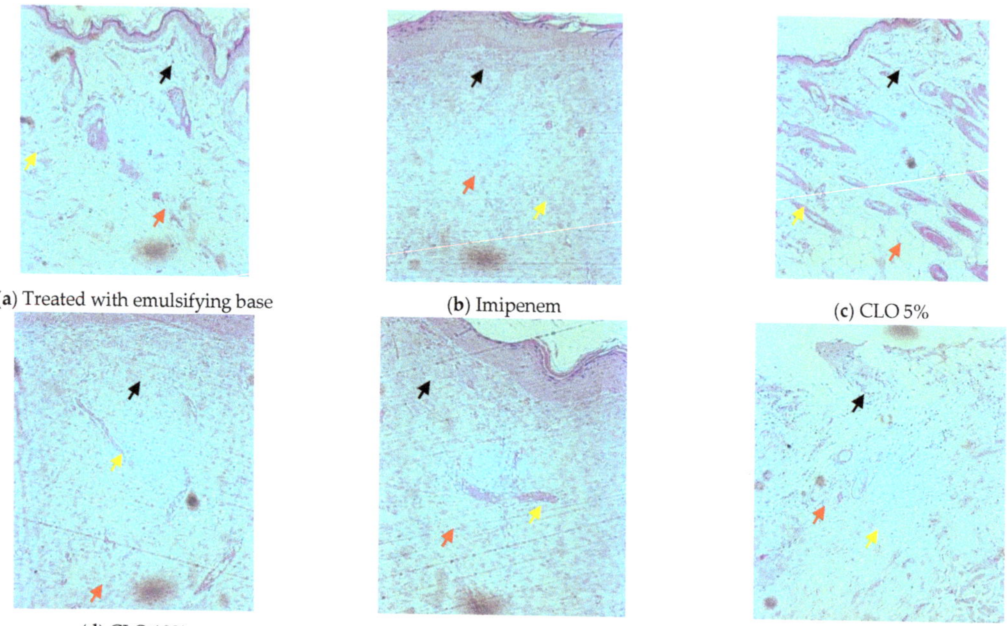

(**a**) Treated with emulsifying base (**b**) Imipenem (**c**) CLO 5% (**d**) CLO 10% (**e**) Imipenem + CLO 10% (**f**) MRSA-INF

Figure 5. Photomicrographs indicate wound healing after different treatments. Epidermis (black arrow); capillaries (yellow arrow); inflammatory cells (red arrow). (CLO—clove oil, MRSA INF—MRSA-infected).

Table 4. The bacterial load (CFU/g) in the skin tissue after treatment.

Treatment	Bacterial Load (Log$_{10}$ CFU/g of Tissue)
MRSA+base	10.34 ± 0.49
MRSA-infected + Imipenem	4.11 ± 0.48 ***
MRSA-infected + 5% CLO	5.89 ± 0.47 ***
MRSA-infected + 10% CLO	3.95 ± 0.33 ***
MRSA-infected + imipenem + 10% clove oil	3.1 ± 0.32 ***
MRSA-infected wounds	11.07 ± 0.41 ^{ns}

Values are mean ± SEM for six animals *** $p < 0.001$, ^{ns} Non-significant change in comparison to MRSA+base (CLO—clove oil).

No noticeable redness (erythema) or inflammatory reaction was seen until 3 days after the application of the clove oil formulation in the skin irritation test.

3. Discussion

The present study examined the outcome of clove oil treatment on different features of wound healing. The determination of the antibacterial action showed its effect against MRSA, which is known to infect wounds. The histological study on the wounded tissues was carried out to support clove oil's wound-healing activity and infection control ability. The GC-MS analysis revealed the various chemical constituents present in the oil. The major constituents of essential oils evaporate above 60 °C [23]. Gas chromatography is used to separate volatile compounds; hence, GC-MS analysis was carried out to identify the different volatile compounds present in the clove essential oil. The silylation procedure was carried out to mask the polar groups that may be present in the components of essential oil. Silylation is the most common procedure used to reduce the polarity of analytes and improve their stability for gas chromatography. Apart from this, other procedures such as alkylation are used. However, there are no reports suggesting that one method is superior to the other for the derivatization of essential oils. The mass spectra obtained were used to identify compounds via the NIST library. The RI values of the marked compounds in Table 1 are for silyl derivatives, as these were derivatized by silylation. The RI values from the literature were taken from the NIST libraries. The RI from the literature was selected by distinguishing the closed match with our calculated RI values.

Clove oil has been reported to possess several biological activities, including antibacterial, anti-inflammatory, analgesic, antioxidant, fungicidal, and antitumor activities [24]. Clove oil contains several active constituents, including pinene (α and β), neral, geranial, γ-terpinene, cis-ocimene, allo-ocimene, 1,8-cineole, linalool, borneol, myrcene, and pinene-2-ol [24]. In the present study, 16 components were identified by GC-MS analysis. Eugenol, caryophyllene, and 2-(octadecyloxy)-ethanol were the major constituents. All these constituents have been previously reported [25].

There are several bases available for the formulation of ointments; the emulsifying base prepared following the British pharmacopeial method was used in the present study [26]. To demonstrate that the base used for the preparation of the ointment is inert, a separate group (MRSA-infected) was established, and this group was compared with the MRSA control, which was administered with the base only. No significant difference in wound healing was observed and this demonstrated the inertness of the base. The diffusion of active antimicrobial compounds from the clove oil ointment formulation was confirmed by studying the passage through the Muller Hinton agar medium (data not shown). The microorganism used for wound infection was chosen after a literature review in order to produce the most challenging infection [27]. Clove oil has been reported to possess antibacterial activity against MRSA [28]. Imipenem served as positive control and the selection was made based on an antibiotic sensitivity test where an intermediate degree of sensitivity was observed for this antibiotic against MRSA. The pathogen showed resistance to amikacin [29], amoxicillin/k clav [30], ampicillin [31], ceftaroline [32], ciprofloxacin [33], clindamycin [34], erythromycin [35], gentamicin [36], levofloxacin [37], oxacillin [38], tetracycline [39], and trimethoprim/sulfa [40].

Earlier studies on essential oils reported antibacterial effects against Gram-negative bacterial strains causing urinary tract infections [41]. The anti-MRSA potential of clove oil has also been reported and has highlighted the need to explore bioactive constituents [42]. The phytochemicals present in clove bud oil are responsible for antimicrobial activity. Eugenol is one of the phytoconstituents that might have significantly contributed to the antimicrobial effects [43]. Eugenol is a primary constituent of clove but it is also found in cinnamon, pepper, and *Ocimum sanctum* [44]. Another constituent that is responsible for both antibacterial and wound-healing properties is caryophyllene [45,46]. Apart from clove, several other plants, such as basil (*Ocimum* spp.), cannabis, lavender, and black pepper, contain caryophyllene [47]. Other constituents such as levomenthol are used in mouthwashes for antimicrobial effects [48]. However, levomenthol is the main constituent of *Mentha piperita* [49], and is found only in smaller amounts in clove. Humulene is another constituent and is a cannabis terpene similar to caryophyllene with antibacterial effects [50]. However, the minor components might also contribute considerably by merging with other major components for controlling the growth of MRSA. The hydrophobic nature of the oil may also help in interacting with the outer cytoplasmic membrane of MRSA and affects the integrity and the functioning of the cell membrane [51].

Each of the parameters, such as wound contraction and the histological studies, indicates a stage of healing. Clove oil effectively increased the healing of excision wounds without causing any noticeable irritation to the skin. Each of the parameters used in the present study discloses factors that contributed to the wound-healing action. The wound contraction indicates improvement in wound closure while the complete healing of the wound is shown by the fading of the scar, which was taken to be the period of epithelization [52]. The clove oil formulation increased the healing of wounds alone as well as in combination with imipenem in the infected wounds, though the effect produced by the combination was not significantly different from either of the treatments alone. Histological studies were carried out to confirm the results of the macroscopic observations. Cellular organelles can be easily visualized after staining with H and E [53]. The epidermis is the outermost layer of the skin, and its thickness shows the regeneration of the epithelial lining. The number of capillaries indicates the increased formation of blood vessels that are required in the wound-healing process. Inflammation is an early reaction to an injury, and it is mediated through various inflammatory cells that include macrophages, neutrophils, and plasma cells. However, the presence of inflammatory cells in the wounded tissue several days after wounding shows incomplete wound healing [54]. An increase in epidermal height and capillary number and a decrease in inflammatory cells after the clove oil treatment showed increased healing of wounds.The fast and rapid wound healing activity of clove oil may be due to the antioxidant and anti-inflammatory activities of its phytochemical contents in addition to its antibacterial effect, especially with respect to phenolic compounds [55,56]. However, there may be additional mechanisms that have to be explored. A study of the individual components of the clove and their effects in combination may provide more insight into the mechanism(s) underlying their wound-healing action, contribution, and/or interaction towards this effect.

4. Materials and Methods

4.1. GC-MS Analysis of Clove Oil

A GC-MS 7890A GC system with 5975C VL MSD from Agilent Technologies (California, USA) was used. Clove oil (100 μL) was mixed with 250 μL of water and 750 μL of ethyl acetate. Following this, the upper layer was separated and concentrated. To this, a 50 μL mixture of N, O-Bis (trimethylsilyl) trifluoroacetamide (49.5 μL) and trimethylchlorosilane (0.5 μL) was added followed by the addition of pyridine (10 μL). This was heated for 30 min at 60 °C and the contents were dried using liquid nitrogen before finally dissolving the dried sample (20 mg) in methanol (5 ml) for analysis. After filtration through a membrane filter (0.22 μm), 3 μL was injected through a capillary column (30 m, 0.25 mm, and 0.25-microns) with an injector temperature of 270 °C and pressure at 80 kPa. The carrier gas was hydrogen and the total time for analysis was 25 min.

For determination of calculated kovat's retention index, a homologous series of n-alkanes were injected into the same GC-MS operating system. The retention indices of isolated constituents were determined using the following formula [57]:

$$I = 100[n + (N - n) \times (\log t_r \text{ (unknown)} - \log t_r \text{ (n)})/\log t_r(N) - \log t_r(n)) \quad (1)$$

where

I = Kovat's index;

n= number of carbon atoms in smaller n-alkane;

N= number of carbon atoms in larger n-alkane;

t_r = retention time.

The referenced Kovat's retention indices were obtained from the NIST library. The calculated Kovat's retention indices were matched with those from the reported literature.

4.2. Antibiotic Sensitivity Test

MRSA isolates available in the laboratory (ATCC43300) were tested for drug resistance and a methicillin-resistant organism was selected for the study. The antibiotic sensitivity test for the determination of multidrug resistance was carried out using the Microscan system. The trays were read by using an autoSCAN-4 reader at 620, 560, 505,470, 440, and 590 nm, and the results were recorded.

4.3. Antibacterial Assay (Well Diffusion Assay)

The antibacterial evaluation of clove oil was carried out by the well diffusion method. The wells were loaded with oil dissolved in dimethylsulfoxide (DMSO) [58]. DMSO

(10 % v/v) served as a control. The zone of inhibition was measured after incubation at 37 °C for 24 h.

4.4. Minimum Inhibitory Concentration (MIC) and Minimum Bactericidal Concentration (MBC)

Muller Hinton broth was inoculated with MRSA containing 1.5×10^8 CFU/mL (0.5 McFarland standard turbidity). The *MIC* and MBC were determined from the final concentration of clove oil (1.25, 2.5, 5, 10, 20, and 40 µL/mL) in broth. The minimum concentration of clove oil that inhibited the visible growth was taken as *MIC*. The minimum concentration at which no bacterial growth was seen after culturing was taken as MBC.

4.5. Interaction of Clove Oil with Imipenem against MRSA

Antagonistic or synergetic assay of the oil with imipenem was determined by broth dilution method using checkerboard assay [59]. Different concentrations of antibiotic and oil were used. The fractional inhibitory concentration (*FIC*) index was determined using the following formulae:

$$FIC\ index = FIC_{\ oil} + FIC_{\ antibiotic}$$

$$FIC_{\ oil} = MIC_{\ oil\ +\ antibiotic}/MIC_{\ oil}$$

$$FIC_{\ antibiotic} = MIC_{\ oil\ +\ antibiotic}/MIC_{\ antibiotic}$$

FIC index of ≤ 0.5 indicates the synergetic effect of the combination; values between 0.5 and ≤ 2 suggest that the combination is indifferent; and an *FIC index* > 2 shows that the combination is antagonistic [60].

4.6. Wound-Healing Activity in Rats (Excision Wound Model)

Albino Wistar rats of either sex weighing between 220 to 232 g and maintained under controlled conditions were utilized for this study. All procedures performed on the animals followed ARRIVE guidelines. The Ethical Research Committee of Shaqra University approved the study (Approval number—53/18910). Persons handling the animals exercised the utmost care to prevent the transmission of MRSA infection. All parameters were measured by individuals unaware of the treatment. Emulsifying bases prepared by mixing soft paraffin (50%), liquid paraffin (20%), and emulsifying wax (30%) were used to formulate clove oil ointment [61]. Two different concentrations (5% w/w and 10% w/w) of clove oil were employed. Physicochemical properties, stability, and diffusion of the prepared ointment were tested [62]. The emulsifying base was applied to control animals.

Animals were divided into six groups (n = 6):

Group I—Control (emulsifying base);

Group II—Imipenem;

Group III—Clove oil (5% w/w in emulsifying base);

Group IV—Clove oil (10% w/w in emulsifying base);

Group V—Clove oil (10% w/w in emulsifying base) + imipenem;

Group VI—Infected control (MRSA).

To inflict the excision wounds, the dorsal thoracic region of the animals was shaved after anesthetizing them with a combination of ketamine (91 mg) and xylazine (9.1 mg) at a dose of 1 mL/kg administered intraperitoneally [63]. A circular area (500 mm^2) was marked and this was excised to full thickness [64]. The broth culture MRSA (1.5×10^8 CFU/mL) was inoculated into the freshly excised wound [65]. Animals were given different treatments, as mentioned above, once per day starting one day after inoculation of MRSA. To prevent the biting of wounds, animals were individually caged. The formulation was prepared and applied to the wounds. The wound area was determined every 4 days until the formed scab fell off. Wound contraction (%) was determined in each group. The day the scabs fell off was considered the day of epithelization. The healed area of the wound was subjected to histological examination and one for determination of CFU/g tissue. The tissue samples (1 g) were collected on the final day of the experiment and homogenized for 5 min using phosphate buffer saline (1 mL) via an aseptic technique. Homogenates were serially diluted (up to 10^9) and plated on nutrient agar. Plates were incubated at 37 °C and the colonies were counted accordingly.

4.7. *Skin Irritation Test*

Clove oil ointment was applied to the shaved skin on the dorsal side of the rats followed by the area's covering with adhesive tape. Skin reactions that manifested as redness or inflammation (if any) were observed every 12 h for three days [66].

4.8. *Statistical Analysis*

The mean $\pm$ SEM values were used to analyze statistical significance by one-way ANOVA with Tukey's post hoc test (SPSS version 20 for Windows).

5. Conclusions

The clove oil showed good antibacterial activity against MRSA in vitro with an *MIC* of 1.25 µL /ml and an MBC of 2.5 µL/mL, though no interaction was observed with imipenem. The results regarding the in vitro antibacterial effects were further confirmed by a significant reduction in microbial load in the excision-wound-healing model rats. The GC-MS analysis of the clove oil showed the presence of 16 volatile components that have been previously reported to possess antibacterial effects.

Supplementary Materials: The following supporting information can be downloaded at: https://www.mdpi.com/article/10.3390/molecules27238551/s1, Table S1: Antibiotic resistance profile of MRSA; Table S2. The fractional inhibitory concentration of clove oil against MRSA.

Author Contributions: Conceptualization, Y.A., M.A. (Mohammed Asad) and B.J.; methodology, B.J., A.K.A., M.H.A. and M.A. (Mohammed Asad); validation, Y.A., M.A. (Mohammed Asad) and F.A.K.; formal analysis, M.A. (Mohammed Asad) and B.J.; investigation, Y.A.; M.H.A. and A.K.A.; resources, Y.A., M.A. (Mohammed Alrouji) and M.A. (Mohammed Asad); data curation, M.H.A., A.K.A. and F.A.K.; writing—original draft preparation, B.J. and M.A. (Mohammed Asad); writing—review and editing, M.A. (Mohammed Alrouji); visualization, B.J. and M.A. (Mohammed Asad); supervision, Y.A. and M.A. (Mohammed Asad); project administration, A.K.A. and F.A.K.; funding acquisition, M.A. (Mohammed Alrouji) and F.A.K. All authors have read and agreed to the published version of the manuscript.

Funding: This research received no external funding.

Institutional Review Board Statement: The research protocol was reviewed and approved by the Ethical Research Committee of Shaqra University (Approval number–53/18910). The methods used in the current study were standard methods and were in accordance with the ARRIVE guidelines.

Informed Consent Statement: Not applicable.

Data Availability Statement: Not applicable.

Acknowledgments: The authors would like to thank the Deanship of Scientific Research at Shaqra University for supporting this work.

Conflicts of Interest: The authors declare no conflict of interest.

References

1. Díez-Pascual, A.M. Antibacterial Action of Nanoparticle Loaded Nanocomposites Based on Graphene and Its Derivatives: A Mini-Review. *Int. J. Mol. Sci.* **2020**, *21*, 3563. [CrossRef] [PubMed]
2. Nandhini, P.; Kumar, P.; Mickymaray, S.; Alothaim, A.S.; Somasundaram, J.; Rajan, M. Recent Developments in Methicillin-Resistant Staphylococcus aureus (MRSA) Treatment: A Review. *Antibiotics* **2022**, *11*, 606. [CrossRef] [PubMed]
3. Monegro, A.F.; Muppidi, V.; Regunath, H. Hospital Acquired Infections. *Cambridge Handb. Psychol. Health Med. Second Ed.* **2022**, 736–738. [CrossRef]
4. Sannathimmappa, M.B.; Nambiar, V.; Aravindakshan, R. Antibiotics at the crossroads-Do we have any therapeutic alternatives to control the emergence and spread of antimicrobial resistance? *J. Educ. Health Promot.* **2021**, *10*, 438. [CrossRef]

5. Yazarlu, O.; Iranshahi, M.; Kashani, H.R.K.; Reshadat, S.; Habtemariam, S.; Iranshahy, M.; Hasanpour, M. Perspective on the application of medicinal plants and natural products in wound healing: A mechanistic review. *Pharmacol. Res.* **2021**, *174*, 105841. [CrossRef]
6. Mohd Israfi, N.A.; Mohd Ali, M.I.A.; Manickam, S.; Sun, X.; Goh, B.H.; Tang, S.Y.; Ismail, N.; Abdull Razis, A.F.; Ch'ng, S.E.; Chan, K.W. Essential oils and plant extracts for tropical fruits protection: From farm to table. *Front. Plant Sci.* **2022**, *13*, 3644. [CrossRef]
7. Oliveira, T.A.S.; Santiago, M.B.; Santos, V.H.P.; Silva, E.O.; Martins, C.H.G.; Crotti, A.E.M. Antibacterial Activity of Essential Oils against Oral Pathogens. *Chem. Biodivers.* **2022**, *19*, e202200097. [CrossRef]
8. Batiha, G.E.S.; Beshbishy, A.M.; Tayebwa, D.S.; Shaheen, H.M.; Yokoyama, N.; Igarashi, I. Inhibitory effects of Syzygium aromaticum and Camellia sinensis methanolic extracts on the growth of Babesia and Theileria parasites. *Ticks Tick. Borne. Dis.* **2019**, *10*, 949–958. [CrossRef]
9. Kumar Pandey, V.; Shams, R.; Singh, R.; Dar, A.H.; Pandiselvam, R.; Rusu, A.V.; Trif, M. A comprehensive review on clove (*Caryophyllus aromaticus* L.) essential oil and its significance in the formulation of edible coatings for potential food applications. *Front. Nutr.* **2022**, *9*, 2114. [CrossRef]
10. Charoonratana, T. Clove (*Syzygium aromaticum*) oleoresins. *Chem. Funct. Appl.* **2022**, 49–65. [CrossRef]
11. Astuti, R.I.; Listyowati, S.; Wahyuni, W.T. Life span extension of model yeast Saccharomyces cerevisiae upon ethanol derived-clover bud extract treatment. *IOP Conf. Ser. Earth Environ. Sci.* **2019**, *299*, 012059. [CrossRef]
12. Kennewell, T.; Mashtoub, S.; Howarth, G.; Cowin, A.; Kopecki, Z. Antimicrobial and healing-promoting properties of animal and plant oils for the treatment of infected wounds. *Wound Pract. Res.* **2019**, *27*, 175–183. [CrossRef]
13. Senatore, F.; Soria, E.U.; Soria, R.U.; Porta, G.D.; Taddeo, R.; De Feo, V. Essential Oil of *Eremocharis triradiata* (Wolff.) Johnston (Apiaceae) Growing Wild in Perú. *Flavour Fragr. J.* **1997**, *12*, 257–259. [CrossRef]
14. Hoskovec, M.; Grygarová, D.; Cvačka, J.; Streinz, L.; Zima, J.; Verevkin, S.P.; Koutek, B. Determining the vapour pressures of plant volatiles from gas chromatographic retention data. *J. Chromatogr. A* **2005**, *1083*, 161–172. [CrossRef]
15. Bouzouita, N.; Kachouri, F.; Hamdi, M.; Chaabouni, M.M. Antimicrobial activity of essential oils from Tunisian aromatic plants. *Flavour Fragr. J.* **2003**, *18*, 380–383. [CrossRef]
16. Mahmood, U.; Kaul, V.K.; Acharya, R. Volatile constituents of *Capillipedium parviflorum*. *Phytochemistry* **2004**, *65*, 2163–2166. [CrossRef]
17. Yu, Y.; Huang, T.; Yang, B.; Liu, X.; Duan, G. Development of gas chromatography–mass spectrometry with microwave distillation and simultaneous solid-phase microextraction for rapid determination of volatile constituents in ginger. *J. Pharm. Biomed. Anal.* **2007**, *43*, 24–31. [CrossRef]
18. De Feo, V.; Della Porta, G.; Soria, E.U.; Soria, R.U.; Senatore, F. Composition of the essential oil of Tagetes filifolia Lag. *Flavour Fragr. J.* **1998**, *13*, 145–147. [CrossRef]
19. Shatar, S. Essential Oil of *Ferula ferulaoides* from Western Mongolia. *Chem. Nat. Compd.* **2005**, *41*, 607–608. [CrossRef]
20. Nagalakshmi, M.A.H.; Thangadurai, D.; Anuradha, T.; Pullaiah, T. Essential oil constituents of Melia dubia, a wild relative of Azadirachta indica growing in the Eastern Ghats of Peninsular India. *Flavour Fragr. J.* **2001**, *16*, 241–244. [CrossRef]
21. Özel, M.Z.; Göğüş, F.; Lewis, A.C. Comparison of direct thermal desorption with water distillation and superheated water extraction for the analysis of volatile components of Rosa damascena Mill. using GCxGC-TOF/MS. *Anal. Chim. Acta* **2006**, *566*, 172–177. [CrossRef]
22. Adams, R.P.; Morris, J.A.; Pandey, R.N.; Schwarzbach, A.E. Cryptic speciation between Juniperus deltoides and Juniperus oxycedrus (Cupressaceae) in the Mediterranean. *Biochem. Syst. Ecol.* **2005**, *33*, 771–787. [CrossRef]
23. Kant, R.; Kumar, A. Review on essential oil extraction from aromatic and medicinal plants: Techniques, performance and economic analysis. *Sustain. Chem. Pharm.* **2022**, *30*, 100829. [CrossRef]
24. Oboh, G.; Akinbola, I.A.; Ademosun, A.O.; Sanni, D.M.; Odubanjo, O.V.; Olasehinde, T.A.; Oyeleye, S.I. Essential Oil from Clove Bud (Eugenia aromatica Kuntze) Inhibit Key Enzymes Relevant to the Management of Type-2 Diabetes and Some Pro-oxidant Induced Lipid Peroxidation in Rats Pancreas in vitro. *J. Oleo Sci.* **2015**, *64*, 775–782. [CrossRef] [PubMed]
25. Alfikri, F.N.; Pujiarti, R.; Wibisono, M.G.; Hardiyanto, E.B. Yield, Quality, and Antioxidant Activity of Clove (*Syzygium aromaticum* L.) Bud Oil at the Different Phenological Stages in Young and Mature Trees. *Scientifica* **2020**, *2020*, 9701701. [CrossRef] [PubMed]
26. SOULSBY, J. The new ointment bases of the British Pharmocopoeia, 1948. *Br. J. Dermatol. Syph.* **1949**, *61*, 206–210. [CrossRef] [PubMed]
27. Turner, N.A.; Sharma-Kuinkel, B.K.; Maskarinec, S.A.; Eichenberger, E.M.; Shah, P.P.; Carugati, M.; Holland, T.L.; Fowler, V.G. Methicillin-resistant Staphylococcus aureus: An overview of basic and clinical research. *Nat. Rev. Microbiol.* **2019**, *17*, 203–218. [CrossRef] [PubMed]
28. Piasecki, B.; Biernasiuk, A.; Skiba, A.; Skalicka-Woźniak, K.; Ludwiczuk, A. Composition, Anti-MRSA Activity and Toxicity of Essential Oils from Cymbopogon Species. *Molecules* **2021**, *26*, 7542. [CrossRef]
29. Caron, J.P.; Bolin, C.A.; Hauptman, J.G.; Johnston, K.A. Minimum inhibitory concentration and postantibiotic effect of amikacin for equine isolates of methicillin-resistant *Staphylococcus aureus* in vitro. *Vet. Surg.* **2009**, *38*, 664–669. [CrossRef]
30. Jamil, S.; Saad, U.; Hafiz, S. Can amoxicillin clavulanate be used for treating MRSA? *J. Pharmacol. Med. Chem.* **2017**, *1*, 21–23.
31. Foxley, M.A.; Friedline, A.W.; Jensen, J.M.; Nimmo, S.L.; Scull, E.M.; King, J.B.; Strange, S.; Xiao, M.T.; Smith, B.E.; Thomas, K.J.; et al. Efficacy of Ampicillin Against Methicillin-Resistant Staphylococcus aureus Restored Through Synergy with Branched Poly(ethylenimine). *J. Antibiot.* **2016**, *69*, 871. [CrossRef] [PubMed]

32. Singh, R.; Almutairi, M.; Alm, R.A.; Lahiri, S.D.; Martin, M.S.; Chen, A.; Ambler, J.E. Ceftaroline efficacy against high-MIC clinical Staphylococcus aureus isolates in an in vitro hollow-fibre infection model. *J. Antimicrob. Chemother.* **2017**, *72*, 2796–2803. [CrossRef] [PubMed]
33. Namysl, E.; Galinski, J.; Szponar, M.; Piechowicz, L. WRAZLIWOSC BAKTERII IZOLOWANYCH Z MATERIALU KLINICZNEGO NA CYPROFLOKSACYNE. *Med. Dosw. Mikrobiol.* **1995**, *47*, 5–9. [PubMed]
34. Hodille, E.; Badiou, C.; Bouveyron, C.; Bes, M.; Tristan, A.; Vandenesch, F.; Lina, G.; Dumitrescu, O. Clindamycin suppresses virulence expression in inducible clindamycin-resistant *Staphylococcus aureus* strains 11 Medical and Health Sciences 1108 Medical Microbiology. *Ann. Clin. Microbiol. Antimicrob.* **2018**, *17*, 1–6. [CrossRef]
35. Yildiz, Ö.; Çoban, A.Y.; Şener, A.G.; Coşkuner, S.A.; Bayramoğlu, G.; Güdücüoğlu, H.; Özyurt, M.; Tatman-Otkun, M.; Karabiber, N.; Özkütük, N.; et al. Antimicrobial susceptibility and resistance mechanisms of methicillin resistant Staphylococcus aureus isolated from 12 Hospitals in Turkey. *Ann. Clin. Microbiol. Antimicrob.* **2014**, *13*, 1–6. [CrossRef]
36. Lemaître, N.; Sougakoff, W.; Masmoudi, A.; Fievet, M.H.; Bismuth, R.; Jarlier, V. Characterization of Gentamicin-Susceptible Strains of Methicillin-Resistant *Staphylococcus aureus* Involved in Nosocomial Spread. *J. Clin. Microbiol.* **1998**, *36*, 81. [CrossRef]
37. Su, Y.L.; Hong, W.F.; Sutherland, C.; DeRyke, A.C.; Nicolau, D.P. Antibacterial effects of moxifloxacin and levofloxacin simulating epithelial lining fluid concentrations against community-acquired methicillin-resistant *Staphylococcus aureus*. *Drugs R. D.* **2007**, *8*, 69–77. [CrossRef]
38. Conceição, T.; Coelho, C.; de Lencastre, H.; Aires-de-Sousa, M. Frequent occurrence of oxacillin-susceptible mecA-positive Staphylococcus aureus (OS-MRSA) strains in two African countries. *J. Antimicrob. Chemother.* **2015**, *70*, 3200–3204. [CrossRef]
39. Trzcinski, K.; Cooper, B.S.; Hryniewicz, W.; Dowson, C.G. Expression of resistance to tetracyclines in strains of methicillin-resistant Staphylococcus aureus. *J. Antimicrob. Chemother.* **2000**, *45*, 763–770. [CrossRef]
40. Harris, T.M.; Bowen, A.C.; Holt, D.C.; Sarovich, D.S.; Stevens, K.; Currie, B.J.; Howden, B.P.; Carapetis, J.R.; Giffard, P.M.; Tong, S.Y.C. Investigation of trimethoprim/sulfamethoxazole resistance in an emerging sequence type 5 methicillin-resistant *Staphylococcus aureus* clone reveals discrepant resistance reporting. *Clin. Microbiol. Infect.* **2018**, *24*, 1027–1029. [CrossRef]
41. Rosarior, V.L.; Lim, P.S.; Wong, W.K.; Yue, C.S.; Yam, H.C.; Tan, S.A. Antioxidant-rich Clove Extract, A Strong Antimicrobial Agent against Urinary Tract Infections-causing Bacteria in vitro. *Trop. Life Sci. Res.* **2021**, *32*, 45. [CrossRef]
42. Wongsawan, K.; Chaisri, W.; Tangtrongsup, S.; Mektrirat, R. Bactericidal Effect of Clove Oil against Multidrug-Resistant Streptococcus suis Isolated from Human Patients and Slaughtered Pigs. *Pathogens* **2020**, *9*, 14. [CrossRef] [PubMed]
43. Gürbüz, M.; İrem Omurtag Korkmaz, B. The anti-campylobacter activity of eugenol and its potential for poultry meat safety: A review. *Food Chem.* **2022**, *394*, 133519. [CrossRef] [PubMed]
44. Taleuzzaman, M.; Jain, P.; Verma, R.; Iqbal, Z.; Mirza, M.A. Eugenol as a Potential Drug Candidate: A Review. *Curr. Top. Med. Chem.* **2021**, *21*, 1804–1815. [CrossRef] [PubMed]
45. Moo, C.L.; Yang, S.K.; Osman, M.A.; Yuswan, M.H.; Loh, J.Y.; Lim, W.M.; Lim, S.H.E.; Lai, K.S. Antibacterial Activity and Mode of Action of β-caryophyllene on Bacillus cereus. *Polish J. Microbiol.* **2020**, *69*, 49–54. [CrossRef] [PubMed]
46. Koyama, S.; Purk, A.; Kaur, M.; Soini, H.A.; Novotny, M.V.; Davis, K.; Cheng Kao, C.; Matsunami, H.; Mescher, A. Beta-caryophyllene enhances wound healing through multiple routes. *PLoS ONE* **2019**, *14*, e0216104. [CrossRef] [PubMed]
47. Baradaran Rahimi, V.; Askari, V.R. A mechanistic review on immunomodulatory effects of selective type two cannabinoid receptor β-caryophyllene. *Biofactors* **2022**, *48*, 857–882. [CrossRef]
48. Yang, S.J.; Han, S.H.; Lee, A.R.; Jun, J.H.; Son, M.W.; Oh, S.H.; Kim, J.; Paik, S.Y. Evaluation of antimicrobial effects of commercial mouthwashes utilized in South Korea. *BMB Rep.* **2015**, *48*, 42. [CrossRef]
49. Hamad Al-Mijalli, S.; Elsharkawy, E.R.; Abdallah, E.M.; Hamed, M.; El Omari, N.; Mahmud, S.; Alshahrani, M.M.; Mrabti, H.N.; Bouyahya, A. Determination of Volatile Compounds of Mentha piperita and Lavandula multifida and Investigation of Their Antibacterial, Antioxidant, and Antidiabetic Properties. *Evid. Based. Complement. Alternat. Med.* **2022**, *2022*, 9306251. [CrossRef]
50. Jang, H.I.; Rhee, K.J.; Eom, Y. Bin Antibacterial and antibiofilm effects of α-humulene against Bacteroides fragilis. *Can. J. Microbiol.* **2020**, *66*, 389–399. [CrossRef]
51. Zhang, X.; Sun, X.; Wu, J.; Wu, Y.; Wang, Y.; Hu, X.; Wang, X. Berberine Damages the Cell Surface of Methicillin-Resistant *Staphylococcus aureus*. *Front. Microbiol.* **2020**, *11*, 621. [CrossRef]
52. Sardari, K.; Kakhki, E.G.; Mohri, M. Evaluation of wound contraction and epithelialization after subcutaneous administration of Theranekron® in cows. *Comp. Clin. Pathol.* **2006**, *16*, 197–200. [CrossRef]
53. El-Tookhy, O.S.; Shamaa, A.A.; Shehab, G.G.; Abdallah, A.N.; Azzam, O.M. Histological Evaluation of Experimentally Induced Critical Size Defect Skin Wounds Using Exosomal Solution of Mesenchymal Stem Cells Derived Microvesicles. *Int. J. Stem Cells* **2017**, *10*, 144–153. [CrossRef]
54. Van De Vyver, M.; Boodhoo, K.; Frazier, T.; Hamel, K.; Kopcewicz, M.; Levi, B.; Maartens, M.; Machcinska, S.; Nunez, J.; Pagani, C.; et al. Histology Scoring System for Murine Cutaneous Wounds. *Stem Cells Dev.* **2021**, *30*, 1141–1152. [CrossRef]
55. Banerjee, K.; Madhyastha, H.; Sandur, V.R.; Manikandanath, N.T.; Thiagarajan, N.; Thiagarajan, P. Anti-inflammatory and wound healing potential of a clove oil emulsion. *Colloids Surf. B Biointerfaces* **2020**, *193*, 111102. [CrossRef]
56. Dimic, D.S.; Kaluderovic, G.N.; Avdovic, E.H.; Milenkovic, D.A.; Živanovic, M.N.; Potocnák, I.; Samolová, E.; Dimitrijevic, M.S.; Saso, L.; Markovic, Z.S.; et al. Synthesis, Crystallographic, Quantum Chemical, Antitumor, and Molecular Docking/Dynamic Studies of 4-Hydroxycoumarin-Neurotransmitter Derivatives. *Int. J. Mol. Sci.* **2022**, *23*, 1001. [CrossRef]

57. Kováts, E. Gas-chromatographische Charakterisierung organischer Verbindungen. Teil 1: Retentionsindices aliphatischer Halogenide, Alkohole, Aldehyde und Ketone. *Helv. Chim. Acta* **1958**, *41*, 1915–1932. [CrossRef]
58. Balouiri, M.; Sadiki, M.; Ibnsouda, S.K. Methods for in vitro evaluating antimicrobial activity: A review. *J. Pharm. Anal.* **2016**, *6*, 71–79. [CrossRef] [PubMed]
59. Fankam, A.G.; Kuete, V.; Voukeng, I.K.; Kuiate, J.R.; Pages, J.M. Antibacterial activities of selected Cameroonian spices and their synergistic effects with antibiotics against multidrug-resistant phenotypes. *BMC Complement. Altern. Med.* **2011**, *11*, 1–11. [CrossRef] [PubMed]
60. Nidhi, P.; Rolta, R.; Kumar, V.; Dev, K.; Sourirajan, A. Synergistic potential of Citrus aurantium L. essential oil with antibiotics against *Candida albicans*. *J. Ethnopharmacol.* **2020**, *262*, 113135. [CrossRef]
61. Nayeem, N.; Mohammed Basheeruddin Asdaq, S.; Alamri, A.S.; Alsanie, W.F.; Alhomrani, M.; Mohzari, Y.; Alrashed, A.A.; Alotaibi, N.; Aalhathal, A.S.; Alharbi, M.A.; et al. Wound healing potential of Dodonaea viscosa extract formulation in experimental animals. *J. King Saud Univ.-Sci.* **2021**, *33*, 101476. [CrossRef]
62. Kolhe, S.S.; Shinde, K.; Jori, R.; Gadhave, M.V.; Jadhav, S.L.; Gaikwad, D.D. Evaluation of polyherbal ointment for wound healing activity in Wistar rats. *J. Drug Deliv. Ther.* **2018**, *8*, 26–31. [CrossRef]
63. Anesthesia (Guideline) | Vertebrate Animal Research. Available online: https://animal.research.uiowa.edu/iacuc-guidelines-anesthesia (accessed on 13 February 2022).
64. Mukherjee, P.K.; Verpoorte, R.; Suresh, B. Evaluation of in-vivo wound healing activity of Hypericum patulum (Family: Hypericaceae) leaf extract on different wound model in rats. *J. Ethnopharmacol.* **2000**, *70*, 315–321. [CrossRef]
65. Fayez, M.S.; Hakim, T.A.; Agwa, M.M.; Abdelmoteleb, M.; Aly, R.G.; Montaser, N.N.; Abdelsattar, A.S.; Rezk, N.; El-Shibiny, A. Topically Applied Bacteriophage to Control Multi-Drug Resistant Klebsiella pneumoniae Infected Wound in a Rat Model. *Antibiotics* **2021**, *10*, 1048. [CrossRef] [PubMed]
66. Ekom, S.E.; Tamokou, J.D.D.; Kuete, V. Methanol extract from the seeds of Persea americana displays antibacterial and wound healing activities in rat model. *J. Ethnopharmacol.* **2022**, *282*, 114573. [CrossRef]

molecules

Article

Discovery of the Active Compounds of the Ethyl Acetate Extract Site of *Ardisia japonica* (Thunb.) Blume for the Treatment of Acute Lung Injury

Shuding Sun [1,2], Xuefang Liu [1,2], Di Zhao [1,2], Lishi Zheng [1,2], Xiaoxiao Han [3], Yange Tian [1,2,*] and Suxiang Feng [1,2,*]

[1] Academy of Chinese Medical Sciences, Henan University of Chinese Medicine, Zhengzhou 450003, China; sunshuding@126.com (S.S.); liuxf0213@163.com (X.L.); zhaodiabcd@sina.com (D.Z.); zhenglishia129@126.com (L.Z.)
[2] Collaborative Innovation Center for Chinese Medicine and Respiratory Diseases Co-Constructed by Henan Province & Education Ministry of China, Zhengzhou 450046, China
[3] College of Pharmacy, Zhejiang Chinese Medical University, Hangzhou 310053, China; hanxx320@163.com
* Correspondence: yange0910@126.com (Y.T.); fengsx221@163.com (S.F.)

Abstract: The objective of this study was to identify and evaluate the pharmacodynamic constituents of Ardisiae Japonicae Herba (AJH) for the treatment of acute lung injury (ALI). To fully analyze the chemical contents of various extraction solvents (petroleum ether site (PE), ethyl acetate site (EA), *n*-butanol site (NB), and water site (WS)) of AJH, the UPLC–Orbitrap Fusion–MS technique was employed. Subsequently, the anti-inflammatory properties of the four extracted components of AJH were assessed using the lipopolysaccharide (LPS)-induced MH-S cellular inflammation model. The parts that exhibited anti-inflammatory activity were identified. Additionally, a technique was developed to measure the levels of specific chemical constituents in the anti-inflammatory components of AJH. The correlation between the "anti-inflammatory activity" and the constituents was analyzed, enabling the identification of a group of pharmacodynamic components with anti-inflammatory properties. ALI model rats were created using the tracheal drip LPS technique. The pharmacodynamic indices were evaluated for the anti-inflammatory active portions of AJH. The research revealed that the PE, EA, NB, and WS extracts of AJH included 215, 289, 128, and 69 unique chemical components, respectively. Additionally, 528 chemical components were discovered after removing duplicate values from the data. The EA exhibited significant anti-inflammatory activity in the cellular assay. A further analysis was conducted to determine the correlation between anti-inflammatory activity and components. Seventeen components, such as caryophyllene oxide, bergenin, and gallic acid, were identified as potential pharmacodynamic components with anti-inflammatory activity. The pharmacodynamic findings demonstrated that the intermediate and high doses of the EA extract from AJH exhibited a more pronounced effect in enhancing lung function, blood counts, and lung histology in a way that depended on the dosage. To summarize, when considering the findings from the previous study on the chemical properties of AJH, it was determined that the EA contained a group of 13 constituents that primarily contributed to its pharmacodynamic effects against ALI. The constituents include bergenin, quercetin, epigallocatechingallate, and others.

Keywords: Ardisiae Japonicae Herba; anti-acute lung injury; UPLC–Orbitrap Fusion–MS; active compounds

Citation: Sun, S.; Liu, X.; Zhao, D.; Zheng, L.; Han, X.; Tian, Y.; Feng, S. Discovery of the Active Compounds of the Ethyl Acetate Extract Site of *Ardisia japonica* (Thunb.) Blume for the Treatment of Acute Lung Injury. *Molecules* **2024**, *29*, 770. https://doi.org/10.3390/molecules29040770

Academic Editors: Yongjun Wei, Lingbo Qu and Xiao-Jun Ji

Received: 30 December 2023
Revised: 27 January 2024
Accepted: 2 February 2024
Published: 7 February 2024

1. Introduction

Acute lung injury (ALI) is a condition triggered by a variety of insults including sepsis, viral or bacterial pneumonia, and mechanical ventilator-induced trauma that frequently progresses to hypoxemic respiratory failure and clinical acute respiratory distress syndrome (ARDS), with a high mortality rate [1–4]. The primary symptoms include a heightened permeability of the pulmonary blood vessels, an exudation of tissue fluid, a buildup of inflammatory cells, and impaired gas exchange [5,6]. These manifestations pose a significant threat to human health [7]. Presently, there remains a deficiency in the availability of efficacious medications for the management of ALI [8]. Traditional Chinese Medicine (TCM) possesses distinct benefits in the management of ALI, including extensive knowledge, notable clinical effectiveness, and few toxic adverse reactions [9]. Extracting anti-ALI medications from TCM is highly important as it serves as a valuable reference for the practical use of TCM in treating ALI [10].

The clinical symptoms of ALI diagnosed by Chinese medicine include weakness, bloating, dizziness, and heat, with weakness and bloating being particularly prominent [11,12]. The disease is caused by the invasion of pathogenic factors, specifically wind-warm and lung heat [13]. It is commonly treated with a prescription aimed at clearing heat, eliminating toxins, expelling phlegm, and resolving stasis [14]. Ardisiae Japonicae Herba (AJH) is a dried whole grass of *Ardisia japonica* (Thunb.) Blume, a member of the Myrsinaceae plant family. The herb has a flat shape and possesses a pungent and slightly bitter taste. It is associated with the lung and liver meridians, and is effective in treating conditions such as phlegm, cough, dampness, heat, blood circulation, and blood stasis. It is commonly used in the treatment of respiratory diseases like carbuncles, coughs, bronchitis, and tuberculosis [15,16]. The herb is classified as a remedy for respiratory conditions characterized by excessive heat and inflammation in the lungs. The classic *Materia Medica* discusses the treatment of diaphragmatic gas during illness and the elimination of wind and phlegm. Extracts from the *Compendium of Materia Medica* include "Addressing hemoptysis and occupational wounds, alleviating death caused by fearfulness, and relieving prolonged coughing during labor".

Initially, the researchers employed UPLC–Orbitrap Fusion–MS technology to methodically analyze the chemical constituents present in AJH [16]. They also developed a quality control approach to concurrently identify and quantify 31 chemical components in various origins and distinct portions of AJH [17]. The study examined the effects of AJH on ALI model rats, focusing on its pharmacodynamics. The results revealed that AJH exhibited significant anti-inflammatory properties and provided protection against LPS-induced ALI in rats. They found that the therapeutic effect of AJH on ALI was related to regulating the AGE-RAGE, PI3K-AKT, and JAK-STAT signaling pathways [18]. However, the specific pharmacodynamic components of AJH remain unclear, necessitating further investigation. The objective of this study was to fully analyze the chemical contents of various extraction solvents of AJH, tracking, identifying, and evaluating the pharmacodynamic constituents of AJH for the treatment of ALI.

2. Results

2.1. Composition Analysis

The chemical composition of PE, EA, NB, and WE of AJH were compared to a database and a control sample to obtain primary and secondary information (Table 1). A total of 215, 289, 128, and 69 chemical constituents were identified in each respective extract. After removing duplicate values, a total of 528 constituents were identified. These constituents mainly included 63 phenylpropanoids, 59 terpenes, 35 flavonoids, 28 steroids, 14 quinones, and 329 phenolic acids, carboxylic acids, and their derivatives. Furthermore, PE and EA exhibited a predominant enrichment of terpenoids and steroids, whereas EA demonstrated a primary enrichment of phenylpropanoids and flavonoids. The identification results for the PE, EA, NB, and WS are presented in the Abbreviations.

Table 1. Compositional enrichment results of PE, EA, NB, and WS.

Sites	Phenylpropanoid	Terpene	Flavonoid	Steroid	Quinone	Phenolic Acids, Carboxylic Acids, and Their Derivatives	Total
PE	23	33	4	16	5	134	215
EA	40	31	31	13	7	167	289
NB	21	13	7	1	5	81	128
WS	11	3	3	2	1	49	69
Total	63	59	35	28	14	329	528

2.2. Active Site Screening

The vitality of MH-S cells was assessed using the CCK-8 assay, and the findings are presented in Figure 1. The concentration range of 3.13–25 μg/mL in the PE group and 12.5–50 μg/mL in the NB group exhibited a more than 85% viability of MH-S cells. However, other concentrations in the PE and the NB had a discernible impact on the growth of MH-S cells. The concentration ranges of 3.13–6.25 μg/mL in the EA group, 12.5–200 μg/mL in the WS group, and 12.5–100 μg/mL in the AJH group had a survival rate exceeding 90% for MH-S cells. This suggested that the concentration ranges mentioned above in the EA, the WS, and the AJH group did not have an impact on the normal growth of MH-S cells and exhibited minimal cytotoxicity. The concentration designed for AJH in the LPS-induced MH-S cell inflammation model is presented in Table 2, based on these findings.

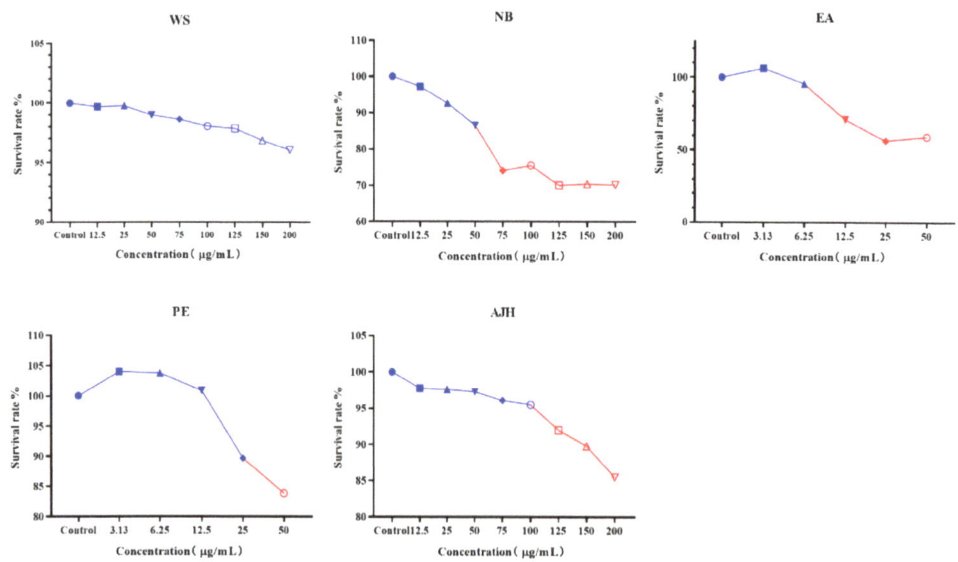

Figure 1. The non-toxic concentration results of different extraction parts of AJH on MH-S cells ($\overline{X} \pm$ SD, n = 8). The color in blue represent the safe concentration and the red illucidate the harmful concentration. The shape of the points represent always the same mean.

Table 2. Design of optimal dosing concentration for each dosing group (μg/mL).

Sites	1	2	3	4
AJH	100	50	25	12.5
PE	25	12.5	6.25	3.13
EA	6.25	3.13	1.56	0.78
NB	50	25	12.5	6.25
WS	200	100	50	25

The levels of IL-6 and TNF-α inflammatory factors in the cell supernatant were measured, and the findings are presented in Figures 2 and 3. The expression levels of IL-6 and TNF-α, which are inflammatory factors, were considered equally important. Therefore, a direct summing method was used to calculate a comprehensive score. The P_0 value and the comprehensive score results were calculated for each group to assess the activity of different components of AJH. The results are presented in Table 3.

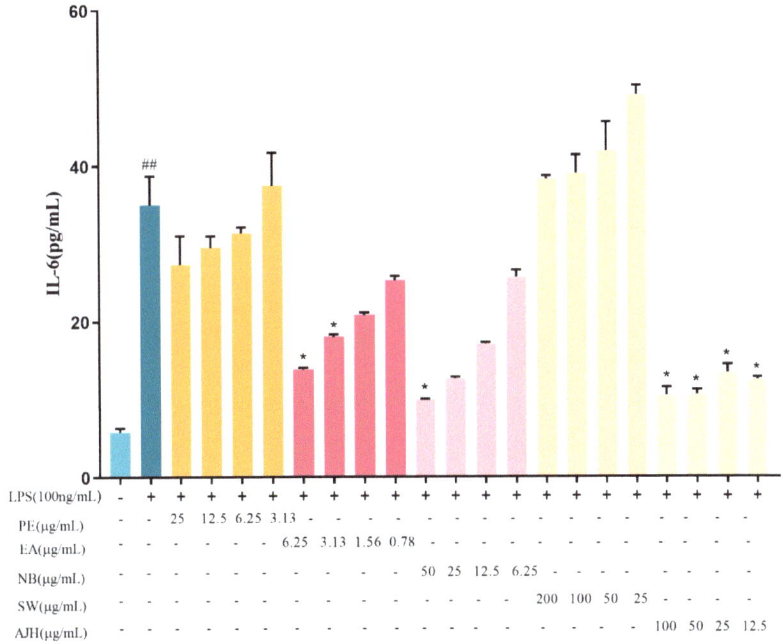

Figure 2. Effects of AJH and different extraction sites on the expression level of IL-6 in MH-S cells ($\overline{X} \pm$ SD, $n = 6$). Note: compared with the blank group: $^{##}$ $p < 0.01$; compared with the model group: * $p < 0.05$. Each color represents a test group, such as PE, EA, NB, WS and AJH.

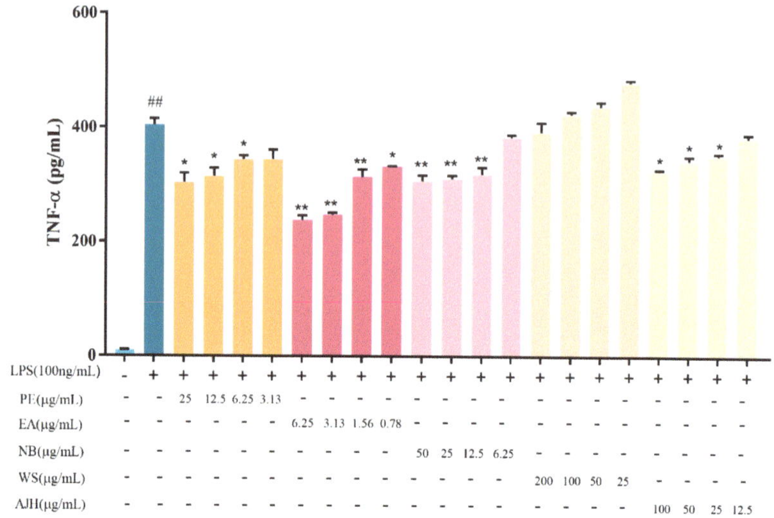

Figure 3. Effects of AJH and different extraction sites on the expression level of TNF-α in MH-S cells ($\overline{X} \pm$ SD, $n = 6$). Note: compared with the blank group: $^{\#\#}$ $p < 0.01$; compared with the model group: * $p < 0.05$, ** $p < 0.01$. Each color represents a test group, such as PE, EA, NB, WS and AJH.

Table 3. Comprehensive scoring results of PE, EA, NB, and WS on the expression levels of IL-6 and TNF-α in AJH.

Sites	IL-6				TNF-α				$P_{0IL\text{-}6} + P_{0TNF\text{-}\alpha}$
	p_1	p_2	p_0	SUMP_0	p_1	p_2	p_0	SUMP_0	
PE 25 μg/mL	0.374	0.039	0.104	0.136	0.01	0.004	0.400	0.806	0.942
PE 12.5 μg/mL	0.475	0.003	0.006		0.01	0.003	0.300		
PE 6.25 μg/mL	0.740	0.000	0.000		0.018	0.001	0.056		
PE 3.13 μg/mL	0.997	0.025	0.025		0.06	0.003	0.050		
EA 6.25 μg/mL	0.031	0.014	0.452	0.525	0.001	0.002	2.000	2.429	2.954
EA 3.13 μg/mL	0.039	0.002	0.051		0.003	0.000	0.000		
EA 1.56 μg/mL	0.059	0.001	0.017		0.007	0.003	0.429		
EA 0.78 μg/mL	0.184	0.001	0.005		0.032	0.000	0.000		
NB 50 μg/mL	0.043	0.001	0.023	0.023	0.003	0.002	0.667	0.952	0.975
NB 25 μg/mL	0.065	0.000	0.000		0.005	0.000	0.000		
NB 12.5 μg/mL	0.091	0.000	0.000		0.007	0.002	0.286		
NB 6.25 μg/mL	0.181	0.000	0.000		0.333	0.000	0.000		
WS 200 μg/mL	−0.807	0.000	0.000	−0.039	−0.017	0.002	0.118	−0.118	−0.157
WS 100 μg/mL	−0.794	0.006	0.008		−0.1	0.000	0.000		
WS 50 μg/mL	−0.483	0.015	0.031		−0.335	0.000	0.000		
WS 25 μg/mL	−0.076	0.000	0.000		−0.976	0.000	0.000		
AJH 100 μg/mL	0.024	0.037	1.542	2.127	0.02	0.000	0.000	0.059	2.186
AJH 50 μg/mL	0.029	0.005	0.172		0.017	0.001	0.059		
AJH 25 μg/mL	0.031	0.012	0.387		0.045	0.000	0.000		
AJH12.5 μg/mL	0.038	0.001	0.026		0.315	0.000	0.000		

Figures 2 and 3 demonstrate that the levels of IL-6 and TNF-α in MH-S cells of the model group were markedly higher compared to the blank group, with statistical significance ($p < 0.01$). This suggested the successful establishment of the LPS-induced inflammatory cell model. The four concentrations of PE, EA, and NB exhibited varying degrees of inhibition on the release of IL-6 and TNF-α in MH-S cells. However, the water concentration did not have a significant inhibitory effect on the release of IL-6 and TNF-α. The results in Table 3 demonstrated varying levels of inhibition in the expression of two inflammatory factors, IL-6 and TNF-α, in LPS-induced MH-S cells. The inhibition strength followed the order EA group > AJH group > NB group > PE group > WS group. Therefore, the EA exhibited a stronger anti-inflammatory effect compared to the NB, PE, and WS. Hence, the anti-inflammatory efficacy of EA surpassed that of NB, PE, and WS.

2.3. Spectral Effect Relationship Analysis

The concentrations of 19 chemical constituents with significant levels in the EA powder location were measured (μg/g EA powder), and the corresponding components in the raw AJH were also measured (μg/g raw AJH). The content of these components in the EA powder was converted (μg/g raw AJH), indicating that these components are mainly enriched in EA, as presented in Table 4. A gray correlation analysis was conducted to assess the relationship between the expression of IL-6 and TNF-α inflammatory factors. The results, presented in Table 5, revealed a strong correlation between the expression of IL-6 and TNF-α with 17 components, including caryophyllene oxide, bergenin, and gallic acid.

Table 4. Results of the determination of the content of AJH and EA ($\overline{X} \pm$ SD, $n = 3$).

Compounds	EA	AJH (Obversion)	AJH
	Content (μg/g EA Powder)	Content (μg/g Raw AJH)	
Isoeugenol	758.18 ± 4.75	13.98 ± 0.11	100.33 ± 0.02
Caryophyllene oxide	453.91 ± 1.84	8.27 ± 0.1	81.32 ± 1.54
Quercetin	6057.59 ± 71.5	112.15 ± 0.08	155.49 ± 1.93
Taxifolin	662.33 ± 8.79	12.07 ± 0.21	11.18 ± 0.07
Myricetin	4521.58 ± 9.77	83.15 ± 0.89	80.89 ± 0.25
Gallic acid	8300.42 ± 23.04	153.48 ± 0.35	174.36 ± 1.01
Ethyl gallate	1398.6 ± 1.67	24.24 ± 1.28	30.86 ± 0.03
Fraxetin	918.08 ± 10.3	16.34 ± 0.09	18.34 ± 0.12
Palmitic acid	15,525.36 ± 39.26	282.86 ± 0.61	269.68 ± 0.06
Kaempferol	23,916.89 ± 42.79	439.05 ± 2.34	476.97 ± 1.09
Eriodictyol	446.7 ± 1.78	8.23 ± 0.04	8.2 ± 0.03
Cianidanol	2897.69 ± 6.14	52.52 ± 0.01	49.02 ± 0.1
Bergenin	285,582.94 ± 2309.97	5064.71 ± 59.58	5150.12 ± 23.68
Cynaroside	17,346.17 ± 88.96	318.37 ± 1.35	311.3 ± 0.36
Astilbin	2343.77 ± 1.24	43.11 ± 0.16	54.9 ± 0.09
Epigallocatechingallate	12,936.53 ± 55.07	238.32 ± 0.14	234.04 ± 0.69
Myricitrin	12,030.85 ± 8.51	220.46 ± 0.28	265.71 ± 4.68
Nictoflorin	397.4 ± 2.15	7.27 ± 0.03	8.71 ± 0.02
Rutin	618.27 ± 0.82	11.28 ± 0.02	6.73 ± 0.03

Note: 2500 g of raw AJH could yield 45.75 g of EA powder.

Table 5. Results of the correlation between the content of 19 chemical components in the EA and anti-inflammatory activity.

Compounds	Correlation	Range	Compounds	Correlation	Range
Caryophyllene oxide	1	1	Taxifolin	0.833	11
Bergenin	0.878	2	Isoeugenol	0.831	12
Gallic acid	0.863	3	Palmitic acid	0.828	13
Quercetin	0.86	4	Luteolin-7-O-glucoside	0.824	14
Epigallocatechin gallate	0.846	5	Myricetin	0.824	15
Rutin	0.84	6	Fraxetin	0.821	16
Catechin	0.84	7	Astilbin	0.82	17
Myricitrin	0.839	8	Eriodictyol	0.753	18
Ethyl gallate	0.836	9	Nicotiflorin	0.626	19
Kaempferide	0.835	10			

2.4. *Pharmacodynamic Evaluation*

2.4.1. Changes in Lung Function

In comparison to the control group, the model group showed significantly higher levels of EF50, PEFb, PAU, and Penh ($p < 0.05$, 0.01). On the other hand, when compared to the model group, the middle- and the high-dose groups of the EA, the 70% alcoholic extract group of AJH, the aqueous extract group of AJH, and the dexamethasone group exhibited significantly lower levels of PEFb ($p < 0.01$). Additionally, the high-dose group of the EA and the dexamethasone group showed reduced levels of EF50, PAU, and Penh ($p < 0.05$, 0.01). The study demonstrated that the EA played a role in influencing the level of bronchoconstriction and airway resistance in rats with LPS-induced ALI. Furthermore, it greatly enhanced the lung function of rats with ALI, as illustrated in Figure 4.

2.4.2. Changes in Blood Routine

In comparison to the control group, the model group showed a significant increase in peripheral blood white blood cells (WBC), neutrophil percentage (NEU%), and monocyte percentage (MONO%), while lymphocyte percentage (LYM%) significantly decreased ($p < 0.05$, 0.01). When compared to the model group, the ethyl acetate high-dose group exhibited a significant decrease in WBC and MONO%. The EA low-, medium-, and high-dose groups, the AJH 70% alcoholic extract group, and the AJH aqueous extract group showed a significant decrease in NEU% and a significant increase in LYM% ($p < 0.01$). The results are illustrated in Figure 5.

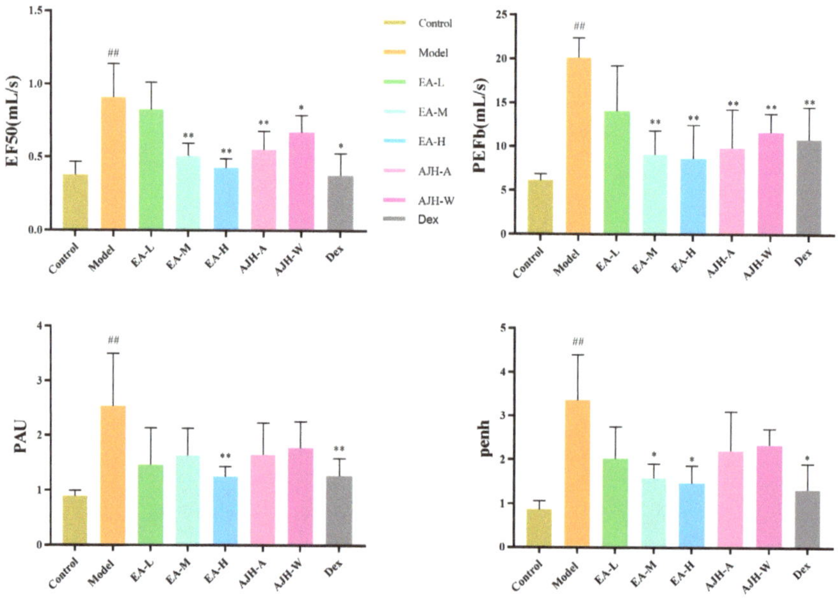

Figure 4. Effects of EA of AJH on the lung function of ALI model rats ($\overline{X} \pm$ SD, $n = 8$). Note: comparison with the blank group: $^{\#\#}$ $p < 0.01$; comparison with the model group: * $p < 0.05$, ** $p < 0.01$. Control: blank group, Model: model group, EA-L: EA low-dose group; EA-M: EA medium-dose group, EA-H: EA high-dose group, AJH-A: AJH 70% alcohol extract group, AJH-W: AJH aqueous extract group, Dex: dexamethasone group.

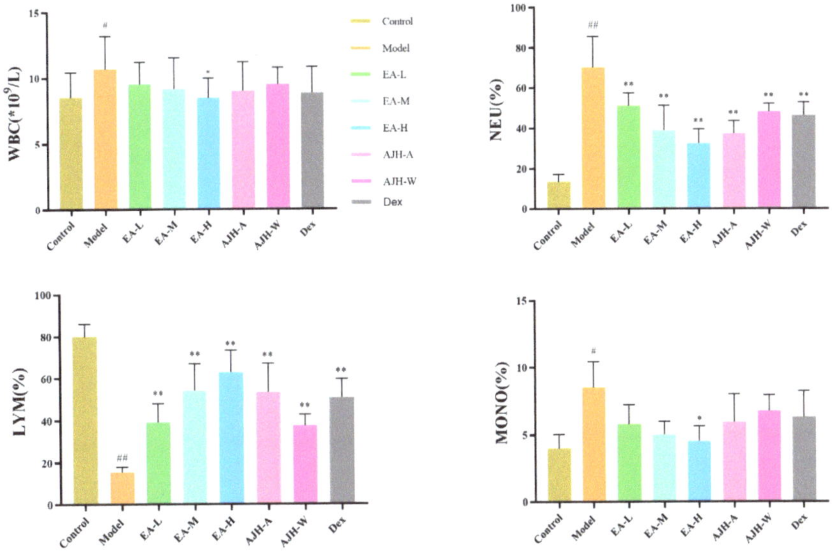

Figure 5. Effects of the EA site of AJH on the blood routine of ALI model rats ($\overline{X} \pm$ SD, $n = 8$). Note: comparison with the blank group: # $p < 0.05$, ## $p < 0.01$; comparison with the model group: * $p < 0.05$, ** $p < 0.01$. Control: blank group, Model: model group, EA-L: EA low-dose group, EA-M: EA medium-dose group, EA-H: EA high-dose group, AJH-A: AJH 70% alcohol extract group, AJH-W: AJH aqueous extract group, Dex: dexamethasone group.

2.4.3. CRP, TNF-α, IL-6, and IL-10 Expression Levels in Serum and Alveolar Lavage Fluid

In comparison to the control group, the model group exhibited a significant increase in the serum levels of TNF-α and IL-6, and a significant decrease in IL-10 ($p < 0.01$). Furthermore, the middle- and high-dose groups of ethyl acetate and the 70% alcoholic extract of the AJH group showed a significant decrease in the serum levels of TNF-α compared to the model group. The high-dose group of ethyl acetate also exhibited a significant decrease in serum levels of IL-6, while IL-10 showed a significant increase ($p < 0.05$, 0.01). There was a tendency to decrease CRP in the serum of the low, medium, and high dose groups of ethyl acetate, the AJH 70% alcoholic extract group, the AJH aqueous extract group, and the dexamethasone group. However, this difference was not statistically significant. These results are illustrated in Figure 6.

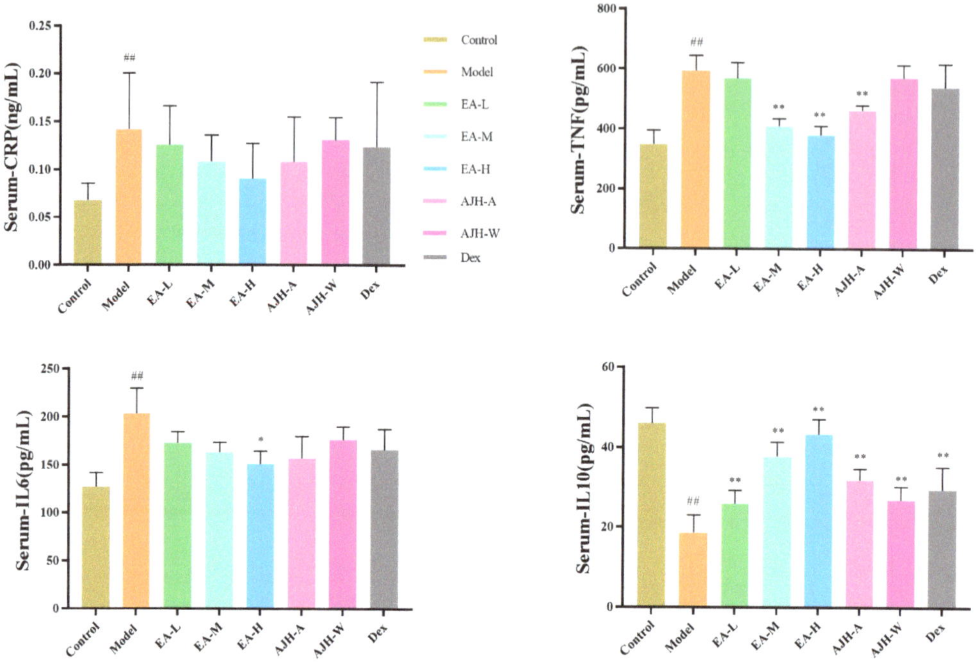

Figure 6. Effects of the EA site of AJH on serum levels of CRP, TNF-α, IL-6, and IL-10 in ALI model rats ($\overline{X} \pm$ SD, $n = 8$). Note: comparison with the blank group: $^{\#\#}$ $p < 0.01$, comparison with the model group: * $p < 0.05$, ** $p < 0.01$. Control: blank group, Model: model group; EA-L: EA low-dose group, EA-M: EA medium-dose group, EA-H: EA high-dose group, AJH-A: AJH 70% alcohol extract group, AJH-W: AJH aqueous extract group, Dex: dexamethasone group.

In comparison to the control group, the levels of CRP and TNF-α in the BALF of the experimental group showed a significant increase, while the level of IL-10 showed a significant decrease ($p < 0.01$). The level of IL-6 showed a tendency to increase, but the difference was not significant. On the other hand, when compared to the experimental group, the high-dose group of the EA, and the group treated with the 70% alcoholic extract of AJH, the dexamethasone group showed a significant decrease in CRP levels in the BALF ($p < 0.01$). Additionally, the CRP level in the BALF of the experimental group also showed a significant decrease ($p < 0.01$). The levels of TNF-α in BALF were significantly reduced in the EA middle- and high-dose groups, the AJH 70% alcohol extract group, and the dexamethasone group.

Conversely, the levels of IL-10 were significantly higher in these groups. There was also a tendency for lower levels of IL-6 in BALF in the EA low-, middle-, and high-dose groups, the AJH 70% alcohol extract group, the AJH aqueous extract group, and the dexamethasone group. However, this difference was not statistically significant. Ultimately, when the dosage of EA increased, the ability of EA to regulate inflammatory variables in ALI model rats also increased in a manner that was dependent on the dosage. These findings are depicted in Figure 7.

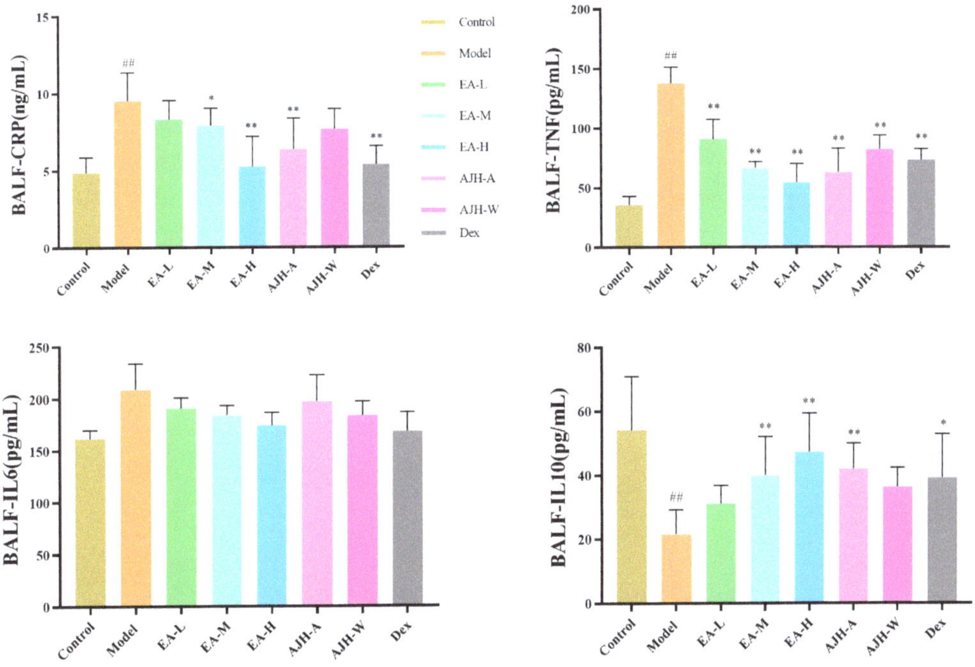

Figure 7. Effects of the EA of AJH on the levels of CRP, TNF-α, IL-6, and IL-10 in the BALF of ALI model rats ($\overline{X} \pm SD$, $n = 8$). Note: comparison with the blank group: ## $p < 0.01$, comparison with the model group: * $p < 0.05$, ** $p < 0.01$. Control: blank group, Model: model group, EA-L: EA low-dose group, EA-M: EA medium-dose group, EA-H: EA high-dose group, AJH-A: AJH 70% alcohol extract group, AJH-W: AJH aqueous extract group, Dex: dexamethasone group.

2.4.4. Changes in Lung Wet–Dry Weight Ratio and Organ Indexes

Compared with the blank group, the thymus and spleen indices of ALI rats were significantly reduced ($p < 0.01$). The lung wet–dry weight ratio of ALI rats was significantly increased ($p < 0.01$), indicating model animals with severe pulmonary edema.

Compared with the model group, the thymus and spleen indices of rats in the middle- and high-dose groups of EA, the 70% alcoholic extract group of AJH, and the dex group were significantly increased ($p < 0.01$). The wet–dry weight ratio of lungs in the middle- and high-dose groups of ethyl acetate, the 70% alcoholic extract of the AJH group, the aqueous extract of the AJH group, and the dexamethasone group were significantly lower ($p < 0.05$, 0.01). With the increase in the dose of EA, the wet–dry weight of lungs was significantly lower, indicating that the EA had a certain protective effect on the damage of the organs of the ALI rats, and had a certain dose-dependence; the results are shown in Figure 8.

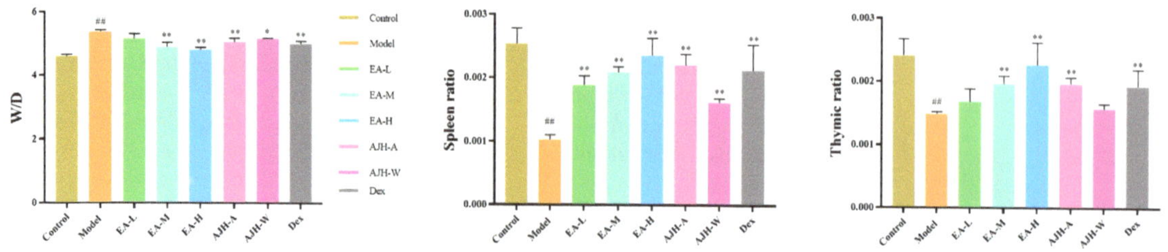

Figure 8. Effects of EA of AJH on lung wet–dry weight ratio and organ index of ALI rats ($\overline{X} \pm$ SD, n = 8). Note: comparison with the blank group: $^{\#\#}$ $p < 0.01$; comparison with the model group: * $p < 0.05$, ** $p < 0.01$. Control: blank group, Model: model group, EA-L: EA low-dose group, EA-M: EA medium-dose group, EA-H: EA high-dose group, AJH-A: 70% alcoholic extract group, AJH-W: aqueous extract group, Dex: dexamethasone group.

2.4.5. Histopathologic Changes in the Lungs

ALI animals exhibited significantly higher levels of inflammatory cell infiltration in the alveolar lumen, a thickening of the alveolar wall, more severe lung injury, and a wider extent of lung injury compared to the control group. In comparison to the model group, the rats in the dexamethasone group and the middle- and high-dose groups of ethyl acetate exhibited significant improvements in lung injury. This suggests that dexamethasone and the middle- and high-dose groups of EA have a protective effect on LPS-induced ALI. The rats in the low-dose group of EA also experienced reduced lung injury, but the effect was relatively weak. These findings are presented in Figure 9.

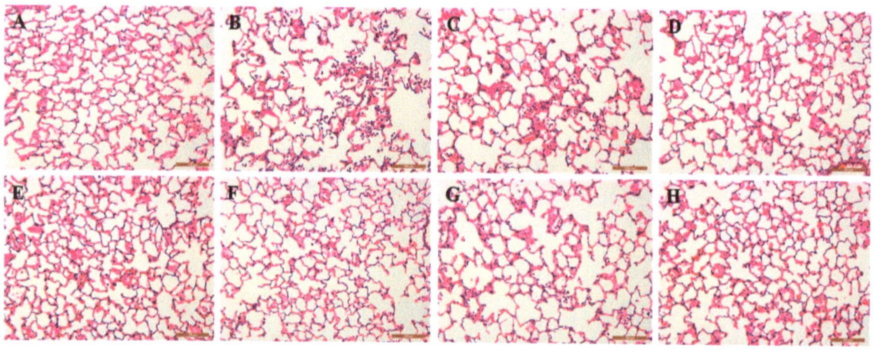

Figure 9. Histopathologic morphology of rat lungs in each group (HE staining, ×200). Note: (**A**): blank group; (**B**): model group; (**C**): low-dose group at EA; (**D**): medium-dose group at EA; (**E**): high-dose group at EA; (**F**): AJH 70% alcoholic extract group; (**G**): AJH aqueous extract group; (**H**): dexamethasone group.

3. Discussion

This study aimed to analyze and characterize the chemical constituents of four extraction sites (PE, EA, NB, and WS) of AJH. The results revealed that the EA had the highest concentration of constituents, followed by the PE, NB, and WS. Phenylpropanoids and flavonoids were predominantly found in the EA, while terpenes and steroids were mainly present in the PE and EA. The NB and WS exhibited relatively lower levels of enrichment compared to the PE and EA. Terpenoids and steroids exhibited higher levels of enrichment in the PE and EA locations, while the NB and WS showed comparatively lower levels of enrichment. Research in the field has demonstrated that flavonoids can mitigate the immune response and diminish inflammatory activation by lowering the levels of expression of inflammatory factors [19,20].

In this study, we employed LPS to stimulate MH-S cells and created a cellular inflammation model [21]. We then examined the impact of AJH, as well as its petroleum ether, ethyl acetate, n-butanol, and water extracts, on the expression levels of IL-6 and TNF-α in these cells. The findings revealed varying degrees of inhibitory effect on the release of IL-6 and TNF-α by AJH and its extracted components. Notably, the ethyl acetate extract exhibited significantly superior inhibitory effects compared to the petroleum ether extract. The anti-inflammatory activity of the EA was markedly superior to that of the PE, NB, and WS. The EA was identified as a potent anti-inflammatory site, which was related to the abundance of phenolpropanoids and flavonoids [22,23].

The chemical constituents of different parts of the AJH were characterized, and the anti-inflammatory activity in vitro was studied. Based on the results and the related literature, 19 components were selected for testing in the ethyl acetate part of AJH. The analysis showed that the contents of these 19 components were not significantly different between the ethyl acetate part and the whole AJH herb. Furthermore, 19 components were found to be mostly concentrated in the EA. The gray correlation analysis revealed the presence of 17 components, such as caryophyllene oxide, bergenin, gallic acid, quercetin, and epigallocatechin gallate, in the EA. These components showed a strong correlation with anti-inflammatory activity, making them suitable for the quality control of AJH. Studies have demonstrated that bergenin could effectively enhance lung injury [24] in mice infected with Klebsiella. It achieved this by suppressing the production of inflammatory cytokines, including IL-6, TNF-α, IL-1β, and PEG2, reducing the levels of MPO and MDA, and increasing the activities of SOD and GSH. These effects are mediated through the NF-κB [25] and MAPKs pathways, resulting in anti-inflammatory effects on Klebsiella-infected mice [26]. Quercetin [27,28], rutin [29,30], and kaempferide [31], all flavonoids, have anti-inflammatory properties to ameliorates acute lung injury. Khazdair Mohammad et al. demonstrated that quercetin can reduce the levels of reactive oxygen species, as well as the expression of inducible nitric oxide synthase and pro-inflammatory mediators such as TNF-α, IL-1α, IL-1β, IL-6, IL-10, and IL-12. These effects suggest that quercetin and kaempferide may have therapeutic potential against SARS-CoV-2 [32].

The pharmacodynamic effects of the ethyl acetate component of AJH on ALI rats were evaluated by examining lung function, blood routine, inflammatory factors, and other pharmacodynamic indexes. The results demonstrated that the ethyl acetate component improved lung injury, suppressed the expression of pro-inflammatory factors, and enhanced the release of anti-inflammatory factors in ALI rats. Furthermore, the intervention effect of the ethyl acetate component on ALI rats was found to be dependent on the dosage. Based on the examination of lung tissue samples, it was shown that the administration of ethyl acetate at high, medium, and low doses resulted in varying degrees of improvement in the infiltration of inflammatory cells surrounding the alveolar lumen and bronchial wall in rats with ALI. The administration of ethyl acetate at low, medium, and high doses had varying degrees of positive effects on the rats' apnea (PAU) and the medium and high doses of AJH ethyl acetate were particularly effective in improving the rats' maximum expiratory flow rate (PEFb) and airway narrowing index (Penh). Furthermore, the high-dose group of AJH ethyl acetate showed superior efficacy compared to the AJH group. The blood analysis showed that administering a high dose of the ethyl acetate component of AJH resulted in a significant decrease in WBC and MONO%, while the low, medium, and high doses reduced the neutrophil ratio (NEU%) and increased the lymphocyte ratio (LYM%) in rats. These findings suggested that the ethyl acetate component had anti-inflammatory properties and enhanced the immune system. Furthermore, the concentration of TNF-α in the serum samples from the EA was significantly decreased in the middle- and high-dosage groups. The concentration of IL-6 in the serum was significantly decreased in the high-dosage group. The concentrations of IL-10 in both the serum and BALF were significantly increased. Additionally, the concentration of CRP in BALF was significantly decreased. Hence, the ethyl acetate compound exhibits a specific therapeutic impact on rats with ALI, demonstrating a clear dependence on dosage. The serum pharmacochemistry of TCM was designed to screen the efficacy material base of TCMs from the constituents absorbed into the blood after oral administration [33,34]. Thus the hypothesis suggests that the pharmacodynamic components of AJH, including bergenin, quercetin, epigallocatechingallate, rutin, cianidanol, myricitrin, ethyl gallate, kaempferol, taxifolin, cynaroside, myricetin, fraxetin, astilbin, and 13 other components (Figure 10), may have an effect when combined with serum medicinal chemistry [18]. However, further research is needed to determine the specific mechanisms of action.

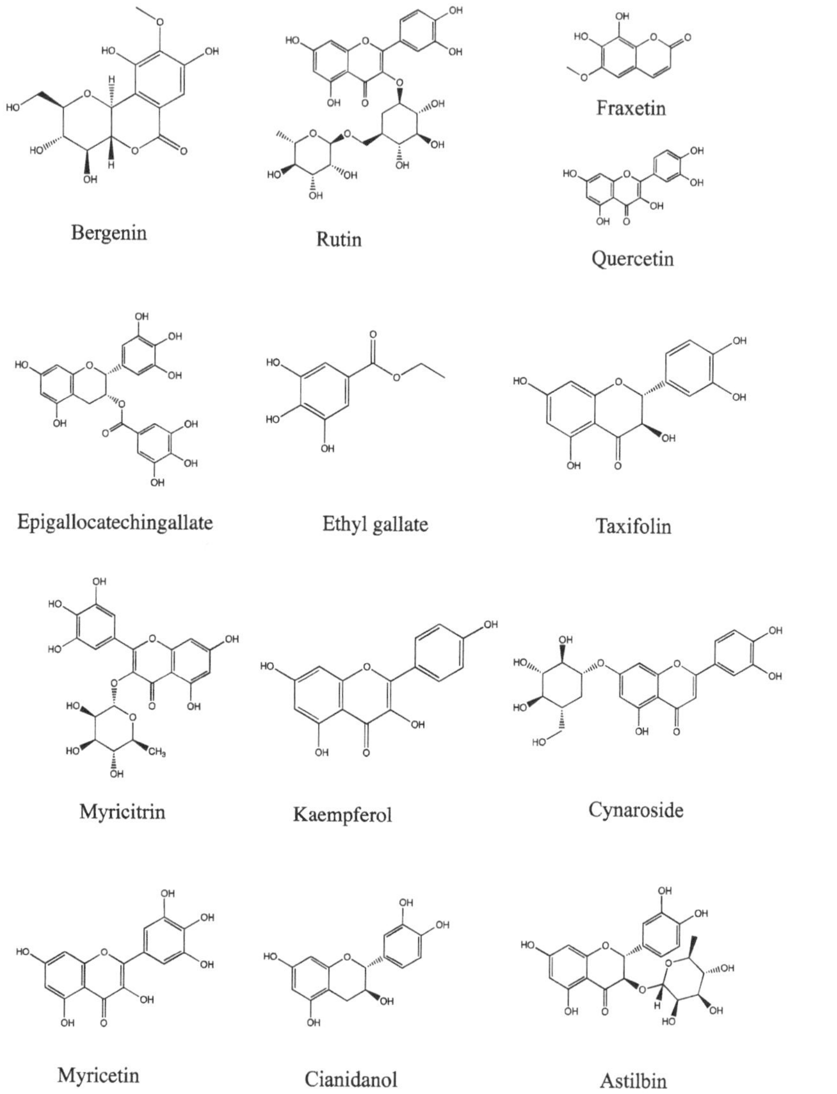

Figure 10. Chemical structure of 13 main compounds (bergenin, quercetin, epigallocatechingallate, rutin, cianidanol, myricitrin, ethyl gallate, kaempferol, taxifolin, cynaroside, myricetin, fraxetin, astilbin).

4. Materials and Methods

4.1. Materials and Instruments

The herbs were purchased from Anhui Sheng'Antang Drug Co., Ltd. (Longhui, Hunan, China, batch no. 210301), and identified by Prof. Chen Suiqing of the Henan University of Traditional Chinese Medicine as the dried whole plant of AJH. Gallic acid, fraxetin, and kaempferide were purchased from Chengdu Manstar Biotechnology Co., Ltd. (Chengdu, China) and Chengdu Kroma Biotechnology Co. (Chengdu, China) Caryophyllene oxide (batch no. J16GB155133, purity $\geq$ 90%) and taxifolin were purchased from Shanghai Yuanye Biotechnology Co. (Shanghai, China) (batch no. C15J11Q107526, purity $\geq$ 98%). Petroleum ether (batch no. 20220108, Tianjin Jindong Tianzheng Fine Chemical Reagent Factory, analytical pure), ethyl acetate (batch no. 20210820, Tianjin Hengxing Chemical Reagent Manufacturing Co., Ltd. (Tianjin, China), analytical pure), n-butanol (batch no. 20210104, Tianjin Hengxing Chemical Reagent Manufacturing Co., Ltd. (Tianjin, China), batch no. F22M6O102), mass spectrometry grade acetonitrile (Thermo Fisher Scientific Co., Ltd. (Shanghai, China) batch no. F22M8L203), mass spectrometry grade formic acid (Thermo Fisher Scientific Co., Ltd. (Shanghai, China), batch no. 214911), lipopolysaccharides (LPS, batch no. 039M4004V, Sigma-Aldrich, St. Louis, MO, USA), Mouse IL-6 reagent kit (batch no. A206H20634, Hangzhou C&T Biotechnology Co., Ltd. (Hangzhou, China)), Mouse TNF-α kit (batch no. UX0448RH0115, Wuhan Elite Bioscience and Technology Co., Ltd. (Wuhan, China)), CCK-8 (batch no. GK100011, Shanghai Hongye Bio-Technology Co., Ltd. (Shanghai, China)), fetal bovine serum (batch no. 21050701, Zhejiang Ltd. (Hangzhou, China)), RPMI 1640 medium (batch no. 20220424, Beijing Solebao Technology Co., Ltd. (Beijing, China)), and mouse alveolar macrophage (MH-S cell) lines were purchased from Procell Life Science&Technology Co., Ltd. (Wuhan, China). We also used UPLC–Orbitrap Fusion–MS (Thermo Scientific, Waltham, MA, USA), MS105DU one-hundred-thousandth analytical balance (Mettler Instruments Ltd., Greifensee, Switzerland), Heraeus Multifuge X_1R refrigerated tabletop high-speed centrifuge (Thermo Scientific, Waltham, MA, USA), Model SPD2010-230 Centrifugal Cryoconcentrator (Thermo Scientific, Waltham, MA, USA), Milli-QPOD Ultrapure Water Preparator (Millipore, Darmstadt, Germany), inverted microscope (ZEISS, Oberkochen, Germany), carbon dioxide incubator (Thermo Scientific, Waltham, MA, USA), and multi-function enzyme labeling instrument (Berten Instruments Ltd., Burlington, VT, USA).

4.2. Analysis of Components in Different Polar Parts of AJH

4.2.1. Sample Preparation

Take 2500 g of dried AJH and add 12 times the amount of 70% ethanol. Let it soak in the cold for 12 h, then heat it three times for 2 h each for the reflux. Filter the mixture and combine the three extracts. Let it sit for 12 h, allowing the ethanol to evaporate and remove any alcohol flavor. Concentrate the mixture under reduced pressure until it reaches the desired volume of 1500 mL. This will yield the extract of the AJH herbs. The AJH was subjected to sequential extraction using three organic solvents, namely petroleum ether, ethyl acetate, and *n*-butanol [35,36], in a 1:1.5 volume ratio. This extraction process was repeated five times. The extracts obtained using the same solvent were combined and concentrated under reduced pressure. The resulting concentrate was then freeze-dried to obtain powdered extracts of different parts of the AJH: petroleum ether site (PE, 26.45 g), ethyl acetate site (EA, 45.75 g), *n*-butanol site (NB, 71.80 g), and water site (WS, 169.40 g). The powder yields of the four components were 1.06%, 1.83%, 2.87%, and 6.78%, respectively. The powders from various components were measured and treated with ultrasound in methanol. The solutions from different components were created, filtered through a 0.22 μm microporous membrane, and examined using mass spectrometry.

4.2.2. Mass Spectrometry

The chromatographic conditions used in this experiment were as follows: a Hypersil GOLD C_{18} column with dimensions of 2.1 mm $\times$ 100 mm and a particle size of 3 μm was employed. The mobile phase consisted of methanol (A) and a 0.1% aqueous solution of formic acid (B). A gradient elution method was employed, with the following time intervals and corresponding changes in solvent composition: 0–2 min (95–75% B), 2–10 min (75–20% B), 10–22 min (20–12% B), 22–24 min (12–11.5% B), and 24–25 min (11.5–0% B). The flow rate was set at 0.2 mL/min. The column temperature was maintained at 25 °C. The sample injection volume was 5 μL.

Mass spectrometry was conducted under the following conditions: electrospray ionization (ESI) was used, with nitrogen as the carrier gas. The sheath gas flow rate was set at 35 Arb. The auxiliary gas flow rate was set at 5 Arb. The spray voltage was 3.50 kv for positive ions and 2.80 kv for negative ions. The ion transfer tube was maintained at a temperature of 300 °C, while the gasification temperature was set at 275 °C. The scanning mode used was Full MS/dd-MS^2 mode, which detected positive and negative ions separately. The first-stage full-scanning resolution was 120,000, with a scanning range of *m/z* 100–1000. The secondary scanning resolution was 60,000, and the HCD collision energy was set at 35%.

4.2.3. Mass Spectrometry Data Analysis

The raw data gathered were integrated into Compound Discoverer version 3.3 software. The procedure of identifying and analyzing unknown compounds was established using databases such as Chemspider (http://www.chemspider.com, accessed on 30 December 2023) and mzCloud (https://www.mzcloud.org/, accessed on 30 December 2023) [37,38]. The raw data underwent peak alignment and peak extraction. The information from the characteristic fragmentation ion peaks was synthesized and analyzed. Xcalibur software version 4.0 was utilized to calculate the possible chemical compositions with an error of less than 5 ppm. Additionally, Mass Frontier software was employed to infer the structure by examining the cleavage pattern [39]. This allowed for the qualitative identification of the chemical compositions of the four extracted parts of AJH.

4.3. *Analysis of the Activity of Different Parts of AJH*

4.3.1. Preparation of Reserve Solution of Different Extracted Parts of AJH

Measure 5 g of AJH powder. Combine it with 100 mL of 70% ethanol for reflux extraction for 2 h. Filter the mixture and then evaporate the resulting filtrate until it becomes a powder. Measure 20 mg of the extracted powder from AJH. Add 400 mL of dimethyl sulfoxide (DMSO) to dissolve the powder. This will result in a concentration of 50 mg/mL for the solution. Measure 20 mg of powdered AJH for each of the following solvents: petroleum ether, ethyl acetate, n-butanol, and water. Add 400 μL of DMSO to each sample to create a stock solution with a concentration of 50 mg/mL.

4.3.2. Mapping of Administration Concentration

The MH-S cells were cultivated in a cell culture incubator at 37 °C, 5% CO_2, and saturated humidity using RPMI-1640 media supplemented with 10% fetal bovine serum. Once the cell fusion degree reached 80–90%, the cells were transferred to a new culture dish, and only cells in the logarithmic growth phase were used for the studies. The cell suspension was diluted to a density of 2.5×10^6 cells/mL using RPMI-1640 media. The cells were then added to 96-well culture plates at a volume of 100 μL per well. Subsequently, the plates were placed in an incubator and grown for 24 h. The initial culture solution was discarded, and the cells were allocated into several groups: a control group, a group treated with varying concentrations of AJH extracts, a group treated with varying concentrations of petroleum ether, a group treated with varying concentrations of acetic acid, and a group treated with varying concentrations of acetate. The control group was supplemented with a medium containing 10% fetal bovine serum. The other groups received varying concentrations of drugs. Six wells were prepared for each experimental group. After drug administration, incubation was carried out for 24 h. Following this, the CCK-8 reagent solution was added and incubated for 3 h in the incubator. Finally, the enzyme marker was introduced into the incubator. Following drug administration, the cells were placed in an incubator for 24 h. Subsequently, a solution of CCK-8 reagent was given to the cells. They were further cultured in the incubator for 3 h. The absorbance of each group was quantified at a wavelength of 450 nm using an enzyme marker. The cell viability of each group was determined, and is shown in Table 6.

Table 6. Different concentration designs of dosing groups (μg/mL).

Sites	1	2	3	4	5	6	7	8
AJH	200	150	125	100	75	50	25	12.5
PE	50	25	12.5	6.25	3.13	-	-	-
EA	50	25	12.5	6.25	3.13	-	-	-
NB	200	150	125	100	75	50	25	12.5
WS	200	150	125	100	75	50	25	12.5

4.3.3. Expression of IL-6 and TNF-α in MH-S Cells

The administration concentration was determined based on the results from screening the administration concentration of each site of MH-S cells using the CCK-8 technique. Cells in the logarithmic growth phase, namely MH-S cells, were used for digestion. The density of the cell suspension was adjusted to 1×10^5 cells/mL. Next, 1 mL of the cell suspension was inoculated into each well of 12-well plates. The plates were then placed in a cell incubator and cultivated for 24 h. The cells were divided randomly into several groups: the control group, the LPS group, the LPS+AJH group, the LPS+ group, the LPS+EA group, the LPS+NB group, and the LPS+WS group. The control group was treated with medium containing 10% fetal bovine serum. The LPS group and the LPS+ administration group were induced to create the MH-S cell inflammation model by adding 100 ng/mL of LPS. After adding the drug, the cells were incubated for 24 h. The cell supernatants were then collected to measure the levels of IL-6 and TNF-α inflammatory factors.

The study also introduced the parameter p_0 value to comprehensively evaluate the anti-inflammatory activity of the four extracted portions of AJH. The equation $p_0 = \frac{P2}{P1}$ represents the relationship between p-values in the context of one-way ANOVA. p_1 refers to the p-value obtained from comparing each dosing group to the model group. A smaller p_1 value indicates a greater difference between each dosing group and the model group. On the other hand, p_2 represents the p-value obtained from comparing each dosing group to the blank group. A larger p_2 value suggests that the inhibition of inflammatory factor release in each dosing group after administration is more like that of the blank group. The expression of IL-6 and TNF-α inflammatory factors is the same. To analyze the active components of AJH, we utilized direct summing to compute comprehensive scores for each group based on the expression of IL-6 and TNF-α inflammatory factors. Additionally, we evaluated the p_0 value and comprehensive scoring results for each group, considering both factors equally important.

4.4. Study on the Spectrum-Effect Relationship between the Anti-Inflammatory Activities of the Active Parts of AJH

4.4.1. Preparation of Test Solution

Weigh precisely 0.1 g of AJH powder. Heat it twice in methanol under reflux for 1 h each time. Filter the mixture and combine the filtrate. Concentrate and evaporate the solution. Redissolve the residue in methanol and adjust the volume to 10 mL using a volumetric flask. Centrifuge the solution at 14,000 r/min for 15 min and collect the supernatant. Filter the supernatant using a 0.22 μm micropore filter membrane to obtain the test solution of AJH. Take approximately 0.009 g of the ethyl acetate component from AJH powder. Use precise weighing techniques. Add 50 mL of methanol and subject it to ultrasonic treatment for 10 min. Compensate for any weight loss by adding more methanol. Filter the mixture and then centrifuge it at a speed of 14,000 revolutions per minute for 15 min. Collect the liquid portion (supernatant) and pass it through a 0.22 μm micropore filter membrane. This will yield the ethyl acetate component of AJH, which could be used as the test solution.

To ensure accuracy, the following substances were carefully weighed: gallic acid, ethyl gallate, fraxetin, palmitic acid, kaempferide, eriodictyol, catechin, bergenin, epicatechin gallate, luteolin-7-*O*-glucoside, astilbin, myricitrin, nicotiflorin, rutin, isoeugenol, caryophyllene oxide, quercetin, taxifolin, and myricetin. Additionally, a control solution was prepared by adding an appropriate amount to a 2 mL volumetric flask and filling it with methanol. The solution was then thoroughly mixed. Thoroughly agitate the control stock solution and store it in a refrigerator at a temperature of 4 °C for future use. Before usage, prepare the mixed standard solution with an appropriate concentration.

4.4.2. Mass Spectrometry

The chromatographic conditions used in this study were as follows: a Hypersil GOLD C_{18} column with dimensions of 2.1 mm × 100 mm and a particle size of 3 μm was employed. The mobile phase consisted of a mixture of methanol (A) and a 0.1% aqueous solution of formic acid (B). A gradient elution method was employed, with the following time intervals and corresponding percentages of B: 0–2 min (95% B), 2–10 min (95–60% B), 10–12 min (60–30% B), 12–18 min (30–5% B), and 18–21 min (30–30% B). The composition of the sample was 5% B for 18–21 min, 5% B for 21–22 min, and 95% B for 22–25 min. The flow rate was 0.2 mL/min, the column temperature was 25 °C, and the injection volume was 1 μL.

Mass spectrometry was performed under the following conditions: electrospray ionization source (ESI), nitrogen as the carrier gas, a sheath gas flow rate of 25 Arb, an auxiliary gas flow rate of 10 Arb, a spray voltage of 3.50 kv (+) and 2.50 kv (−), an ion transfer tube temperature of 325 °C, and a gasification temperature of 325 °C. The scanning mode used was SIM mode, which involved the switching of positive and negative ions. The resolution of the scanning mode was 60,000. Table 7 displayed the ion information of the components that were measured for quantitative analysis.

Table 7. Ion information of the components to be measured for quantitative analysis.

Compounds	Test Mode	Test Ion	*m/z*	Rt (min)
Isoeugenol	$[M + H]^+$	$C_{10}H_{13}O_2{}^+$	165.0910	15.73
Caryophyllene oxide	$[M + H]^+$	$C_{15}H_{25}O^+$	221.1900	17.01
Quercetin	$[M + H]^+$	$C_{15}H_{11}O_7{}^+$	303.0499	8.67
Taxifolin	$[M + H]^+$	$C_{15}H_{13}O_7{}^+$	305.0656	9.19
Myricetin	$[M + H]^+$	$C_{15}H_{11}O_8{}^+$	319.0448	8.86
Gallic acid	$[M - H]^-$	$C_7H_5O_5{}^-$	169.0142	2.34
Ethyl gallate	$[M - H]^-$	$C_9H_9O_5{}^-$	197.0455	8.89
Fraxetin	$[M - H]^-$	$C_{10}H_7O_5{}^-$	207.0299	8.15
Palmiticacid	$[M - H]^-$	$C_{16}H_{31}O_2{}^-$	255.2330	20.25
Kaempferol	$[M - H]^-$	$C_{15}H_9O_6{}^-$	285.0405	9.3
Eriodictyol	$[M - H]^-$	$C_{15}H_{11}O_6{}^-$	287.0561	11.37
Cianidanol	$[M - H]^-$	$C_{15}H_{13}O_6{}^-$	289.0718	6.95
Bergenin	$[M - H]^-$	$C_{14}H_{15}O_9{}^-$	327.0722	6.42
Cynaroside	$[M - H]^-$	$C_{21}H_{19}O_{11}{}^-$	447.0933	8.98
Astilbin	$[M - H]^-$	$C_{21}H_{21}O_{11}{}^-$	449.1089	9.24
Epigallocatechingallate	$[M - H]^-$	$C_{22}H_{17}O_{11}{}^-$	457.0776	7.86
Myricitrin	$[M - H]^-$	$C_{21}H_{19}O_{12}{}^-$	463.0882	8.8
Nictoflorin	$[M - H]^-$	$C_{27}H_{29}O_{15}{}^-$	593.1512	9.09
Rutin	$[M - H]^-$	$C_{27}H_{29}O_{16}{}^-$	609.1461	8.66

4.4.3. Correlation Analysis of "Anti-Inflammatory Activity-Constituents" in the EA of AJH

A gray correlation study was conducted using the SPSS PRO program [40]. The raw data were transformed into a dimensionless format. The expression levels of IL-6 and TNF-α inflammatory factors were designated as the parent sequence Y(k), while the results of determining the content of the 19 components in the ethyl acetate part of AJH were designated as the subsequence Xi(k), where i represents the component. The absolute difference sequence Δi(k) was defined as the absolute value of the difference between the elements of the parent sequence Y(k) and the subsequence Xi(k). The calculation of the correlation coefficient is as follows:

$$Ai(k) = [\Delta i(k)min + \rho \times \Delta i(k)max]/[\Delta i(k) + \rho \times \Delta i(k)max]$$

The discrimination coefficient, denoted as ρ, is defined in this experiment as a value between 0 and 1. In this case, ρ was set to 0.5. A correlation coefficient greater than 0.8 indicated a strong relationship between the subsequence and the parent sequence. A correlation coefficient between 0.6 and 0.8 indicated a moderate relationship, while a correlation coefficient less than 0.6 indicated a weak relationship [41].

4.5. Pharmacological Activities of the Ethyl Acetate Parts of AJH

4.5.1. Preparation of AJH and Its Related Medicinal Fluid

Take 500 g of AJH herbs, combine with 12 times the volume of water, and perform a heating and reflux extraction process. The filtrate is then concentrated under reduced pressure three times, with each concentration lasting for 2 h. This process results in a final concentration of 2 g of AJH per milliliter, which is the water extract of AJH. Take 500 g of AJH herbs and add 12 times the volume of 70% ethanol. Heat the mixture and carry out a reflux extraction. The filtrate is subjected to decompression and concentration three times, with each iteration lasting for 2 h, to remove any alcohol flavor. The resulting concentrate will have a concentration of 2 g of AJH per milliliter, equivalent to a 70% alcohol extract of AJH. To prepare the ethyl acetate portion of AJH powder, follow the same procedure as described in Section 4.2.1. Weigh the powder and then add a 0.5% solution of CMC-Na to create a suspension equivalent to 2 g of AJH/mL. This suspension will serve as the ethyl acetate portion of the AJH liquid.

4.5.2. Animal Administration and Sample Collection

Male Sprague Dawley rats were acclimated and provided with food for 7 days (the Approval Code is DWLLGZR202303103 and the Approval Date is 1 March 2023. All animal experiments also complied with ethical guidelines for the use of animals in research according to policies of the UK). The participants were divided into 8 groups ($n = 8$) through a random process. These groups included a blank group, a model group, a low-dose group receiving 5 g of AJH raw drug/kg/d of ethyl acetate, a medium-dose group receiving 10 g of AJH raw drug/kg/d of ethyl acetate, a high-dose group receiving 20 g of AJH raw drug/kg/d of ethyl acetate, a group receiving AJH 70% alcoholic extract at a dose of 20 g/kg/d, a group receiving AJH aqueous extract at a dose of 20 g/kg/d, and a dexamethasone group receiving a dose of 2.3×10^{-4} g/kg/d. The low, medium, and high doses of EA, 70% alcoholic extract of the AJH group, aqueous extract of the AJH group, and the dexamethasone administration group were orally administered with their respective doses. The blank and model groups were orally administered with the same dose of distilled water for 7 consecutive days.

After 1 h following the final dose, the rats were rendered unconscious by injecting 20% ursodiol into their peritoneal cavity. The ALI rat model was induced by administering lipopolysaccharide (LPS) by a tracheal drip at a dosage of 2 mg/kg in each experimental group, except for the control group, which received a tracheal drip of saline. Following the rats' natural awakening and a 24 h fasting period, blood samples were obtained from the abdominal aorta. One portion of the blood was collected using a lithium heparin negative-pressure tube to assess routine pharmacodynamic indexes of blood, including white blood cell count (WBC), neutrophil ratio (NEU%), lymphocyte ratio (LYM%), and monocyte ratio (MONO%). The other portion of the blood serum was collected in a negative-pressure tube without sodium heparin and left in the tube for 2 h. Subsequently, the serum was subjected to centrifugation at a temperature of 4 °C for 15 min. Following this, it was transferred to a negative-pressure tube devoid of sodium heparin and left undisturbed for 2 min. The other portion was placed in a negative pressure tube without sodium heparin to collect serum.

It was then left for 2 h and centrifuged at 4 °C for 15 min at a speed of 3000 revolutions per minute. The resulting supernatant was used to detect the levels of IL-6, IL-10, CRP, and TNF-α inflammatory factors. The trachea and whole lungs of the rats were removed by opening the chest. The bronchoalveolar lavage fluid (BALF) from the left lungs was withdrawn after ligating the right main bronchus, and was also used to detect the levels of IL-6, IL-10, CRP, and TNF-α. The left lungs were then perfused with 4% paraformaldehyde for 30 min. The tissue samples were treated with paraformaldehyde for 30 min and then fixed in a 4% paraformaldehyde solution after perfusion. These samples were intended for pathological examination. The upper lobes of the right lung, spleen, and thymus were rinsed with saline solution. Excess surface water was removed using filter paper. The samples were then weighed to determine the lung wet/dry weight ratio and organ indexes.

4.5.3. Testing Indicators

Lung Function

After 24 h following the end of modeling, the animal whole-body volume scanning detection system non-invasive (WBP) was used to detect the lung function parameters of each group of rats, including airway narrowing index (Penh), maximal expiratory flow rate (PEF), expiratory flow rate in mid-expiration (EF50), and apnea (PAU) [42].

Blood Routine

Blood was taken from the abdominal aorta for routine blood tests to count white blood cell count (WBC), neutrophil ratio (NEU%), lymphocyte percentage (LYM%), and monocyte percentage (MONO%).

Measurement of IL-10, IL-6, CRP, and TNF-α Levels in Serum and BALF

An enzyme-linked immunosorbent assay (ELISA) was used to determine IL-10, IL-6, CRP, and TNF-α expression according to the respective manual in serum and BALF.

Lung Wet/Dry Weight Ratio and Organ Index

The upper lobes of the right lung of the rats in each group were excised and weighed to obtain the wet weight (W), then placed in the oven at 60 °C for 48 h and weighed to obtain the dry weight (D). The degree of edema of lung tissue was evaluated by calculating the ratio of W/D. The thymus and spleen of rats in each group were taken and weighed to calculate the rat organ index, and the calculated results were statistically analyzed.

Histopathological Analysis of Lungs

The left lung was fixed with 4% paraformaldehyde, routinely dehydrated, paraffin-embedded, sectioned at 4 μm, and routinely HE-stained, and the pathological changes in the lung tissues of rats in each group were observed under the microscope.

4.6. Data Analysis

All data were presented as mean ± SD ($\overline{X} \pm$ SD) and analyzed by a one-way analysis of variance, followed by Tukey's multiple comparisons as a post hoc test. All statistical analyses and graphs were performed using GraphPad Prism version 7.0 software (GraphPad Software, Inc., San Diego, CA, USA). p values less than 0.05 were considered statistically significant.

5. Conclusions

This study determined the chemical compositions of several solvent extraction methods in AJH. There were a total of 528 chemical components examined. Within the petroleum ether, ethyl acetate, *n*-butanol, and aqueous environments, a total of 215, 289, 128, and 69 components were identified, respectively. The research revealed that the ethyl acetate extraction site consisted of a high concentration of flavonoids, phenylpropanoids, and quinones. Furthermore, the anti-inflammatory activities of the four solvents of AJH were evaluated using the cellular tests conducted. The findings demonstrated that the EA site of AJH shown significant efficacy in anti-inflammatory treatment. Hence, a comprehensive analysis was conducted on a total of 19 chemical constituents at the EA site. The gray relational analysis was used to examine the correlation between 19 constituents and their anti-inflammatory activity. A total of 17 substances have been identified as pharmacodynamic components with anti-inflammatory properties. Relying on prior studies on serum medicinal chemistry, a total of 13 components were ultimately recognized as having biological activity in AJH for the purpose of treating acute lung injury. Additionally, the intermediate and high doses of the EA site of AJH exhibited a more significant effect in enhancing lung function, blood routine, and lung histology.

Supplementary Materials: The supplementary materials for this study can be downloaded at: https://www.mdpi.com/article/10.3390/molecules29040770/s1, Compositional enrichment results for different sites of AJH.

Author Contributions: S.S. and X.L. designed the study, analyzed the data, and drafted the manuscript. D.Z. and L.Z. performed animal experiments. Y.T. and X.H. designed the study and analyzed the corresponding data. S.F. designed, supervised, and reviewed the manuscript. All authors have read and agreed to the published version of the manuscript.

Funding: This paper was supported by the Key scientific research projects in higher education institutions in Henan Province (24B360002).

Institutional Review Board Statement: The animal study protocol was approved by the Experimental Animal Welfare Ethics Committee of Henan University of Traditional Chinese Medicine (protocol code DWLLGZR202303103 and date of approval 1 March 2023).

Data Availability Statement: The data presented in this study are available in the article and Supplementary Material.

Conflicts of Interest: The authors declare no conflict of interest.

Abbreviations

AJH: Ardisiae Japonicae Herba, ALI: acute lung injury, PE: petroleum ether site, EA: ethyl acetate site, NB: *n*-butanol site, WS: water site, WBC: white blood cells, NEU%: neutrophil percentage, MONO%: monocyte percentage, LYM%: lymphocyte percentage, EF50: mid-expiration, PAU: the rats' apnea, PEFb: the rats' maximum expiratory flow rate, Penh: airway narrowing index, NEU%: neutrophil ratio, LYM%: lymphocyte ratio, DMSO: dimethyl sulfoxide, ESI: electrospray ionization, TIC: total ion current, LPS: lipopolysaccharide, BALF: bronchoalveolar lavage fluid.

References

1. Shah, D.; Das, P.; Acharya, S.; Agarwal, B.; Christensen, D.J.; Robertson, S.M.; Bhandari, V. Small Immunomodulatory Molecules as Potential Therapeutics in Experimental Murine Models of Acute Lung Injury (Ali)/Acute Respiratory Distress Syndrome (Ards). *Int. J. Mol. Sci.* **2021**, *22*, 52573. [CrossRef]
2. Zhang, C.Y.; Lin, W.; Gao, J.; Shi, X.; Davaritouchaee, M.; Nielsen, A.E.; Mancini, R.J.; Wang, Z. PH-Responsive Nanoparticles Targeted to Lungs for Improved Therapy of Acute Lung Inflammation/Injury. *ACS Appl. Mater. Interfaces* **2019**, *11*, 16380–16390. [CrossRef]
3. Maloney, J.P. Genetic Factors Impacting Therapy in Acute Lung Injury/Acute Respiratory Distress Syndrome. *J. Investig. Med.* **2009**, *57*, 865–869. [CrossRef] [PubMed]
4. Song, L.; Li, G.; Guan, W.; Zeng, Z.; Ou, Y.; Zhao, T.; Li, J.; He, D.; Fang, X.; Zhang, Y.; et al. Design, Synthesis and Anti-Inflammatory Activity Study of Lansiumamide Analogues for Treatment of Acute Lung Injury. *Biomed. Pharmacother.* **2023**, *166*, 115412. [CrossRef] [PubMed]
5. Butt, Y.; Kurdowska, A.; Allen, T.C. Acute Lung Injury: A Clinical and Molecular Review. *Arch. Pathol. Lab. Med.* **2016**, *140*, 345–350. [CrossRef] [PubMed]
6. Peiris, J.S.M.; Guan, Y.; Yuen, K.Y. Severe Acute Respiratory Syndrome. *Nat. Med.* **2004**, *10*, S88–S97. [CrossRef] [PubMed]
7. Xu, D.; Sun, X.; Zhang, Y.; Cao, L. Ketamine Alleviates HMGB1-Induced Acute Lung Injury through TLR4 Signaling Pathway. *Adv. Clin. Exp. Med.* **2020**, *29*, 813–817. [CrossRef] [PubMed]

8. Lian, J.; Lin, J.; Zakaria, N.; Yahaya, B.H. Acute Lung Injury: Disease Modelling and the Therapeutic Potential of Stem Cells. *Adv. Exp. Med. Biol.* **2020**, *1298*, 149–166. [CrossRef] [PubMed]
9. Ding, Z.; Zhong, R.; Xia, T.; Yang, Y.; Xing, N.; Wang, W.; Wang, Y.; Yang, B.; Sun, X.; Shu, Z. Advances in Research into the Mechanisms of Chinese Materia Medica against Acute Lung Injury. *Biomed. Pharmacother.* **2020**, *122*, 109706. [CrossRef]
10. Chao, X.; Li, H.; Wang, T.S.; Pan, Y.S. Research Progress on Active Ingredients of Animal-Derived Traditional Chinese Medicine in Treatment of Allergic Rhinitis. *Molecules* **2023**, *48*, 4884–4892. [CrossRef]
11. Zian, Z.; Xiaoying, W.; Wenjing, W.; Yun, T.A.N.; Junxia, W.; Jinglian, Q.U. Discussion the Importance of Phlegm Retention in the Pathogenesis and Treatment of Acute Lung Injury. *J. Guizhou Univ. Tradit. Chin. Med.* **2021**, *43*, 1–3.
12. Wang, P.; Ding, Q.; Qu, S.; Li, Z.; Yu, Q.; Zhu, H.; Shi, Y.; Pan, J. The Effects of Qingfei Litan Fang on Lung Function, Lung Tissue Pathology, and Inflammatory Factors in Vivo in a Mouse Model of Acute Lung Injury Induced by Klebsiella Pneumoniae. *J. Tradit. Chin. Med.* **2021**, *62*, 1991–2002.
13. Xiaolong, M.; Yueru, C.; Feiyan, S.; Sunqi, Y.; Zhihui, Z.; Huajun, Z. Research Progress on Therapeutic Effect and Mechanism of Traditional Chinese Medicine in Acute Lung Injury. *Chin. J. Mod. Appl. Pharm.* **2022**, *39*, 269–276. [CrossRef]
14. Lingyang, Z.; Ziyi, W.; Cuiwei, H. Research Progress on Chemical Composition and Pharmacological Action of Ardisia Japonic. *Guangxi Sci.* **2019**, *26*, 484–489.
15. Yu, K.Y.; Gao, W.; Li, S.Z.; Wu, W.; Li, P.; Dou, L.L.; Wang, Y.Z.; Liu, E.H. Qualitative and Quantitative Analysis of Chemical Constituents in Ardisiae Japonicae Herba. *J. Sep. Sci.* **2017**, *40*, 4347–4356. [CrossRef] [PubMed]
16. Feng, S.; Yuan, J.; Zhao, D.; Li, R.; Liu, X.; Tian, Y.; Li, J. Systematic Characterization of the Effective Constituents and Molecular Mechanisms of Ardisiae Japonicae Herba Using UPLC-Orbitrap Fusion MS and Network Pharmacology. *PLoS ONE* **2022**, *17*, e0269087. [CrossRef] [PubMed]
17. Feng, S.X.; Han, X.X.; Zhao, D.; Li, R.R.; Liu, X.F.; Tian, Y.G.; Li, J.S. Simultaneous Quantitation of 31 Bioactive Components in Different Parts of Ardisiae Japonicae Herba from Different Regions by UPLC-Orbitrap Fusion MS. *J. Liq. Chromatogr. Relat. Technol.* **2021**, *44*, 649–662. [CrossRef]
18. Han, X.X.; Tian, Y.G.; Liu, W.J.; Zhao, D.; Liu, X.F.; Hu, Y.P.; Feng, S.X.; Li, J.S. Metabolomic Profiling Combined with Network Analysis of Serum Pharmacochemistry to Reveal the Therapeutic Mechanism of Ardisiae Japonicae Herba against Acute Lung Injury. *Front. Pharmacol.* **2023**, *14*, 1131479. [CrossRef]
19. Yi, Y.S. Regulatory Roles of Flavonoids on Inflammasome Activation during Inflammatory Responses. *Mol. Nutr. Food Res.* **2018**, *62*, e1800147. [CrossRef]
20. Zhang, Y.L.; Zhang, W.X.; Yan, J.Q.; Tang, Y.L.; Jia, W.J.; Xu, Z.W.; Xu, M.J.; Chattipakorn, N.; Wang, Y.; Feng, J.P.; et al. Chalcone Derivatives Ameliorate Lipopolysaccharide-Induced Acute Lung Injury and Inflammation by Targeting MD2. *Acta Pharmacol. Sin.* **2022**, *43*, 76–85. [CrossRef]
21. Han, F.; Wu, G.; Han, S.; Li, Z.; Jia, Y.; Bai, L.; Li, X.; Wang, K.; Yang, F.; Zhang, J.; et al. Hypoxia-Inducible Factor Prolyl-Hydroxylase Inhibitor Roxadustat (FG-4592) Alleviates Sepsis-Induced Acute Lung Injury. *Respir. Physiol. Neurobiol.* **2020**, *281*, 103506. [CrossRef]
22. Al-Khayri, J.M.; Sahana, G.R.; Nagella, P.; Joseph, B.V.; Alessa, F.M.; Al-Mssallem, M.Q. Flavonoids as Potential Anti-Inflammatory Molecules: A Review. *Molecules* **2022**, *27*, 2901. [CrossRef]
23. Maleki, S.J.; Crespo, J.F.; Cabanillas, B. Anti-Inflammatory Effects of Flavonoids. *Food Chem.* **2019**, *299*, 125124. [CrossRef]
24. Lin, H.; Wang, P.; Zhang, W.; Yan, H.; Yu, H.; Yan, L.; Chen, H.; Xie, M.; Shan, L. Novel Combined Preparation and Investigation of Bergenin-Loaded Albumin Nanoparticles for the Treatment of Acute Lung Injury: In Vitro and In Vivo Evaluations. *Inflammation* **2022**, *45*, 428–444. [CrossRef] [PubMed]
25. Yang, S.; Yu, Z.; Wang, L.; Yuan, T.; Wang, X.; Zhang, X.; Wang, J.; Lv, Y.; Du, G. The Natural Product Bergenin Ameliorates Lipopolysaccharide-Induced Acute Lung Injury by Inhibiting NF-KappaB Activition. *J. Ethnopharmacol.* **2017**, *200*, 147–155. [CrossRef] [PubMed]
26. Tang, Q.; Wang, Q.; Sun, Z.; Kang, S.; Fan, Y.; Hao, Z. Bergenin Monohydrate Attenuates Inflammatory Response via MAPK and NF-KB Pathways against Klebsiella pneumonia Infection. *Front. Pharmacol.* **2021**, *12*, 651664. [CrossRef] [PubMed]
27. Nassef, N.A.A.; Abd-El Hamid, M.S.; Abusikkien, S.A.; Ahmed, A.I. Quercetin Ameliorates Acute Lung Injury in a Rat Model of Hepatopulmonary Syndrome. *BMC Complement. Med. Ther.* **2022**, *22*, 320. [CrossRef]
28. Wang, X.F.; Song, S.D.; Li, Y.J.; Hu, Z.Q.; Zhang, Z.W.; Yan, C.G.; Li, Z.G.; Tang, H.F. Protective Effect of Quercetin in LPS-Induced Murine Acute Lung Injury Mediated by CAMP-Epac Pathway. *Inflammation* **2018**, *41*, 1093–1103. [CrossRef]
29. Chen, S.; Bai, Y.; Xia, J.; Zhang, Y.; Zhan, Q. Rutin Alleviates Ventilator-Induced Lung Injury by Inhibiting NLRP3 Inflammasome Activation. *iScience* **2023**, *26*, 107866. [CrossRef]
30. Yeh, C.H.; Yang, J.J.; Yang, M.L.; Li, Y.C.; Kuan, Y.H. Rutin Decreases Lipopolysaccharide-Induced Acute Lung Injury via Inhibition of Oxidative Stress and the MAPK-NF-KB Pathway. *Free Radic. Biol. Med.* **2014**, *69*, 249–257. [CrossRef] [PubMed]
31. Ren, J.; Lu, Y.; Qian, Y.; Chen, B.; Wu, T.; Ji, G. Recent Progress Regarding Kaempferol for the Treatment of Various Diseases. *Exp. Ther. Med.* **2019**, *18*, 2759–2776. [CrossRef]
32. Khazdair, M.R.; Anaeigoudari, A.; Agbor, G.A. Anti-Viral and Anti-Inflammatory Effects of Kaempferol and Quercetin and COVID-2019: A Scoping Review. *Asian Pac. J. Trop. Biomed.* **2021**, *11*, 327–334. [CrossRef]

33. Wang, D.; Liu, R.; Zeng, J.; Li, C.; Xiang, W.; Zhong, G.; Xia, Z. Preliminary Screening of the Potential Active Ingredients in Traditional Chinese Medicines Using the Ussing Chamber Model Combined with HPLC-PDA-MS. *J. Chromatogr. B Anal. Technol. Biomed. Life Sci.* **2022**, *1189*, 123090. [CrossRef]
34. Liu, X.; Liu, J.; Fu, B.; Chen, R.; Jiang, J.; Chen, H.; Li, R.; Xing, L.; Yuan, L.; Chen, X.; et al. DCABM-TCM: A Database of Constituents Absorbed into the Blood and Metabolites of Traditional Chinese Medicine. *J. Chem. Inf. Model.* **2023**, *63*, 4948–4959. [CrossRef]
35. Šavikin, K.; Živković, J.; Alimpić, A.; Zdunić, G.; Janković, T.; Duletić-Laušević, S.; Menković, N. Activity Guided Fractionation of Pomegranate Extract and Its Antioxidant, Antidiabetic and Antineurodegenerative Properties. *Ind. Crops Prod.* **2018**, *113*, 142–149. [CrossRef]
36. Qureshi, M.N.; Numonov, S.; Aisa, H.A. Chemical and Pharmacological Evaluation of Hulls of Prunus Dulcis Nuts. *Int. J. Anal. Chem.* **2019**, *2019*, 5861692. [CrossRef]
37. Chaowongdee, S.; Malichan, S.; Pongpamorn, P.; Paemanee, A.; Siriwan, W. Metabolic Profiles of Sri Lankan Cassava Mosaic Virus-Infected and Healthy Cassava (Manihot Esculenta Crantz) Cultivars with Tolerance and Susceptibility Phenotypes. *BMC Plant Biol.* **2023**, *23*, 178. [CrossRef] [PubMed]
38. Oulkar, D.; Singh, K.; Narayan, B. Characterization of Different Parts of Litchi Fruit Using UHPLC-QExactive Orbitrap. *J. Food Sci. Technol.* **2022**, *59*, 4889–4906. [CrossRef] [PubMed]
39. Li, Q.; Zhang, C.; Xilin, T.; Ji, M.; Meng, X.; Zhao, Y.; Siqin, B.; Zhang, N.; Li, M. Effects of Koumiss on Intestinal Immune Modulation in Immunosuppressed Rats. *Front. Nutr.* **2022**, *9*, 765499. [CrossRef] [PubMed]
40. Guanya, B.; Pan, H.; Yuanyuan, L.; Yong-hua, Y.; Yan, N. Spectrum Effect Relationship of Anti-Inflammatory Effect of Different Polar Parts of Unripe Forsythiae Fructus. *Chin. J. Exp. Tradit. Med. Formulae* **2017**, *23*, 1–6.
41. Linghang, Q.; Guosheng, C.; Ji Yuan, T.; Kun, S.; Jie, W.; Yiyi, X.; Yanju, L. Spectrum-Effect Relationship of Dryness of Atractylodis Rhizoma Volatile Oil Based on Grey Relational Grade and Orthogonal Partial Least Squares Analysis. *Chin. Tradit. Herb. Drugs* **2019**, *50*, 150–156. [CrossRef]
42. Tian, Y.; Li, Y.; Li, J.; Xie, Y.; Wang, M.; Dong, Y.; Li, L.; Mao, J.; Wang, L.; Luo, S. Bufei Yishen Granule Combined with Acupoint Sticking Improves Pulmonary Function and Morphormetry in Chronic Obstructive Pulmonary Disease Rats. *BMC Complement. Altern. Med.* **2015**, *15*, 266. [CrossRef] [PubMed]

molecules

MDPI

Article

Ginsenoside Rb1, Compound K and 20(S)-Protopanaxadiol Attenuate High-Fat Diet-Induced Hyperlipidemia in Rats via Modulation of Gut Microbiota and Bile Acid Metabolism

Kang-Xi Zhang [1,†], Yue Zhu [2,†], Shu-Xia Song [1,2], Qing-Yun Bu [1,2,3], Xiao-Yan You [1,2,*], Hong Zou [4,*] and Guo-Ping Zhao [2,4,5,*]

1 Henan Engineering Research Center of Food Microbiology, College of Food and Bioengineering, Henan University of Science and Technology, Luoyang 471023, China; zhangkx@tib.cas.cn (K.-X.Z.); asongshuxia@163.com (S.-X.S.); buqy@tib.cas.cn (Q.-Y.B.)

2 Master Lab for Innovative Application of Nature Products, National Center of Technology Innovation for Synthetic Biology, Tianjin Institute of Industrial Biotechnology, Chinese Academy of Sciences, Tianjin 300308, China; zhuyue@tib.cas.cn

3 Haihe Laboratory of Synthetic Biology, Tianjin 300308, China

4 CAS Engineering Laboratory for Nutrition, Shanghai Institute of Nutrition and Health, Chinese Academy of Sciences, Shanghai 200031, China

5 CAS-Key Laboratory of Synthetic Biology, CAS Center for Excellence in Molecular Plant Sciences, Shanghai Institute of Plant Physiology and Ecology, Chinese Academy of Sciences, Shanghai 200032, China

* Correspondence: xiaoyanyou@haust.edu.cn (X.-Y.Y.); hzhou01@sinh.ac.cn (H.Z.); gpzhao@sibs.ac.cn (G.-P.Z.); Tel.: +86-135-2595-2624 (X.-Y.Y.); +86-139-1761-2639 (H.Z.); +86-139-0169-1631 (G.-P.Z.)

† These authors contributed equally to this work.

Abstract: Hyperlipidemia, characterized by elevated serum lipid concentrations resulting from lipid metabolism dysfunction, represents a prevalent global health concern. Ginsenoside Rb1, compound K (CK), and 20(S)-protopanaxadiol (PPD), bioactive constituents derived from Panax ginseng, have shown promise in mitigating lipid metabolism disorders. However, the comparative efficacy and underlying mechanisms of these compounds in hyperlipidemia prevention remain inadequately explored. This study investigates the impact of ginsenoside Rb1, CK, and PPD supplementation on hyperlipidemia in rats induced by a high-fat diet. Our findings demonstrate that ginsenoside Rb1 significantly decreased body weight and body weight gain, ameliorated hepatic steatosis, and improved dyslipidemia in HFD-fed rats, outperforming CK and PPD. Moreover, ginsenoside Rb1, CK, and PPD distinctly modified gut microbiota composition and function. Ginsenoside Rb1 increased the relative abundance of *Blautia* and *Eubacterium*, while PPD elevated *Akkermansia* levels. Both CK and PPD increased *Prevotella* and *Bacteroides*, whereas *Clostridium-sensu-stricto* and *Lactobacillus* were reduced following treatment with all three compounds. Notably, only ginsenoside Rb1 enhanced lipid metabolism by modulating the PPARγ/ACC/FAS signaling pathway and promoting fatty acid β-oxidation. Additionally, all three ginsenosides markedly improved bile acid enterohepatic circulation via the FXR/CYP7A1 pathway, reducing hepatic and serum total bile acids and modulating bile acid pool composition by decreasing primary/unconjugated bile acids (CA, CDCA, and β-MCA) and increasing conjugated bile acids (TCDCA, GCDCA, GDCA, and TUDCA), correlated with gut microbiota changes. In conclusion, our results suggest that ginsenoside Rb1, CK, and PPD supplementation offer promising prebiotic interventions for managing HFD-induced hyperlipidemia in rats, with ginsenoside Rb1 demonstrating superior efficacy.

Keywords: ginsenoside Rb1; ginsenoside CK; ginsenoside PPD; hyperlipidemia; gut microbiota; bile acid metabolism

check for updates

Citation: Zhang, K.-X.; Zhu, Y.; Song, S.-X.; Bu, Q.-Y.; You, X.-Y.; Zou, H.; Zhao, G.-P. Ginsenoside Rb1, Compound K and 20(S)-Protopanaxadiol Attenuate High-Fat Diet-Induced Hyperlipidemia in Rats via Modulation of Gut Microbiota and Bile Acid Metabolism. *Molecules* **2024**, *29*, 1108. https://doi.org/10.3390/molecules29051108

Academic Editor: Yun-Bae Kim

Received: 3 February 2024
Revised: 25 February 2024
Accepted: 27 February 2024
Published: 1 March 2024

1. Introduction

Hyperlipidemia is a critical factor in the onset of numerous diseases, including non-alcoholic fatty liver disease (NAFLD), obesity, and cardiovascular disease (CVD) [1]. Strategies to mitigate lipid metabolism dysfunction are pivotal in curbing the prevalence of hyperlipidemia [2]. The liver is central to systemic lipid regulation, overseeing processes such as cholesterol metabolism, fatty acid (FA) uptake and export, de novo lipogenesis, and β-oxidation [3]. Excessive cholesterol accumulation is detrimental, posing risks for various health conditions [4]. A primary pathway for cholesterol clearance involves its conversion into bile acids (BAs), which also function as signaling molecules and metabolic regulators, influencing hepatic lipid, glucose, and energy balance to maintain metabolic equilibrium [5,6]. The gut microbiota exerts a significant influence on cholesterol homeostasis and BA metabolism, transforming primary BAs into secondary BAs via several biochemical reactions [7]. Notably, the FXR/Fgf15 signaling pathway plays a key role in the reabsorption and secretion of intestinal BAs [8].

Panax ginseng, a perennial herb, is renowned for its diverse bioactive and medicinal properties, primarily attributed to its ginsenosides, polysaccharides, and peptides [9]. Ginsenosides, particularly the dammarane and oleanolic acid types, are the herb's principal bioactive constituents [10]. Among these, ginsenoside Rb1 stands out for its potential to mitigate HFD-induced hepatic steatosis by modulating peroxisome proliferator-activated receptor γ (PPARγ) expression and intestinal microbial structure and function [11,12]. Compound K (CK) exhibits enhanced bioavailability and water solubility using nanocarriers and cyclodextrin for delivery, with its anti-diabetic and anti-hyperlipidemic effects the subjects of recent research [13,14]. Meanwhile, 20(S)-protopanaxadiol (PPD), a deglycosylated metabolite produced by gut microbiota, has demonstrated immunostimulatory effects [15]. However, the specific contributions of ginsenoside Rb1, CK, and PPD to hyperlipidemia progression in HFD-fed rats remain to be elucidated.

The interaction between ginsenosides and gut microbiota post-oral administration is crucial for human health [16]. The structural complexity of ginsenosides, including their polar surface area and the presence of sugar moieties, limits their direct pharmacological action [17]. The gut microbiota facilitates the transformation of ginsenosides into more bioavailable metabolites, such as CK and PPD, enhancing their pharmacological activity [18]. Despite existing research, comparative studies on the effects of ginsenoside Rb1, CK, and PPD on lipid metabolism in HFD-fed rats and the mechanisms involved are scarce. Our study aims to fill this gap by investigating the impact of these ginsenosides on hyperlipidemia and delving into the mechanisms related to gut microbiota composition and function, as well as the enterohepatic circulation of BAs. This research could unveil new natural product-based strategies for improving lipid metabolism disorders.

2. Results

2.1. *Effects of Ginsenoside Rb1, CK, and PPD on Body Weight, Serum Biochemistry, Hepatic Adipose Accumulation in HFD-Fed Rats*

The animal experimental design was shown in Figure 1A. Compared with rats fed with standard diet (Con group), rats fed with a high-fat diet (HFD group) showed significantly increased BW and body weight gain (Supplementary Materials Figure S1A,B). Compared with the HFD group, Rb1 treatment dramatically reduced BW and body weight gain, while there was no significant difference in body weight gain among the PPD group, the CK group and the HFD group, indicating that the effect of ginsenoside Rb1 on improving weight gain was superior to that of ginsenoside CK and PPD treatment (Figure 1B,C). Further, eWAT/BW, liver/BW and fasting blood glucose (FBG) levels were decreased in the Rb1 group and the PPD group. Ginsenoside CK also reduced the eWAT/BW and FBG levels, whereas there was no significant difference in liver/BW between the CK group and the HFD group (Figure 1D–F). Furthermore, compared to the Con group, serum TC, TG, and LDL-C levels were higher and the serum HDL-C level was lower in the HFD group (Supplementary Materials Figure S1D). The Rb1 group showed significantly reduced serum TC, TG, and LDL-C levels compared to the HFD group. Although the serum LDL-C level was lower, there was little effect on decreasing serum TG and TC levels after ginsenoside CK and PPD treatment (Figure 1G–J). In this study, atorvastatin treatment decreased FBG, TC and LDL-C levels in HFD-fed rats, while atorvastatin treatment failed to reduce body weight gain, eWAT/BW, liver/BW, TG and HDL-C levels. To further evaluate the effects of the ginsenoside Rb1, CK, and PPD on alleviating lipid metabolism disorders, HE staining was performed on liver and eWAT in HFD-fed rats. As shown in Supplementary Materials Figure S1D and Figure 1K, ginsenoside Rb1 improved hepatic steatosis and lipid accumulation in eWAT. Similarly, ginsenoside CK and PPD treatment could also slightly reduce hepatic fat droplets and adipocyte size in HFD-fed rats. However, atorvastatin treatment failed to ameliorate adipose deposition and lipid accumulation in liver and eWAT. Furthermore, in the HFD group, serum TNF-α, IL-6, and IL-1β levels of the HFD group were significantly increased compared with the Con group. Ginsenoside Rb1 treatment observably reduced serum TNF-α, IL-6, and IL-1β levels. Ginsenoside CK and PPD treatment decreased serum TNF-α and IL-6 secretion, respectively. In addition, the serum TNF-α level was lower, whereas the serum IL-6 level was higher in the Ato group compared with the HFD group (Figure 1L). These results indicate that ginsenoside Rb1 was more effective than ginsenoside CK and PPD in improving obesity, hyperlipidemia and hepatic steatosis in HFD-fed rats.

2.2. Effects of Ginsenoside Rb1, CK, and PPD on the Composition of Gut Microbiota in HFD-Fed Rats

As shown in Figure 2A, the Chao1 index, the richness index, Shannon's index and Simpson's index estimate species richness and species evenness, which reflect the diversity of microbial communities. After ginsenoside Rb1 treatment, the Chao1 index, the richness index and Shannon's index were significantly enhanced and Simpson's index was reduced, while there was no significant difference in the Chao1 index, the richness index, Simpson's index and Shannon's index between the CK group, the PPD group and the HFD group. In addition, we observed a distinct clustering of microbiota composition for the HFD group, the Rb1 group, the CK group and the PPD group by UniFrac-based principal coordinates analysis (PCoA) (Figure 2B), indicating that there were significant differences in gut microbiota structure after ginsenoside Rb1, CK, and PPD treatment. Next, we analyzed the relative abundance of gut microbiota at the phylum and genus levels in HFD-fed rats. At the phylum level, the relative abundance of *Actinobacteria* was lower in the Rb1 group. Moreover, the relative abundance of *Firmicutes* and *Actinobacteria* was lower, while the relative abundance of *Bacteroidetes* was higher in the CK group compared with the HFD group. In the PPD group, the relative abundance of *Verrucomicrobia* and *Bacteroidetes* was higher compared with the HFD group (Figure 2C). At the genus level, the relative abundance of the top 30 genera was shown in Figure 2D. Among them, the relative abundance of *Blautia, Eisenbergiella,* and *Anaerostipes* was higher, whereas the relative abundance of *Clostridium subcluster XIVa, Clostridium cluster XVIII, Eubacterium, Lactobacillus, Turicibacter, Bifidobacterium* and *Clostridium-sensu-stricto* was lower in the Rb1 group. The relative abundance of *Clostridium-sensu-stricto, Lactobacillus* was observably reduced, while the relative abundance of *Prevotella, Clostridium subcluster XIVa, Sutterella, Parabacteroides, Flavonifractor, Ruminococcus* and *Bacteroides* was enhanced after ginsenoside CK and PPD treatment (Figure 2E). Particularly, further analysis at the genus level showed that the relative abundance of *Blautia* and *Eubacterium* was changed only after ginsenoside Rb1 treatment. Ginsenoside PPD treatment markedly increased the relative abundance of *Akkermansia* in HFD-fed rats. Ginsenoside Rb1, CK and PPD treatment simultaneously decreased the relative abundance of *Clostridium-sensu-stricto* and *Lactobacillus* in the HFD-fed rats (Figure 2F).

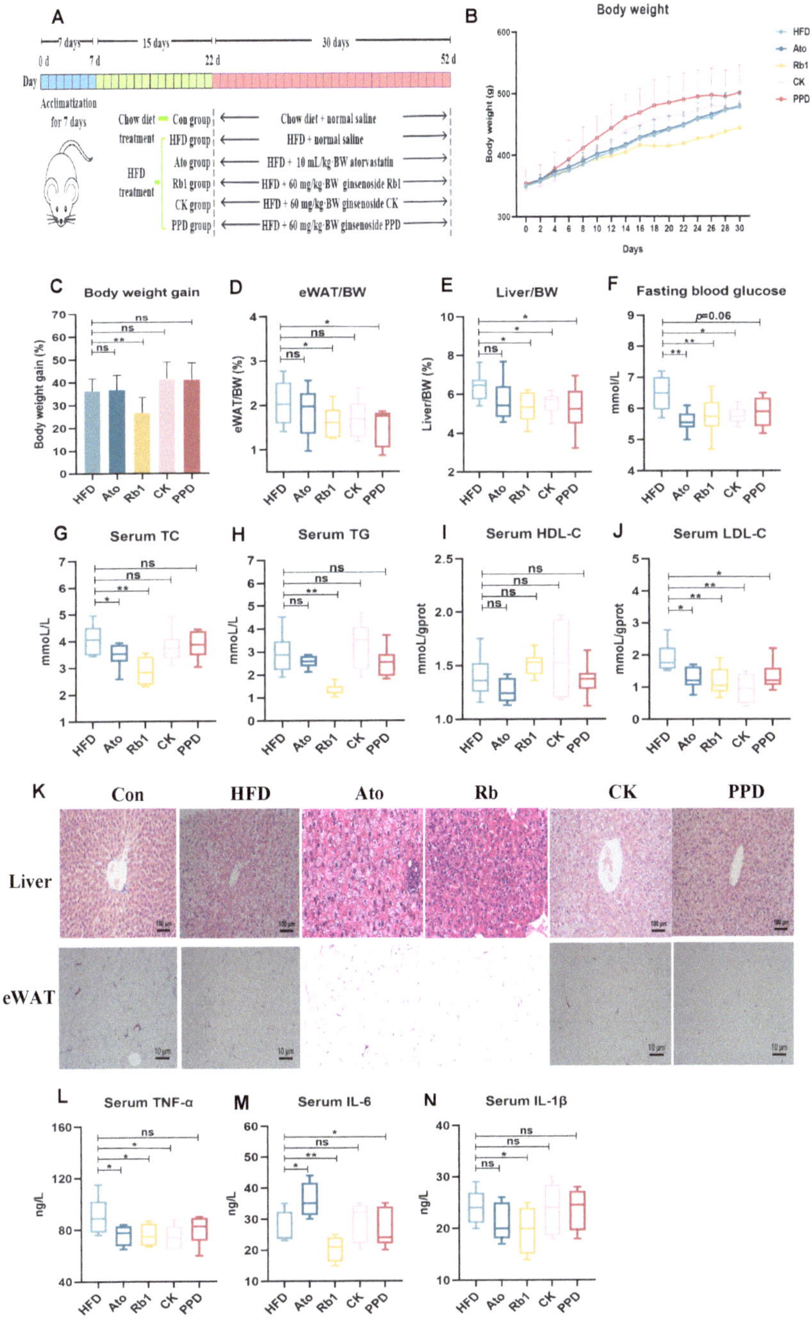

Figure 1. Effect of ginsenoside Rb1, CK and PPD treatment on body weight, hyperlipidemia and liver steatosis in HFD-fed rats. (**A**) The experimental design, (**B**) body weight (g), (**C**) body weight gain (%), (**D**) eWAT/BW (%), (**E**) liver/BW (%), (**F**) fasting blood glucose (mmol/L), (**G**–**J**) serum concentrations of TC, TG, HDL-C and LDL-C (mmol/L), (**K**) histological analysis for liver and epididymal adipose tissues (eWAT), and (**L**–**N**) serum concentrations of TNF-α, IL-6 and IL-1β (ng/L). Values are the means ± SEMs. * $p < 0.05$, ** $p < 0.01$ and versus the HFD group, ns means no significantly difference.

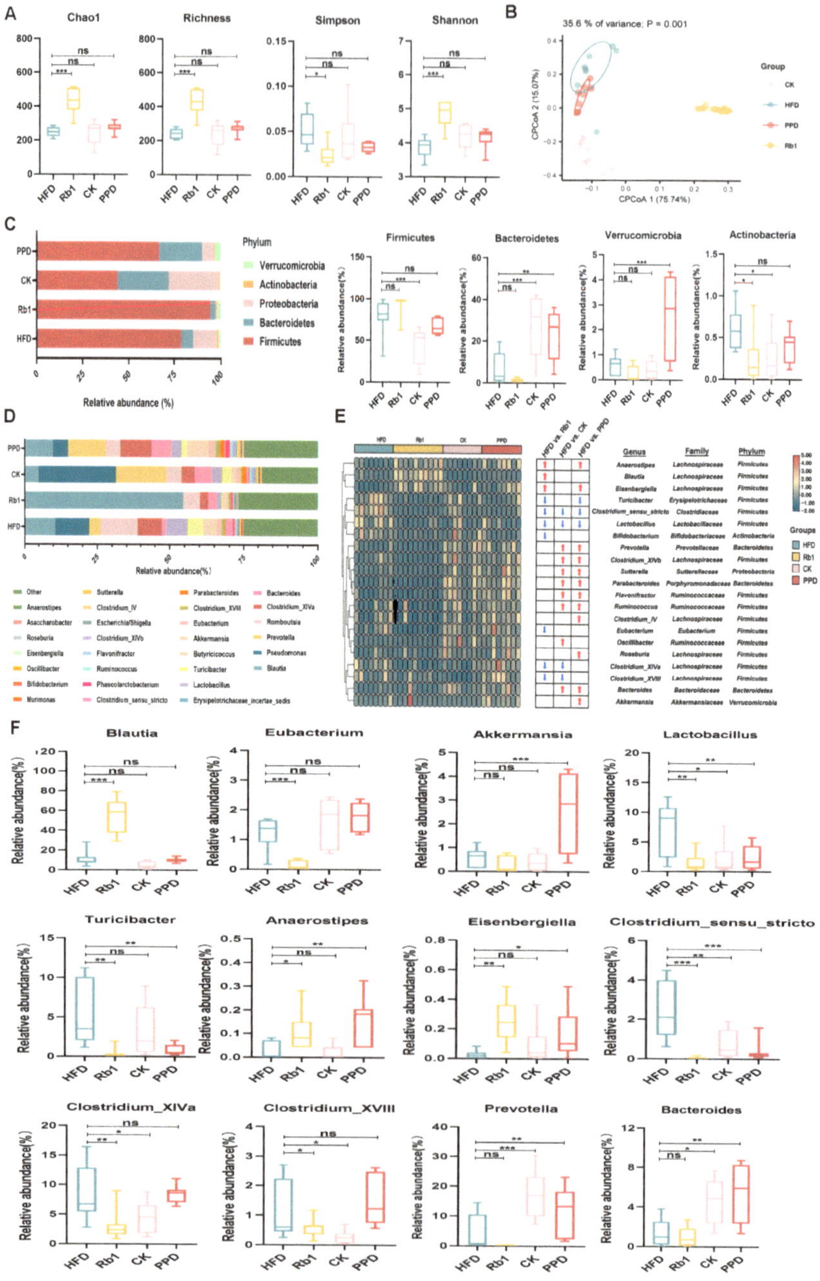

Figure 2. Effects of ginsenoside Rb1, CK and PPD on gut microbiota structure in HFD-fed rats. (**A**) α diversity analysis, (**B**) PCoA analysis, (**C**) phylum-level analysis, (**D**) top 30 genus-level analysis, (**E**) heatmap showing the relative abundance of 21 genera significantly altered after ginsenoside Rb1, CK and PPD treatment, ⬆ representing more abundance in the Rb1, CK and PPD groups compared with the HFD group, ⬇ representing less abundance in the Rb1, CK and PPD groups compared with the HFD group, (**F**) change in the relative abundance of representative genera. Values are the means ± SEMs. * $p < 0.05$, ** $p < 0.01$ and *** $p < 0.001$ versus the HFD group, ns means no significantly difference.

LEfSe analysis also demonstrated significant taxonomic differences at the bacterial genus level among the Rb1 group, the PPD group, the CK group and the HFD group (Figure 3A–C). The relative abundance of *Firmicutes* and *Blautia* was higher, and the relative abundance of *Clostridium subcluster XlVa* and *Lactobacillus* was lower in the Rb1 group compared to the HFD group. In addition, the relative abundance of *Prevotella* and *Bacteroides* was increased and the relative abundance of *Clostridium subcluster XlVa* and *Lactobacillus* was reduced after ginsenoside CK treatment. Similar to the ginsenoside CK treatment, the relative abundance of *Prevotella*, *Bacteroides* and *Akkermansia* was higher and the relative abundance of *Clostridium subcluster XlVa* was lower after ginsenoside PPD treatment. These results were in accordance with the genus-level analysis.

2.3. Effects of Ginsenoside Rb1, CK, and PPD on the Function of Gut Microbiota in HFD-Fed Rats

PICRUSt analysis was applied to predict the function of microbial genes involved in the metabolism pathway in the HFD group, the Rb1 group, the CK group, and the PPD group. Compared with the HFD group, multiple pathways related to lipid metabolism were significantly improved after ginsenoside Rb1 treatment, including the superpathway of phospholipid biosynthesis I, fatty acid salvage, fatty acid β-oxidation, CDP-diacylglycerol biosynthesis I, and CDP-diacylglycerol biosynthesis II. In addition, metabolism pathways involved in glycometabolism, including the glycolysis II (from fructose 6-phosphate), D-fructuronate degradation, D-glucuronide and D-glucuronate degradation pathways, were altered in the Rb1 group compared with the HFD group (Figure S2A). Moreover, the fatty acid elongation-saturated, superpathway of fatty acid biosynthesis, GDP-mannose biosynthesis, sucrose degradation III, peptidoglycan maturation and sucrose degradation IV were changed after ginsenoside CK treatment (Figure S2B). Further, lipid IVA biosynthesis was markedly altered after ginsenoside PPD treatment (Figure S2C). Overall, these studies suggested that ginsenoside Rb1, CK, and PPD could regulate the composition and function of gut microbiota in HFD-fed rats in different ways.

2.4. Effects of Ginsenoside Rb1, CK, and PPD on FA and Cholesterol Metabolism in HFD-Fed Rats

To examine the molecular mechanism of ginsenoside Rb1, CK, and PPD in lipid metabolism, we measured the relative expression of genes related to FA and cholesterol metabolism in the liver, including farnesoid X receptor (*Fxr*), 3-hydroxy-3-methylglutaryl-coenzyme A (CoA) reductase (*Hmgcr*), cytochrome P450 (CYP) isoform 7A1 (*Cyp7a1*), *Cyp7b1*, *Cyp27a1*, perixisome proliferation-activated receptor alpha (*Pparα*), *Pparγ*, acetyl-CoA carboxylase (*Acc*), fatty acid synthase (*Fas*) and hormone-sensitive triglyceride lipase (*Hsl*). As shown in Figure 4A, genes involved in cholesterol synthesis (*Hmgcr*, *Cyp7a1*, *Cyp27a1*) were down-regulated and genes involved in cholesterol catabolism (*Fxr*, *Cyp7b1*) were up-regulated in the HFD group, whereas ginsenoside Rb1, CK, and PPD treatment markedly up-regulated the mRNA relative expression of *Hmgcr*, *Cyp7a1*, and *Cyp27a1* and down-regulated the mRNA relative expression of *Fxr* and *Cyp7b1*. In addition, only ginsenoside Rb1 supplementation significantly down-regulated the relative expression of genes related to FA synthesis *Pparγ*, *Acc* and *Fas* and up-regulated the relative expression of genes related to FA β-oxidation *Pparα* and *Hsl* in HFD-fed rats. Ginsenoside CK and PPD treatment failed to change the relative expression of genes related to FA metabolism. Further, hepatic protein expression related to FA metabolism (ACC, HSL, FAS, PPARα and PPARγ) and cholesterol metabolism (FXR, CYP7A1) was measured among the HFD group, the Rb1 group, the CK group and the PPD group. As shown in Figure 4B, ginsenoside Rb1 supplementation down-regulated the relative expression of PPARγ, ACC and FAS and up-regulated the relative expression of PPARα and HSL in HFD-fed rats. Compared to the HFD group, the relative expression of PPARγ and FXR was lower, whereas the relative expression of CYP7A1 was higher in the CK group. Likewise, the relative expression of PPARγ, FXR was down-regulated, whereas the relative expression of HSL and CYP7A1 was up-regulated in the PPD group. These studies suggested that ginsenoside Rb1, CK, and PPD observably improved cholesterol catabolism (BA metabolism) in HFD-fed rats.

Meanwhile, only ginsenoside Rb1 supplementation could alleviate FA metabolism disorder through regulating FA synthesis and β-oxidation.

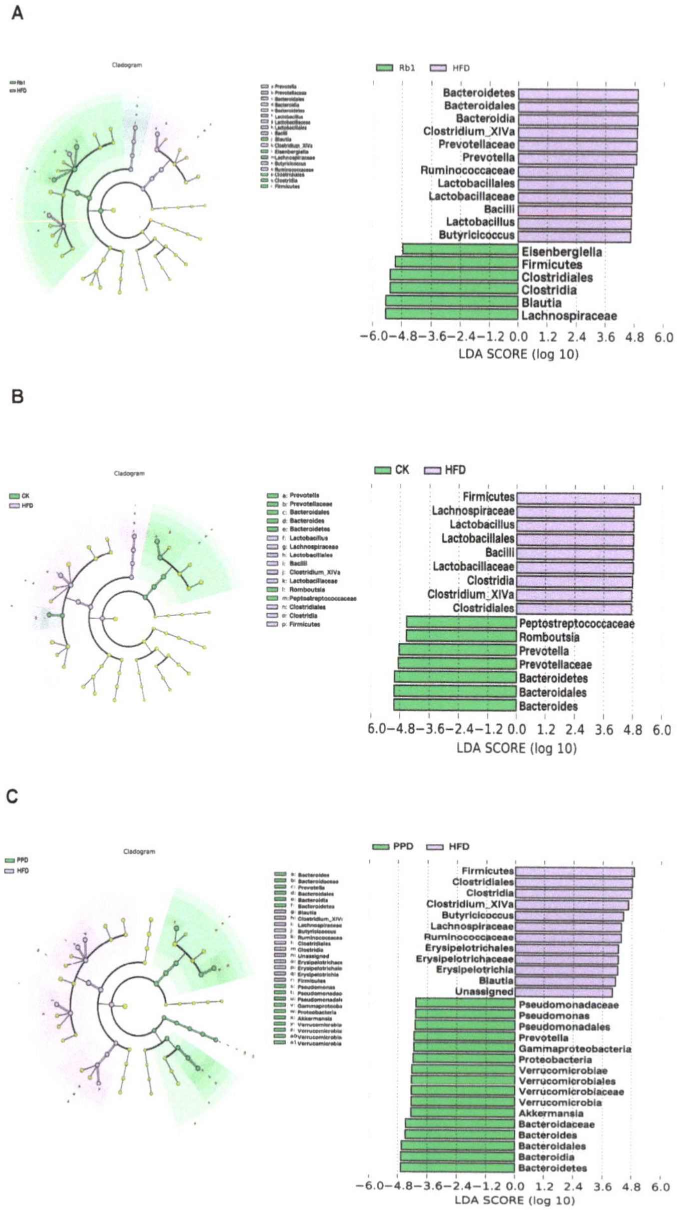

Figure 3. Effects of ginsenoside Rb1, CK and PPD on the gut microbial community in HFD-fed rats. The cladogram of the linear discriminant analysis effect size (LEfSe) analysis for (**A**) Rb1 vs. HFD, (**B**) CK vs. HFD, and (**C**) PPD vs. HFD.

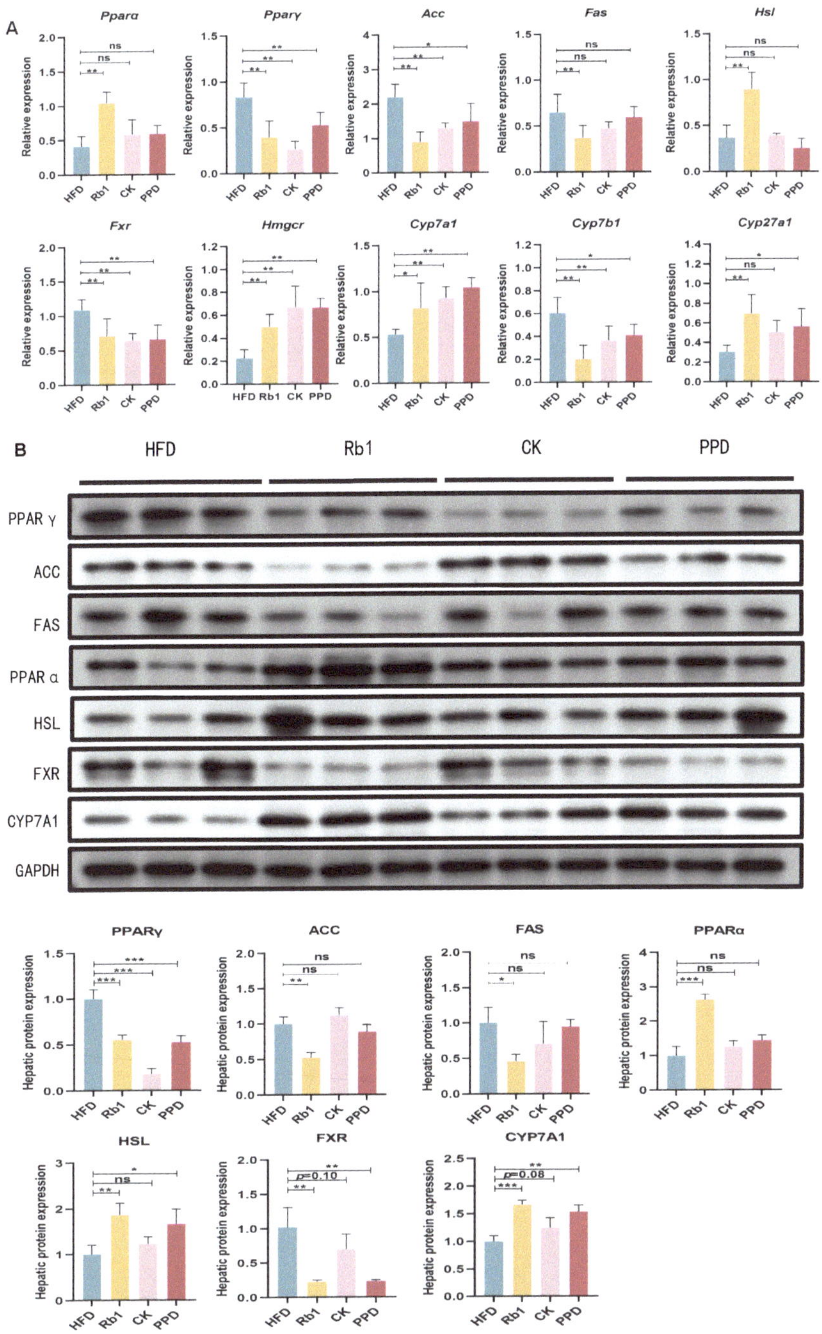

Figure 4. Effects of ginsenoside Rb1, CK and PPD on the relative expression of genes and proteins related to cholesterol metabolism and fatty acid metabolism in HFD-fed rats. (**A**) The relative expression of genes related to cholesterol metabolism and fatty acid metabolism. (**B**) The protein expression of PPARγ, ACC, FAS, PPARα, HSL, FXR and CYP7A1 in liver. Values are the means ± SEMs. * $p < 0.05$, ** $p < 0.01$ and *** $p < 0.001$ versus the HFD group, ns means no significantly difference.

2.5. *Effects of Ginsenoside Rb1, CK, and PPD on BA Metabolism in HFD-Fed Rats*

Furthermore, to further clarify the effect of ginsenoside Rb1, CK, and PPD treatment on BA metabolism, we measured serum and hepatic 12 BA levels by HPLC-MS/MS. As shown in Figure 5A, in serum, the total BA level was significantly decreased in the Rb1, CK, and PPD groups compared with the HFD group. Among them, the relative abundance of primary/unconjugate BA (CA and β-MCA) was lower after ginsenoside Rb1, CK, and PPD treatment. Ginsenoside Rb1 and PPD treatment decreased the relative abundance of CDCA and α-MCA, which was not altered after ginsenoside CK treatment. Further, ginsenoside Rb1 treatment notably enhanced the relative abundance of primary/conjugate BA, including TCA, TCDCA, GCDCA and α/β-TMCA, while ginsenoside PPD treatment did not change the relative abundance of TCDCA and GCDCA. Similarly, ginsenoside CK treatment increased the relative abundance of TCA, TCDCA and α/β-TMCA. Meanwhile, the relative abundance of secondary/conjugate BA, including TDCA, GDCA, TUDCA and GUDCA, was significantly reduced in the Rb1, CK and PPD groups compared with the HFD group. In liver, the BA profile was similar to that in serum. The total BA level was significantly decreased after ginsenoside Rb1, CK and PPD treatment. The relative abundance of primary/unconjugate BA, CA, CDCA and β-MCA was lower in the Rb1 group, the CK group and the PPD group than that in the HFD group. Furthermore, TCDCA, GCDCA, GDCA and TUDCA, belonging to conjugate BA, were observably increased after ginsenoside Rb1, CK and PPD treatment. Collectively, these results indicated that ginsenoside Rb1, CK, and PPD treatment could decrease the accumulation of hepatic and serum total BAs and regulate the profile of the BA pool in HFD-fed rats.

2.6. *Correlation Analysis of Gut Microbiota and BA Metabolism*

To clarify the correlation between gut microbiota and the levels of BAs in serum, Spearman correlation analysis was further performed among the four groups. The serum 12 BAs were correlated with the changes in the top 20 genus. As shown in Figure 6A, in the HFD group, *Clostridium_sensu_stricto*, *Romboutsia*, *Turicibacter*, *Pseudomonas*, *Clostridium_XlVb*, and *Eubacterium* were negatively correlated with conjugate BA (TUDCA, GCDCA, GUDCA, TCA, TCDCA and TDCA), positively correlated with the primary/unconjugate BA (CDCA, CA, α-MCA, β-MCA). Conversely, *Blautia*, *Anaerostipes*, *Flavonifractor*, *Prevotella*, *Akkermansia*, *Phascolarctobacterium*, *Eisenbergiella* and *Butyricicoccus* were positively correlated with conjugate BA (TUDCA, GCDCA, GUDCA, TCA, TCDCA and TDCA), and *Blautia*, *Clostridium_XVIII*, *Anaerostipes*, *Clostridium_XlVa* and *Butyricicoccus* were negatively correlated with primary/unconjugate BA, including α-MCA and CDCA. In Rb1 group, most BAs (except CA and TCDCA) were positively correlated with *Bacteroides*, *Clostridium_XlVa*, *Phascolarctobacterium*, *Prevotella*, *Akkermansia* and *Eisenbergiella*, and negatively correlated with *Clostridium_sensu_stricto*, *Pseudomonas*, *Lactobacillus* and *Anaerostipes* (Figure 6B). Moreover, primary BAs (CA, CDCA, β-MCA, TCDCA and GCDCA) were negatively correlated with most genera (except *Pseudomonas* and *Lactobacillus*), whereas GUDCA, TUDCA, TDCA and GDCA were positively correlated with *Eubacterium*, *Bacteroides*, *Flavonifractor*, *Romboutsia*, *Clostridium_sensu_stricto*, *Akkermansia* and *Parabacteroides* in the CK group (Figure 6C). In addition, *Lactobacillus*, *Akkermansia*, *Clostridium_sensu_stricto*, *Butyricicoccus* and *Pseudomonas* were negatively correlated with primary/unconjugate BA (CDCA, CA, α-MCA, β-MCA), and positively correlated with GDCA and GUDCA. On the contrary, *Parabacteroides*, *Bacteroides*, *Flavonifractor*, *Eisenbergiella* and *Clostridium_XlVb* were negatively correlated with TUDCA, GDCA and GUDCA, and positively correlated with α-MCA and TCA (Figure 6D). These results suggested that key-phylotypes in the gut microbiota after ginsenoside Rb1, CK, and PPD treatment were strongly correlated with the serum BAs in HFD-fed rats.

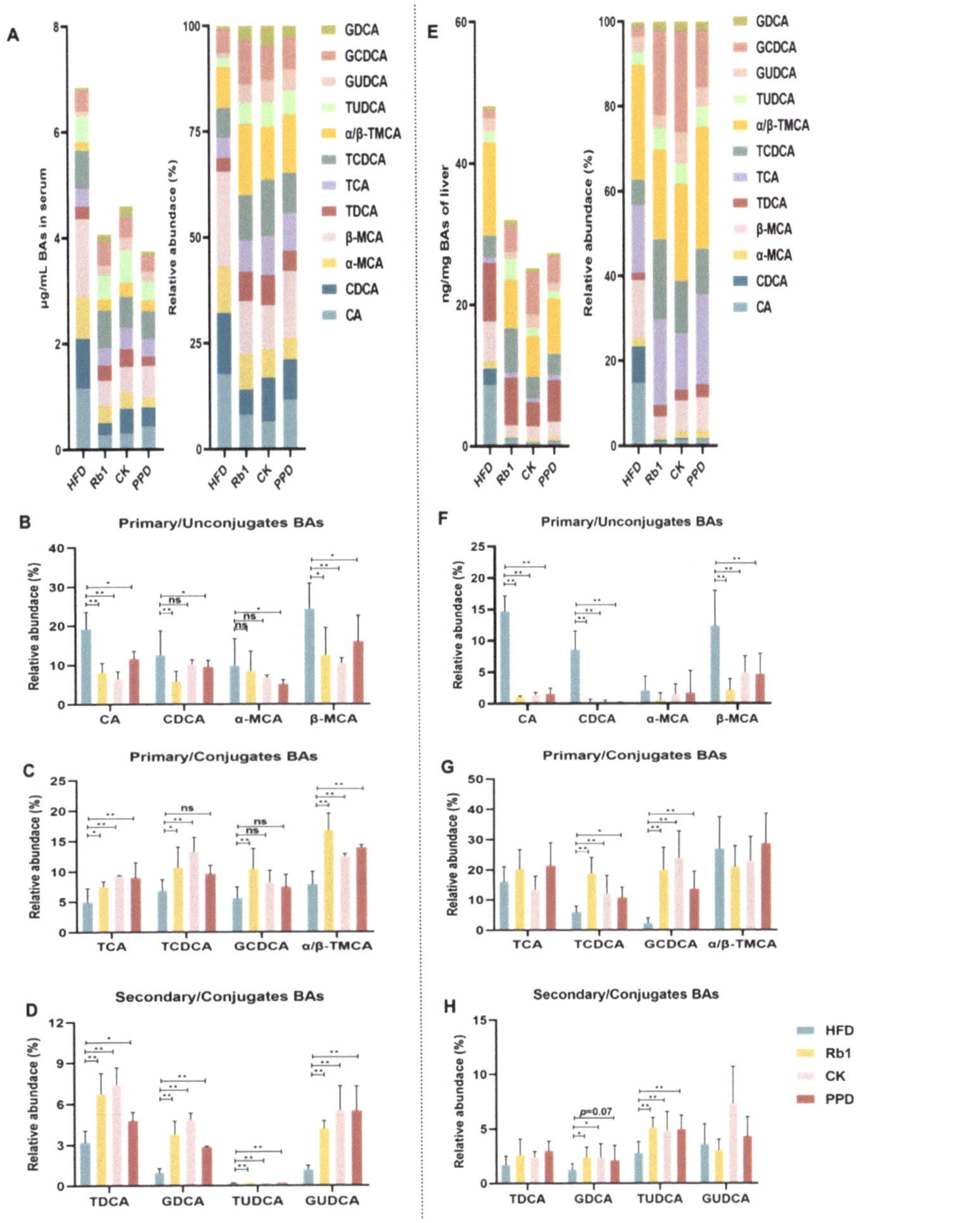

Figure 5. Effects of ginsenoside Rb1, CK and PPD on BA metabolism in HFD-fed rats. (**A**) The total BA level in serum; (**B**) primary/unconjugate BA level in serum; (**C**) primary/conjugate BA in serum; (**D**) secondary/conjugate BA in serum; (**E**) the total BA level in liver; (**F**) primary/unconjugate BA level in liver; (**G**) primary/conjugate BA in liver; (**H**) secondary/conjugate BA in liver. Values are the means ± SEMs. * $p < 0.05$, ** $p < 0.01$ versus the HFD group, ns means no significantly difference.

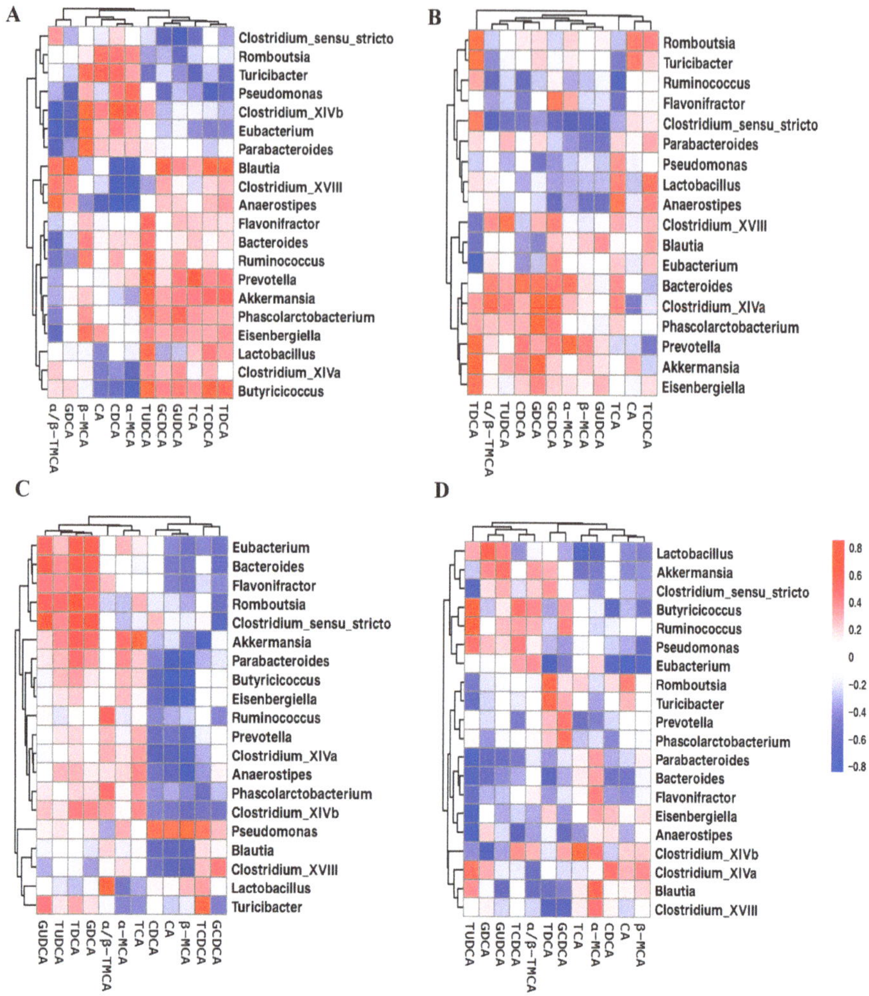

Figure 6. The heatmap showed Pearson's correlation coefficient with $p < 0.05$ between key gut microbes and BAs in the HFD, Rb1, CK and PPD groups. (**A**) The correlation between the top 20 genus and serum BAs in the HFD group, (**B**) the correlation between the top 18 genus and serum BAs in the Rb1 group, (**C**) the correlation between the top 20 genus and serum BAs in the CK group, and (**D**) the correlation between the top 20 genus and serum BAs in the PPD group. Red cells represent positive correlation and blue cells represent negative correlation.

3. Discussion

Many studies have reported that the gut microbiota has been shown to be associated with cardiovascular disease (CVD) development such as obesity, type 2 diabetes (T2DM), metabolic syndrome, and hyperlipidemia. Dietary intervention plays critical roles in maintaining a healthy cardiovascular function and contributes to alleviating insulin sensitivity and lipid metabolism dysfunction through regulating gut microbiota composition [19,20]. Ginsenoside has been attracting attention as a dietary phytochemical, which decrease TC and TG levels in HFD-induced mice [21]. However, whether the beneficial effects of ginsenoside Rb1, CK, and PPD treatment has HFD-induced hyperlipidemia could be attributed to the regulation of gut microbiota and BA metabolism has been rarely reported. In this study, we aimed to investigate the effects of ginsenoside Rb1, CK, and PPD treatment on hyperlipidemia in HFD-fed rats and further explore their differences and underlying mechanisms by focusing on the gut microbiota and BA metabolism. Specifically, ginsenoside Rb1 treatment reduced body weight gain and serum TC and TG levels, and alleviated hepatic steatosis and inflammation, whereas ginsenoside CK and PPD treatment had no such effect. Serum FBG and LDL-C levels were significantly reduced and the serum HDL-C level was not altered after ginsenoside Rb1, CK, and PPD treatment. Moreover, ginsenoside Rb1, CK, and PPD treatment changed the structure and function of gut microbiota through different ways, which were closely connected with BA metabolism in HFD-fed rats.

Previous studies have indicated that ginsenoside Rb1 supplementation improved HFD-induced insulin resistance and obesity in mice [22]. Similarly, ginsenoside CK and PPD was shown to reduce body weight and fat accumulation and improve sensitivity to insulin in mice [23,24]. Our results were consistent with previous studies that ginsenoside Rb1 treatment observably reduced body weight, body weight gain, eWAT/BW and liver/BW in HFD-fed rats (Figure 1B–E). However, ginsenoside CK and PPD failed to affect the increased body weight after HFD treatment in rats. In addition, ginsenoside Rb1 improved hyperlipidemia through decreasing serum TC, TG and LDL-C levels, while ginsenoside CK and PPD treatment partly alleviated hyperlipidemia by reducing the LDL-C level. There was no significant difference in the HDL-C level after ginsenoside Rb1, CK and PPD supplementation (Figure 1G–J). Moreover, ginsenoside Rb1 treatment could ameliorate hepatic steatosis though reducing adipose deposition in liver and eWAT, which was better than ginsenoside CK and PPD treatment (Figure 1K). Statins, such as 3-hydroxy-3-methyl-glutaryl coenzyme A (HMG-CoA) reductase inhibitor, suppressed the cholesterol biosynthetic pathway, resulting in decreased TC and LDL-C levels in serum [25]. In our study, atorvastatin treatment did not reduce body weight or relieve hepatic steatosis, which could improve hyperlipidemia by reducing serum TC and LDL-C levels in HFD-fed rats.

A growing body of research has underscored the pivotal role of inflammatory cytokines in the initiation and advancement of hyperlipidemia and obesity. Lai et al. have elucidated that various inflammatory processes exhibit a strong correlation with hyperlipidemia and its concomitant cardiovascular diseases (CVDs). Notably, the circulating levels of cytokines associated with inflammation, namely IL-1β, IL-6, and TNF-α, were found to be significantly higher in patients afflicted with CVDs [26]. Furthermore, a link between obesity and elevated serum levels of inflammatory cytokines has been established. The surplus of macronutrients within adipose tissues acts as a catalyst for the secretion of inflammatory mediators, such as TNF-α and IL-6, while concurrently diminishing the production of adiponectin. This imbalance fosters a pro-inflammatory state and induces oxidative stress [27]. The escalation of cytokines in individuals with obesity is implicated in the progression of several disorders, including cardiovascular disease, hypertension, and insulin resistance [28].

Targeting inflammatory pathways offers a promising strategy for managing obesity and mitigating the risk of cardiovascular diseases. Ginsenosides, in particular, have been identified as potent agents in dampening inflammation. Oral administration of ginsenosides has been shown to significantly reduce inflammatory markers. Specifically, ginsenoside Rb1 has emerged as a potential anti-inflammatory agent, demonstrating the ability to inhibit the activation of NF-κB, a pivotal regulator of inflammation and a key factor in controlling TNF-α production in LPS-activated mouse peritoneal macrophages [29]. Furthermore, Compound K (CK) exerts its anti-inflammatory effects primarily by diminishing the levels of inducible nitric oxide synthase (iNOS), cyclooxygenase (COX)-2, and pro-inflammatory cytokines [29–31]. CK achieves this by down-regulating the activity of IRAK-1, MAPKs, IKK-α, and NF-κB in the peritoneal macrophages of LPS-treated mice [32]. In our study, we observed a notable reduction in serum TNF-α, IL-6, and IL-1β levels following treatment with ginsenoside Rb1, in comparison to the high-fat diet (HFD) group. Additionally, treatments with ginsenoside CK and PPD individually led to a decrease in serum TNF-α and IL-6 levels, respectively (Figure 1L). Our findings confirm the anti-inflammatory properties of ginsenoside.

The importance of the gut microbiota in human health has been revealed. The imbalance of gut microbiota was related to many metabolic diseases, such as hyperlipidemia, non-alcoholic liver disease and atherosclerosis [33,34]. The effect mechanism of microbiota-related metabolites such as BAs, lipopolysaccharide (LPS), and short-chain fatty acids (SCFAs) in the regulation of hyperlipidemia has been reported [35]. In our study, ginsenoside Rb1 improved the diversity of gut microbiota through increasing the Chao1 index, the richness index and Shannon's index, whereas ginsenoside CK and PPD failed to change the diversity index in HFD-fed rats (Figure 2A). Furthermore, *Blautia*, as a dominant genus in the gut microbiota, has revealed a negative correlation with obesity-related metabolic disorders [36]. After ginsenoside Rb1 treatment, the relative abundance of *Blautia* was significantly enhanced (58.75%) in HFD-fed rats (Figure 2F). The genus *Akkermansia*, belonging to the *Verrucomicrobia* phylum, is considered to be a promising candidate as a probiotic, which regulates metabolic functions and protects mice from high-fat diets [37,38]. The relative abundance of *Akkermansia* was increased after ginsenoside PPD treatment. In addition, some studies indicated that *Lactobacillus* have shown inhibitory effects on HFD-induced obesity, whereas Aline et al.'s study indicated that the dysbiosis of gut microbiota seems be related to increases in *Lactobacillus* [19,39]. *Clostridium_sensu_stricto* is regarded as the true *Clostridium* genus, which is associated with lipid metabolism disorders and insulin resistance [40]. Our results indicated that the relative abundance of *Lactobacillus* and *Clostridium_sensu_stricto* was decreased after ginsenoside Rb1, CK, and PPD treatment. Meanwhile, previous studies had shown that the relative abundance of *Clostridium_XIVa* was enhanced in feces of 34 obese subjects with metabolic syndrome than without metabolic syndrome [41] and *Clostridium_XVIII* was positively related to obese phenotypes in mice [42]. Ginsenoside Rb1 and CK treatment reduced the relative abundance of *Clostridium_XIVa* and *Clostridium_XVIII* in HFD-fed rats. In addition, *Prevotella* and *Bacteroides* were the dominant genera after ginsenoside CK and PPD treatment. Previous studies have shown that ginsenoside Rb1 has only three more molecules of glucose (Glc) than ginsenoside CK, which is transformed into ginsenoside CK and PPD under the action of *Prevotella* and *Bacteroides* by removing Glc at C-3 [17,43]. In our study, the relative abundance of *Prevotella* and *Bacteroides* was significantly increased after ginsenoside CK and PPD treatment, while ginsenoside Rb1 treatment had no similar effect. These results suggested that ginsenoside Rb1, CK, and PPD selectively increased and reduced intestinal bacterial genera, which contributed to the anti-hyperlipemia effects in HFD-fed rats.

It is well known that *lactobacilli* are common probiotics that play a major role in ameliorating several metabolic disorders. However, our study found a reduction in *Lactobacillus* abundance following treatment with all three ginsenosides. This finding is indeed intriguing, considering the well-documented beneficial effects of various *Lactobacillus* strains on hyperlipidemia [44–46]. While many studies highlight the positive effects of *Lactobacillus* on hyperlipidemia, the interaction between ginsenosides and *Lactobacillus* is less well understood. Some research suggests that ginsenosides can have prebiotic-like effects, promoting the growth of beneficial gut bacteria [47]. However, these effects may not be uniform across all bacterial strains, and our findings underscore the need for a nuanced understanding of how specific ginsenosides interact with specific strains of *Lactobacillus*. In addition, it is also possible that ginsenosides may exert lipid-lowering effects through mechanisms independent of *Lactobacillus* modulation.

Concomitant with gut microbial composition alteration, we observed a regulation in bacterial gene functions after ginsenoside Rb1, CK, and PPD supplementation. Through PICRUSt KEGG function analysis, we found that the gene abundance involved in lipid metabolism pathways (the superpathway of phospholipid biosynthesis I, fatty acid salvage, fatty acid β-oxidation, CDP-diacylglycerol biosynthesis I, and CDP-diacylglycerol biosynthesis II) and glycometabolism (glycolysis II (from fructose 6-phosphate), D-fructuronate degradation, D-glucuronide and D-glucuronate degradation pathway) changed after ginsenoside Rb1 supplementation. Further, ginsenoside CK intervention altered glycometabolism pathways (sucrose degradation III, peptidoglycan maturation and sucrose degradation IV) to a greater extent compared to lipid metabolism pathways (fatty acid elongation-saturated and superpathway of fatty acid biosynthesis). Only the genes involved in the lipid IVA biosynthesis pathway were enriched in the PPD group compared with the HFD group (Figure S2A–C).

Next, we measured gene expression involved in lipid metabolism pathways in the HFD group, the Rb1 group, the CK group and the PPD group. Peroxisome proliferator-activated receptors (PPARs) have been considered as integrators of inflammatory and metabolic signaling networks [48]. PPARα controls lipid oxidation by regulating the expression of genes related to fatty acid transport and β-oxidation, thus reducing plasma levels of TG [49]. In contrast, PPARγ accelerates fatty acid uptake, TG formation and storage in lipid droplets, resulting in increased insulin sensitivity and glucose metabolism. The activation of PPARγ continues to regulate the downstream genes ACC1 and FAS, thus accelerating fatty acid synthesis [50]. In adipose tissues, stored TGs were hydrolyzed to free fatty acids by HSL [51]. In our study, the relative expression of *Pparα* and *Hsl* was up-regulated and the relative expression of *Pparγ*, *Acc* and *Fas* was down-regulated after ginsenoside Rb1 treatment. Likewise, hepatic protein expression of PPARα and HSL was higher and that of PPARγ, ACC and FAS was lower in the Rb1 group compared with the HFD group. These results indicated that ginsenoside Rb1 could improve lipid metabolism through regulating the PPARγ/ACC/FAS signaling pathway and stimulating FA β-oxidation in HFD-fed rats, consistent with the function analysis of gut microbiota. However, ginsenoside CK and PPD treatment down-regulated the relative expression of PPARγ, which failed to alter the relative expression of PPARα, HSL ACC and FAS. This result was in agreement with our studies that ginsenoside Rb1, but not ginsenoside CK and PPD, could reduce serum TC and TG levels in HFD-fed rats. Furthermore, FXR, as a BA synthesis and enterohepatic circulation mediator, plays a key role in BAs and cholesterol homeostasis, which regulated the rate-limiting enzyme (CYP7A1) for BA synthesis [52]. HMGCR, as the rate-limiting enzyme of cholesterol biosynthesis, contributes to cholesterol biosynthesis and uptake [53]. However, the relative expression of *Hmgcr* was enhanced in the Rb1, CK and PPD groups compared with the HFD group, which still needs further research. Notably, whether ginsenoside Rb1 or CK and PPD treatment, gene and protein relative expression of FXR and CYP7A1 were significantly changed in HFD-fed rats. Therefore, on the one hand, we assumed that ginsenoside Rb1 treatment inhibited lipogenesis via the PPARγ/ACC1/FAS pathway and promoted FA β-oxidation by down-regulating the mRNA expression of PPARα and HSL. On the other hand, ginsenoside Rb1, CK, and PPD could act as potent FXR agonists that ameliorate the TC level through regulating BA metabolism.

Further, to clarify the regulatory effect of ginsenoside Rb1, CK, and PPD treatment on BA metabolism, we measured hepatic and serous 12 BA levels in HFD-fed rats. The hepatic and serous total BA levels were markedly decreased after ginsenoside Rb1, CK, and PPD treatment in HFD-fed rats. Previous studies have indicated serum total BAs can be used to evaluate hepatic dysfunction and diagnose the various hepatobiliary diseases [54]. Increased serum BA levels induced metabolic dysfunctions and positively correlated obesity and type 2 diabetes [55,56]. In our study, compared with the HFD group, the serum primary/unconjugate BA (CA and β-MCA) level was lower in the Rb1 group, the CK group and the PPD group. Further, only ginsenoside PPD treatment could reduce the serum α-MCA levels and there was no significant difference in hepatic α-MCA levels among these groups. The ginsenoside Rb1 and PPD treatment decreased the serum CDCA level, whereas ginsenoside CK had no such effect. In liver, the primary/unconjugate BA (CA, CDCA and β-MCA) were markedly decreased after ginsenoside Rb1, CK, and PPD treatment, which was consistent with serum primary/unconjugate BA profile. FXR regulates hepatic BA synthesis via a negative feedback mechanism. Activated FXR suppresses the expression of CYP7A1 and thus inhibits hepatic BA synthesis. CDCA and CA were shown to be the agonists of FXR function, inhibiting BA synthesis [57]. In the HFD group, increased CA and CDCA levels could activate FXR and inhibit hepatic BA synthesis, leading to more cholesterol accumulation in the liver. After ginsenoside Rb1, CK, and PPD treatment, the hepatic and serous primary/unconjugate BA (CA, CDCA and β-MCA) were decreased in the HFD-fed rats, indicating that ginsenoside Rb1, CK, and PPD treatment promoted hepatic BA synthesis and reduced hepatic cholesterol levels. Furthermore, serum conjugate BA, including TCA, α/β-TMCA, TDCA, GDCA, TUDCA and GUDCA, were enhanced in the Rb1, CK, and PPD groups. Likewise, the hepatic conjugate BA (TCDCA, GUDCA, TDCA, GDCA, TUDCA and GUDCA) increased after ginsenoside Rb1, CK, and PPD treatment. Jiang et al. research suggested that alteration of the gut microbiota changed BA composition, notably alteration of conjugated BAs that regulate the intestinal FXR [58]. In liver, ginsenoside Rb1, CK, and PPD contributed to the conjugation of primary BAs with taurine (T) and glycine (G), thereby increasing BA hydrophilicity, reducing the hepatotoxicity of BAs and inhibiting cholestatic liver injury [59]. Our results suggested that ginsenoside Rb1, CK, and PPD regulated BA enterohepatic circulation through enhancing hepatic FXR-CYP7A1 signaling, resulting in decreased total BAs and BA profile alteration in the liver and serum.

Notably, the gut microbiota has been demonstrated to be a central regulator in BA metabolism. BAs are metabolized by enzymes derived from gut microbiota and play a critically important role in balanced lipid metabolism, insulin sensitivity and innate immunity [57]. In our study, the correlation between serum BAs and gut microbiota was analyzed after ginsenoside Rb1, CK, and PPD treatment. We found that ginsenoside Rb1, CK, and PPD treatment significantly reduced the abundance of BSH-producing *Clostridium* spp., which resulted in increased levels of conjugated BAs [60]. Furthermore, GUDCA, TUDCA, TDCA and GDCA were positively correlated with *Bacteroides* and *Akkermansia* in the Rb1 and CK groups (Figure 6B,C). In addition, *Parabacteroides*, *Bacteroides*, *Flavonifractor*, *Eisenbergiella* and *Clostridium_XlVb* were negatively correlated with TUDCA, GDCA and GUDCA, and positively correlated with α-MCA and TCA (Figure 6D). This notion was supported by previous research [61].

4. Materials and Methods

4.1. Animal Study Design

Four-week-old male SD rats were purchased from Beijing Vital River Laboratory Animal Technology Co., Ltd. (Beijing, China). They were raised under specific pathogen-free (SPF) conditions in an animal facility with a temperature of 20–25 °C, a relative humidity of (55 ± 15)%, and an air exchange rate of 10–20 times/h. Artificial lighting was controlled to provide a 12 h day/night cycle. Four rats were placed in each cage and had free access to food and water. After 7 days of acclimation, the rats were randomly divided into two groups (n = 10 and n = 50) and a fed standard diet (1010086) and a high-fat diet (XT19004), respectively. The standard diet and high-fat diet were purchased from Jiangsu Synergetic Biotechnology Co., Ltd. (Nanjing, China). Ginsenoside CK (CAS No. 39262-14-1) was purchased from BBI Life Sciences (Shanghai, China). Ginsenoside Rb1 (CAS No. 41753-43-9) was purchased from Chengdu Pusen Biotechnology Co., Ltd. (Chengdu, China). Ginsenoside PPD (CAS No. 30636-90-9) was provided by Shanghai GsynBioT Bio-Technology Co., Ltd. (Shanghai, China). All three ginsenosides were dissolved in 0.5% CMC-Na and were orally administered by gavage daily.

After 15 days of standard diet and high-fat diet treatment, rats were divided into six groups: (1) the Con group (n = 10), fed with a standard diet and administered with 2 mL/kg·BW of saline via gavage; (2) the HFD group (n = 10), fed with a high-fat diet and administered with 2 mL/kg·BW of saline via gavage; (3) the Ato group (n = 10), fed with a high-fat diet and administered with atorvastatin at a dose of 10 mg/kg·BW via gavage; (4) the Rb1 group (n = 10), fed with a high-fat diet and administered with ginsenoside Rb1 at a dose of 60 mg/kg·BW via gavage; (5) the CK group (n = 10), fed with a high-fat diet and administered with ginsenoside CK at a dose of 60 mg/kg·BW via gavage; (6) the PPD group (n = 10), fed with a high-fat diet and administered with ginsenoside PPD at a dose of 60 mg/kg·BW via gavage. All of these treatments were orally administered and lasted for 30 days. The body weight (BW) of rats was measured every 2 days. After 30 days, animals were euthanized by CO_2 asphyxiation. Dissected tissues, feces and serum were immediately frozen in liquid nitrogen and stored at −80 °C for further analysis.

4.2. Biochemical Analysis

Serum total cholesterol (TC), triglyceride (TG), high-density lipoprotein (HDL-C) and low-density lipoprotein (LDL-C) were measured using commercial assay kits according to the manufacturer's instructions (Nanjing Jiancheng Bioengineering Institute (Nanjing, China)). Serum tumor necrosis factor-α (TNF-α), interleukin-6 (IL-6) and interleukin-1β (IL-1β) were measured using commercial enzyme-linked immunosorbent assay kits (Beijing Baoruyi Biotechnology Co., LTD Co. Ltd., Beijing, China).

4.3. Histological Analysis

Liver and epididymal white adipose tissue (eWAT) were fixed in 10% formalin and stained with hematoxylin and eosin. The slices were examined under a light microscope (Olympus, BH-2, Tokyo, Japan) for tissue pathological analysis.

4.4. Real-Time Quantitative PCR (RT-qPCR) Analysis

Total RNA was extracted from liver tissue using a RNAprep Pure Tissue Kit (TianGen, Beijing, China) and the RNA samples concentration was measured by Nanodrop one (Thermo Scientific, Waltham, MA, USA). The RNA was then converted to cDNA using a Fastking RT Kit (TianGen, Beijing, China). Reverse transcription and quantitative PCR were performed using SuperReal PreMix Plus Kit (TianGen, Beijing, China) and LightCycler®96 System (Roche, Basel, Switzerland). The target genes' expression was analyzed using the ΔΔCt method with GAPDH as the internal reference gene. The primers used are listed in Supplementary Materials Table S1.

4.5. Western Blot Analysis

Liver tissues were homogenized in RIPA buffer (P0013B, Beyotime, Shanghai, China) with protease and phosphatase inhibitors. Proteins were separated on SDS-PAGE and then transferred to PVDF membranes. TBS-Tween containing 5% skimmed milk powder was used to immerse the PVDF membranes at room temperature on a shaker for 2 h. Primary antibodies were incubated overnight at 4 °C. After blotting, the membranes were washed and incubated with secondary antibody for 1 h at room temperature. Target proteins were normalized against β-actin. The protein bands were examined by using the ChemiDoc MP Imaging System (Bio-Rad, Hercules, CA, USA) and the intensities of these proteins were analyzed by the Image J 1.5.4 software (Madison, WI, USA).

4.6. Extraction and Sequencing of Fecal DNA Using 16S rRNA Gene

According to the manufacturer's protocol, the Tiangen Magnetic Stool DNA Kit was used to extract total genomic DNA from fecal samples. The V3–V4 hypervariable regions of 16S rRNA genes were amplified with specific primers. The PCR products were separated on a 2% agarose gel and visualized under UV light. Samples with bright main bands between 400 and 450 bp were selected for further experiments. The PCR products were purified using the Qiagen Gel Extraction Kit (Qiagen, Hong Kong) and then subjected to library preparation using the TruSeq® DNA PCR-Free Sample Preparation Kit (San Diego, CA, USA) according to the manufacturer's recommendations, including the addition of index codes. The quality and quantity of the libraries were assessed using the Qubit@ 2.0 Fluorometer (Thermo Scientific) and the Agilent (Santa Clara, CA, USA) Bioanalyzer 2100 system. Finally, the libraries were sequenced on the Illumina NovaSeq6000 platform, generating paired-end reads of 250 bp. 16S rRNA gene sequencing analyses were performed using Usearch (v11.0.1). Alpha diversity (number of observed OTUs, the Chao1 index, the richness index, Shannon's index, Simpson's index) was visualized and principal coordinates analysis (PCoA, Tucson, AZ, USA) was performed to measure beta diversity. Linear discriminant analysis Effect Size (LEfSe) was used to analyze the biomarkers within different groups. PICRUSt2 was performed to predict the gut microbiota function.

4.7. UHPLC-MS/MS Analysis

The BA concentrations in the serum were quantified using the ACQUITY Ultra performance liquid chromatography (UPLC) quaternary system coupled with a 6500 QTRAP mass spectrometer (AB SCIEX, Toronto, ON, Canada). A Waters UPLC BEH C18 (2.5 μm × 100 mm, 2.1 mm) was used for the separation. The mobile phase consisted of solvent A (water:formic acid (100:0.1, *v*/*v*)) and solvent B (methanol: formic acid (100:0.1, *v*/*v*)). The gradient was as follows: 60–65% B from 0 to 13 min, 65–80% B from 13 to 28 min, 80–90% B from 28 to 28.5 min, 90–100% B from 28.5 to 30 min, and 100–60% B from 30 to 40 min. The column temperature was 50 °C and the sample injection volume was 2 μL. Optimization of multiple reaction monitoring (MRM) parameters was performed on a ESI-QqQ-MS operated in the negative ion mode for BA analysis. The optimal MS conditions were as follows: ion spray voltage: 4500 V; temperature: 550 °C; curtain gas (CUR):40; Ion Source Gas1 (Gas1):55; Ion Source Gas2 (Gas2):55. Acquired data were processed using Analyst Software (version 1.6.1, AB Sciex).

4.8. Statistical Analysis

Graphpad Prism 9.2.0 software (GraphPad Software Inc., San Diego, CA, USA) was used to perform the statistical analyses and visualize the data. All data were presented as the mean ± SD. All p values were corrected for Dunnett's multiple comparisons test and subsequent significant differences established were recorded as follows: * $p < 0.05$, ** $p < 0.01$ and *** $p < 0.01$. Correlations were assessed with the Spearman nonparametric rank test.

5. Conclusions

In conclusion, these findings revealed that ginsenoside Rb1 significantly attenuated HFD-induced body weight gain and improved hepatic steatosis and hyperlipidemia, whereas ginsenoside CK and PPD did not show a similar effect. Meanwhile, gut microbiota sequencing studies indicated that ginsenoside Rb1, CK and PPD treatment could reshape the structure and function of gut microbiota through regulating the relative abundance of different genera in HFD-fed rats, including *Blautia*, *Akkermansia*, *Clostridium_XIVa* and *Clostridium_XVIII*, *Lactobacillus* and *Clostridium_sensu_stricto*. Furthermore, ginsenoside Rb1 supplement improved lipid metabolism through regulating the PPARγ/ACC/FAS signaling pathway and stimulating FA β-oxidation in HFD-fed rats, which was consistent with the function analysis of gut microbiota. In addition, ginsenoside Rb1, CK and PPD treatment all regulated BA metabolism through the FXR/CYP7A1 signaling pathway, decreasing total BA levels and altering BA enterohepatic circulation in HFD-fed rats. Our findings provide evidence supporting the beneficial effects of ginsenoside Rb1, CK and PPD intake in ameliorating hyperlipidemia in HFD-fed rats.

Supplementary Materials: The following supporting information can be downloaded at: https://www.mdpi.com/article/10.3390/molecules29051108/s1.

Author Contributions: G.-P.Z., X.-Y.Y. and Y.Z. conceived and designed the experiments. K.-X.Z., S.-X.S., Q.-Y.B. and Y.Z. performed the animal experiments. Y.Z. analyzed the data. Y.Z. and K.-X.Z. draft the manuscript. X.-Y.Y., H.Z. and G.-P.Z. revised the manuscript. All authors have read and agreed to the published version of the manuscript.

Funding: This research was supported by the Tianjin Synthetic Biotechnology Innovation Capacity Improvement Projects (TSBICIP-CXRC-008), and the Major Project of Haihe Laboratory of Synthetic Biology (E2M9560201).

Institutional Review Board Statement: The animal study protocol was approved by the Ethics Committee of Nankai University, China (2022-SYDWLL-000570), approval date: 22 November 2022.

Informed Consent Statement: Not applicable.

Data Availability Statement: The data presented in this study are available in article and Supplementary Materials.

Acknowledgments: We would like to extend our appreciation to all the people who participated in the experiments.

Conflicts of Interest: There are no conflicts of interest to declare.

References

1. Leuti, A.; Fazio, D.; Fava, M.; Piccoli, A.; Oddi, S.; Maccarrone, M. Bioactive lipids, inflammation and chronic diseases. *Adv. Drug Deliv. Rev.* **2020**, *159*, 133–169. [CrossRef]
2. Deprince, A.; Haas, J.T.; Staels, B. Dysregulated lipid metabolism links NAFLD to cardiovascular disease. *Mol. Metab.* **2020**, *42*, 101092. [CrossRef] [PubMed]
3. Badmus, O.O.; Hillhouse, S.A.; Anderson, C.D.; Hinds, T.D.; Stec, D.E. Molecular mechanisms of metabolic associated fatty liver disease (MAFLD): Functional analysis of lipid metabolism pathways. *Clin. Sci.* **2022**, *136*, 1347–1366. [CrossRef] [PubMed]
4. Lu, X.Y.; Shi, X.J.; Hu, A.; Wang, J.Q.; Ding, Y.; Jiang, W.; Sun, M.; ZAhao, X.; Luo, J.; Qi, W.; et al. Feeding induces cholesterol biosynthesis via the mTORC1–USP20–HMGCR axis. *Nature* **2020**, *588*, 479–484. [CrossRef] [PubMed]
5. Chiang, J.Y.L. Bile Acid Metabolism and Signaling. *Compr. Physiol.* **2013**, *3*, 1191–1212. [PubMed]
6. de Aguiar Vallim, T.Q.; Tarling, E.J.; Edwards, P.A. Pleiotropic Roles of Bile Acids in Metabolism. *Cell Metab.* **2013**, *17*, 657–669. [CrossRef] [PubMed]
7. Cai, J.; Rimal, B.; Jiang, C.; Chiang, J.Y.; Patterson, A.D. Bile acid metabolism and signaling, the microbiota, and metabolic disease. *Pharmacol. Ther.* **2022**, *237*, 108238. [CrossRef] [PubMed]
8. Guzior, D.V.; Quinn, R.A. Review: Microbial transformations of human bile acids. *Microbiome* **2021**, *9*, 140. [CrossRef]
9. Li, X.; Liu, J.; Zuo, T.-T.; Hu, Y.; Li, Z.; Wang, H.-D.; Xu, X.-Y.; Yang, W.-Z.; Guo, D.-A. Advances and challenges in ginseng research from 2011 to 2020: The phytochemistry, quality control, metabolism, and biosynthesis. *Nat. Prod. Rep.* **2022**, *39*, 875–909. [CrossRef] [PubMed]
10. Shi, Z.; Zeng, J.; Wong, A.S.T. Chemical Structures and Pharmacological Profiles of Ginseng Saponins. *Molecules* **2019**, *24*, 2443. [CrossRef] [PubMed]
11. Song, B.; Sun, Y.; Chu, Y.; Wang, J.; Zheng, H.; Liu, L.; Cai, W.; Zhang, H. Ginsenoside Rb1 Alleviated High-Fat-Diet-Induced Hepatocytic Apoptosis via Peroxisome Proliferator-Activated Receptor γ. *BioMed Res. Int.* **2020**, *2020*, 2315230. [CrossRef]
12. Yang, X.; Dong, B.; An, L.; Zhang, Q.; Chen, Y.; Wang, H.; Song, Z. Ginsenoside Rb1 ameliorates Glycemic Disorder in Mice with High Fat Diet-Induced Obesity via Regulating Gut Microbiota and Amino Acid Metabolism. *Front. Pharmacol.* **2021**, *12*, 756491. [CrossRef] [PubMed]
13. Sharma, A.; Lee, H.-J. Ginsenoside Compound K: Insights into Recent Studies on Pharmacokinetics and Health-Promoting Activities. *Biomolecules* **2020**, *10*, 1028. [CrossRef]
14. Lee, S.J.; Ko, W.G.; Kim, J.H.; Sung, J.H.; Lee, S.J.; Moon, C.K.; Lee, B.H. Induction of apoptosis by a novel intestinal metabolite of ginseng saponin via cytochrome c-mediated activation of caspase-3 protease. *Biochem. Pharmacol.* **2000**, *60*, 677–685. [CrossRef] [PubMed]
15. Kim, M.-Y.; Cho, J.Y. 20S-dihydroprotopanaxadiol, a ginsenoside derivative, boosts innate immune responses of monocytes and macrophages. *J. Ginseng Res.* **2013**, *37*, 293–299. [CrossRef]
16. de Vos, W.M.; Tilg, H.; Van Hul, M.; Cani, P.D. Gut microbiome and health: Mechanistic insights. *Gut* **2022**, *71*, 1020–1032. [CrossRef] [PubMed]
17. Kim, D.-H. Gut microbiota-mediated pharmacokinetics of ginseng saponins. *J. Ginseng Res.* **2018**, *42*, 255–263. [CrossRef] [PubMed]
18. Zhang, X.; Chen, S.; Duan, F.; Liu, A.; Li, S.; Zhong, W.; Sheng, W.; Chen, J.; Xu, J.; Xiao, S. Prebiotics enhance the biotransformation and bioavailability of ginsenosides in rats by modulating gut microbiota. *J. Ginseng Res.* **2021**, *45*, 334–343. [CrossRef] [PubMed]
19. Tang, W.W.; Kitai, T.; Hazen, S.L. Gut microbiota in cardiovascular health and disease. *Circ. Res.* **2017**, *120*, 1183–1196. [CrossRef] [PubMed]
20. Wang, Z.; Zhao, Y. Gut microbiota derived metabolites in cardiovascular health and disease. *Protein Cell* **2018**, *9*, 416–431. [CrossRef] [PubMed]
21. Ratan, Z.A.; Haidere, M.F.; Hong, Y.H.; Park, S.H.; Lee, J.-O.; Lee, J.; Cho, J.Y. Pharmacological potential of ginseng and its major component ginsenosides. *J. Ginseng Res.* **2021**, *45*, 199–210. [CrossRef]
22. Zou, H.; Zhang, M.; Zhu, X.; Zhu, L.; Chen, S.; Luo, M.; Xie, Q.; Chen, Y.; Zhang, K.; Bu, Q.; et al. Ginsenoside Rb1 Improves Metabolic Disorder in High-Fat Diet-Induced Obese Mice Associated with Modulation of Gut Microbiota. *Front. Microbiol.* **2022**, *13*, 826487. [CrossRef]
23. Yang, S.; Liu, T.; Hu, C.; Li, W.; Meng, Y.; Li, H.; Song, C.; He, C.; Zhou, Y.; Fan, Y. Ginsenoside Compound K Protects against Obesity through Pharmacological Targeting of Glucocorticoid Receptor to Activate Lipophagy and Lipid Metabolism. *Pharmaceutics* **2022**, *14*, 1192. [CrossRef] [PubMed]

24. Liu, C.; Li, H.; Zhou, Z.; Li, J.; Chen, H.; Liu, Y.; Huang, C.; Fan, S. Protopanaxadiol alleviates obesity in high-fat diet-fed mice via activation of energy-sensing neuron in the paraventricular nucleus of hypothalamus. *Biochem. Biophys. Res. Commun.* **2019**, *513*, 1092–1099. [CrossRef] [PubMed]
25. Willey, J.Z.; Elkind, M.S. 3-Hydroxy-3-methylglutaryl-coenzyme A reductase inhibitors in the treatment of central nervous system diseases. *Arch. Neurol.* **2010**, *67*, 1062–1067. [CrossRef] [PubMed]
26. Lai, M.; Peng, H.; Wu, X.; Chen, X.; Wang, B.; Su, X. IL-38 in modulating hyperlipidemia and its related cardiovascular diseases. *Int. Immunopharmacol.* **2022**, *108*, 108876. [CrossRef]
27. Ellulu, M.S.; Patimah, I.; KhazáAi, H.; Rahmat, A.; Abed, Y. Obesity and inflammation: The linking mechanism and the complications. *Arch. Med. Sci.* **2017**, *13*, 851–863. [CrossRef] [PubMed]
28. Moghbeli, M.; Khedmatgozar, H.; Yadegari, M.; Avan, A.; Ferns, G.A.; Mobarhan, M.G. Cytokines and the immune response in obesity-related disorders. *Adv. Clin. Chem.* **2021**, *101*, 135–168.
29. Joh, E.-H.; Lee, I.-A.; Jung, I.-H.; Kim, D.-H. Ginsenoside Rb1 and its metabolite compound K inhibit IRAK-1 activation—The key step of inflammation. *Biochem. Pharmacol.* **2011**, *82*, 278–286. [CrossRef] [PubMed]
30. Cuong, T.T.; Yang, C.-S.; Yuk, J.-M.; Lee, H.-M.; Ko, S.-R.; Cho, B.-G.; Jo, E.-K. Glucocorticoid receptor agonist compound K regulates dectin-1-dependent inflammatory signaling through inhibition of reactive oxygen species. *Life Sci.* **2009**, *85*, 625–633. [CrossRef] [PubMed]
31. Park, J.-S.; Shin, J.A.; Jung, J.-S.; Hyun, J.-W.; Van Le, T.K.; Kim, D.-H.; Park, E.-M.; Kim, H.-S. Anti-Inflammatory Mechanism of Compound K in Activated Microglia and Its Neuroprotective Effect on Experimental Stroke in Mice. *J. Pharmacol. Exp. Ther.* **2012**, *341*, 59–67. [CrossRef] [PubMed]
32. Rhule, A.; Navarro, S.; Smith, J.R.; Shepherd, D.M. Panax notoginseng attenuates LPS-induced pro-inflammatory mediators in RAW264.7 cells. *J. Ethnopharmacol.* **2006**, *106*, 121–128. [CrossRef] [PubMed]
33. Dabke, K.; Hendrick, G.; Devkota, S. The gut microbiome and metabolic syndrome. *J. Clin. Investig.* **2019**, *129*, 4050–4057. [CrossRef] [PubMed]
34. Schoeler, M.; Caesar, R. Dietary lipids, gut microbiota and lipid metabolism. *Rev. Endocr. Metab. Disord.* **2019**, *20*, 461–472. [CrossRef] [PubMed]
35. Jia, X.; Xu, W.; Zhang, L.; Li, X.; Wang, R.; Wu, S. Impact of Gut Microbiota and Microbiota-Related Metabolites on Hyperlipidemia. *Front. Cell. Infect. Microbiol.* **2021**, *11*, 634780. [CrossRef]
36. Liu, X.; Mao, B.; Gu, J.; Wu, J.; Cui, S.; Wang, G.; Zhao, J.; Zhang, H.; Chen, W. Blautia—A new functional genus with potential probiotic properties? *Gut Microbes* **2021**, *13*, 1875796. [CrossRef]
37. Cani, P.D.; Depommier, C.; Derrien, M.; Everard, A.; de Vos, W.M. *Akkermansia muciniphila*: Paradigm for next-generation beneficial microorganisms. *Nat. Rev. Gastroenterol. Hepatol.* **2022**, *19*, 625–637. [CrossRef]
38. Zhang, T.; Li, Q.; Cheng, L.; Buch, H.; Zhang, F. *Akkermansia muciniphila* is a promising probiotic. *Microb. Biotechnol.* **2019**, *12*, 1109–1125. [CrossRef]
39. Kang, Y.; Kang, X.; Yang, H.; Liu, H.; Yang, X.; Liu, Q.; Tian, H.; Xue, Y.; Ren, P.; Kuang, X.; et al. *Lactobacillus* acidophilus ameliorates obesity in mice through modulation of gut microbiota dysbiosis and intestinal permeability. *Pharmacol. Res.* **2022**, *175*, 106020. [CrossRef]
40. Gao, X.; Du, L.; Randell, E.; Zhang, H.; Li, K.; Li, D. Effect of different phosphatidylcholines on high fat diet-induced insulin resistance in mice. *Food Funct.* **2021**, *12*, 1516–1528. [CrossRef]
41. Jamar, G.; Santamarina, A.B.; Dias, G.C.; Masquio, D.C.L.; de Rosso, V.V.; Pisani, L.P. Relationship between fatty acids intake and Clostridium coccoides in obese individuals with metabolic syndrome. *Food Res. Int.* **2018**, *113*, 86–92. [CrossRef] [PubMed]
42. Huang, S.; Zou, Y.; Tang, H.; Zhuang, J.; Ye, Z.; Wei, T.; Lin, J.; Zheng, Q. Cordyceps militaris polysaccharides modulate gut microbiota and improve metabolic disorders in mice withdiet-induced obesity. *J. Sci. Food Agric.* **2023**, *103*, 1885–1894. [CrossRef]
43. Chen, Z.; Zhang, Z.; Liu, J.; Qi, H.; Li, J.; Chen, J.; Huang, Q.; Liu, Q.; Mi, J.; Li, X. Gut Microbiota: Therapeutic Targets of Ginseng Against Multiple Disorders and Ginsenoside Transformation. *Front. Cell. Infect. Microbiol.* **2022**, *12*, 853981. [CrossRef]
44. Lv, X.-C.; Chen, M.; Huang, Z.-R.; Guo, W.-L.; Ai, L.-Z.; Bai, W.-D.; Yu, X.-D.; Liu, Y.-L.; Rao, P.-F.; Ni, L. Potential mechanisms underlying the ameliorative effect of Lactobacillus paracasei FZU103 on the lipid metabolism in hyperlipidemic mice fed a high-fat diet. *Food Res. Int.* **2021**, *139*, 109956. [CrossRef] [PubMed]
45. Lew, L.-C.; Hor, Y.-Y.; Jaafar, M.-H.; Lau, A.-S.; Lee, B.-K.; Chuah, L.-O.; Yap, K.-P.; Azlan, A.; Azzam, G.; Choi, S.-B.; et al. *Lactobacillus* Strains Alleviated Hyperlipidemia and Liver Steatosis in Aging Rats via Activation of AMPK. *Int. J. Mol. Sci.* **2020**, *21*, 5872. [CrossRef] [PubMed]
46. Teng, Y.; Guan, W.-Y.; Wang, C.; Yu, H.-S.; Li, X.; Wang, Y.-H. Effect of Lactobacillus plantarum LP104 on hyperlipidemia in high-fat diet induced C57BL/6N mice via alteration of intestinal microbiota. *J. Funct. Foods* **2022**, *95*, 105176. [CrossRef]
47. Zhuang, T.; Li, W.; Yang, L.; Wang, Z.; Ding, L.; Zhou, M. Gut Microbiota: Novel Therapeutic Target of Ginsenosides for the Treatment of Obesity and Its Complications. *Front. Pharmacol.* **2021**, *12*, 731288. [CrossRef] [PubMed]
48. Gross, B.; Pawlak, M.; Lefebvre, P.; Staels, B. PPARs in obesity-induced T2DM, dyslipidaemia and NAFLD. *Nat. Rev. Endocrinol.* **2017**, *13*, 36–49. [CrossRef]
49. Peters, J.M.; Hennuyer, N.; Staels, B.; Fruchart, J.-C.; Fievet, C.; Gonzalez, F.J.; Auwerx, J. Alterations in Lipoprotein Metabolism in Peroxisome Proliferator-activated Receptor α-deficient Mice. *J. Biol. Chem.* **1997**, *272*, 27307–27312. [CrossRef]

50. Montaigne, D.; Butruille, L.; Staels, B. PPAR control of metabolism and cardiovascular functions. *Nat. Rev. Cardiol.* **2021**, *18*, 809–823. [CrossRef] [PubMed]
51. Lorente-Cebrián, S.; Kulyté, A.; Hedén, P.; Näslund, E.; Arner, P.; Rydén, M. Relationship between Site-Specific HSL Phosphorylation and Adipocyte Lipolysis in Obese Women. *Obes. Facts* **2011**, *4*, 365–371. [CrossRef]
52. Pushpass, R.-A.G.; Alzoufairi, S.; Jackson, K.G.; Lovegrove, J.A. Circulating bile acids as a link between the gut microbiota and cardiovascular health: Impact of prebiotics, probiotics and polyphenol-rich foods. *Nutr. Res. Rev.* **2022**, *35*, 161–180. [CrossRef] [PubMed]
53. Guo, Y.; Zhao, M.; Bo, T.; Ma, S.; Yuan, Z.; Chen, W.; He, Z.; Hou, X.; Liu, J.; Zhang, Z.; et al. Blocking FSH inhibits hepatic cholesterol biosynthesis and reduces serum cholesterol. *Cell Res.* **2019**, *29*, 151–166. [CrossRef] [PubMed]
54. Lefebvre, P.; Cariou, B.; Lien, F.; Kuipers, F.; Staels, B. Role of Bile Acids and Bile Acid Receptors in Metabolic Regulation. *Physiol. Rev.* **2009**, *89*, 147–191. [CrossRef] [PubMed]
55. Desai, M.S.; Mathur, B.; Eblimit, Z.; Vasquez, H.; Taegtmeyer, H.; Karpen, S.J.; Penny, D.J.; Moore, D.D.; Anakk, S. Bile acid excess induces cardiomyopathy and metabolic dysfunctions in the heart. *Hepatology* **2016**, *65*, 189–201. [CrossRef] [PubMed]
56. Song, Y.; Sun, L.; Ma, P.; Xu, L.; Xiao, P. Dihydromyricetin prevents obesity via regulating bile acid metabolism associated with the farnesoid X receptor inob/ob mice. *Food Funct.* **2022**, *13*, 2491–2503. [CrossRef] [PubMed]
57. Jia, W.; Xie, G.; Jia, W. Bile acid–microbiota crosstalk in gastrointestinal inflammation and carcinogenesis. *Nat. Rev. Gastroenterol. Hepatol.* **2018**, *15*, 111–128. [CrossRef] [PubMed]
58. Jiang, C.; Xie, C.; Li, F.; Zhang, L.; Nichols, R.G.; Krausz, K.W.; Cai, J.; Qi, Y.; Fang, Z.-Z.; Takahashi, S.; et al. Intestinal farnesoid X receptor signaling promotes nonalcoholic fatty liver disease. *J. Clin. Investig.* **2015**, *125*, 386–402. [CrossRef] [PubMed]
59. Chiang, J.Y.L.; Ferrell, J.M. Bile Acid Metabolism in Liver Pathobiology. *Gene Expr.* **2018**, *18*, 71–87. [CrossRef] [PubMed]
60. Münzker, J.; Haase, N.; Till, A.; Sucher, R.; Haange, S.-B.; Nemetschke, L.; Gnad, T.; Jäger, E.; Chen, J.; Riede, S.J.; et al. Functional changes of the gastric bypass microbiota reactivate thermogenic adipose tissue and systemic glucose control via intestinal FXR-TGR5 crosstalk in diet-induced obesity. *Microbiome* **2022**, *10*, 96. [CrossRef] [PubMed]
61. Lin, H.; An, Y.; Tang, H.; Wang, Y. Alterations of Bile Acids and Gut Microbiota in Obesity Induced by High Fat Diet in Rat Model. *J. Agric. Food Chem.* **2019**, *67*, 3624–3632. [CrossRef] [PubMed]

molecules

MDPI

Article

Increasing 1,4-Diaminobutane Production in *Escherichia coli* by Optimization of Cofactor PLP and NADPH Synthesis

Tong Sun [1,2], Yongcan Zhao [1,2], Jinjin Wang [1,2], Wenke Kang [1,2], Xiangxiang Sun [1,2], Yanling Sun [1,2], Meixue Chu [1,2], Zhengyu Liu [1,2], Fuping Lu [1,2,*] and Ming Li [1,2,*]

1 Key Laboratory of Industrial Fermentation Microbiology (Tianjin University of Science and Technology), Ministry of Education, Tianjin 300457, China; aatong9@163.com (T.S.); cc13053767790@163.com (Y.Z.); jinjw1437@163.com (J.W.); k17526620717@163.com (W.K.); sunxx99@outlook.com (X.S.); 13686327051@163.com (Y.S.); chumeixue@163.com (M.C.); liuzhengyu397@163.com (Z.L.)

2 College of Biotechnology, Tianjin University of Science and Technology, Tianjin 300457, China

* Correspondence: lfp@tust.edu.cn (F.L.); liming09@tust.edu.cn (M.L.)

Abstract: 1,4-diaminobutane is widely used in the industrial production of polymers, pharmaceuticals, agrochemicals and surfactants. Owing to economic and environmental concerns, there has been a growing interest in using microbes to produce 1,4-diaminobutane. However, there is lack of research on the influence of cofactors pyridoxal phosphate (PLP) and NADPH on the synthesis of 1,4-diaminobutane. PLP serves as a cofactor of ornithine decarboxylase in the synthesis of 1,4-diaminobutane. Additionally, the synthesis of 1 mol 1,4-diaminobutane requires 2 mol NADPH, thus necessitating consideration of NADPH balance in the efficient synthesis of 1,4-diaminobutane by *Escherichia coli*. The aim of this study was to enhance the synthesis efficiency of 1,4-diaminobutane through increasing production of PLP and NADPH. By optimizing the expression of the genes associated with synthesis of PLP and NADPH in *E. coli*, cellular PLP and NADPH levels increased, and the yield of 1,4-diaminobutane also increased accordingly. Ultimately, using glucose as the primary carbon source, the yield of 1,4-diaminobutane in the recombinant strain NAP19 reached 272 mg/L·DCW, by increased 79% compared with its chassis strain.

Keywords: PLP; NADPH; 1,4-diaminobutane; cofactor

check for updates

Citation: Sun, T.; Zhao, Y.; Wang, J.; Kang, W.; Sun, X.; Sun, Y.; Chu, M.; Liu, Z.; Lu, F.; Li, M. Increasing 1,4-Diaminobutane Production in *Escherichia coli* by Optimization of Cofactor PLP and NADPH Synthesis. *Molecules* **2024**, *29*, 3094. https://doi.org/10.3390/molecules29133094

Academic Editors: Marina DellaGreca and Yongjun Wei

Received: 14 May 2024
Revised: 12 June 2024
Accepted: 27 June 2024
Published: 28 June 2024

1. Introduction

The compound 1,4-diaminobutane, also known as putrescine, is not only involved in cellular metabolism and growth [1–3], but also a monomer material used in industry to produce PA46, PA410 and so on [4]. The market demand for 1,4-diaminobutane is growing rapidly due to its high industrial value. At present, 1,4-diaminobutane is mainly produced from petroleum. Industrial scale production of 1,4-diaminbutane mainly relies on chemical synthesis by the addition of hydrogen cyanide to acrylanitrile to obtain succinonitrile, and then hydrogenation of succinonitrile under harsh reaction conditions. The process requires non-renewable petrochemicals as raw materials, as well as expensive catalysts, which can put pressure on environmental protection [5,6]. Therefore, it is necessary to develop a new green production process for 1,4-diaminobutane. By constructing high-yield 1,4-diaminobutane engineered strains, production of 1,4-diaminobutane via microbial fermentation has environmental advantages and represents the future direction of its industrialization.

Engineered strains for the production of 1,4-diaminobutane have been developed and made some progress. At present, the engineered strains were constructed generally by increasing the expression levels of genes in the 1,4-diaminobutane synthesis pathway [7–9], knocking out branching metabolic pathways and the degradation pathways of 1,4-diaminobutane [10,11], and screening for high-activity ornithine decarboxylase (ODC) to enhance the synthesis flux of 1,4-diaminobutane [12,13]. In our laboratory, we constructed a chassis strain, *E. coli* PUT11 (DE3), which knocked out 11 genes related to the branching pathway, degradation pathway, and transporter pathway of 1,4-diaminobutane from *E. coli* MG1655 and integrated the T7 RNA polymerase expression cassette (DE3) from *E. coli* BL21(DE3) into the genome. Compared with the wild strain, the 1,4-diaminobutane synthesis was increased by 1888% [14]. However, the yields are insufficient for industrial production requirements. Therefore, other measures are needed to increase the yield of 1,4-diaminobutane.

In the synthetic pathway of 1,4-diaminobutane from glucose, the synthesis of 1 mol 1,4-diaminobutane requires 2 mol NADPH, which are consumed by $NADP^{-}$ -dependent glutamate dehydrogenase (Gdh) and $NADP^{-}$ -dependent N-acetyl-glutaminyl phosphate reductase (ArgC) [15]. Overexpression the genes of the pentose phosphate (PP) pathway are a common method to increase NADPH levels [16–18]. In addition to the PP pathway, by overexpression of pyridine nucleotide transhydrogenase gene (*pntAB*) and NAD^{+} kinase gene (*ppnK*) are also a major pathway to increase NADPH concentration in the strain [19–21]. Increasing NADPH synthesis in the 1,4-diaminobutane synthesis pathway is crucial.

PLP is involved in various enzymatic reactions in *E. coli*, such as decarboxylation, transamination, and racemization in amino acid synthesis [22–24]. The de novo pathway of PLP synthesis proceeds from E4P to synthesize 4-phosphohydroxy-l-serine (4PHT), followed by the condensation of 4PHT with 1-deoxy-D-xylulose-5-phosphate (DOXP) catalyzed by pyridoxine 5′-phosphate synthase (PdxJ) to form pyridoxamine 5′-phosphate (PNP), which is further oxidized to PLP [25,26] (Figure 1). Many studies have shown that optimizing the synthesis of PLP can effectively increase the yield of the cadaverine [27–29]. PLP is used as a cofactor for ODC, a key enzyme in the synthesis of the 1,4-diaminobutane [30,31]. A moderate increase in PLP supply is important for 1,4-diaminobutane synthesis.

Currently, the construction processes of recombinant strains for 1,4-diaminobutane biosynthesis lack the researches on PLP and NADPH. In this study, overexpression of the genes related to NADPH synthesis, including *pntAB*, *ppnK* and the genes in PP pathway, was used to increase NADPH concentration in the strains to study the impact of NADPH on 1,4-diaminobutane synthesis (Figure 1). The relationship between PLP synthesis and 1,4-diaminobutane production was investigated by increasing PLP synthesis in *E. coli* through overexpression of the genes *pdxJ* and *dxs*, which are involved in the PLP synthesis pathway, and *tktA* and *talB*, which are the genes of the E4P synthesis pathway, a precursor of PLP (Figure 1). Furthermore, the impact of increasing NADPH and PLP synthesis in the chassis strain *E. coli* PUT11 (DE3) on 1,4-diaminobutane production was explored.

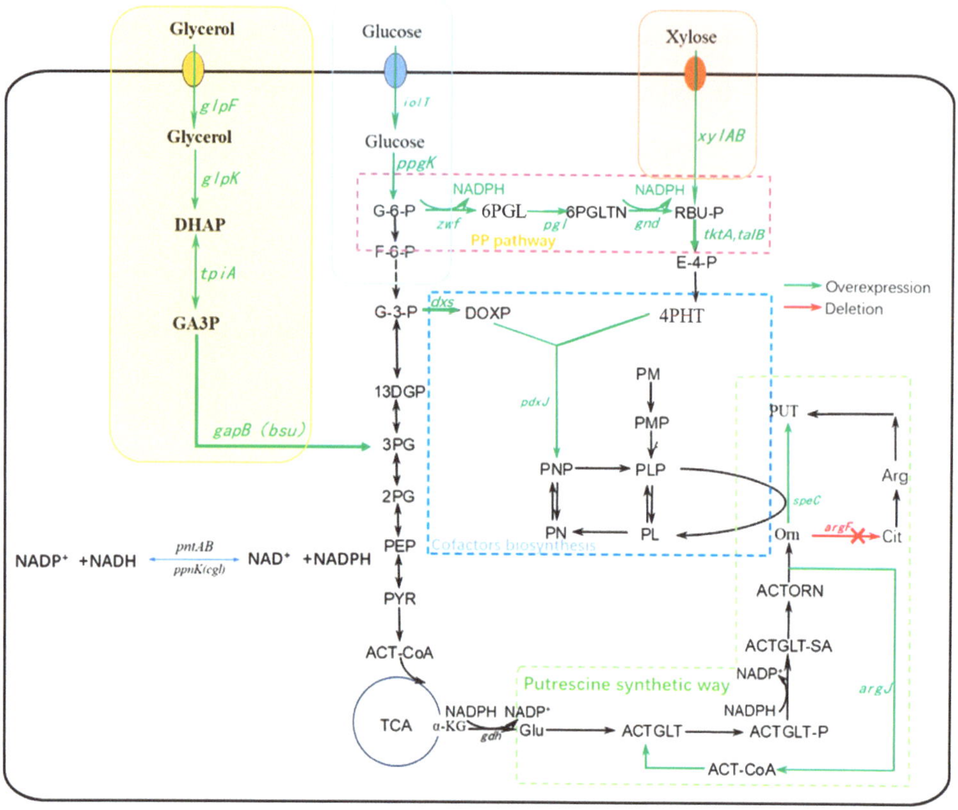

Figure 1. Metabolic pathway design and key metabolic engineering strategies for 1,4-diaminobutane production in *E. coli*. Green arrows indicate overexpressed genes. Red arrows indicate relevant gene deletions. Yellow boxes indicate genes related to glycerol utilization pathway. Blue boxes indicate genes related to glucose utilization pathway and red boxes indicate genes related to D-xylose utilization pathway. *glpF*: glycerol facilitator; *glpK*: glycerol kinase; *tpiA:* triose-phosphate isomerase; *gapB*: glyceraldehyde-3-phosphate dehydrogenase; *iolT*: myoinositol permease; *ppgK*: polyphosphate glucokinase; *xylAB*: xylose isomerase and xylulokinase; *zwf*: ADP(+)-dependent glucose-6-phosphate dehydrogenase; *pgl*: 6-phosphogluconolactonase; *gnd*: 6-phosphogluconate dehydrogenase; *tktA*: transketolase 1; *talB*: transaldolase B; *pdxJ*: pyridoxine 5′-phosphate synthase; *dxs*: 1-deoxy-D-xylulose-5-phosphate synthase; *pntAB*: pyridine nucleotide transhydrogenase subunit alpha and beta; *ppnK*: inorganic polyphosphate/ATP-NAD kinase; *speC*: constitutive ornithine decarboxylase; *argF*: ornithine carbamoyltransferase; *gdh*: Glutamate dehydrogenase; *argC*: N-acetylglutamylphosphate reductase; *argJ*: bifunctional glutamate N-acetyltransferase/amino-acid acetyltransferase; *cgl*: from *Corynebacterium glutamicum*; *bsu*: from *Bacillus subtilis*.

2. Results

2.1. Optimisation of NADPH Synthesis in the 1,4-Diaminobutane Synthetic Pathway

In the 1,4-diaminobutane synthesis pathway, 2 mol NADPH is required to synthesize 1,4-diaminobutane. PntAB and PpnK catalyze $NADP^+$ and NADH to generate NADPH. In addition, overexpression of PP pathway could effectively increase NADPH concentration in strains. Therefore, in this study, the key genes involved in NADPH synthesis, *pntAB* and *ppnK*, as well as the genes in the PP pathway, were overexpressed to increase the production of NADPH and thus increased the production of 1,4-diaminobutane (Figure 1).

2.1.1. Overexpression of *ppnK* and *pntAB* Increased NADPH Supply

The cloned *ppnK* and *pntAB* genes were ligated into pETM6 plasmid and transformed into *E. coli* BL21(DE3) to form recombinant strains NAP1 and NAP2. Recombinant strains NAP1 and NAP2 could catalyze NADP$^+$ to NADPH by expressing *pntAB* and *ppnK*. NADPH synthesis and 1,4-diaminobutane yield were shown in Figure 2. The NADPH yield of NAP1 and NAP2 was significantly higher than that of the control strain *E. coli* BL21(DE3), and similarly, the yield of 1,4-diaminobutane was also increased companied with the elevation of NADPH, showing that improvement of NADPH can enhance the synthesis of 1,4-diaminobutane. The titer of 1,4-diaminobutane in NAP1 and NAP2 was 77 mg/L·DCW and 83 mg/L·DCW, respectively, which was about 27% and 37% higher than that in the control strain *E. coli* BL21(DE3). In order to furtherly increase the synthesis of NADPH, co-expression plasmid of *pntAB* and *ppnK* was transformed into *E. coli* BL21(DE3) to generate recombinant strain NAP3. The NADPH yield of NAP3 strain was higher, and the 1,4-diaminobutane titer was 95 mg/L·DCW, which was 57% greater than the control strain *E. coli* BL21(DE3) (Figure 2), indicating that the synthesis of 1,4-diaminobutane was positively correlated with NADPH.

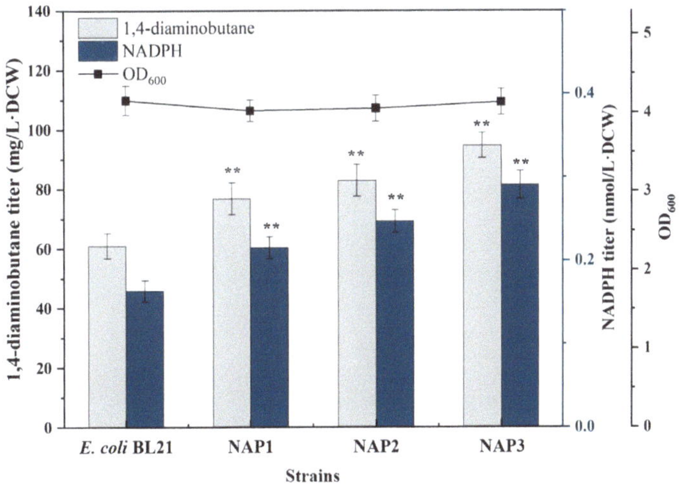

Figure 2. Determination of NADPH and 1,4-diaminobutane titer of NAP1, NAP2 and NAP3. Strain genotypes: NAP1: pETM6-ppnK; NAP2: pETM6-pntAB; NAP3: pETM6-ppnK-pntAB. Data were expressed as means, and error bars indicate standard deviation (n = 3 independent experiments. The ** symbol represented $p < 0.01$ compared with their controls).

2.1.2. Overexpression of PP Pathway Genes Increased NADPH Supply

In the PP pathway, NADP$^+$-dependent glucose-6-phosphate dehydrogenase (Zwf) and NADP$^+$-dependent glucose-6-phosphate dehydrogenase (Gnd) can reduce NADP$^+$ to NADPH [32,33] (Figure 1). The *zwf*, 6-phosphogluconolactonase (*pgl*) and *gnd* genes of this pathway were cloned into the pETM6 plasmid, and transformed into *E. coli* BL21(DE3) to construct recombinant strains NAP4, NAP5 and NAP6. Recombinant strains NAP4, NAP5 and NAP6 increased NADPH synthesis by expressing *zwf*, *pgl* and *gnd* genes of PP pathway. The result was shown in Figure 3, overexpression of the *zwf*, *pgl* and *gnd* genes could all promote NADPH synthesis, thereby increased the yield of 1,4-diaminobutane. The 1,4-diaminobutane titer of NAP4, NAP5 and NAP6 was 75 mg/L·DCW, 70 mg/L·DCW and 83 mg/L·DCW, respectively, which was 24%, 17% and 38% higher than those of the control strain *E. coli* BL21(DE3). Subsequently, the expression plasmids harbouring *zwf-gnd* and *zwf-gnd-pgl* were transformed into *E. coli* BL21(DE3) to construct recombinant strains NAP7 and NAP8. As shown in Figure 3, with the further improvement of NADPH yield, the synthesis levels of 1,4-diaminobutane were furtherly increased. The titer of 1,4-diaminobutane of NAP7 and NAP8 was 88 mg/L·DCW and 98 mg/L·DCW, respectively, which were 47% and 63% higher than the control strain *E. coli* BL21(DE3). It was proved that overexpression of the genes in the PP pathway could effectively improve NADPH level and 1,4-diaminobutane synthesis.

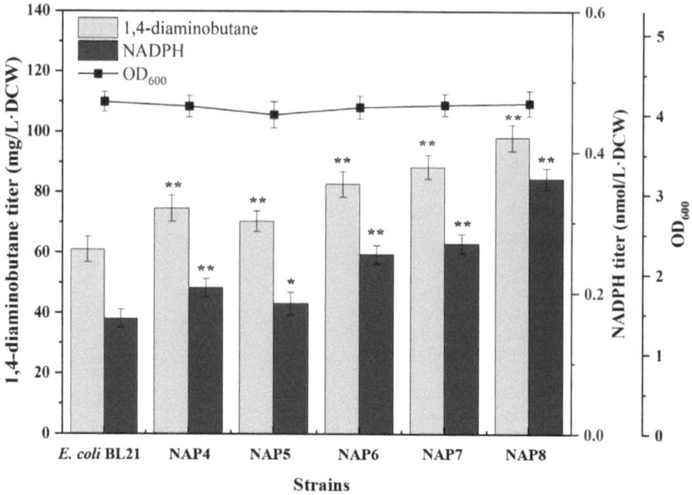

Figure 3. Determination of NADPH and 1,4-diaminobutane titer in NAP4, NAP5, NAP6, NAP7 and NAP8. Strain genotypes: NAP4: pETM6-zwf; NAP5: pETM6-pgl; NAP6: pETM6-gnd; NAP7: pETM6-zwf-pgl; NAP8: pETM6-zwf-gnd-pgl. Data were expressed as means, and error bars indicated standard deviation ($n = 3$ independent experiments. The * symbol represented $0.01 < p < 0.05$, and ** represented $p < 0.01$ compared with their controls).

2.2. *Optimization of PLP Supply in the 1,4-Diaminobutane Synthesis Pathway*

2.2.1. Effect of Exogenous Supplementation of PLP on the Synthesis of 1,4-Diaminobutane in *E. coli*

Excessive accumulation of PLP in cells can cause certain toxicity and affect cell growth. Therefore, it is first necessary to determine the effect of PLP concentration on the synthesis of 1,4-diaminobutane. PLP of 100, 200, 300, 400 and 500 μmol/L were added to the medium, and the yield of 1,4-diaminobutane in the fermentation supernatant and OD_{600} was measured after 24 h of fermentation culture, as shown in Figure 4. When PLP is not added, the titer of 1,4-diaminobutane of *E. coli* BL21(DE3) is 59 mg/L·DCW, while when PLP is added exogenously at 100 μmol/L, the titer of 1,4-diaminobutane of *E. coli* BL21(DE3) and OD_{600} were the highest, respectively reaching 66 mg/L·DCW and 4. However, with the increase of PLP concentration, both the production of 1,4-diaminobutane and OD_{600} decreased. These results indicated that a reasonable increase of PLP synthesis is beneficial to the synthesis of 1,4-diaminobutane, but not too much, otherwise it will affect the growth of strains and the synthesis of 1,4-diaminobutane.

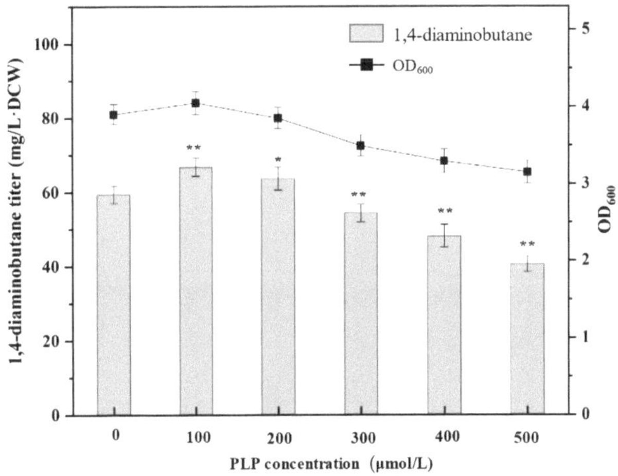

Figure 4. Production of 1,4-diaminobutane and OD_{600} after 24 h fermentation culture with different concentrations of PLP. The concentration of PLP was 100, 200, 300, 400 and 500 μmol/L, and the fermentation supernatant was used to measure the concentration of 1,4-diaminobutane. Data were expressed as means, and error bars indicated standard deviation ($n = 3$ independent experiments. The * symbol represented $0.01 < p < 0.05$, and ** represented $p < 0.01$ compared with their controls).

2.2.2. Enhancing the Synthesis of PLP

The genes *pdxJ* and *dxs*, involved in the PLP synthesis pathway, and the genes *tktA* and *talB*, responsible for the synthesis of PLP precursors E4P, were cloned into the pETM6 plasmid and transformed into *E. coli* BL21(DE3) to generate recombinant strains NAP9, NAP10, NAP11 and NAP12. These recombinant strains all could increase PLP synthesis due to expressing *pdxJ*, *dxs*, *tktA* and *talB* genes related to synthesis of PLP (Figure 1). The results are shown in Figure 5. Overexpression of these genes could increase the synthesis of 1,4-diaminobutane. Overexpression of *pdxJ* and *dxs* genes in the PLP synthesis pathway resulted in 1,4-diaminobutane accumulation levels of 67 mg/L·DCW and 69 mg/L·DCW, respectively, with an increase of 11% and 15% compared with the control strain *E. coli* BL21(DE3). Overexpression of the genes *tktA* and *talB* resulted in 1,4-diaminobutane accumulation levels of 86 mg/L·DCW and 80 mg/L·DCW, with an increase of 44% and 35%, respectively, compared with the control strain *E. coli* BL21(DE3). This showed that increasing the PLP significantly increased 1,4-diaminobutane production.

So as to furtherly enhance PLP synthesis, *pdxJ*, *dxs*, *tktA* and *talB* were cloned into pETM6 to construct the the PLP synthesis module plasmid pETM6-pdxJ-dxs-tktA-talB. This plasmid was transformed into *E. coli* BL21(DE3) to create the recombinant strain NAP13. The results showed that when enhancing the expression of PLP synthesis module, the accumulation of 1,4-diaminobutane not only did not increase, on the contrary, the yield of 1,4-diaminobutane decreased by approximately 22% compared with the control strain *E. coli* BL21(DE3) (Figure 5). At the same time, OD_{600} also decreased. It was possible that after enhancing the expression of PLP synthesis module, PLP concentration was too high, which led to toxicity to the cells and a reduction in 1,4-diaminobutane production. This was in line with the previous findings that an excessively high PLP concentration did not improve 1,4-diaminobutane production, inversely, it reduced its accumulation.

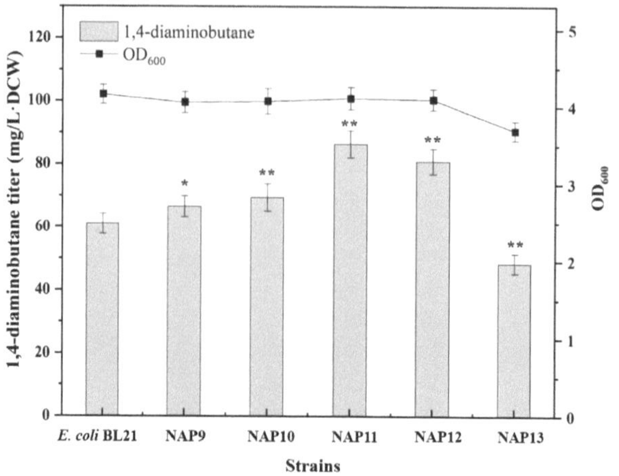

Figure 5. Determination of 1,4-diaminobutane titer in NAP9, NAP10, NAP11, NAP12 and NAP13. Strain genotypes: NAP9: pETM6-pdxJ; NAP10: pETM6-dxs; NAP11: pETM6-tktA; NAP12: pETM6-talB; NAP13: pETM6-pdxJ-dxs-tktA-talB. Data were expressed as means, and error bars indicated standard deviation ($n = 3$ independent experiments. The * symbol represented $0.01 < p < 0.05$, and ** represented $p < 0.01$ compared with their controls).

2.3. *Overexpression of 1,4-Diaminobutane Synthesis Module*

The bifunctional N-acetyltransferase enzyme encoded by *argJ* in *C. glutamicum* catalyzes the conversion of N-acetylornithine to ornithine through deacetylation, and simultaneously transfers the acetyl group produced from this reaction to glutamic acid to form N-acetylglutamic acid, thus facilitating the recycling of the acetyl group [34–36]. Therefore, the key gene *speC* gene encoding ODC and *argJ* were cloned into the pRSM3 plasmid to construct a 1,4-diaminobutane synthesis module plasmid pRSM3-speC-argJ.

The strain NAP14 was constructed by co-transforming the 1,4-diaminobutane synthesis module plasmid pRSM3-speC-argJ and the PLP synthesis module plasmid pETM6-pdxJ-dxs-tktA-talB into *E. coli* BL21(DE3), and the yield of 1,4-diaminobutane was detected. The results showed that NAP14 produced 86 mg/L·DCW of 1,4-diaminobutane, with a 44% increase compared to the control strain *E. coli* BL21(DE3) (Figure 6). As ODC is a crucial gene in the synthesis of 1,4-diaminobutane, increasing the expression of ODC in the strain allows for the extensive utilization of PLP and decreases PLP's toxicity.

Afterward, the NADPH synthesis module plasmid pCDM4-pntAB-ppnK and pACM4-zwf-gnd-pgl were constructed and transferred into NAP14 to obtain strain NAP15. The results are shown in Figure 6, the 1,4-diaminobutane titer of NAP15 was 91 mg/L·DCW, with an increase of 51% compared with the control strain *E. coli* BL21(DE3), which proved that with increasing synthesis of PLP and NAPDH, overexpression of the 1,4-diaminobutane synthesis module could lead to an increase in 1,4-diaminobutane production. At the same time, the density of NAP15 strain decreased, speculating that the four plasmids in the same strain increased the metabolic burden of the strain and inhibited the growth of the strains.

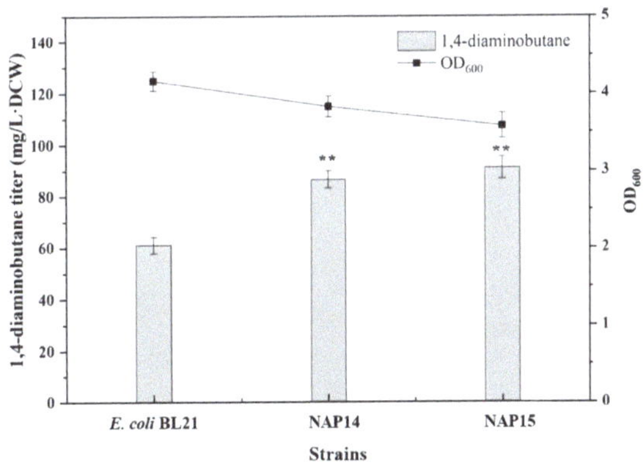

Figure 6. Determination of 1,4-diaminobutane titer in NAP14 and NAP15. Strain genotypes: NAP14: pETM6-pdxJ-dxs-tktA-talB and pRSM3-speC-argJ; NAP15: NAP14 harbouring the pCDM4-pntAB-ppnK and pACM4-zwf-gnd-pgl. Data were expressed as means, and error bars indicated standard deviation (n = 3 independent experiments. The ** represented $p < 0.01$ compared with their controls).

2.4. Effect of Increasing NADPH and PLP Synthesis on 1,4-Diaminobutane Yield When Utilizing Different Carbon Sources

Overexpression of Carbon Source Utilization Pathway Genes

Carbon source is one of the factors that must be considered in fermentation production [37–39]. In this study, glycerol utilization pathway, glucose utilization pathway, and xylose utilization pathway plasmids were constructed, respectively, and transformed into *E. coli* BL21(DE3) to construct recombinant strains NAP16, NAP17 and NAP18. The strains were cultured in SOB medium containing 1 g/L glycerol, 1 g/L glucose and 1 g/L xylose, respectively. As shown in Figure 7, the 1,4-diaminobutane titer of NAP16, NAP17 and NAP18 were 93 mg/L·DCW, 102 mg/L·DCW and 78 mg/L·DCW with an increase of 30%, 40% and 24% compared with the control strain *E. coli* BL21(DE3). These results showed that increasing utilization of the carbon source by overexpressing their genes could significantly increase the yield of the target products, and the recombinant strains overexpressing the glucose pathway genes had the most obvious effect.

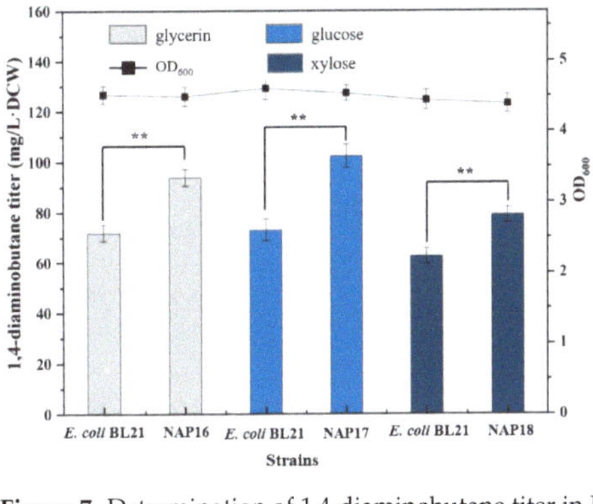

Figure 7. Determination of 1,4-diaminobutane titer in NAP16, NAP17 and NAP18. Strain genotypes: NAP16: pETM6-glpFK-tpiA-gapB; NAP17: pETM6-iolT-ppgK; NAP18: pETM6-xylAB. Grey represented adding 1 g/L of glycerol to the medium; Light blue represented adding 1 g/L of glucose to the medium; Dark blue represented adding 1 g/L of xylose to the medium. Data were expressed as means and error bars indicated standard deviation (n = 3 independent experiments. The ** represented $p < 0.01$ compared with their controls).

2.5. Fermentation Results of Recombinant Strains

The constructed glucose utilization pathway plasmid, NADPH and PLP synthesis plasmid, and ornithine decarboxylase plasmid were co-transformed into *E. coli* PUT11(DE3) to construct the strain NAP19. *E. coli* PUT11(DE3) has been knocked out 11 genes which were associated with the pathways for 1,4-diaminobutane degradation and transport and knocked in DE3 sequence from *E. coli* BL21(DE3) in PuuR operon site. The fermentation results are shown in Figure 8. NAP19 finally achieved a 1,4-diaminobutane titer of 272 mg/L·DCW, there was a 79% increase compared to the control strain *E. coli* PUT11(DE3). The results showed that in the strain overexpressing the genes of glucose pathway, enhancing the PLP and NADPH synthesis could increase the yield of 1,4-diaminobutane. However, NAP19 exhibited slower growth compared with the control strain, *E. coli* PUT11(DE3), which was attributed to the metabolic burden imposed by the presence of four free plasmids containing multi-gene expression cassettes, ultimately impacting the strain's growth efficiency.

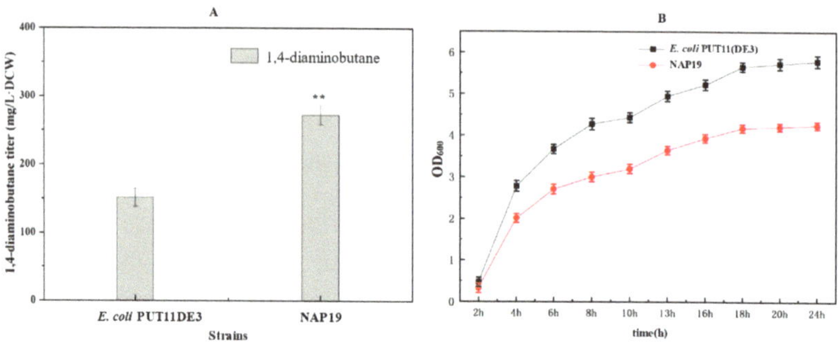

Figure 8. Determination of 1,4-diaminobutane titer in NAP19. (**A**) 1,4-diaminobutane titer in NAP19; (**B**) Cell growth of the recombinant NAP19 strain and the control strain. Strain genotypes NAP19: *E. coli* PUT11(DE3) harbouring the pRSM3-iolT-ppgk, pTrc99A-speC-argJ, pACM4-zwf-gnd-pgl, pCDM4-pdxJ-dxs-tktA-talB-pntAB-ppnk. Data were expressed as means and error bars indicated standard deviation (n = 3 independent experiments. The ** represented $p < 0.01$ compared with their controls).

3. Discussion

NADPH and PLP are the crucial cofactors in the biosynthetic pathway of 1,4-diaminobutane. The construction process of the reported high-yield 1,4-diaminobutane recombinant strains lacked researches on the synthesis of NADPH and PLP. Furthermore, in this study, the effect on the yield of 1,4-diaminobutane was determined by optimizing the PLP and NADPH synthesis pathways in *E. coli*.

Synthesizing 1 mol 1,4-diaminobutane in the pathway requires the consumption of 2 mol NADPH. Overexpressing key genes involved in NADPH synthesis, such as *pntAB* and *ppnK*, along with genes in the PP pathway, could enhance NADPH production and increased the yield of 1,4-diaminobutane. Ultimately, the accumulated amounts in the recombinant strains NAP3 and NAP8 were 95 mg/L·DCW and 98 mg/L·DCW, respectively, showing increases of 57% and 63% compared to the control strains *E. coli* BL21(DE3). This demonstrated that boosting NADPH synthesis in *E. coli* could effectively enhance the production of 1,4-diaminobutane.

PLP is the cofactor of ODC. Within a certain range of PLP concentrations, the synthesis of 1,4-diaminobutane is positively correlated with the PLP content in the cells. When the PLP content exceeds a certain range, cell growth is inhibited, leading to a decrease in 1,4-diaminobutane synthesis. Hence, optimizing the PLP content is essential when constructing high-yield 1,4-diaminobutane strains. Overexpressing genes involved in PLP synthesis, such as *pdxJ* and *dxs*, and genes involved in the synthesis of PLP precursors E4P, like *tktA* and *talB*, could increase the synthesis of 1,4-diaminobutane. However, when *pdxJ*, *dxs*, *tktA* and *talB* were co-expressed in the recombinant strain NAP13, the accumulation of

1,4-diaminobutane inversely decreased. At the same time, OD_{600} also decreased because enhancing the expression of PLP synthesis led to toxicity to the cells. This showed that PLP could promote the biosynthesis of 1,4-diaminobutane in a certain concentration range, otherwise, the biosynthesis of 1,4-diaminobutane was inhibited.

The fermentation of *E. coli* using glucose, glycerol and xylose was optimized, and the fermentation results showed that the 1,4-diaminobutane produced by the strain using glucose was the highest. Therefore, NADPH and PLP synthesis pathway plasmids, glucose utilization plasmids, and 1,4-diaminobutane synthesis plasmids were transferred to *E. coli* PUT11(DE3), which knocked out 11 genes associated with 1,4-diaminobutane degradation and transport pathways and incorporated the T7 RNA polymerase of *E. coli* BL21(DE3). Finally, the yield of the recombinant strain NAP19 reached 272 mg/L·DCW, which was 79% higher than that of the control strain *E. coli* PUT11(DE3). This suggested that the production of 1,4-diaminobutane could be increased by enhancing the synthesis of NADPH and PLP in the engineering strain. The effect of NADPH and PLP concentrations on 1,4-diaminobutane yield should also be considered.

This study revealed the importance of NADPH and PLP synthesis for the production of 1,4-diaminobutane. Although the yield of 1,4-diaminobutane in recombinant strains was not high, these results proved that the synthesis of NADPH and PLP in recombinant strains could effectively improve the yield of 1,4-diaminobutane, which provided a theoretical basis for the construction of high-yielding 1,4-diaminobutane recombinant strains.

4. Materials and Methods

4.1. Strain, Plasmid and Culture Conditions

E. coli str. K12 substr. MG1655 was used as the starting strain for genome manipulation. *E. coli* BL21(DE3) and *E. coli* PUT11(DE3) were used for the synthesis of 1,4-diaminobutane. The constructed strains and genetic phenotypes are shown in Table 1.

Table 1. List of strains.

Strains	Genotypes	Source
E. coli str. K12 substr. MG1655	Template for gene from *E.coli* cloning	Kept in our lab
C. glutamicum 13032	Wild type, donor of *ppnk, argJ* gene	Kept in our lab
B. subtilis subsp. subtilis str. 168	Wild type, donor of *gapB* gene	Kept in our lab
E. coli PUT11(DE3)	K12 MG1655△*argR*△*patA*△*puuA*△*speED*△*speG*△*puuP*△*argF*△*ydcSTUV*△*potFGHI*△*plaP*::DE3	Kept in our lab
NAP1	K12 harbouring pETM6-ppnk, Amp	This study
NAP2	K12 harbouring pETM6-pntAB, Amp	This study
NAP3	K12 harbouring pETM6-ppnk -pntAB, Amp	This study
NAP4	BL21(DE3) harbouring pETM6-zwf, Amp	This study
NAP5	BL21(DE3) harbouring pETM6-pgl, Amp	This study
NAP6	BL21(DE3) harbouring pETM6-gnd, Amp	This study
NAP7	BL21(DE3) harbouring pETM6-zwf-pgl, Amp	This study
NAP8	BL21(DE3) harbouring pETM6-zwf-pgl-gnd, Amp	This study
NAP9	BL21(DE3) harbouring pETM6-pdxJ, Amp	This study
NAP10	BL21(DE3) harbouring pETM6-dxs, Amp	This study
NAP11	BL21(DE3) harbouring pETM6-tktA, Amp	This study
NAP12	BL21(DE3) harbouring pETM6-talB, Amp	This study
NAP13	BL21(DE3) harbouring pETM6-pdxJ-dxs-tktA-talB, Amp	This study
NAP14	NAP13 harbouring pRSM3-speC-argJ, Kan	This study
NAP15	NAP14 harbouring pCDM4-pntAB-ppnK and pACM4-zwf-gnd-pgl, Cl	This study
NAP16	BL21(DE3) harbouring pETM6-glpFK-tpiA-gapB, Amp	This study
NAP17	BL21(DE3) harbouring pETM6-iolT-ppgK, Amp	This study
NAP18	BL21(DE3) harbouring pETM6-xylAB, Amp	This study
NAP19	PUT11(DE3) harbouring the pRSM3-glpFK-tpiA-gapB, pTrc99A-speC-argJ, pACM4-zwf-gnd-pgl, pCDM4-pdxJ-dxs-tktA-talB-pntAB-ppnk. Kan, Amp, Cl, Str	This study

During plasmid and strain construction, cells were grown on LB solid plates containing appropriate antibiotics. When needed, antibiotics were added at the following concentrations: 50 mg/mL kanamycin (Kan), 100 mg/mL ampicillin (Amp), 25 mg/mL chloramphenicol (Cl) and 50 mg/mL streptomycin (Str).

The target genes in this study were amplified using matched primers using relevant host genomic DNA as a template. Heterologous genes were amplified using primers from *E. coli* MG1655, *C. glutamicum* 13032 and *B. subtilis* genomes as templates, respectively. All of the primers used in this study were listed in Supplementary Table S1.

4.2. Plasmids Construction

The pETM6 plasmid and interest genes were digested using *Spe*I and *Nde*I and subsequently ligated with a ligase to construct the pETM6 expression plasmid.

The pACM4, pCDM4, pRSM3 and pETM6 plasmids have three major cleavage sites: *Bln*I, *Nhe*I and *Sal*I, and after cleavage they will produce the same CTAG sticky end. Firstly, the interest gene was cloned into the *Nde*I and *Spe*I of pETM6. Then, interest gene expression cassette was digested using *Bln*I and *Sal*I from a constructed plasmid and ligated into between *Nhe*I and *Sal*I of the other plasmid by isocaudamers for achieving a tandem linkage of genes (Supplementary Figure S1). The same method was used to tandem genes on pACM4, pCDM4 and pRSM3. The constructed plasmids are shown in Supplementary Table S2.

4.3. Shake Flask Culture and Fermentation

The constructed engineering strains were inoculated from solid medium into LB liquid medium containing corresponding antibiotics, and incubated at 37 °C, 220 r/min for 14–16 h as seed culture. The seed was inoculated into 500 mL shake flasks containing fermentation medium according to 1% inoculum volume, and the corresponding antibiotics was added.

Fermentation medium:

(1) SOB medium (Super Optimal Broth): 20 g/L tryptone, 5 g/L yeast extract, 0.5 g/L NaCl, add 10 mL of 250 mmol/L KCl solution, and adjust pH to 7.0. Supplemented with 1 g/L carbon source. The solution was autoclaved at 115 °C for 20 min. 5 mL of sterilised 2 mol/L $MgCl_2$ was added before use.
(2) R/2 medium: 5 g/L yeast extract, 1 g/L peptone, 2 g/L $(NH_4)_2HPO_4$, 6.75 g/L KH_2PO_4, 0.85 g/L citric acid, 0.7 g/L $MgSO_4{\cdot}7H_2O$, 10 g/L glucose, 3 g/L $(NH_4)_2SO_4$ and 5 mL/L trace metal solution. The pH was adjusted to 6.8 and autoclaved at 115 °C for 20 min. The trace metal solution contained 5 mol/L HCl, 10 g/L $FeSO_4{\cdot}7H_2O$. 2.25 g/L $ZnSO_4{\cdot}7H_20$, 1 g/L $CuSO_4{\cdot}5H_2O$, 0.5 g/L $MnSO_4{\cdot}5H_2O$, 1.23 g/L $Na_2B_4O_7{\cdot}10H_2O$, 2 g/L$CaCl_2{\cdot}2H_2O$ and 0.1 g/L $(NH_4)_6Mo_7O_{24}$.

4.4. Analytical Method

Strain growth was monitored by measuring absorbance at 600 nm (OD_{600}) using a UV spectrophotometer. The dry cell weights (DCW) were determined by heating the samples to a constant weight at 80 °C.

NADPH concentrations were determined based on the glucose dehydrogenase cycle reaction using the CheKine^{TM} Micro Coenzyme II NADP(H) Assay Kit (Abbkine Scientific, CHN. https://www.abbkine.com/product/chekine-micro-coenzyme-%E2%85%A1-nadph-assay-kit-ktb1010/, accessed on 5 June 2024). The strains samples were harvested by centrifugaed and washed with PBS. The pellet cells were added 100 µL of NADPH Extraction Buffer and broken by ultrasonic crushing (power 200 W, ultrasonic 3 s, interval 10 s, repeated 30 times). And then, the extracts were heated at 60 °C for 5 min, 20 µL of Assay Buffer was added, followed by the addition of an equal volume of reverse extraction solution to neutralize it. Vortexed briefly, the samples were centrifuged at 14,000 r/min for 5 min at 4 °C. The supernatant was transferred to a new microcentrifuge tube, and the NADPH content was measured at 450 nm.

1,4-diaminobutane was derived with dansyl chloride so that it can be detected by ultraviolet at 254 nm [40]. After 1 mL sample was centrifuged, 500 µL supernatant was taken into a microcentrifuge tube. Added 500 µL saturated $NaHCO_3$, and adjusted the pH to 10 with saturated NaOH. Added 1 mL dansyl chloride (5 g/L) for derivative reaction.

Incubated the tubes at 60 °C for 30 min, and then 2 mL anhydrous ether was added for extraction. Took the top organic phase into the microcentrifuge tube, dried, and added of 500 µL acetonitrile to dissolve. The concentration of 1,4-diaminobutane was detected by HPLC. The chromatographic column was Agilent porpshell EC-C18 column (4.6 mm × 250 mm, particle size 4 µm). The detection method was as follows: 30 °C, UV wavelength 254 nm, sample size 10 uL, mobile phase A as ultra-pure water, mobile phase B as acetonitrile, flow rate 0.7 mL/min, gradient elution: 0–5 min: 55% to 70% B; 5–10 min: 70% B; 10–15 min: 70 to 95% B; 15–20 min: 95% B; 20-25 min: 95 to 55% B. The retention time of 1,4-diaminobutane is about 14 min. Chromatographic peak of 1,4-diaminobutane is shown in Supplementary Materials Figure S2.

4.5. Statistical Analysis

The data in this study represented the mean values and standard deviations of three independent experiments. The significant differences in data were determined via a one-way analysis of variance. The value of $0.01 < p < 0.05$ indicated a significant difference compared with their controls, and a p value < 0.01 indicated a highly significant difference compared with their controls.

5. Conclusions

This study optimized the supply of NADPH and PLP in the 1,4-diaminobutane biosynthesis pathway of *E. coli* and confirmed the positive effects of increased NADPH and PLP synthesis on 1,4-diaminobutane production. Ultimately transferring NADPH and PLP synthetic plasmids and glucose utilization plasmids into the *E. coli* PUT11(DE3) strain, the resulting strain NAP19 accumulated 272 mg/L·DCW of 1,4-diaminobutane. Increasing the supply of the cofactors PLP and NADPH was shown to be of significant importance for the synthesis of 1,4-diaminobutane. It laid a foundation for the construction of high-yielding 1,4-diaminobutane recombinant strains.

Supplementary Materials: The following supporting information can be downloaded at: https://www.mdpi.com/article/10.3390/molecules29133094/s1, Table S1: Primers used in this study. Table S2: List of plasmids. Figure S1: Flow chart of tandem carrier constructed by isocaudamers. Figure S2: Determination of 1,4-diaminobutane by HPLC. (A) the chromatographic peak of 1,4-diamino-butane standard, which peaked in about 14 min. (B) the chromatographic peak of the sample, which peaked in about 14 min.

Author Contributions: Experiment method design, M.L., F.L. and T.S.; experiment performance, T.S., Y.Z., J.W., W.K., X.S. and Z.L.; writing—original draft preparation, M.L., F.L. and T.S.; writing—review and editing, T.S., Y.Z., J.W., Y.S. and M.C. All authors have read and agreed to the published version of the manuscript.

Funding: This work was supported by the National Key Research and Development Program of China (No. 2021YFC2101800, No. 2022YFC2105500), and the National Natural Science Foundation of China (No. 22278321, No. 32072161). Tianjin Postgraduate Research Innovation Program (2022SKYZ061).

Institutional Review Board Statement: Not applicable.

Informed Consent Statement: Not applicable.

Data Availability Statement: The data are available from the corresponding author upon request.

Conflicts of Interest: The authors declare no conflicts of interest.

References

1. Zhang, X.; Zhao, P.; Kuai, J.; Chang, C.; Yuan, Q. Spermidine alleviates lipopolysaccharide-induced myocardial injury in mice by suppressing apoptosis, ROS production and ferroptosis. *J. South. Med. Univ.* **2024**, *44*, 166–172.
2. Holbert, C.E.; Cullen, M.T.; Casero, R.A., Jr.; Stewart, T.M. Polyamines in cancer: Integrating organismal metabolism and antitumour immunity. *Nat. Rev. Cancer* **2022**, *22*, 467–480. [CrossRef]

3. Pérez-Pertejo, Y.; García-Estrada, C.; Martínez-Valladares, M.; Murugesan, S.; Reguera, R.M.; Balaña-Fouce, R. Polyamine Metabolism for Drug Intervention in Trypanosomatids. *Pathogens* **2024**, *13*, 79. [CrossRef]
4. Lee, P.C.; Kim, S.Y.; Ko, Y.K.; Ha, J.U.; Jeoung, S.K.; Shin, D.; Kim, J.H.; Kim, M.G. Tribological Properties of Polyamide 46/Graphene Nanocomposites. *Polymers* **2022**, *14*, 1139. [CrossRef]
5. Li, Z.; Liu, J.Z. Transcriptomic Changes in Response to Putrescine Production in Metabolically Engineered *Corynebacterium glutamicum*. *Front. Microbiol.* **2017**, *8*, 1987. [CrossRef]
6. Sanders, J.; Scott, E.; Weusthuis, R.; Mooibroek, H. Biorefinery as the bioinspired process to bulk chemicals. *Macromol. Biosci.* **2007**, *7*, 105–117. [CrossRef]
7. Son, J.; Sohn, Y.J.; Baritugo, K.A. Recent advances in microbial production of diamines, aminocarboxylic acids, and diacids as potential platform chemicals and bio-based polyamides monomer. *Biotechnol. Adv.* **2023**, *62*, 108070. [CrossRef]
8. Yang, S.-C.; Ting, W.-W.; Ng, I.-S. Effective whole cell biotransformation of arginine to a four-carbon diamine putrescine using engineered *Escherichia coli*. *Biochem. Eng. J.* **2022**, *185*, 108502.
9. Qian, Z.G.; Xia, X.X.; Lee, S.Y. Metabolic engineering of *Escherichia coli* for the production of putrescine: A four carbon diamine. *Biotechnol. Bioeng.* **2009**, *104*, 651–662. [CrossRef]
10. Li, Z.; Shen, Y.P.; Jiang, X.L.; Feng, L.S.; Liu, J.Z. Metabolic evolution and a comparative omics analysis of *Corynebacterium glutamicum* for putrescine production. *J. Ind. Microbiol. Biotechnol.* **2018**, *45*, 123–139. [CrossRef]
11. Nguyen, A.Q.; Schneider, J.; Wendisch, V.F. Elimination of polyamine N-acetylation and regulatory engineering improved putrescine production by *Corynebacterium glutamicum*. *J. Biotechnol.* **2015**, *201*, 75–85. [CrossRef] [PubMed]
12. Hong, E.Y.; Kim, J.Y.; Upadhyay, R.; Park, B.J.; Lee, J.M.; Kim, B.G. Rational engineering of ornithine decarboxylase with greater selectivity for ornithine over lysine through protein network analysis. *J. Biotechnol.* **2018**, *281*, 175–182. [CrossRef] [PubMed]
13. Tan, Q.; Gou, L.; Fan, T.P.; Cai, Y. Enzymatic properties of ornithine decarboxylase from Clostridium aceticum DSM1496. *Biotechnol. Appl. Biochem.* **2024**, *71*, 525–535. [CrossRef] [PubMed]
14. Li, M.; Lu, F.; Sun, X. Chassis Strain of Escherichia coli Producing 1,4-Diaminobutane and Its Application. CN202311595033.0, 27 March 2024.
15. Schneider, J.; Wendisch, V.F. Biotechnological production of polyamines by bacteria: Recent achievements and future perspectives. *Appl. Microbiol. Biotechnol.* **2011**, *91*, 17–30. [CrossRef] [PubMed]
16. Noh, M.; Yoo, S.M.; Kim, W.J.; Lee, S.Y. Gene Expression Knockdown by Modulating Synthetic Small RNA Expression in *Escherichia coli*. *Cell Syst.* **2017**, *5*, 418–426. [CrossRef] [PubMed]
17. Jensen, J.V.; Eberhardt, D.; Wendisch, V.F. Modular pathway engineering of *Corynebacterium glutamicum* for production of the glutamate-derived compounds ornithine, proline, putrescine, citrulline, and arginine. *J. Biotechnol.* **2015**, *214*, 85–94. [CrossRef] [PubMed]
18. Kim, S.Y.; Lee, J.; Lee, S.Y. Metabolic engineering of *Corynebacterium glutamicum* for the production of L-ornithine. *Biotechnol. Bioeng.* **2015**, *112*, 416–421. [CrossRef] [PubMed]
19. Sauer, U.; Canonaco, F.; Heri, S.; Perrenoud, A.; Fischer, E. The soluble and membrane-bound transhydrogenases UdhA and PntAB have divergent functions in NADPH metabolism of *Escherichia coli*. *J. Biol. Chem.* **2004**, *279*, 6613–6619. [CrossRef] [PubMed]
20. Kawai, S.; Mori, S.; Mukai, T.; Suzuki, S.; Yamada, T.; Hashimoto, W.; Murata, K. Inorganic Polyphosphate/ATP-NAD kinase of Micrococcus flavus and Mycobacterium tuberculosis H37Rv. *Biochem. Biophys. Res. Commun.* **2000**, *276*, 57–63. [CrossRef]
21. Shi, F.; Li, Y.; Li, Y.; Wang, X. Molecular properties, functions, and potential applications of NAD kinases. *Acta Biochim. Biophys. Sin.* **2009**, *41*, 352–361. [CrossRef]
22. Okazaki, S.; Suzuki, A.; Mizushima, T.; Kawano, T.; Komeda, H.; Asano, Y.; Yamane, T. The novel structure of a pyridoxal 5′-phosphate-dependent fold-type I racemase, alpha-amino-epsilon-caprolactam racemase from *Achromobacter obae*. *Biochemistry* **2009**, *48*, 941–950. [CrossRef] [PubMed]
23. Fitzpatrick, T.B.; Amrhein, N.; Kappes, B.; Macheroux, P.; Tews, I.; Raschle, T. Two independent routes of de novo vitamin B6 biosynthesis: Not that different after all. *Biochem. J.* **2007**, *407*, 1–13. [CrossRef] [PubMed]
24. Schiroli, D.; Peracchi, A. A subfamily of PLP-dependent enzymes specialized in handling terminal amines. *Biochim. Biophys. Acta* **2015**, *1854*, 1200–1211. [CrossRef] [PubMed]
25. Bisercić, M.; Feutrier, J.Y.; Reeves, P.R. Nucleotide sequences of the gnd genes from nine natural isolates of *Escherichia coli*: Evidence of intragenic recombination as a contributing factor in the evolution of the polymorphic gnd locus. *J. Bacteriol.* **1991**, *173*, 3894–3900. [CrossRef]
26. Moccand, C.; Kaufmann, M.; Fitzpatrick, T.B. It takes two to tango: Defining an essential second active site in pyridoxal 5′-phosphate synthase. *PLoS ONE* **2011**, *6*, e16042. [CrossRef] [PubMed]
27. Kim, J.H.; Kim, J.; Kim, H.J.; Sathiyanarayanan, G.; Bhatia, S.K.; Song, H.S.; Choi, Y.K.; Kim, Y.G.; Park, K.; Yang, Y.H. Biotransformation of pyridoxal 5′-phosphate from pyridoxal by pyridoxal kinase (pdxY) to support cadaverine production in *Escherichia coli*. *Enzym. Microb. Technol.* **2017**, *104*, 9–15. [CrossRef] [PubMed]
28. Han, H.; Xu, B.; Zeng, W.; Zhou, J. Regulating the biosynthesis of pyridoxal 5′-phosphate with riboswitch to enhance L-DOPA production by *Escherichia coli* whole-cell biotransformation. *J. Biotechnol.* **2020**, *321*, 68–77. [CrossRef] [PubMed]
29. Ma, W.; Cao, W.; Zhang, B.; Chen, K.; Liu, Q.; Li, Y.; Ouyang, P. Engineering a pyridoxal 5′-phosphate supply for cadaverine production by using *Escherichia coli* whole-cell biocatalysis. *Sci. Rep.* **2015**, *5*, 15630. [CrossRef] [PubMed]

30. Momany, C.; Ernst, S.; Ghosh, R.; Chang, N.L.; Hackert, M.L. Crystallographic structure of a PLP-dependent ornithine decarboxylase from Lactobacillus 30a to 3.0 A resolution. *J. Mol. Biol.* **1995**, *252*, 643–655. [CrossRef]
31. Oliveira, E.F.; Cerqueira, N.M.; Fernandes, P.A.; Ramos, M.J. Mechanism of formation of the internal aldimine in pyridoxal 5′-phosphate-dependent enzymes. *J. Am. Chem. Soc.* **2011**, *133*, 15496–15505. [CrossRef]
32. Seol, E.; Sekar, B.S.; Raj, S.M.; Park, S. Co-production of hydrogen and ethanol from glucose by modification of glycolytic pathways in *Escherichia coli*—from Embden-Meyerhof-Parnas pathway to pentose phosphate pathway. *Biotechnol. J.* **2016**, *11*, 249–256. [CrossRef]
33. Moritz, B.; Striegel, K.; De Graaf, A.A.; Sahm, H. Kinetic properties of the glucose-6-phosphate and 6-phosphogluconate dehydrogenases from *Corynebacterium glutamicum* and their application for predicting pentose phosphate pathway flux in vivo. *Eur. J. Biochem.* **2000**, *267*, 3442–3452. [CrossRef]
34. Hao, N.; Mu, J.; Hu, N.; Xu, S.; Shen, P.; Yan, M.; Li, Y.; Xu, L. Implication of ornithine acetyltransferase activity on l-ornithine production in *Corynebacterium glutamicum*. *Biotechnol. Appl. Biochem.* **2016**, *63*, 15–21. [CrossRef]
35. Hao, N.; Mu, J.; Hu, N.; Xu, S.; Yan, M.; Li, Y.; Guo, K.; Xu, L. Improvement of L-citrulline production in *Corynebacterium glutamicum* by ornithine acetyltransferase. *J. Ind. Microbiol. Biotechnol.* **2015**, *42*, 307–313. [CrossRef]
36. Caldovic, L.; Tuchman, M. N-acetylglutamate and its changing role through evolution. *Biochem. J.* **2003**, *372 Pt 2*, 279–290. [CrossRef]
37. Zhang, B.; Gao, G.; Chu, X.H.; Ye, B.C. Metabolic engineering of *Corynebacterium glutamicum* S9114 to enhance the production of l-ornithine driven by glucose and xylose. *Bioresour. Technol.* **2019**, *284*, 204–213. [CrossRef]
38. Kim, S.M.; Choi, B.Y.; Ryu, Y.S.; Jung, S.H.; Park, J.M.; Kim, G.H.; Lee, S.K. Simultaneous utilization of glucose and xylose via novel mechanisms in engineered *Escherichia coli*. *Metab. Eng.* **2015**, *30*, 141–148. [CrossRef]
39. Shams Yazdani, S.; Gonzalez, R. Engineering *Escherichia coli* for the efficient conversion of glycerol to ethanol and co-products. *Metab. Eng.* **2008**, *10*, 340–351. [CrossRef]
40. Hawi, A.A.; Yip, H.; Sullivan, T.S.; Digenis, G.A. Development of an HPLC assay for the analysis of tetrafluoroputrescine—A putrescine analog. *Anal. Biochem.* **1988**, *172*, 235–240. [CrossRef]

molecules

Review

Challenges and Advances in the Bioproduction of L-Cysteine

Daniel Alejandro Caballero Cerbon [1], Leon Gebhard [2], Ruveyda Dokuyucu [3], Theresa Ertl [3], Sophia Härtl [3], Ayesha Mazhar [3] and Dirk Weuster-Botz [1,*]

1 Chair of Biochemical Engineering, TUM School of Engineering and Design, Technical University of Munich, Boltzmannstraße 15, D-85748 Garching, Germany; daniel.caballero@tum.de
2 TUM School of Engineering and Design, Technical University of Munich, Boltzmannstraße 15, D-85748 Garching, Germany
3 TUM Campus Straubing for Biotechnology and Sustainability, Technical University of Munich, Petersgasse 5, D-94315 Straubing, Germany; ruveyda.dokuyucu@tum.de (R.D.); theresa.maria.ertl@tum.de (T.E.); ge73xob@mytum.de (S.H.)
* Correspondence: dirk.weuster-botz@tum.de; Tel.: +49-89-289-15712; Fax: +49-89-289-15714

Abstract: L-cysteine is a proteogenic amino acid with many applications in the pharmaceutical, food, animal feed, and cosmetic industries. Due to safety and environmental issues in extracting L-cysteine from animal hair and feathers, the fermentative production of L-cysteine offers an attractive alternative using renewable feedstocks. Strategies to improve microbial production hosts like *Pantoea ananatis*, *Corynebacterium glutamicum*, *Pseudomonas* sp., and *Escherichia coli* are summarized. Concerning the metabolic engineering strategies, the overexpression of feedback inhibition-insensitive L-serine O-acetyltransferase and weakening the degradation of L-cysteine through the removal of L-cysteine desulfhydrases are crucial adjustments. The overexpression of L-cysteine exporters is vital to overcome the toxicity caused by intracellular accumulating L-cysteine. In addition, we compiled the process engineering aspects for the bioproduction of L-cysteine. Utilizing the energy-efficient sulfur assimilation pathway via thiosulfate, fermenting cheap carbon sources, designing scalable, fed-batch processes with individual feedings of carbon and sulfur sources, and implementing efficient purification techniques are essential for the fermentative production of L-cysteine on an industrial scale.

Keywords: L-cysteine; biosynthesis; metabolic engineering; fermentation; fed-batch process; downstream processing

Citation: Caballero Cerbon, D.A.; Gebhard, L.; Dokuyucu, R.; Ertl, T.; Härtl, S.; Mazhar, A.; Weuster-Botz, D. Challenges and Advances in the Bioproduction of L-Cysteine. *Molecules* **2024**, *29*, 486. https://doi.org/10.3390/molecules29020486

Academic Editors: Yongjun Wei, Xiao-Jun Ji and Lingbo Qu

Received: 21 November 2023
Revised: 15 January 2024
Accepted: 17 January 2024
Published: 18 January 2024

1. Introduction

L-cysteine is a nucleophilic amino acid with a sulfhydryl group, which enables the formation of disulfide bridges. It can be produced by most eukaryotic and prokaryotic organisms. This amino acid is semi-essential in humans, since it can be synthesized endogenously or directly absorbed from an L-cysteine-rich diet. In cells, cysteine can be produced via the reverse transsulfuration pathway from L-methionine [1] or from L-serine through acetylation and subsequent sulfhydration [2]. Figure 1 presents a step-by-step diagram for these pathways. The cells tightly control the intracellular L-cysteine concentration because already-low concentrations can be cytotoxic [3,4]. Intracellular L-cysteine plays a vital role as a precursor for protein synthesis, and for the production of glutathione, hydrogen sulfide, and taurine [5]. In humans and animals, L-cysteine is primarily found in the hair, nails, and skin, especially in collagen. Prokaryotic cells use L-cysteine mostly in proteins, in enzymes for the formation of iron–sulfur clusters, or as a source of sulfur for other macromolecules [6].

L-cysteine is produced for applications in the pharmaceutical, food, animal feed, and cosmetics industries, with an estimated yearly production of roughly 3000 tons [7,8]. In the pharmaceutical and food industries, L-cysteine is a precursor for the production of drugs, food supplements, and nutraceuticals [9]. In animal feed, L-cysteine is used as a sulfur supplement for wool production [10]. In the cosmetic industry, L-cysteine is applied as a substitute for thioglycolic acid in hair applications for breaking the disulfide bridges of keratin to achieve permanent hair waving [10].

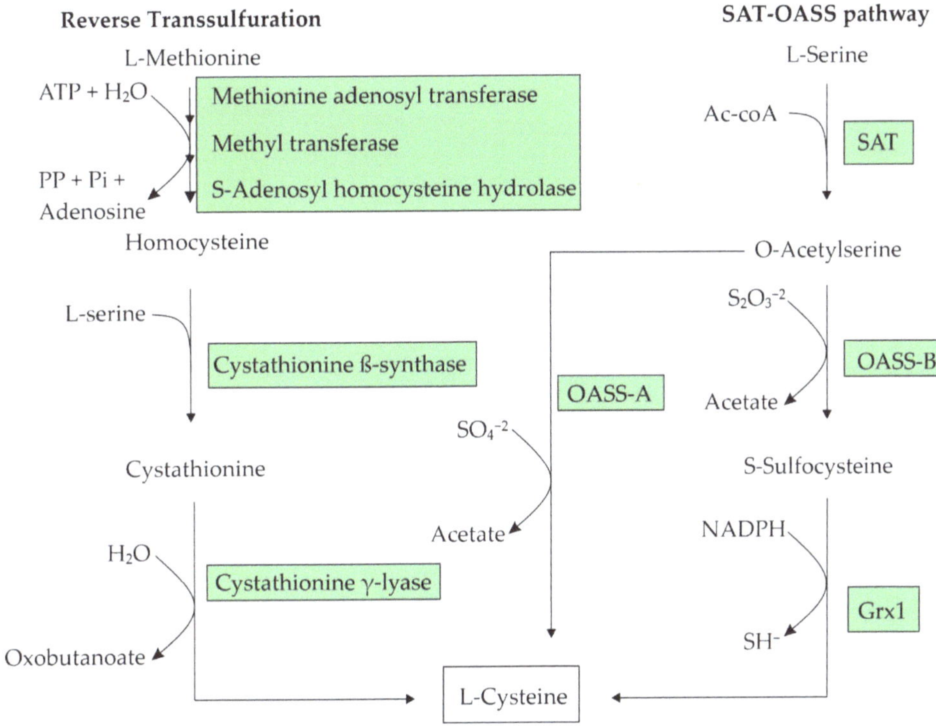

Figure 1. Metabolic pathways for L-cysteine biosynthesis in microorganisms. In the L-serine O-acetyltransferase-O-acetyl-L-serine (SAT-OASS) pathway, there are two branches for the transformation of O-acetyl-L-serine (OAS) into L-cysteine, depending on the sulfur source available to the organism. These biotransformations are catalyzed by two different OASSs: OASS-A in the case of sulfate as the S-source and OASS-B for thiosulfate. The transformation through OASS-B requires a subsequent reduction of S-sulfocysteine to L-cysteine catalyzed by glutathioredoxin (Grx1).

Animal feathers and human hair are natural sources of L-cysteine, which is the main component of keratin. First, cystine is extracted with activated charcoal after the acidic hydrolysis of feathers and hair. After desorption from the activated charcoal matrix, the isolated and purified L-cystine is then transformed into L-cysteine via electrolytic reduction [11]. This is the most common production method applied so far, although the L-cysteine process yield is low, and unpleasant odors may occur. However, the most important disadvantage is the environmental and safety issues caused by the disposal of the organic waste, which contains hydrochloric acid [12].

An alternative approach is the biotechnological production of L-cysteine from renewable resources [13]. Since the 1980s, microorganisms of the genus *Pseudomonas* have been used for the asymmetric whole-cell biocatalysis of DL-2-amino-Δ^2-thiazoline-4-carboxylic acid (DL-ATC) to generate L-cysteine (Figure 2).

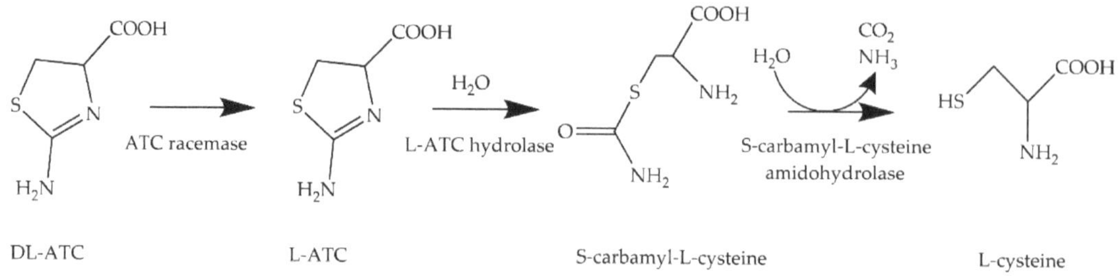

Figure 2. Transformation of DL-ATC to L-cysteine. In this enzymatic three-step process, DL-ATC is isomerized to L-ATC by ATC racemase. Then, L-ATC is hydrolyzed to S-carbamyl-L-cysteine (SCC) by L-ATC hydrolase. Finally, SCC is cleaved into L-cysteine, CO_2, and ammonia by S-cabamyl-L-cysteine amidohydrolase.

Pseudomonas is a genus of gram-negative bacteria belonging to the family Pseudomonadaceae in the class of γ-proteobacteria. From the many bacteria, especially in the genus *Pseudomonas*, that have an activity to convert DL-ATC to L-cysteine, two biosynthetic pathways have been described [14–16]. As an intermediate, one pathway produces N-carbamoyl cysteine, and the other, S-carbamoyl cysteine [17]. However, in the *Pseudomonas* cells, L-cysteine is converted to pyruvate, hydrogen sulfide, and ammonia by L-cysteine desulfhydrase (CD). The L-cysteine desulfhydrase gene could not be knocked out to increase the production of L-cysteine during the bioconversion of DL-ATC by *Pseudomonas* sp., since the degradation of L-cysteine is required to supply the cells with carbon and nitrogen [18]. Furthermore, the difficulty of procurement of the substrate DL-ATC and the low enzymatic yields of this process limited its application as an industrial process [12].

Through the understanding of the complex regulation system of the L-cysteine synthesis in bacteria, and with the advancements in synthetic biology, the direct fermentative L-cysteine production from glucose became feasible and has been exploited by global amino acid producers to supply an increasing portion of the global L-cysteine demand [19].

In this review, the metabolic engineering strategies for enhancing the L-cysteine yield and cell-specific productivities will be compiled, e.g., improving the metabolic availability of precursors, deregulating controlling enzymes, knocking out genes coding for L-cysteine-degrading enzymes, and overexpressing transporter genes to transport L-cysteine out of the cell. The final section of this review focuses on the process engineering aspects like sulfur sources, fermentation process designs, and L-cysteine purification strategies.

2. Metabolic Engineering Strategies

Microorganisms use two main pathways for the biosynthesis of L-cysteine. In enteric bacteria such as *E. coli*, the precursor of L-cysteine is L-serine [20]. L-cysteine is synthesized from L-serine in two steps, which involve the substitution of the β-hydroxyl group with a thiol group (Figure 1). The first step involves the acetylation of the β-hydroxyl of L-serine, resulting in an O-acetyl-L-serine (OAS) catalyzed by L-serine O-acetyltransferase (SAT), which is encoded by the *cysE* gene. The second step involves the α,β-elimination of acetate from OAS and the addition of H_2S to give an L-cysteine catalyzed by O-acetyl-L-serine sulfhydrylase-A (OASS-A), which is encoded by the *cysK* gene. *Corynebacterium glutamicum* and *Pantoea ananantis* synthesize L-cysteine through the same pathway as *E. coli* [12]. Other microorganisms, such as *Pseudomonas putida* [21], *Saccharomyces cerevisiae* [22], and *Lactococcus lactis* [23] synthesize L-cysteine from L-methionine through the reverse transsulfuration pathway that consists in the cleavage of a methyl group from L-methionine to generate a homocysteine catalyzed by a multienzyme complex, and the homocysteine's subsequent transformation to cystathionine by the enzyme cystathionine beta-synthase (CBS). L-cysteine can be produced from cystathionine through the enzyme cystathionine gamma-lyase (CSE) (Figure 1).

The choice of the starting strain for metabolic engineering has an impact on the potential enhancements achieved through the modification of metabolic steps [24]. This will be shown in an example with *E. coli*: It has been established that the transcription levels of key metabolic genes are the major modulators of different glucose utilization pathways in *E. coli* [25]. Also, the genetic difference in the catabolite repression protein (CRP) among the *E. coli* K-12 strains affects sulfur metabolism [26]. During the last 20 years, four *E. coli* K-12 strains (BW25113, JM109, W3110, and MG1655, respectively) have been utilized in L-cysteine production [27]. Therefore, the selection of a suitable chassis for L-cysteine production has been deemed the foundation for the successful implementation of metabolic engineering strategies [28]. To reach a more suitable chassis cell, the L-cysteine-producing plasmid pLH03, containing P-*cysE* (coding for serine acetyltransferase) and P-*ydeD* (a cysteine exporter), was transformed into the four aforementioned *E. coli* K-12 hosts. The results of shake flask experiments (minimal medium with thiosulfate as the sulfur source) showed that the best chassis for the plasmid pLH03 was BW25113, since it produced 345.8 mg L^{-1} of L-cysteine, followed by JM109 (190.3 mg L^{-1}). By contrast, neither W3110 nor MG1655 were deemed suitable base strains for this plasmid, since they each produced less than 70 mg L^{-1} product. [28]

Henceforth, the advances in the metabolic engineering of a variety of organisms will be discussed. In general, the fermentative production of L-cysteine can be enhanced by utilizing a combination of three different metabolic engineering strategies, which are as follows:

- Enhancing the biosynthesis of L-cysteine by expressing genes encoding feedback inhibition-insensitive phosphoglycerate dehydrogenase (PGDH) L-serine O-acetyltransferase (SAT);
- Weakening the degradation of L-cysteine by knocking out genes encoding L-cysteine desulfhydrase (CD);
- Enhancing and strengthening the efflux system while weakening and deleting the genes responsible for the import of L-cysteine.

Table 1 at the end of this chapter summarizes the maximal L-cysteine concentrations achieved by applying these metabolic engineering approaches, as well as the organism and scale of the process by which they were reached.

2.1. Biosynthesis of L-Cysteine

The acetylation of L-serine catalyzed by serine acetyltransferase (SAT) is the rate-limiting step of L-cysteine biosynthesis in many microorganisms. The activity of SAT is suppressed by L-cysteine via feedback inhibition (Figure 2). An approach to enhance the fermentative production of L-cysteine utilizes feedback inhibition-insensitive SATs. These SATs are obtained by the following means:

- Engineering SAT from *E. coli* through site-directed mutagenesis or random mutagenesis [29,30];
- Utilizing natural SAT from higher plants insensitive to feedback inhibition [31,32].

Nakamori et al. [33] obtained feedback inhibition-insensitive SATs by engineering SAT from *E. coli* through site-directed mutagenesis using PCR. In this study, the *cysE* gene coding for SAT was altered by replacing the methionine at position 256 with other amino acids. The transformed microorganisms, having the altered *cysE* gene, produced 50–300 mg L^{-1} of L-cysteine plus L-cystine after 72 h of cultivation in shake flasks containing a minimal medium at 30 °C. Furthermore, the transformation of the altered *cysE* gene in a mutant strain of *E. coli*, JM39-8, with decreased cysteine degradation abilities (10% CD activity) resulted in an increased production to a maximum of 790 mg L^{-1} of L-cysteine plus L-cystine in shake flasks under the same conditions as above. Takagi et al. [30] isolated feedback inhibition-insensitive SAT through a random mutagenesis into the *cysE* gene by using the error-prone PCR. The production test of L-cysteine and the enzymatic analysis of the mutant SAT suggested that the carboxy-terminal region of SAT is involved in the desensitization to feedback inhibition and the substantial production of L-cysteine.

Another approach to obtain feedback inhibition-insensitive SATs could be to use natural SATs from higher plants. In *Arabidopsis thaliana*, three cDNA clones encoding SAT isoforms (SAT-c, SAT-p, and SAT-m) have been isolated. Noji et al. [33] analyzed these SATs and reported that SAT-p and SAT-m were feedback inhibition-insensitive isozymes. However, the activity of SAT-c was feedback inhibited by a low concentration of L-cysteine. Takagi et al. [31] expressed two cDNAs from *A. thaliana* encoding feedback inhibition-insensitive SAT (SAT-p and SAT-m) in the *E. coli* strain JM39-8, and reported an increased production of L-cysteine (1.6–1.7 g L^{-1}) when cultivated for 96 h in shaking flasks containing a minimal medium at 30 °C. Wirtz and Hell [32] isolated three new cDNAs from *Nicotiana tabacum* encoding SATs and reported that the SAT4 isoform has a lower feedback sensitivity than the previously reported SATs. The results showed that the expression of SAT4 in *E. coli* causes a 50-fold increase in the accumulation of L-cysteine in the medium.

Kai et al. [34] analyzed the crystal structure and reaction mechanism of SAT and introduced various amino acid substitutions to change the position of Asp92. As a result, they designed a mutant SAT having a high enzymatic activity and low sensitivity to inhibition by L-cysteine. For one promising mutant, the inhibitory L-cysteine concentration that halved the enzyme's activity went from the wild type's 0.8 μM to 410 μM, while simultaneously increasing the initial enzymatic activity by 220 μM mg^{-1} min^{-1}.

In another study, a strain of *P. ananatis* that carried a mutant of the gene *cysE (cysEX)* coding for a feedback-insensitive SAT was constructed. A second strain of *P. ananatis*, additionally harboring a plasmid with the gene *cefA*, which codes for an L-cysteine efflux pump, was generated. A strain of *P. ananatis* carrying the gene *cefA* was able to produce 900 mg L^{-1} of L-cysteine in agitated test tubes, which is roughly four times the amount produced by the *P. ananatis* strain carrying only the feedback-insensitive SAT [6].

C. glutamicum is a nonpathogenic and high G+C gram-positive bacterium industrially used for decades for the fermentative production of amino acids such as L-glutamic acid and L-lysine [35,36]. Joo et al. [37] investigated the effect of the overexpression of genes encoding SAT, OASS, and a transcriptional regulator in *C. glutamicum*. The results showed that L-cysteine production was enhanced by approximately three-fold, with the recombinant strain reaching a maximal L-cysteine concentration of approximately 60 mg L^{-1} in shake flasks at 30 °C after 15 h of cultivation. Wei et al. [7] further engineered *C. glutamicum* for L-cysteine production by enhancing the L-serine biosynthetic pathway, expressing the gene encoding feedback inhibition-insensitive SAT, deleting genes encoding CD, and increasing the L-cysteine export. Their engineered strain, cultivated for 48 h at 30 °C in shake flasks, produced 947.9 ± 46.5 mg L^{-1} of L-cysteine, with a yield of 27.27 mg g^{-1} of glucose and a volumetric productivity of 26.33 mg L^{-1} h^{-1}. Kondoh and Hirasawa [38] achieved approximately 200 mg L^{-1} of L-cysteine production with *C. glutamicum* by overexpressing the mutant *cysE* and *serA* genes to enhance the biosynthesis, disrupting *aecD* gene coding for CD to prevent degradation and disrupting the NCgl2463 gene coding for a membrane transporter to prevent L-cystine import.

One of the key enzymes in the L-serine biosynthesis pathway, 3-phosphoglycerate dehydrogenase (PGDH), coded by the gene *serA*, is feedback inhibited by L-serine [7,13]. The authors of [39] converted the 344th and 346th amino acids of *E. coli*'s PGDH to alanine, which created a feedback-insensitive *serA*. Liu et al. [13] overexpressed the feedback-resistant *serA* as well as wild-type variants of the two subsequent enzymes of the pathway, phosphoserine transferase (*serB*) and phosphoserine phosphate-lyase (*serC*), to produce L-serine. The *E. coli* strain accumulated 11.7 g L^{-1} of L-serine after 48 h in shake flask experiments with glucose as the carbon source, which was the highest concentration so far.

From these results, it was concluded that the overexpression of *serA* along with *serC* and *serB* in the L-serine synthesis pathway was effective in increasing the production of L-cysteine, which resulted in L-cysteine concentrations in shake flask experiments of about 492.0 mg L^{-1}, with a yield of about 47 mg of L-cysteine g^{-1} glucose in 48 h [13].

2.2. Weakening the Degradation of L-Cysteine

It is well known that disrupting the degradation pathway and enhancing metabolic fluxes for the desired amino acids are effective approaches to improve their production [19,39–41]. The degradation of L-cysteine to pyruvate, ammonia, and sulfide is mainly catalyzed by L-cysteine desulfhydrase (CD). Wada et al. [42] purified and characterized CD from *C. glutamicum*. After comparing the partial amino acid sequence, it was identified that the enzyme CD is a C-S lyase with α,β-elimination activity, and it is encoded by the *aecD* gene [43]. They showed that the *aecD* gene product is involved in L-cysteine degradation, and the disruption of the *aecD* gene resulted in an increased production of L-cysteine (approximately 290 mg L^{-1} in shake flasks).

Awano et al. [44] investigated the enzyme having CD activity in *E. coli* using native-PAGE and CD activity staining. An analysis with gene-disrupted mutants showed that the tryptophanase (TNase) encoded by *tnaA* and cystathionine β-lyase (CBL) encoded by *metC* catalyze the degradation of L-cysteine in *E. coli*. By using a plasmid gene library, Awano et al. [45] identified three additional proteins as O-acetylserine sulfhydrylase-A (OASS-A), O-acetylserine sulfhydrylase-B (OASS-B), and the bifunctional ß-cystathionase (MalY), all having CD activity in *E. coli*. The disruption of these CD genes in *E. coli* resulted in the increased production of L-cysteine and L-cystine after 72 h by factors of 1.8 to 2.3 in shake flasks. The five above-mentioned proteins are pyridoxal 5′-phosphate (PLP)-dependent enzymes, and cysteine desulfhydrase activity might just be a side reaction.

The transcriptional activator CysB, belonging to the LysR family, induces the transcription of genes in the sulfate pathway [46]. Kawano et al. [46] found a most plausible sequence for the CysB-binding motif upstream of the *yciW* gene. YciW encodes a soluble protein of 401 amino acids, although its function and structural information are still unknown. The CysB-binding motif is located at the −35 region of the predicted promotor, which is similar to that of *cysJ* coding for a sulfite reductase. The *yciW*-disrupted strain exhibited a much higher sensitivity to L-cysteine than the native strain. This indicates that *yciW* confers tolerance to cysteine in *E. coli* cells, and the authors suggest that *yciW* may convert L-cysteine to L-methionine or glutathione. Based on the deduced amino acid sequence, *yciW* is predicted to encode an oxidoreductase-like protein. YciW is also involved in the degradation of intracellular L-cysteine, leading to its detoxification. However, further investigation is needed at the protein level [46].

L-cysteine desulfidases are alternative L-cysteine-degrading enzymes that are less well characterized as compared to L-cysteine desulfhydrases. The activity of this enzyme was first observed in *Methanocaldococcus jannaschii* [47]. Nonaka and Takumi [48] identified the *yhaM* gene, which encodes cysteine desulfidase and decomposes L-cysteine into hydrogen sulfide, pyruvate, and ammonium in *E. coli*. They reported that *yhaM* is L-cysteine inducible; therefore, it could serve as the main target for metabolic engineering to enhance L-cysteine production. Loddeke et al. [49] reported that L-cysteine desulfidase (CyuA) is the major enzyme that degrades L-cysteine during anaerobic growth in both *E. coli* and *Salmonella enterica*. In their research, Δ*cyuA* mutants of both species grown with L-cysteine anaerobically presented a reduced growth capacity and a 57% reduction in their sulfur production. The authors postulated that the presence of *cyuA* is a vestige predating the separation of bacteria and archaea and has three possible functions in bacteria: the use of L-cysteine as a carbon source, the detoxification of L-cysteine at growth-inhibiting concentrations, and the modulation of intracellular L-cysteine levels at nontoxic concentrations.

Takumi et al. [6] identified a major L-cysteine desulfhydrase gene (*ccdA*) involved in L-cysteine degradation in *Pantoea ananatis*. In their study, the *E. coli* wild-type strain MG1655 was transformed with a multicopy genomic library prepared from *P. ananatis* genomic DNA. From the investigated genes, the genes *ccdA*, which presents a cysteine degradation activity, and *cefA* and *cefB*, which are putative efflux pumps, were found to have a significant impact on the L-cysteine production from glucose and thiosulfate in experiments performed with *E. coli* in agitated test tubes. The overexpression of *ccdA* in *P. ananatis* led to a decrease of 58% in the maximal achieved L-cysteine concentration in test tubes. Conversely, the deletion of *ccdA* resulted in an increase in the achieved L-cysteine concentration in test tubes from 202 mg L^{-1} with the wild type to 516 mg L^{-1}.

For the accumulation of L-cysteine in *C. glutamicum*, the deletion of the enzyme CD, encoded by *aecD*, and the overexpression of the native CysE, encoded by *cysE*, were performed [7]. In contrast to the wild-type strain that produced 7.1 ± 1.5 mg L^{-1} of L-cysteine on a shake flask scale, the CD-deleted *C. glutamicum* strain accumulated 10.6 ± 2.3 mg L^{-1} of L-cysteine, thus showing no significant improvement. After the overexpression of a *CysE* mutant from *E. coli*, coding for a feedback-insensitive serine acetyltransferase in the CD deletion strain, an L-cysteine concentration of 346.1 ± 33.1 mg L^{-1} was measured. This result indicates that the combination of the overexpression of a feedback-insensitive CysE and deletion of CD is necessary to enhance L-cysteine production with *C. glutamicum* [7,50].

2.3. *Regulation of L-Cysteine Transport*

The bacterial production of metabolites depends on the equilibrium between secretion and reconsumption. High concentrations of L-cysteine have been reported to be toxic and inhibitory for microbial cells [3]. In *E. coli*, concentrations of 1 mM L-cysteine were enough to cause growth limitations [46]. Therefore, microorganisms have an exporter for L-cysteine within the L-cysteine regulon in order to avoid product-associated cytotoxicity.

Daßler et al. [51] identified the transporter protein YdeD in *E. coli*, which is reported to be one of the main facilitators involved in the export of L-cysteine. The corresponding gene encoding for the transporter protein involves an open reading frame coding for 299 amino acids. The substrate range of the YdeD protein is wide; therefore, it secretes other metabolites along with L-cysteine. In addition, YdeD acts as a regulator by exporting O-acetylserine (OAS), therefore lowering the intracellular NAS concentration. This stops the L-cysteine regulon activation [52]. The combination of YdeD with the strong promoter from the gene cassette coding for glyceraldehyde-phosphate dehydrogenase GAPDH in a plasmid generated a higher yield of L-cysteine with *E. coli*, proving the transporter's vital role in the accumulation of the metabolite [13].

CydDC is an ATP-binding transporter reported to be involved in L-cysteine export from the cytoplasm to the periplasm in *E. coli* [53]. In a study, CydDC was compared to the gene product of *ydeD* and was found to more effectively remove high L-cysteine concentrations in intracellular compartments than *ydeD*. In shake flask experiments at 37 °C, *E. coli* mutants with deletions of *cydDC* accumulated 1.6 times higher L-cysteine concentrations than the wild type, while mutants with *ydeD* deletions accumulated 1.3 times higher L-cysteine concentrations. A double-deletion *E. coli* mutant of both *cydDC* and *ydeD* accumulated 2.2 times more L-cysteine than the wild type. [53]

yfiK was discovered to be a gene that augmented L-cysteine production when it was overexpressed in an *E. coli* production strain [52]. The gene product is an integral membrane protein with about six predicted transmembrane helices and belongs to the resistance to homoserine/threonine (RhtB) family of export proteins. YfiK overproduction from a plasmid leads to the drastic and parallel secretion of O-acetyl serine and L-cysteine into the medium, but only when the *E. coli* strain possesses an L-serine transacetylase that is feedback insensitive to L-cysteine. Externally provided, OAS obviated this requirement for L-cysteine secretion, both in the *yfiK*-carrying transformant and in the wild type. A Δ*yfiK* mutant did not show any phenotype, and this mutant exported OAS and L-cysteine when transformed with a plasmid carrying *ydeD*, a previously characterized, alternate OAS/L-cysteine exporter. Since a *ydeD–yfiK* double mutant showed the same pattern, it appears that YfiK and YdeD act independently. The necessity for the cell to regulate the size of the internal pool of OAS via the synthesis of exporter proteins could be connected to the fact that this compound inhibits growth when supplied externally. The overexpression of either *ydeD* or *yfiK* leads to alleviation of this inhibition parallel by an increased resistance to azaserine, which is an analog of OAS [52].

In another study, different drug transporter genes were tested, and *Bcr* (a multidrug transporter) was found to be involved in L-cysteine export with an increased selectivity compared to other amino acids [54].

Besides these cytoplasmic membrane transporters, a novel gene called *tolC*, which encodes the outer membrane channel TolC, was reported to be involved in L-cysteine export from the periplasm in the medium [55].

The genes *cefA* and *cefB* were identified as cysteine efflux pumps. In a first experiment, the overexpression of *cefA* and *cefB* in *P. ananatis* resulted in L-cysteine resistance when 200 μM of L-cysteine was added to the growth medium. When the *P. ananatis* strains overexpressing either *cefA* or *cefB* were grown in aerated test tubes in the presence of glucose and thiosulfate, the maximum final L-cysteine concentrations were increased by 0.2 g L^{-1} and 1.9 g L^{-1}, to maximal concentrations of 0.4 g L^{-1} and 2.3 g L^{-1}, respectively, compared to *P. ananatis*, without the overexpression of these efflux pump genes [6].

The genes NCgl0580 and NCgl2566 were identified as potential L-cysteine exporters in *C. glutamicum*, and it was found that their overexpression in a *C. glutamicum* strain increased the L-cysteine formation from glucose and thiosulfate. This strain achieved a maximal L-cysteine concentration of 280 mg L^{-1} in shake flasks, which is a 1.4-fold improvement compared with the original *C. glutamicum* strain [56].

Wei et al. [7] evaluated the effectivity of three heterologous proteins responsible for L-cysteine transport in *C. glutamicum*. The transporters YdeD and Bcr from *E. coli* and CefA from *P. ananatis* were individually overexpressed. While all three strains overexpressing L-cysteine transporters have shown improvements in L-cysteine production, the best results were observed with Bcr, followed by YdeD and CefA. The strain with Bcr achieved an 88.9% increase in the L-cysteine concentration in shake flasks (634.4 mg L^{-1}) compared to the strain without additional transporters. The results indicated that the overexpression of these proteins enhanced the secretion of L-cysteine, preventing its accumulation in the cytoplasm of *C. glutamicum* cells and, therefore, improving cell growth. The levels of L-cysteine obtained from the engineered strains were still approximately five times lower in comparison with the results of [54], whose authors performed the same alteration on the *Bcr* and *YdeD* genes in *E. coli* cells. This difference is believed to be caused by the different cell membrane structures of the two bacteria, with *C. glutamicum* being gram positive and *E. coli* gram negative [7].

Among L-cysteine transport proteins, there are few that are solely responsible for the import of L-cysteine back into the cytoplasm. This is disadvantageous for the industrial biosynthesis of L-cysteine, since it doesn't allow for the accumulation of the product and causes backflow. In a recent study, the *yeaN* gene product, a known 2-nitroimidazole exporter, was found to take part in L-cysteine import in *E. coli* strains [57]. The deletion of the *yeaN* gene in an L-cysteine producer strain increased the overall L-cysteine production by 50% in shake flasks.

A recent study by Du et al. [58] has shown how to regulate the L-cysteine transport system in *C. glutamicum* effectively to establish intracellular homeostasis, avoiding toxicity. This was achieved by the deletion of *NCgl2463*, an L-cysteine importer gene [37], the upregulation of the export gene *eamA*, and the enhancement of the NADPH pool through the overexpression of glucose-6-phosphate dehydrogenase. These genetic modifications resulted in a producer strain that, in a 5 L stirred-tank reactor, was able to accumulate 5.92 g L^{-1} of L-cysteine with a sulfur conversion rate of nearly 75%.

Table 1. Summary of metabolic engineering approaches increasing the fermentative L-cysteine production. The studies presented in this table are ordered by the kind of approach each investigation followed, as well as the date of the publication. Only the highest L-cysteine concentration of each publication is listed, together with the molecular biology step applied in order to reach that specific concentration.

Source	Organism	Method	Scale	L-Cysteine Concentration
Enhancing biosynthesis of L-cysteine				
[28]	*E. coli* JM39-8	Feedback inhibition-insensitive SAT from site-directed mutagenesis	Shake flasks, 72 h	0.79 g L^{-1}
[30]	*E. coli* JM39-8	Insensitive SAT from *Arabidopsis thaliana*	Shake flasks, 96 h	1.66 g L^{-1}
[31]	*E. coli* MW1	Insensitive SAT from *Nicotiana tabacum*	Shake flasks, 72 h	0.30 g L^{-1}
[6]	*P. ananatis*	Insensitive SAT with L-cysteine efflux pump	Test tubes, 24 h	0.90 g L^{-1}
[36]	*C. glutamicum*	Overexpression of native SAT, OAS, and transcription regulator	Shake flasks, 15 h	0.06 g L^{-1}
[7]	*C. glutamicum*	Insensitive SAT, deletion of CD, and increased transport	Shake flasks, 48 h	0.95 g L^{-1}
[37]	*C. glutamicum*	Overexpression of SAT and PGDH with disruption of CD and L-cysteine import	Shake flasks, 12 h	0.20 g L^{-1}
[13]	*E. coli*	Overexpression of PGDH, PSERT, and PSL	Shake flasks, 48 h	0.49 g L^{-1}
Weakening degradation of L-cysteine				
[41]	*C. glutamicum*	Disruption of *aecD* gene (CD)	Shake flasks, 72 h	0.29 g L^{-1}
[44]	*E. coli*	Disruption of TNase, CBL, OASS-A, OASS-B, and MalY	Shake flasks, 72 h	1.36 g L^{-1} (per OD_{562})
[46]	*E. coli*	Disruption of the *yciW* gene	Shake flasks, 72 h	0.31 g L^{-1} (per OD_{562})
[6]	*P. ananatis*	Deletion of CD *ccdA*	Test tubes, 28 h	0.52 g L^{-1}
[7]	*C. glutamicum*	Disruption of *aecD* (CD) and overexpression of insensitive SAT	Shake flasks, 48 h	0.35 g L^{-1}
Enhancing efflux of L-cysteine				
[51]	*E. coli*	Overexpression of exporter YdeD	Shake flasks, 48 h	0.07 g L^{-1}
[52]	*E. coli*	Overexpression of exporter YfiK	Shake flasks, 20 h	0.15 g L^{-1}
[54]	*E. coli*	Overexpression of exporter bcr and deletion of TNase	Shake flasks, 48 h	0.50 g L^{-1}
[55]	*E. coli*	Overexpression of exporters YdeD and TolC	Shake flasks, 24 h	0.12 g L^{-1}
[6]	*P. ananatis*	Overexpression of transporter *cefB*	Test tubes, 28 h	2.30 g L^{-1}
[56]	*C. glutamicum*	Overexpression of transport genes NCgl0580 and NCgl2566	Shake flasks, 20 h	0.28 g L^{-1}
[7]	*C. glutamicum*	Overexpression of exporter bcr	Shake flasks, 48 h	0.63 g L^{-1}
[57]	*E. coli*	Deletion of importer *yeaN*	Shake flasks, 20 h	1.20 g L^{-1}
[58]	*C. glutamicum*	Deletion of importer *NCgl2463*, overexpression of exporter *eamA* and G6PDH	5 L stirred tank, 72 h	5.92 g L^{-1}

OD_{562} is the optical density of the fermentation broth measured at 562 nm.

2.4. Limitations of the Metabolic Engineering Approaches

Liu et al. [13] developed an integrated approach, which used many of the previously presented metabolic engineering strategies in tandem. They overexpressed PSERT, PSL, and a feedback-insensitive version of PGDH, disrupted genes for the degradation of L-serine and L-cysteine, and overexpressed the L-cysteine exporter YdeD. Even though this strategy showed a cumulative improvement in the L-cysteine concentration, the final yield of L-cysteine on glucose was about 4.7% (w/w). This means that most of the fed substrate was not directly utilized for the L-cysteine production. Nonetheless, the authors indicate that this is the highest yield on glucose found in the literature at the time of its publication.

This exemplifies one of the major challenges of the metabolic engineering approaches. Most of the methods described above are applied directly on the L-cysteine synthesis pathway, which has been extensively characterized so that the bottlenecks are extensively identified. However, most metabolic pathways are thoroughly interconnected with regulation interactions that have not been yet characterized [59]. These interactions must then first be mapped out in order to identify further bottlenecks that can be tackled through rational metabolic engineering methods [59].

Furthermore, this interconnectivity of metabolic pathways may also result in unintended consequences from seemingly straightforward molecular biology interventions, like the lack of growth observed in *Pseudomonas* sp. by Huai et al. [18] when an enzyme responsible for L-cysteine degradation was deleted, or the increase in the exogenous L-cysteine concentration leading to lower levels of c-type cytochromes and increased penicillin resistance [53].

3. Process Engineering Aspects

Metabolic engineering of the microorganisms used for L-cysteine biosynthesis has been applied to enable L-cysteine production by fermentation. However, the economic L-cysteine production by fermentation must also take into account the process engineering viewpoint. Emphasizing energy-efficient sulfur pathways [7,60], selecting economically viable carbon sources [61], designing scalable fed-batch processes with individual feedings of carbon and sulfur sources [59], and implementing efficient purification techniques [62] are important for the industrial-scale production of L-cysteine.

3.1. Sulfur Assimilation Pathways in Bacteria

E. coli has two sulfur assimilation pathways for L-cysteine biosynthesis (Figure 2). One is the sulfate (SO_4^{2-}) pathway, which consumes two molecules of ATP and four molecules of NADPH as a reducing power to produce L-cysteine from sulfate. The other is the thiosulfate ($S_2O_3^{2-}$) pathway, which spends only one molecule of NADPH from thiosulfate [63]. Thiosulfate is receiving attention as a sulfur source due to its advantageous effects on bacterial growth and L-cysteine production in comparison to sulfate [64].

Both sulfur assimilation pathways utilize OAS as a carbon skeleton to which they may incorporate sulfur. In the thiosulfate pathway, O-acetyl-L-serine sulfhydrylase B (CysM) catalyzes the conversion of OAS and thiosulfate into S-sulfocysteine, which is reduced into L-cysteine and sulfite by glutaredoxin (Grx1), and the glutaredoxin-like protein NrdH [63,65]. In comparison, sulfate is first reduced to sulfide in a four-step reaction catalyzed in subsequent order by sulfate adenylyltransferase (CysDN), adenylylsulfate kinase (CysC), 3′-phospho-adenylylsulfate reductase (CysH), and sulfite reductase (CysIJ). The subsequent incorporation of sulfide into OAS is enabled by O-acetyl-L-serine sulfhydrylase A (CysK) [60].

There is strong evidence suggesting that *E. coli* may possess an alternative pathway for assimilating thiosulfate, independent of the conventional pathway through the CysM enzyme [64]. This evidence comes from an experiment where ΔCysM *E. coli* cells were grown using thiosulfate as the only sulfur source. The ΔCysM cells showed a slower growth compared to the wild-type cells. However, when L-cystine was added to the growth media, the impaired growth was restored, indicating a deficiency in L-cysteine. The growth rate in the growth media without L-cystine was then increased by overexpressing thiosulfate sulfurtransferase (GlpE). Based on these findings, researchers postulated that the CysM-independent pathway involves the direct breakdown of thiosulfate into sulfite by GlpE, followed by assimilation through the sulfate pathway [64]. The proposed sulfur assimilation pathway, as summarized in Table 2, is more energy efficient than the sulfate pathway, but less efficient than the traditional thiosulfate-based sulfur assimilation pathway. The energy efficiency of the sulfur assimilation pathway influences the fermentative production of L-cysteine significantly. This has been shown as increasing the levels of GlpE and CysK, which are responsible for the CysM-independent pathway, enhances L-cysteine overproduction only in wild-type *E. coli* strains, while the same modifications worsen L-cysteine overproduction in strains with elevated expressions of CysM and nrdH, responsible for the traditional, energy-efficient thiosulfate γ pathway [66].

Table 2. ATP and NADPH demand and catalyzing enzymes for different sulfur assimilation pathways [59,63].

Pathway	Steps	Enzymes	ATP/NADPH Demand
Thiosulfate	1. Conversion of thiosulfate to S-sulfocysteine	CysM	-
	2. Lyase of S-sulfocysteine and sulfite	NrdH, Grx1	1 NADPH
Thiosulfate (postulated)	1. Conversion of thiosulfate to sulfite ($SO_3{}^{-2}$)	GlpE	-
	2. Reduction of sulfite to sulfide (S^{-2})	CysIJ, CysG	3 NADPH
	3. Reaction with O-acetylserine to L-cysteine	CysK	-
Sulfate	1. Activation to adenosine 5′-phosphosulfate	CysDN	1 ATP
	2. Phosphorylation to 3′-phosphoadenosine 5′-phosphosulfate	CysC	1 ATP
	3. Conversion to sulfite	CysH	1 NADPH
	4. Reduction of sulfite to sulfide	CysIJ, CysG	3 NADPH
	5. Reaction with O-acetylserine to L-cysteine	CysK	-

3.2. Glycerol as Carbon Source

Although glucose remains the predominant carbon source for L-cysteine biosynthesis, glycerol has gained increasing attention as an alternative carbon source due to its abundant availability as a by-product in biodiesel production [66]. Utilizing glycerol as a carbon source is crucial for reducing the costs of bioproduction and improving resource utilization. Moreover, the metabolic pathway from glycerol to L-cysteine is shorter and exhibits better carbon economy compared to glucose [67].

Significant progress has been made towards the bioproduction of L-cysteine from glycerol, as its essential precursor, L-serine, has already been produced effectively using glycerol [67]. The highest recorded concentration of L-cysteine from glycerol achieved so far is 313.4 mg L^{-1} with *E. coli* in shake flasks [67]. First, L-serine production was enhanced by making use of an L-serine biosensor based on the transcriptional regulator NCgl0581 for coupling L-serine biosynthesis to the growth rate of *E. coli* and using adaptive laboratory evolution (ALE). Following this, the L-cysteine degradation pathway (encoded by *tnaA*) was eliminated, while serine acetyltransferase (encoded by *cysE*) and the L-cysteine exporter (encoded by *ydeD*) were overexpressed [67].

3.3. Fed-Batch Processes in Stirred-Tank Bioreactors

Up to this point, most research presented had been carried out as simple batch processes in shake flasks, which allow for easier preparation and higher throughput than the scalable stirred-tank bioreactor. Investigations in fed-batch operated and fully controlled stirred-tank bioreactors using the observations provided by the studies in shake flasks and optimizing the reaction conditions may lead to an L-cysteine production scenario that is more in line with the expectations of the industry.

Liu et al. [13] utilized a rational approach which combined previously discussed metabolic engineering approaches to enhance the L-cysteine production of the *E. coli* strain JM109, including the overexpression of the enzymes in the L-serine biosynthetic pathway, feedback-insensitive phosphoglycerate dehydrogenase and serine acetyltransferase, the ydeD exporter, and the deletion of L-cysteine and serine deaminases. These modifications lead to a total L-cysteine production in shake flasks of 600 mg L^{-1}. They then cultivated this modified strain in a 5 L stirred-tank reactor in a fed-batch operation, where a glucose solution was fed constantly into the reactor, but thiosulfate (the sulfur source) was only supplied at the beginning of the process. This led to an increase in the maximal achieved L-cysteine concentration to 5.1 g L^{-1}.

Caballero et al. [59] transferred the shake flask experiments of Daßler et al. [51] with *E. coli* W3110, carrying L-cysteine production plasmids to a stirred-tank bioreactor on a 15 L scale using continuous but independent glucose and ammonium thiosulfate feedings (presented in Figure 3a), thereby increasing the maximal achieved L-cysteine concentration from 72 mg L^{-1} in shake flasks to 16 g L^{-1} in the fed-batch process. An in vivo metabolic control analysis (MCA) was applied after rapid media transition [68] and short-term perturbation experiments [69,70] for the identification of bottlenecks in the L-cysteine synthesis pathway of producer cells withdrawn from this fed-batch process in the production phase. Synthetic biology efforts to increase the expression of OASS-A based on the results of the in vivo MCA led to an increase in the L-cysteine product concentration by 46% in the dual-feeding, fed-batch process, reaching maximal L-cysteine concentrations of 22 g L^{-1}, which are the highest reported in the literature so far [59]. The L-cysteine concentration profiles of this process before and after the strain optimization are presented in Figure 3b.

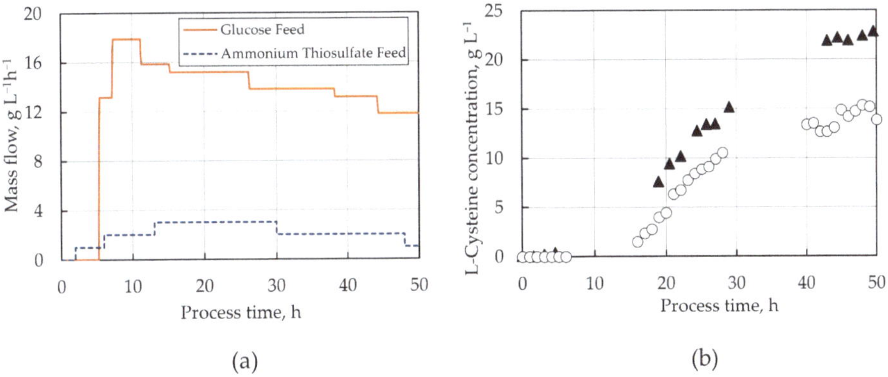

Figure 3. (**a**) Substrate feeding profiles followed by [59] in the fed-batch L-cysteine production process. The thiosulfate feeding started two hours after the process was inoculated in order to avoid unnecessary oxidative stress on the cells in the initial lag phase. The glucose feeding started after five hours, once the initially added glucose was completely consumed. (**b**) L-cysteine concentration profiles from the fermentation processes carried out with the *E. coli* strain W3110, with the L-cysteine production plasmid and without the overexpression of OASS-A (circles), and the same strain and plasmid but additional overexpression of OASS-A (triangles).

3.4. Purification of L-Cysteine from the Fermentation Broth

L-cysteine, due to its high solubility in water and the vulnerability of its thiol group to oxidation, presents challenges in its purification for industrial applications. Several approaches have been proposed to address this issue [62].

Most approaches follow a two-stage process, which involves converting L-cysteine into L-cystine by treating the clarified solution with an oxidizing agent. The reduced solubility and increased chemical stability of L-cystine allow for its purification through crystallization. The resulting crystals are then dissolved in water and reduced back to L-cysteine using electrolysis [71].

More recently, Wacker AG (Munich, Germany) has developed a method for the direct purification of L-cysteine from a clarified fermentation broth [62]. This approach selectively isolates L-cysteine while eliminating its derivatives, emphasizing the importance of preventing the oxidation of L-cysteine to L-cystine. The oxidation of L-cysteine's sulfhydryl (SH) groups by oxygen, the primary oxidizing agent present in fermentation broths, is particularly favored under alkaline conditions. Additionally, thiosulfate, which serves as the preferred sulfur source for the fermentative production of L-cysteine, oxidizes L-cysteine at a pH of 5 and below. It is recommended to maintain a pH of 5–7 during fermentation to minimize any loss in yield due to the formation of L-cystine as a by-product [62].

The first step in the direct purification process (Figure 4) involves ion exchange chromatography. In this step, L-cysteine, as well as L-cystine and other cations present in the clarified fermentation broth, are adsorbed onto an ion exchanger containing sulfonic and phosphonic acid groups in their protonated (H^+) form. The absorbed compounds are subsequently eluted using ammonia solution (1–2 molar NH_4^+) as the eluant. The use of a gradually increasing ammonia concentration and fractioned separation of the product solution further improves the purity of L-cysteine. It is important to note that although the L-cysteine-containing solution is highly acidic, the chromatography process effectively removes all oxidizing agents, mitigating concerns about the oxidation of L-cysteine [62]. Subsequently, to further purify L-cysteine from the solution, fractioned crystallization is performed as described previously [72]. This method involves the addition of hydrochloric acid at 20 °C to facilitate the crystallization of remaining impurities. The resulting deposits are then removed through filtration, and the filtrate is cooled to −20 °C to induce the crystallization of L-cysteine.

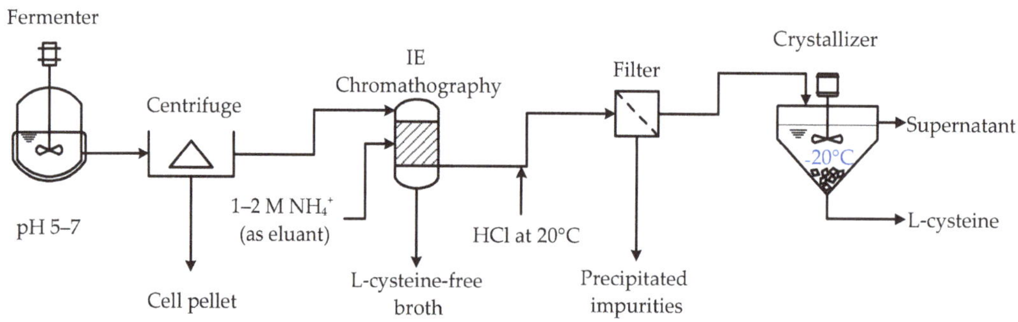

Figure 4. Flow diagram representation of an L-cysteine purification process according to Boehm [62]. This process relies on the L-cysteine remaining in an unoxidized state. Hence, the oxidants in the fermentation broth are removed during the anion-exchange chromatography, and the pH of the L-cysteine-rich solution must be closely monitored and adjusted throughout the purification line.

4. Conclusions

In general, the most widely researched microorganism for L-cysteine bioproduction is *E. coli*. Nonetheless, other production hosts like *C. glutamicum* and *P. ananatis* have been the focus of strain optimization, with promising improvements in L-cysteine production.

The combination of three metabolic engineering strategies has been successful, irrespective of the production host: The first one is enhancing the L-cysteine biosynthesis by overexpressing the genes encoding for feedback-insensitive L-serine O-acetyltransferase (SAT) and O-acetyl-L-serine sulfhydrylase. In *E. coli*, the feedback inhibition of SAT was bypassed through the introduction of an SAT from *Arabidopsis thaliana*. *C. glutamicum* showed a high L-cysteine production with the homologous overexpression of the *cysE* and *serA* genes.

The second metabolic engineering strategy is to reduce L-cysteine degradation by the microorganisms. Therefore, the L-cysteine desulfhydrase gene *aecD* was deleted in *C. glutamicum*, which increased the production of L-cysteine. In *E. coli*, the gene products of *tnaA* and *metC* were identified to catalyze the degradation of L-cysteine, and after the deletion of these genes, the L-cysteine production was improved.

The third metabolic engineering strategy is to enhance the export of L-cysteine and weaken the import. As L-cysteine is cytotoxic, it has to be transported out of the cell. In *E. coli*, two membrane proteins called YdeD and YfiK are responsible for the export of L-cysteine. But by overexpression of *ydeD*, OAS is simultaneously secreted, which causes the L-cysteine regulon activation to stop. In contrast, the gene product of *yfiK* enables L-cysteine export without depleting the intracellular OAS concentrations. In addition, the gene product of *yeaN* is a recently discovered L-cysteine uptake protein found in *E. coli*, which may be a promising target to avoid the reconsumption of L-cysteine by the cells.

Taking into account the process engineering aspects beyond metabolic engineering of the L-cysteine biosynthesis pathway enables economic L-cysteine production by fermentation. Emphasizing energy-efficient sulfur pathways, selecting economically viable carbon sources, designing scalable fed-batch processes with individual feedings of carbon and sulfur sources, and implementing efficient purification techniques are important for the production of L-cysteine on an industrial scale. The utilization of thiosulfate instead of sulfates requires fewer reduction equivalents, and thus, enhances L-cysteine production rates. Even though the highest L-cysteine concentrations are still obtained with glucose as a carbon source, microbial L-cysteine production with glycerol has been shown. Glycerol is abundant as a by-product of the biodiesel industry, and thus, its low cost might improve the industrial-scale L-cysteine production in the future.

The different ways to enhance the bioproduction of L-cysteine show a high potential for success. A holistic approach which combines the presented methodologies may uncover synergies that may result in higher L-cysteine yields and increased productivity in the future.

Author Contributions: D.A.C.C., R.D., T.E., S.H., L.G., A.M. and D.W.-B. contributed to the conception and design of the manuscript; T.E. redacted the introduction and conclusions; S.H. compiled the information on microbial L-cysteine producers; A.M. performed the literature research on metabolic engineering strategies; L.G. carried out the process engineering literature research. D.A.C.C. coordinated the literature research and evaluation; A.M. and D.A.C.C. designed the figures. D.A.C.C., R.D., T.E., S.H., L.G., A.M. and D.W.-B. wrote the manuscript. All the authors contributed to the critical revision and final approval of the manuscript. All authors have read and agreed to the published version of the manuscript.

Funding: This research received no external funding.

Institutional Review Board Statement: Not applicable.

Informed Consent Statement: Not applicable.

Data Availability Statement: The databases utilized in this review were Scopus and Web of Science with the keywords "cysteine biosynthesis", "cysteine production", and "cysteine fermentation" as search terms. The search timespan was originally set to publications between 2012 and 2023 but was extended further to cover additional metabolic approaches.

Acknowledgments: The authors would like to thank Rose Jacobs and Natasha Wellershausen (both from TUM Language Center), Anna-Lena Heins (Biochemical Engineering, TUM), Franziska Glasl (Center for Key Competences, TUM), and Kristin Preuß (Library, TUM) for their support in academic writing, collaborative scientific editing, virtual teamwork and digital collaboration, and literature research, respectively. The support of Daniel Caballero of the TUM Graduate School is acknowledged as well.

Conflicts of Interest: The authors declare no conflicts of interest.

References

1. Paul, B.D.; Sbodio, J.I.; Snyder, S.H. Cysteine metabolism in neuronal redox homeostasis. *Trends Pharmacol. Sci.* **2018**, *39*, 513–524. [CrossRef] [PubMed]
2. Kredich, N.M. Regulation of l-Cysteine Biosynthesis in *Salmonella typhimurium*: I. Effects of growth on varying sulfur sources and O-acetyl-L-serine on gene expression. *J. Biol. Chem.* **1971**, *246*, 3474–3484. [CrossRef] [PubMed]
3. Sørensen, M.A.; Pedersen, S. Cysteine, even in low concentrations, induces transient amino acid starvation in *Escherichia coli*. *J. Bacteriol.* **1991**, *173*, 5244–5246. [CrossRef]
4. Takumi, K.; Nonaka, G. Bacterial Cysteine-Inducible Cysteine Resistance Systems. *J. Bacteriol.* **2016**, *198*, 1384–1392. [CrossRef]
5. Yin, J.; Ren, W.; Yang, G.; Duan, J.; Huang, X.; Fang, R.; Li, C.; Li, T.; Yin, Y.; Hou, Y.; et al. L-Cysteine metabolism and its nutritional implications. *Mol. Nutr. Food Res.* **2016**, *60*, 134–146. [CrossRef] [PubMed]
6. Takumi, K.; Ziyatdinov, M.K.; Samsonov, V.; Nonaka, G. Fermentative production of cysteine by *Pantoea ananatis*. *Appl. Environ. Microbiol.* **2017**, *83*, e02502-16. [CrossRef]
7. Wei, L.; Wang, H.; Xu, N.; Zhou, W.; Ju, J.; Liu, J.; Ma, Y. Metabolic engineering of *Corynebacterium glutamicum* for l-cysteine production. *Appl. Microbiol. Biotechnol.* **2019**, *103*, 1325–1338. [CrossRef]
8. Wendisch, V.F. Metabolic engineering advances and prospects for amino acid production. *Metab. Eng.* **2020**, *58*, 17–34. [CrossRef]
9. Clemente Plaza, N.; Reig García-Galbis, M.; Martínez-Espinosa, R.M. Effects of the Usage of l-Cysteine (l-Cys) on Human Health. *Molecules* **2018**, *23*, 3. [CrossRef]
10. Ismail, N.I.; Yumi Zuhanis Has-Yun, H.; Parveen, J. Production of Cysteine: Approaches, Challenges and Potential Solution. *Int. J. Biotechnol. Wellness Ind.* **2014**, *3*, 95–101.
11. Bichon, E.; Sitthisack, P.; Sérée-Vattier, L.; Prévost, S.; Schiphorst, A.-M.; Pouponneau, K.; Monteau, F.; Le Bizec, B. Determination of l-cysteine origin on the basis of its δ15N values. *Food Chem.* **2018**, *260*, 283–288. [CrossRef]
12. Takagi, H.; Ohtsu, I. L-Cysteine Metabolism and Fermentation in Microorganisms. In *Amino Acid Fermentation*; Yokota, A., Ikeda, M., Eds.; Springer: Tokyo, Japan, 2017; pp. 129–151. [CrossRef]
13. Liu, H.; Fang, G.; Wu, H.; Li, Z.; Ye, Q. L-Cysteine Production in *Escherichia coli* Based on Rational Metabolic Engineering and Modular Strategy. *Biotechnol. J.* **2018**, *13*, 1700695. [CrossRef] [PubMed]
14. Sano, K.; Yokozeki, K.; Tamura, F.; Yasuda, N.; Noda, I.; Mitsugi, K. Microbial conversion of DL-2-amino-delta2-thiazoline-4-carboxylic acid to L-cysteine and L-cystine: Screening of microorganisms and identification of products. *Appl. Environ. Microbiol.* **1977**, *34*, 806–810. [CrossRef] [PubMed]
15. Sano, K.; Mitsugi, K. Enzymatic Production of l-Cysteine from dl-2-Amino-Δ2-thiazoline-4-carboxylic Acid by *Pseudomonas thiazolinophilum*: Optimal Conditions for the Enzyme Formation and Enzymatic Reaction. *Agric. Biol. Chem.* **1978**, *42*, 2315–2321. [CrossRef]
16. Tamura, Y.; Nishino, M.; Ohmachi, T.; Asada, Y. N-Carbamoyl-L-Cysteine as an Intermediate in the Bioconversion from D,L-2-Amino-Δ2-Thiazoline-4-Carboxylic Acid to L-Cysteine by *Pseudomonas* sp. ON-4a. *Biosci. Biotechnol. Biochem.* **1998**, *62*, 2226–2229. [CrossRef] [PubMed]
17. Shiba, T.; Takeda, K.; Yajima, M.; Tadano, M. Genes from *Pseudomonas* sp. Strain BS Involved in the Conversion of l-2-Amino-Δ2-Thiazolin-4-Carbonic Acid to l-Cysteine. *Appl. Environ. Microbiol.* **2002**, *68*, 2179–2187. [CrossRef]
18. Huai, L.; Chen, N.; Yang, W.; Bai, G. Metabolic control analysis of L-cysteine producing strain TS1138 of *Pseudomonas* sp. *Biochemistry* **2009**, *74*, 288–292. [CrossRef]
19. Wendisch, V.F. Microbial production of amino acids and derived chemicals: Synthetic biology approaches to strain development. *Curr. Opin. Biotechnol.* **2014**, *30*, 51–58. [CrossRef]
20. Pizer, L.I. The pathway and control of serine biosynthesis in *Escherichia coli*. *J. Biol. Chem.* **1963**, *238*, 3934–3944. [CrossRef]
21. Vermeij, P.; Kertesz, M.A. Pathways of assimilative sulfur metabolism in *Pseudomonas putida*. *J. Bacteriol.* **1999**, *181*, 5833–5837. [CrossRef]
22. Ono, B.-I.; Hazu, T.; Yoshida, S.; Kawato, T.; Shinoda, S.; Brzvwczy, J.; Paszewski, A. Cysteine biosynthesis in *Saccharomyces cerevisiae*: A new outlook on pathway and regulation. *Yeast* **1999**, *15*, 1365–1375. [CrossRef]
23. Sperandio, B.; Polard, P.; Ehrlich, D.S.; Renault, P.; Guédon, E. Sulfur amino acid metabolism and its control in *Lactococcus lactis* IL1403. *J. Bacteriol.* **2005**, *187*, 3762–3778. [CrossRef] [PubMed]
24. Byrne, C.R.; Monroe, R.S.; Ward, K.A.; Kredich, N.M. DNA sequences of the cysK regions of *Salmonella typhimurium* and *Escherichia coli* and linkage of the cysK regions to ptsH. *J. Bacteriol.* **1988**, *170*, 3150–3157. [CrossRef] [PubMed]

25. Phue, J.-N.; Shiloach, J. Transcription levels of key metabolic genes are the cause for different glucose utilization pathways in *E. coli* B (BL21) and *E. coli* K (JM109). *J. Biotechnol.* **2004**, *109*, 21–30. [CrossRef]
26. Tanaka, N.; Hatano, T.; Saito, S.; Wakabayashi, Y.; Abe, T.; Kawano, Y.; Ohtsu, I. Generation of hydrogen sulfide from sulfur assimilation in *Escherichia coli*. *J. Gen. Appl. Microbiol.* **2019**, *65*, 234–239. [CrossRef] [PubMed]
27. Liu, H.; Hou, Y.; Wang, Y.; Li, Z. Enhancement of Sulfur Conversion Rate in the Production of l-Cysteine by Engineered *Escherichia coli*. *J. Agric. Food Chem.* **2020**, *68*, 250–257. [CrossRef]
28. Liu, H.; Wang, Y.; Hou, Y.; Li, Z. Fitness of Chassis Cells and Metabolic Pathways for L -Cysteine Overproduction in *Escherichia coli*. *J. Agric. Food Chem.* **2020**, *68*, 14928–14937. [CrossRef]
29. Nakamori, S.; Kobayashi, S.I.; Kobayashi, C.; Takagi, H. Overproduction of L-cysteine and L-cystine by *Escherichia coli* strains with a genetically altered serine acetyltransferase. *Appl. Environ. Microbiol.* **1998**, *64*, 1607–1611. [CrossRef]
30. Takagi, H.; Kobayashi, C.; Kobayashi, S.; Nakamori, S. PCR random mutagenesis into *Escherichia coli* serine acetyltransferase: Isolation of the mutant enzymes that cause overproduction of L-cysteine and L-cystine due to the desensitization to feedback inhibition. *FEBS Lett.* **1999**, *452*, 323–327. [CrossRef]
31. Takagi, H.; Awano, N.; Kobayashi, S.; Noji, M.; Saito, K.; Nakamori, S. Overproduction of l-cysteine and l-cystine by expression of genes for feedback inhibition-insensitive serine acetyltransferase from *Arabidopsis thaliana* in *Escherichia coli*. *FEMS Microbiol. Lett.* **1999**, *179*, 453–459. [CrossRef]
32. Wirtz, M.; Hell, R. Production of cysteine for bacterial and plant biotechnology: Application of cysteine feedback-insensitive isoforms of serine acetyltransferase. *Amino Acids* **2003**, *24*, 195–203. [CrossRef] [PubMed]
33. Noji, M.; Inoue, K.; Kimura, N.; Gouda, A.; Saito, K. Isoform-dependent Differences in Feedback Regulation and Subcellular Localization of Serine Acetyltransferase Involved in Cysteine Biosynthesis from *Arabidopsis thaliana*. *J. Biol. Chem.* **1998**, *273*, 32739–32745. [CrossRef] [PubMed]
34. Kai, Y.; Kashiwagi, T.; Ishikawa, K.; Ziyatdinov, M.K.; Redkina, E.I.; Kiriukhin, M.Y.; Gusyatiner, M.M.; Kobayashi, S.; Takagi, H.; Suzuki, E. Engineering of *Escherichia coli* L-serine O-acetyltransferase on the basis of crystal structure: Desensitization to feedback inhibition by l-cysteine. *Protein Eng. Des. Sel.* **2006**, *19*, 163–167. [CrossRef] [PubMed]
35. Hirasawa, T.; Shimizu, H. Glutamic Acid Fermentation: Discovery of Glutamic Acid-Producing Microorganisms, Analysis of the Production Mechanism, Metabolic Engineering, and Industrial Production Process. In *Industrial Biotechnology*; John Wiley & Sons, Ltd.: Hoboken, NJ, USA, 2017; pp. 339–360. [CrossRef]
36. Kinoshita, S.; Udaka, S.; Shimono, M. Studies on the Amino Acid Fermentation. *J. Gen. Appl. Microbiol.* **1957**, *3*, 193–205. [CrossRef]
37. Joo, Y.-C.; Hyeon, J.E.; Han, S.O. Metabolic Design of *Corynebacterium glutamicum* for Production of l-Cysteine with Consideration of Sulfur-Supplemented Animal Feed. *J. Agric. Food Chem.* **2017**, *65*, 4698–4707. [CrossRef]
38. Kondoh, M.; Hirasawa, T. L-Cysteine production by metabolically engineered *Corynebacterium glutamicum*. *Appl. Microbiol. Biotechnol.* **2019**, *103*, 2609–2619. [CrossRef]
39. Mundhada, H.; Schneider, K.; Christensen, H.B.; Nielsen, A.T. Engineering of high yield production of L-serine in *Escherichia coli*. *Biotechnol. Bioeng.* **2016**, *113*, 807–816. [CrossRef]
40. Becker, J.; Wittmann, C. Bio-based production of chemicals, materials and fuels—*Corynebacterium glutamicum* as versatile cell factory. *Curr. Opin. Biotechnol.* **2012**, *23*, 631–640. [CrossRef]
41. Park, J.H.; Lee, S.Y. Towards systems metabolic engineering of microorganisms for amino acid production. *Curr. Opin. Biotechnol.* **2008**, *19*, 454–460. [CrossRef]
42. Wada, M.; Awano, N.; Haisa, K.; Takagi, H.; Nakamori, S. Purification, characterization and identification of cysteine desulfhydrase of *Corynebacterium glutamicum*, and its relationship to cysteine production. *FEMS Microbiol. Lett.* **2002**, *217*, 103–107. [CrossRef]
43. Rossol, I.; Pühler, A. The *Corynebacterium glutamicum* aecD gene encodes a C-S lyase with alpha, beta-elimination activity that degrades aminoethylcysteine. *J. Bacteriol.* **1992**, *174*, 2968–2977. [CrossRef] [PubMed]
44. Awano, N.; Wada, M.; Kohdoh, A.; Oikawa, T.; Takagi, H.; Nakamori, S. Effect of cysteine desulfhydrase gene disruption on l-cysteine overproduction in *Escherichia coli*. *Appl. Microbiol. Biotechnol.* **2003**, *62*, 239–243. [CrossRef] [PubMed]
45. Awano, N.; Wada, M.; Mori, H.; Nakamori, S.; Takagi, H. Identification and functional analysis of *Escherichia coli* cysteine desulfhydrases. *Appl. Environ. Microbiol.* **2005**, *71*, 4149–4152. [CrossRef]
46. Kawano, Y.; Ohtsu, I.; Takumi, K.; Tamakoshi, A.; Nonaka, G.; Funahashi, E.; Ihara, M.; Takagi, H. Enhancement of l-cysteine production by disruption of yciW in *Escherichia coli*. *J. Biosci. Bioeng.* **2015**, *119*, 176–179. [CrossRef] [PubMed]
47. Tchong, S.-I.; Xu, H.; White, R.H. l-Cysteine Desulfidase: An [4Fe-4S] Enzyme Isolated from *Methanocaldococcus jannaschii* That Catalyzes the Breakdown of l-Cysteine into Pyruvate, Ammonia, and Sulfide. *Biochemistry* **2005**, *44*, 1659–1670. [CrossRef]
48. Nonaka, G.; Takumi, K. Cysteine degradation gene yhaM, encoding cysteine desulfidase, serves as a genetic engineering target to improve cysteine production in *Escherichia coli*. *AMB Express* **2017**, *7*, 90. [CrossRef] [PubMed]
49. Loddeke, M.; Schneider, B.; Oguri, T.; Mehta, I.; Xuan, Z.; Reitzer, L. Anaerobic cysteine degradation and potential metabolic coordination in *Salmonella enterica* and *Escherichia coli*. *J. Bacteriol.* **2017**, *199*, e00117-17. [CrossRef]
50. Haitani, Y.; Awano, N.; Yamazaki, M.; Wada, M.; Nakamori, S.; Takagi, H. Functional analysis of l-serine O-acetyltransferase from *Corynebacterium glutamicum*. *FEMS Microbiol. Lett.* **2006**, *255*, 156–163. [CrossRef]
51. Daßler, T.; Maier, T.; Winterhalter, C.; Böck, A. Identification of a major facilitator protein from *Escherichia coli* involved in efflux of metabolites of the cysteine pathway. *Mol. Microbiol.* **2000**, *36*, 1101–1112. [CrossRef]

52. Franke, I.; Resch, A.; Dassler, T.; Maier, T.; Böck, A. YfiK from *Escherichia coli* promotes export of O-acetylserine and cysteine. *J. Bacteriol.* **2003**, *185*, 1161–1166. [CrossRef]
53. Pittman, M.S.; Corker, H.; Wu, G.; Binet, M.B.; Moir, A.J.G.; Poole, R.K. Cysteine Is Exported from the *Escherichia coli* Cytoplasm by CydDC, an ATP-binding Cassette-type Transporter Required for Cytochrome Assembly *. *J. Biol. Chem.* **2002**, *277*, 49841–49849. [CrossRef]
54. Yamada, S.; Awano, N.; Inubushi, K.; Maeda, E.; Nakamori, S.; Nishino, K.; Yamaguchi, A.; Takagi, H. Effect of Drug Transporter Genes on Cysteine Export and Overproduction in *Escherichia coli*. *Appl. Environ. Microbiol.* **2006**, *72*, 4735–4742. [CrossRef]
55. Wiriyathanawudhiwong, N.; Ohtsu, I.; Li, Z.-D.; Mori, H.; Takagi, H. The outer membrane TolC is involved in cysteine tolerance and overproduction in *Escherichia coli*. *Appl. Microbiol. Biotechnol.* **2009**, *81*, 903–913. [CrossRef]
56. Kishino, M.; Kondoh, M.; Hirasawa, T. Enhanced L-cysteine production by overexpressing potential L-cysteine exporter genes in an L-cysteine-producing recombinant strain of *Corynebacterium glutamicum*. *Biosci. Biotechnol. Biochem.* **2019**, *83*, 2390–2393. [CrossRef] [PubMed]
57. Yamazaki, S.; Ziyatdinov, M.K.; Nonaka, G. Fermentative production of sulfur-containing amino acid with engineering putative l-cystathionine and l-cysteine uptake systems in *Escherichia coli*. *J. Biosci. Bioeng.* **2020**, *130*, 14–19. [CrossRef] [PubMed]
58. Du, H.; Qiao, J.; Qi, Y.; Li, L.; Xu, N.; Shao, L.; Wei, L.; Liu, J. Reprogramming the sulfur recycling network to improve L-cysteine production in *Corynebacterium glutamicum*. *Green Chem.* **2023**, *25*, 3152–3165. [CrossRef]
59. Caballero Cerbon, D.A.; Widmann, J.; Weuster-Botz, D. Metabolic control analysis enabled the improvement of the L-cysteine production process with *Escherichia coli*. *Appl. Microbiol. Biotechnol.* **2024**, *108*, 1–13. [CrossRef]
60. Kawano, Y.; Suzuki, K.; Ohtsu, I. Current understanding of sulfur assimilation metabolism to biosynthesize l-cysteine and recent progress of its fermentative overproduction in microorganisms. *Appl. Microbiol. Biotechnol.* **2018**, *102*, 19. [CrossRef]
61. Zhang, X.; Sun, Z.; Bian, J.; Gao, Y.; Zhang, D.; Xu, G.; Zhang, X.; Li, H.; Shi, J.; Xu, Z. Rational Metabolic Engineering Combined with Biosensor-Mediated Adaptive Laboratory Evolution for l-Cysteine Overproduction from Glycerol in *Escherichia coli*. *Fermentation* **2022**, *8*, 7. [CrossRef]
62. Boehm, A. Process for Purifying L-Cysteine. U.S. Patent US8088949B2, 3 January 2012. Available online: https://patents.google.com/patent/US8088949B2/en (accessed on 22 September 2023).
63. Nakatani, T.; Ohtsu, I.; Nonaka, G.; Wiriyathanawudhiwong, N.; Morigasaki, S.; Takagi, H. Enhancement of thioredoxin/glutaredoxin-mediated L-cysteine synthesis from S-sulfocysteine increases L-cysteine production in *Escherichia coli*. *Microb. Cell Factories* **2012**, *11*, 62. [CrossRef]
64. Kawano, Y.; Onishi, F.; Shiroyama, M.; Miura, M.; Tanaka, N.; Oshiro, S.; Nonaka, G.; Nakanishi, T.; Ohtsu, I. Improved fermentative l-cysteine overproduction by enhancing a newly identified thiosulfate assimilation pathway in *Escherichia coli*. *Appl. Microbiol. Biotechnol.* **2017**, *101*, 6879–6889. [CrossRef] [PubMed]
65. Kredich, N.M.; Tomkins, G.M. The enzymic synthesis of L-cysteine in *Escherichia coli* and *Salmonella typhimurium*. *J. Biol. Chem.* **1966**, *241*, 4955–4965. [CrossRef] [PubMed]
66. Zhu, Z.; Zhang, J.; Ji, X.; Fang, Z.; Wu, Z.; Chen, J.; Du, G. Evolutionary engineering of industrial microorganisms-strategies and applications. *Appl. Microbiol. Biotechnol.* **2018**, *102*, 4615–4627. [CrossRef]
67. Zhang, X.; Zhang, D.; Zhu, J.; Liu, W.; Xu, G.; Zhang, X.; Shi, J.; Xu, Z. High-yield production of l-serine from glycerol by engineered *Escherichia coli*. *J. Ind. Microbiol. Biotechnol.* **2019**, *46*, 221–230. [CrossRef] [PubMed]
68. Link, H.; Anselment, B.; Weuster-Botz, D. Rapid media transition: An experimental approach for steady state analysis of metabolic pathways. *Biotechnol. Prog.* **2010**, *26*, 1–10. [CrossRef]
69. Weiner, M.; Albermann, C.; Gottlieb, K.; Sprenger, G.A.; Weuster-Botz, D. Fed-batch production of l-phenylalanine from glycerol and ammonia with recombinant *Escherichia coli*. *Biochem. Eng. J.* **2014**, *83*, 62–69. [CrossRef]
70. Schoppel, K.; Trachtmann, N.; Korzin, E.J.; Tzanavari, A.; Sprenger, G.A.; Weuster-Botz, D. Metabolic control analysis enables rational improvement of E. coli L-tryptophan producers but methylglyoxal formation limits glycerol-based production. *Microb. Cell Factories* **2022**, *21*, 201. [CrossRef]
71. Doering, W. Verfahren zur Isolierung von Cystin. German Patent DE10040176A1, 8 September 2002. Available online: https://patents.google.com/patent/DE10040176A1/en/en17 (accessed on 13 September 2023).
72. Miyahara, S.; Kamiguchi, T.; Hashimukai, T.; Nitta, K. Method of Separating Monohydrate of l-Cysteine Hydrochloride. European. Union Patent EP0250987B1, 27 December 1990. Available online: https://patents.google.com/patent/EP0250987B1/en (accessed on 17 August 2023).

molecules

Article

Effect of Nano-Selenium on Nutritional Quality of Cowpea and Response of ABCC Transporter Family

Li Li [1,2,*], Yuzhou Xiong [1], Yuan Wang [1,2], Shuai Wu [1], Chunmei Xiao [1], Shiyan Wang [1], Shuiyuan Cheng [1,2] and Hua Cheng [1,2,*]

1 School of Modern Industry for Selenium Science and Engineering, Wuhan Polytechnic University, Wuhan 430023, China
2 National R&D Center for Se-rich Agricultural Products Processing, Wuhan Polytechnic University, Wuhan 430023, China
* Correspondence: lily7819@whpu.edu.cn (L.L.); 12621@whpu.edu.cn (H.C.)

Abstract: It is an important way for healthy Selenium (Se) supplement to transform exogenous Se into organic Se through crops. In the present study, *Vigna unguiculata* was selected as a test material and sprayed with biological nano selenium (SeNPs) and Na_2SeO_3, and its nutrient composition, antioxidant capacity, total Se and organic Se content were determined, respectively. Further, the response of ABC transporter family members in cowpea to different exogenous Se treatments was analyzed by transcriptome sequencing combined with different Se forms. The results show that the soluble protein content of cowpea increased after twice Se treatment. SeNPs treatment increased the content of cellulose in cowpea pods. Na_2SeO_3 treatment increased the content of vitamin C (Vc) in cowpea pods. Se treatments could significantly increase the activities of Peroxidase (POD), polyphenol oxidase (PPO) and catalase (CAT) in cowpea pods and effectively maintain the activity of Superoxide dismutase (SOD). SeNPs can reduce the content of malondialdehyde (MDA) in pods. After Se treatment, cowpea pods showed a dose-effect relationship on the absorption and accumulation of total Se, and Na_2SeO_3 treatment had a better effect on the increase of total Se content in cowpea pods. After treatment with SeNPs and Na_2SeO_3, the Se species detected in cowpea pods was mainly SeMet, followed by MeSeCys. Inorganic Se can only be detected in the high concentration treatment group. Analysis of transcriptome data of cowpea treated with Se showed that ABC transporters could play an active role in response to Se stress and Se absorption, among which ABCB, ABCC and ABCG subfamilies played a major role in Se absorption and transportation in cowpea. Further analysis by weighted gene co-expression network analysis (WGCNA) showed that the content of organic Se in cowpea treated with high concentration of SeNPs was significantly and positively correlated with the expression level of three transporters ABCC11, ABCC13 and ABCC10, which means that the ABCC subfamily may be more involved in the transmembrane transport of organic Se in cells.

Keywords: biological nano selenium; nutritional composition; antioxidant stress; organic selenium; ABC transporter family

check for updates

Citation: Li, L.; Xiong, Y.; Wang, Y.; Wu, S.; Xiao, C.; Wang, S.; Cheng, S.; Cheng, H. Effect of Nano-Selenium on Nutritional Quality of Cowpea and Response of ABCC Transporter Family. *Molecules* **2023**, *28*, 1398. https://doi.org/10.3390/molecules28031398

Academic Editors: Yongjun Wei, Lingbo Qu and Xiao-Jun Ji

Received: 21 November 2022
Revised: 27 January 2023
Accepted: 30 January 2023
Published: 1 February 2023

1. Introduction

Selenium (Se) is one of the essential trace elements for human body and has very important physiological functions [1]. At present, a total of 25 seleno-proteins have been identified in human proteome, usually oxidoreductases, including SeCys as catalytic residues [2]. Human Se deficiency is associated with thyroid dysfunction, irreversible brain damage, peripheral vascular disease, chronic and degenerative osteoarthropathy (Kashin Beck disease), impaired immune response to viral infection, male infertility, pre-eclampsia in women, heart disease and increased risk of several cancers [3]. In addition, excessive selenium intake can also lead to poisoning. Recent studies have shown that even subtoxic doses may have negative effects, such as an increased risk of type II diabetes [1]. People living in the USA and Canada usually do not have problems related to Se deficiency. On the contrary, people living in China, New Zealand, parts of Europe and occasionally Russia have insufficient intake of micronutrients due to the low Se content in soil and food [4]. According to the research of the Institute of Medicine of the USA National Academy, the recommended dietary intake of Se for adults is 55 μg/day, while the upper tolerable intake for adults is 400 μg/day [5]. Selenate (SeO_4^{2-}) and Selenite (SeO_3^{2-}) are the two major forms of bioavailable Se in soil, with selenate predominating in oxic soils and selenite predominating in anoxic soils [6]. Organic seleno-compounds such as seleno-amino acids are also present in significant concentrations in some soils and can be imported by plant roots [7]. In its early stage, the cultivation technology of increasing Se content in plants is called Se-biofortification, and has been developed for several plants, including trees, crops and vegetables [8–12].

Vigna unguiculata (Linn.) is an important leguminous plant with rich nutritional value and adaptability [13]. Due to its resistance to water loss, it is suitable to hot arid regions [14]. Cowpea is also used as a dietary protein and mineral source as a major vegetable crop in the diets of developing countries in Africa, Latin America and Asia [15]. Se concentrations in plants grown under acidic and weathering conditions (such as tropical soils) are often very low. It is estimated that more than 1 billion people worldwide are Se deficient [13,16]. Therefore, supplementing Se through crops may be an important strategy to increase the daily intake of Se by humans and animals [17]. Although Se has important nutritional value and is beneficial to plant growth, the range between Se deficiency and poisoning is very narrow [16]. High concentrations of Se can affect the normal growth of crops, such as cowpea [18]. High doses of Se in plants can cause stress response and increase of reactive oxygen species, and finally cause crop damage. The rise of antioxidative enzymes such as peroxidase (POD), superoxide dismutase (SOD), and catalase (CAT) in response to high levels of heavy metal exposure also indicates abiotic stress in plants [13,17].

The existing studies indicate that ABC transporters may be involved in the process of Se absorption and transport in plants [19,20]. The ABC transporter is one of the largest and oldest protein families found at present, which widely exists in eukaryotes and prokaryotes. According to the phylogenetic analysis of ABC transporters, the similarity of nucleotide binding region sequences and the organizational form of domains, the family can be divided into eight subfamilies, ABCA to ABCG and ABCI [21]. Studies have shown that phosphate transporters and silicon transporters participate in the absorption of Na_2SeO_3 by plants [22,23]. The rice phosphorus transporter *OsPT2* is mainly expressed in the root, which can promote the effective absorption of Se in crops. After the Se absorbed from the outside is converted into organic Se, it is transported to the upper part of the root. Overexpression of *OsPT2* gene can improve the tolerance of crops to Se [19]. The expression changes of 28 ABC transporters in *Lolium perenne* may be related to the accumulation of selenides after selenate saline culture [20]. After *Oryza sativa* was treated with SeNPs and Na_2SeO_3, seven differentially expressed ABC transporters were screened, of which five were up-regulated and two were down-regulated, indicating that ABC transporters may also be involved in the absorption and transport of Se by rice [24]. These results indicated that ABC transporters are closely related to Se accumulation in plants.

In present study, cowpeas were treated with SeNPs and Na_2SeO_3 to explore the effects of Se on physiological indexes, antioxidant capacity, total Se content and Se species of cowpeas. Further analysis of transcriptome data revealed that ABCC transporter family may participate in the transport of Se and the mechanism of organic Se transformation. The research results are expected to provide a basis for the establishment of efficient Se enriched cultivation method of cowpea, determine the best Se concentration, times of Se application and harvest time of Se enriched cowpea, and provide new insights for leguminous plants to cope with Se stress and the role of ABC transporter in Se transport of legume.

2. Results

2.1. Physiological Indexes of Cowpea Pods

After SeNPs treatment, the content of soluble sugar in cowpea pods decreased first and then increased with the increase of concentration, and there was no significant difference compared with the control group (Figure 1a). Compared with the control group, 0.1–2.5 mmol/L treatment reduced the soluble sugar content of pods, and the lowest concentration of 1.0 mmol/L was 11.64 mg/g. Compared with the control group, the content of soluble sugar increased at 3.0 mmol/L concentration, and the content of soluble sugar was 14.15 mg/g. After Na_2SeO_3 spraying treatment, the content of soluble sugar in cowpea pods had no significant difference compared with the control group and was lower than the control group. The lowest value was 10.07 mg/g under 0.1 mmol/L treatment concentration, and the highest value was 12.24 mg/g under 3.0 mmol/L.

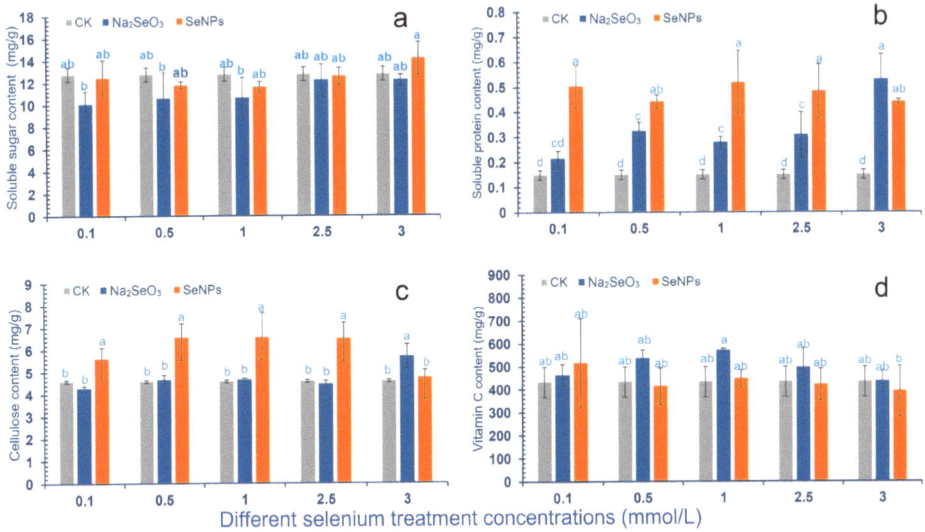

Figure 1. Effects of SeNPs and Na_2SeO_3 with different concentrations on the nutrient content of cowpea. The letter a–d represents the different significance under different treatments, p-value < 0.05. (**a**) Changes of soluble sugar content in cowpea pods; (**b**) Changes of soluble protein content in cowpea pods; (**c**) Change of cellulose content in pods; (**d**) Changes of Vc content in pods.

After SeNPs treatment, the content of soluble protein in cowpea pods was significantly increased compared with the control group: the content was 0.53 mg/g, 0.44 mg/g, 0.52 mg/g, 0.48 mg/g, 0.44 mg/g, respectively, which was 253.3%, 193.3%, 246.7%, 220.0%, 193.3% higher than the control group. After Na_2SeO_3 treatment, the soluble protein content in cowpea pods increased compared with the control group. The concentrations of 0.5 mol/L and 3.0 mmol/L showed significant difference, with the contents of 0.32 mg/g and 0.53 mg/g, respectively, which were 113.3% and 253.3% higher than those of the control group (Figure 1b).

After the cowpea leaves were treated with SeNPs, the cellulose content in the pod was higher than that in the control group. The concentration of 1.0–2.5 mmol/L was the most significant, and the cellulose content was 5.62 mg/g, 6.56 mg/g, 6.56 mg/g and 6.51 mg/g, respectively. After 3.0 mmol/L Na_2SeO_3 treatment, the cellulose content in cowpea pods was significantly increased compared with the control group, the content was 5.71 mg/g, and the increase was 24.5%. There was no significant difference in the other Na_2SeO_3 treatment group compared with the control (Figure 1c).

After spraying cowpea leaves with SeNPs, the Vc content in the pod had no significant change compared with the control group. After spraying cowpea with Na_2SeO_3, the content of Vc in cowpea pods increased first and then decreased with the increase of Na_2SeO_3 concentration, and both were higher than the control group, but there was no significant difference (Figure 1d).

2.2. Effect of Se on Antioxidant System

After SeNPs treatment, the content of malondialdehyde (MDA) in cowpea pods was significantly lower than that in the control group (Figure 2a). Under the treatment concentration of 3.0 mmol/L, the content of MDA reached a minimum of 0.23 mg/g, with a decrease of 42.5%. After Na_2SeO_3 treatment, the content of MDA in cowpea pods of all treatment groups increased, and the content of MDA in cowpea pods was significantly increased under 0.1 mmol/L treatment, which was 41.0% higher than that of the control group.

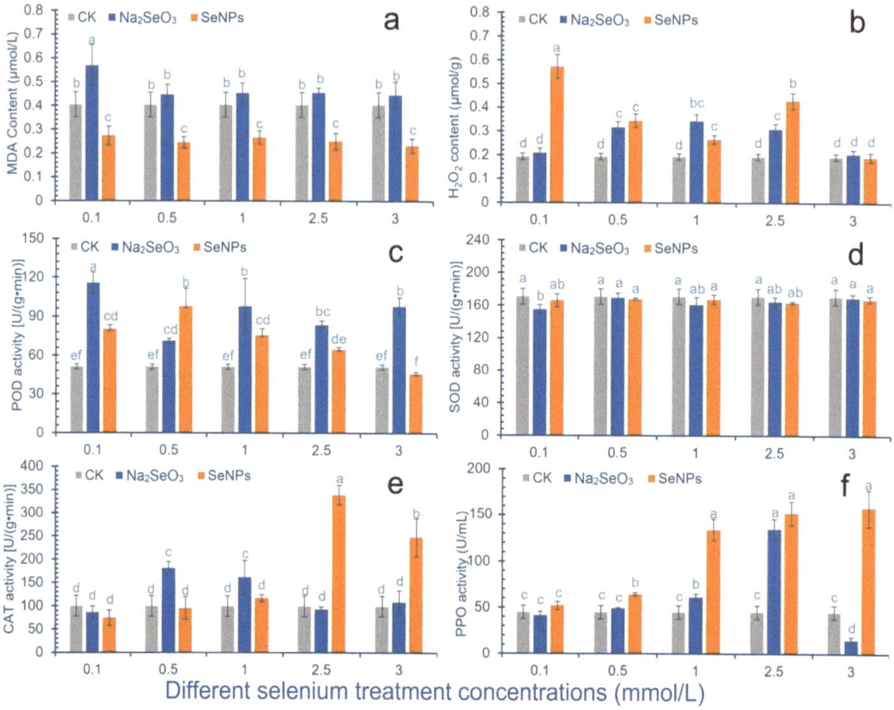

Figure 2. Effect of different Se species treatments on antioxidant activity of cowpea. The letter a–f represents the difference significance under different treatments, p-value < 0.05. (**a**) Change of malondialdehyde (MDA) content in pods; (**b**) Change of hydrogen peroxide solution (H_2O_2) content in pods; (**c**) Change of peroxidase (POD) activity; (**d**) Changes of superoxide dismutase (SOD) activity; (**e**) Change of catalase (CAT) activity; (**f**) Changes of polyphenol oxidase (PPO) activity.

After cowpea was treated with 0.1 mmol/L, 0.5 mmol/L and 2.5 mmol/L SeNPs, the content of H_2O_2 in the pod was significantly higher than that of the control group, which was increased by 195%, 78%, 122%, respectively. After cowpea was treated with 3.0 mmol/L SeNPs, the content of hydrogen peroxide in the pod was slightly lower than that in the control group. After Na_2SeO_3 treatment, the content of hydrogen peroxide in pods first increased and then decreased, and all treatments were higher than the control group, among which 0.5–1.0 mmol/L concentration had a significant increase (Figure 2b).

After SeNPs treatment, POD activity in cowpea pods increased first and then decreased. After 0.1–2.5 mmol/L treatment, the POD activity was improved compared with the control group. After 0.1–1.0 mmol/L treatment, the POD activity increased by 56.6%, 91.4% and 47.9%, respectively, compared with the control group. However, the POD activity of the treatment group with 3.0 mmol/L concentration was lower than that of the control group. The results showed that the high concentration of SeNPs would reduce the POD activity of cowpea. After Na_2SeO_3 treatment, the activity of POD in cowpea pods was improved compared with the control group. Except that there was no significant difference at the concentration of 0.5 mmol/L, the activity of POD in other treatment groups was increased by 126.2%, 39.1%, 91.4%, 63.0% and 91.4%, respectively, compared with the control group (Figure 2c).

After 0.1 mmol/L Na_2SeO_3 treatment, the SOD activity in cowpea pods was significantly lower than that in the control group, which was 91% of the control group. After treatment with 0.1 mmol/L and 2.5 mmol/L SeNPs, the SOD activity in pods decreased compared with the control group, which was 97% and 96% of the control group, respectively, with no significant difference (Figure 2d).

With the increase of SeNPs concentration, the CAT activity in cowpea pods decreased first and then increased. After 0.1–0.5 mmol/L treatment, the CAT activity in cowpea pods was lower than that in the control group. When the concentration was 2.5 and 3 mmol/L, the CAT activity was significantly higher than that of the control group, increased by 239.2% and 147.9%, respectively. After Na_2SeO_3 treatment, the CAT activity in the 0.5 and 1.0 mmol/L treatment groups was significantly increased compared with the control group, with the increase amplitude of 81.3% and 62.1%, respectively (Figure 2e).

The SeNPs treatment can significantly improve the PPO enzyme activity of cowpea. When the concentration of the treatment is greater than 0.5 mmol/L, the PPO activity is increased by 45.45%, 198.73%, 238.46% and 251.44%, respectively, compared with the control. In the Na_2SeO_3 treatment group, only 1 and 2.5 mmol/L treatment groups had significant improvement, which were 36.33% and 204.89% higher than the control, respectively (Figure 2f).

2.3. *Effect of Exogenous Se Treatment on Total Se Content in Cowpea Pods*

Figure 3 shows that the Se content in cowpea pods increases significantly with the increase of exogenous Se concentration. When the concentration of SeNPs is more than 1.0 mmol/L, the Se content in the pod increases significantly. When the treatment concentration is 3.0 mmol/L, the Se content in the pods reaches the maximum of 12.81 μg/g. At low concentration (<0.5 mmol/L), the effect of Na_2SeO_3 on the increase of Se content in cowpea pods was not as good as that of SeNPs. When the concentration of exogenous Se is more than 1.0 mmol/L, the effect of Na_2SeO_3 is more obvious. Compared with the control, the Se accumulation in cowpea pods was significantly increased by 3.22 times, 4.47 times and 8.00 times after the application of 0.5~3.0 mmol/L Na_2SeO_3, and the difference between the treatment groups was significant.

Under the condition of the same Se source and the same treatment concentration, different picking time has a greater impact on the total Se content in cowpea pods. Among them, in the 0.5 mmol/L treatment group, the total Se content in the pods picked the second time was slightly lower than that in the first time, while in the high concentration (2.5 mmol/L) treatment group, the total Se content in the pods picked the second time was higher than that in the first time. In the 2.5 mmol/L SeNPs treatment group, the Se content in the second harvest pods was significantly higher than that in the first harvest, about 2.16 times (Figure 4a).

Figure 4b shows that different Se application times will affect the total Se content of cowpea. The total Se content of soybean pods with the second application of Se was higher than that with the first application of Se. Among them, the treatment concentration of 0.5 mmol/L has a significant increase effect, which is about 3.05 times higher. In the 0.5 mmol/L Na_2SeO_3 treatment group, the total Se content in the pods with the second application of Se was slightly higher than that with the first application of Se, while the total Se content in the pods with the second application of Se in the 2.5 mmol/L treatment group decreased significantly (Figure 4b).

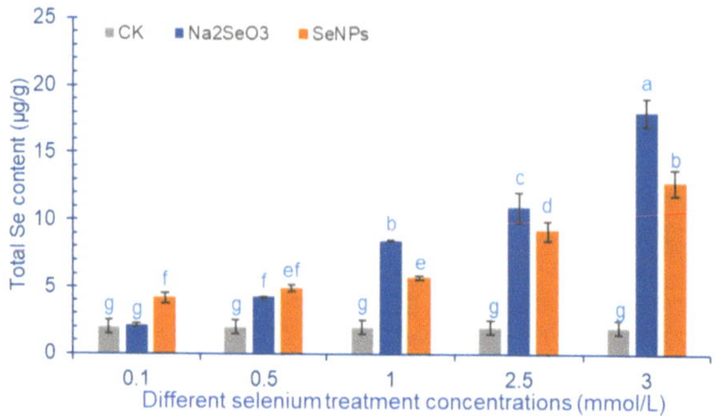

Figure 3. Effect of exogenous Se treatment with different concentrations on total Se content in cowpea. The letter a–g represents the difference significance under different treatments, p-value < 0.05.

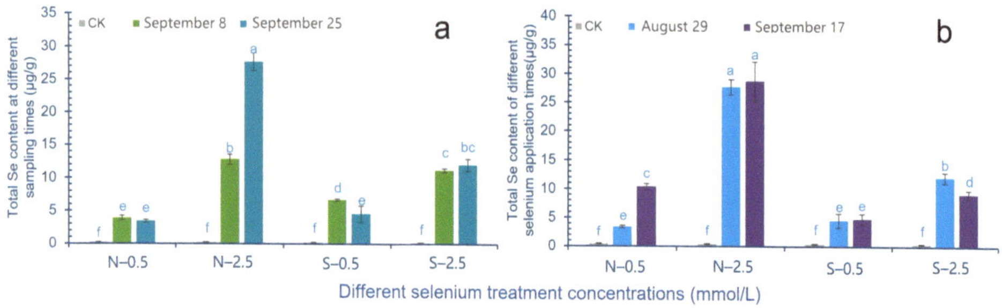

Figure 4. Effects of Se treatment times and picking time on total Se content in cowpea pods. The letter a–f represents the difference significance under different treatments, p-value < 0.05. (**a**) Changes of total Se content at different sampling times after Se treatment; (**b**) Change of total Se content in different treatment time.

2.4. *Effects of Se Application on the Forms and Contents of Se in Cowpea Pods*

In all treatments, SeMet was the main form of Se in cowpea pods, accounting for 55–100%. When the concentration of SeNPs and Na_2SeO_3 was 0~0.1 mmol/L, only the unique Se form, SeMet, was detected in the pods; with the increase of exogenous Se concentration, the detected Se species increased gradually. They all exist in the form of organic Se; when the concentration of exogenous Se is 3.0 mmol/L, inorganic Se (Se^{4+}, Se^{6+}) can be detected (Table 1).

When the concentration of SeNPs treatment is 0.5 mmol/L, the Se forms in cowpea pods are mainly MeSeCys and SeMet, accounting for 28.65% and 71.35%, respectively. When the concentration of SeNPs was 1.0 and 2.5 mmol/L, the Se in cowpea pods was mainly SeCys2, MeSeCys and SeMet, accounting for 2.17%, 26.93%, 70.90% and 4.45%, 32.31% and 63.24%, respectively. When the concentration of SeNPs treatment was 3.0 mmol/L, five Se forms were detected, including Se^{4+}, Se^{6+}, SeCys2, MeSeCys and SeMet, of which inorganic Se accounted for 13.88% and organic Se accounted for 86.12%. Table 1 shows that the content of various Se forms detected in cowpea pods increases with the increase of SeNPs concentration applied, and there are significant differences among the treatments.

Different from SeNPs treatment, after 0.5 mmol/L Na_2SeO_3 treatment, three forms of organic Se can be detected: SeCys2, MeSeCys and SeMet, which account for 1.21%, 14.82% and 83.97%, respectively. When the concentration of Na_2SeO_3 increases to 3.0 mmol/L, four Se forms, namely Se^{4+}, SeCys2, MeSeCys and SeMet, can be detected, accounting for 3.12%, 6.28%, 18.61% and 71.99%, respectively (Table 2). The content of various Se forms detected in cowpea pods gradually increased with the increase of Na_2SeO_3 concentration.

Table 1. The content and proportion of Se forms in cowpea pods after SeNPs treatment.

SeNP Treatment Concentration (mM)	Content of Different Se Forms $SeCys_2$ (μg/g) DW	MeSeCys (μg/g) DW	SeMet (μg/g) DW	Se^{4+} (μg/g) DW	Se^{6+} (μg/g) DW
CK	-	-	0.1187 ± 0.0059 [e] 100.00%	-	-
0.1	-	-	0.2666 ± 0.0028 [e] 100.00%	-	-
0.5	-	0.1509 ± 0.0116 [d] 28.65%	0.3758 ± 0.0061 [d] 71.35%	-	-
1.0	0.0183 ± 0.0057 [c] 2.17%	0.2274 ± 0.0195 [c] 26.93%	0.5988 ± 0.0084 [c] 70.90%	-	-
2.5	0.0587 ± 0.0066 [b] 4.45%	0.4261 ± 0.0167 [b] 32.31%	0.8338 ± 0.0749 [b] 63.24%	-	-
3.0	0.1478 ± 0.0095 [a] 6.88%	0.5206 ± 0.0127 [a] 24.24%	1.1813 ± 0.0272 [a] 55.00%	0.0084 ± 0.0032 [a] 0.39%	0.2897 ± 0.0254 [a] 13.49%

Note: Small letters a–e indicate the significance analysis of each Se form after treatment with the same Se source and different concentrations, p-value < 0.05. The percentage in the table is the proportion of each Se form under the treatment of Se source of this concentration. The short horizontal line represents that the content of the substance is not detected in the sample.

Table 2. The content and proportion of Se forms in cowpea pods after Na_2SeO_3 treatment.

Na_2SeO_3 Treatment Concentration (mM)	Content of Different Se Forms $SeCys_2$ (μg/g) DW	MeSeCys (μg/g) DW	SeMet (μg/g) DW	Se^{4+} (μg/g) DW	Se^{6+} (μg/g) DW
CK	-	-	0.1035 ± 0.0059 [e] 100.00%	-	-
0.1	-	-	0.1545 ± 0.0119 [e] 100.00%	-	-
0.5	0.0136 ± 0.0016 [c] 1.21%	0.1672 ± 0.0073 [c] 14.82%	0.9473 ± 0.0799 [c] 83.97%	-	-
1.0	0.0459 ± 0.0083 [c] 3.53%	0.2872 ± 0.0348 [bc] 22.09%	0.9671 ± 0.0967 [c] 74.38%	-	-
2.5	0.1323 ± 0.0213 [b] 4.68%	0.4095 ± 0.0304 [b] 14.50%	2.2833 ± 0.1101 [b] 80.82%	-	-
3.0	0.2990 ± 0.0121 [a] 6.28%	0.8859 ± 0.0517 [a] 18.61%	3.4273 ± 0.2747 [a] 71.99%	0.1485 ± 0.0412 [a] 3.12%	-

Note: Small letters a–e indicate the significance analysis of each Se form after treatment with the same Se source and different concentrations, p-value < 0.05. The percentage in the table is the proportion of each Se form under the treatment of Se source of this concentration. The short horizontal line represents that the content of the substance is not detected in the sample.

2.5. Analysis of Transcriptome Data of Cowpea Treated with Se

The sequencing results showed that the contrast rate of all treatment groups remained at 82.99–94.89%, indicating that the reference genome assembly could meet the needs of information analysis and the reliability of later data analysis (Figure S1, Supplementary Materials). A sample correlation analysis was carried out between each two of all treatment groups, the Pearson correlation coefficient calculated between all gene samples and a heat map was made. The results showed that the p-value value was between 0.938 and 0.995, indicating that the similarity within the sample group was very high and the number of different genes between samples was small (Figure S1). The transcriptome of the treated samples was analyzed. It was found that the relative expression value of FPKM (Log10) was between 0–1.5, indicating that the samples had high repeatability and reliability (Figure S2, Supplementary Materials).

All genes were analyzed for differential expression (the screening threshold was q value <0.05 and ∣ log2FoldChange ∣>1). A total of 3778 differentially expressed genes (DEGs) were screened, of which 3680 were annotated, accounting for 11.5% of the total genes. GO enrichment analysis showed that DEGs are mainly enriched in molecular function, kinase activity and DNA binding subclasses; cell components are enriched in cellular component and protein complex subclasses; and molecular biological processes are mainly enriched in response to stress, transmembrane transport and transport (Figure S3, Supplementary Materials).

The cowpea transcriptome data were analyzed, and the transcriptome databases of *Arabidopsis thaliana*, *Nicotiana tabacum* and *O. sativa* were used for comparative annotation. A total of 31,948 genes were annotated, and the number of genes annotated by *Arabidopsis*, tobacco and rice was 17,464, 9522 and 17,733, respectively. Only 2552, 1391 and 2852 genes were compared and annotated separately by *A. thaliana*, *N. tabacum* and *O. sativa*. A total of 6328 genes were compared and annotated by the three species (Figure 5a).

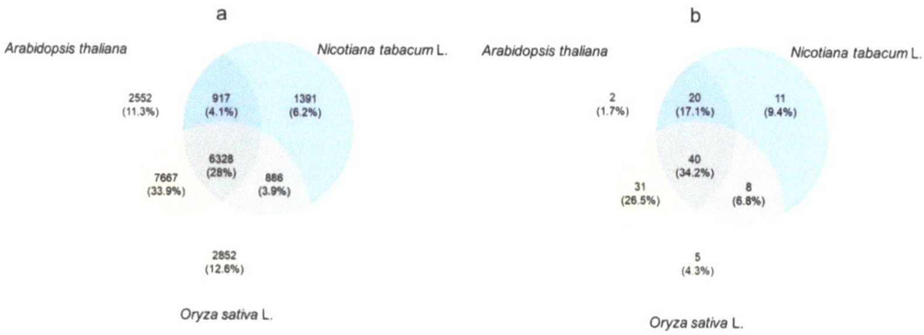

Figure 5. Venn diagram of differentially expressed ABC transporter genes in cowpea transcriptome. (**a**) Annotation of cowpea transcriptome genes with transcriptome data of *A. thaliana*, *Nicotiana tabacum* and *Oryza sativa*; (**b**) The number of ABC transporter genes of cowpea transcriptome annotated to the above three species.

Using ABC transporter as the key word, 107 genes encoding ABC transporter were screened from cowpea transcriptome data (Figure 5b). The comparison results showed that 93, 79 and 84 ABC transporter genes were annotated into the databases of arabidopsis, tobacco and rice, respectively. Among them, there were 2, 11 and 5 genes annotated by three species alone, accounting for 1.7%, 9.4% and 4.3%, respectively. The number of ABC protein genes annotated by the three species was 40, accounting for 34.2%.

The results of comparison and screening of DEGs according to different treatment groups are shown in Figure 6. Among them, the control group has the largest number of DEGs compared with the SeNPs treatment group, indicating that a large number of genes in cowpea leaves respond to Se stress after SeNPs treatment, which may play a role in promoting the absorption of SeNPs in cowpea leaves. SeNPs treatment showed that the number of DEGs between high concentration treatment groups was significantly less, while the number of DEGs between low concentration and high concentration treatment groups (N-100 vs. N-2500, N-100 vs. N-3000, N-500 vs. N-3000) was more. It is speculated that with the increase of SeNPs concentration, more genes respond to the stress of SeNPs treatment on cowpea leaves. The number of DEGs between different Se species treatments (N-1000 vs. S-1000) is also high, which may be related to the different ways of absorbing SeNPs and Na_2SeO_3 in leaves (Figure 6).

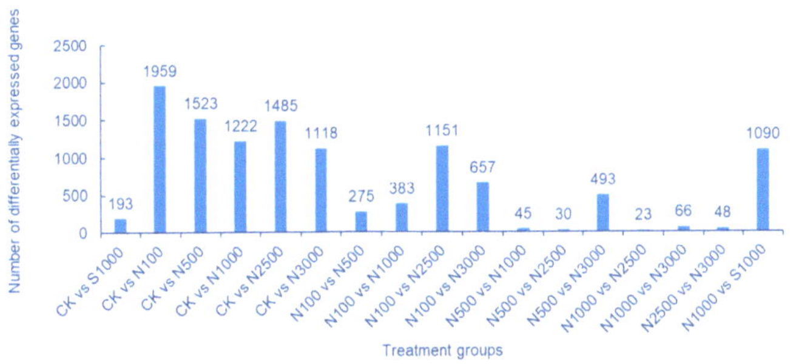

Figure 6. Grouping of DEGs in cowpea transcriptome data.

2.6. Analysis of Transcriptome ABC Transporter Gene

Seven ABC transporter subfamilies were identified in the Se treated cowpea transcriptome data, of which ABCB, ABCC, and ABCG have a large number of 32, 15, and 47, respectively. The three subfamilies account for 80.3% of the total ABC transporter genes. The number of ABCA subfamilies is 14. In cowpea ABC transporters, the three subfamilies of ABCD, ABCE, and ABCF are relatively small. No ABCH subfamily genes were detected in cowpea transcriptome (Figure 7a).

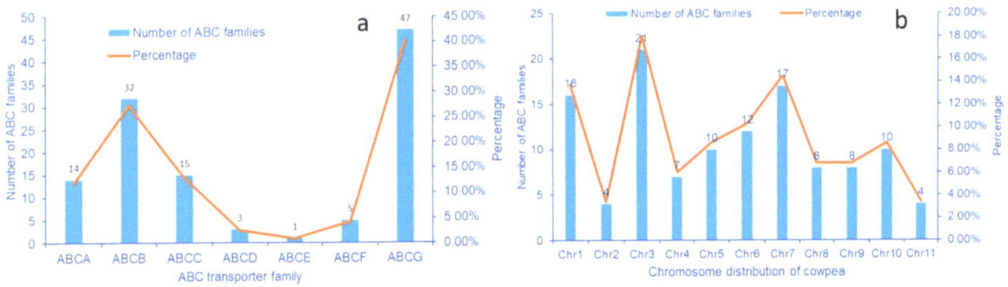

Figure 7. Quantitative distribution of ABC transporter gene subfamilies in *V. unguiculata*. (**a**) Quantitative distribution of ABC transporter family subclasses in cowpea; (**b**) Distribution of ABC transporter subfamily in cowpea genome.

Cowpea has 11 pairs of chromosomes (2n = 22). ABC transporter genes are unevenly distributed on cowpea chromosomes. The number of ABC transporter genes on chromosome No. 3 is up to 21, accounting for 18.0%; The second is chromosome No. 7 and No. 1, which contain 17 and 16 genes, respectively. There are four ABC transporter genes distributed on chromosomes No.2 and No.11, respectively (Figure 7b).

2.7. Differential Expression Analysis of ABC Transporter Gene

Among the DEGs of transcriptome data, 107 ABC transporter genes were screened from cowpea pods. The ABCG transporter subfamily has 44 genes at most, followed by ABCB and ABCC transporter subfamilies, with 27 and 14 DEGs, respectively. It indicated that the ABCG, ABCB and ABCC subfamilies had an obvious response to stress after exogenous Se treatment (Figure 8).

The differential expression analysis of cowpea ABC transporter gene was conducted among groups. The expression level of the gene was based on the fold changes value. When log2FoldChange > 0, it was considered that the DEGs was up-regulated. On the contrary, when log2FoldChange < 0, it was considered that the DEGs was down-regulated (Table S1, Supplementary Materials). There was no significant difference in the number of differentially expressed ABC transporter genes among the control groups. After 1.0 mmol/L SeNPs treatment, the number of differentially expressed ABC transporter genes was the largest, with 55 up-regulated and 49 down-regulated; after 0.5 mmol/L SeNPs treatment, the number of differentially expressed ABC transporter genes was the least, and the number of up/down regulated genes was 49. The differential expression analysis of ABCB, ABCC and ABCG subfamilies showed that the DEGs in ABCC subfamily were identical. The change of Se concentration had little effect on the number of DEGs in the three subfamilies, and the three subfamilies accounted for a large proportion of differentially expressed ABC transporter genes, accounting for 77.6–77.9%. These results indicate that ABCB, ABCC and ABCG subfamilies are the main transporters family in response to Se stress.

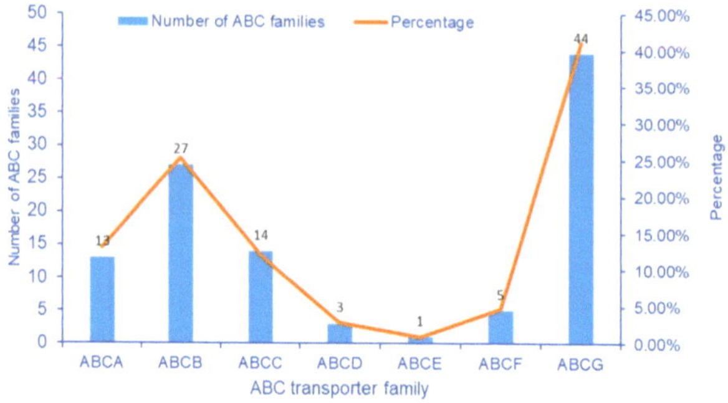

Figure 8. Number distribution of ABC transporter gene family in *V. unguiculata* after Se treatment.

Seventeen groups of DEGs were screened, and a total of 40 highly expressed ABC transporter differentially expressed genes were screened. The number of ABCB, ABCC and ABCG subfamilies were 9, 7 and 21 genes, respectively, accounting for 92.5% of all differentially expressed ABC transporters. These data support that these three subfamilies are the main genes for ABC transporters to cope with Se stress. After increasing the screening threshold, compared with the control, the number of differentially expressed ABC transporter genes in pods of 100 μmol/L SeNPs treatment group was the largest, of which 17 genes were up-regulated and 9 genes were down regulated, indicating that ABC transporters could respond to SeNPs stress after low concentration Se application. The number of differentially expressed ABC transporter genes decreased with the increase of SeNPs concentration, which may be related to the different reversals of cowpea leaves caused by excessive SeNPs concentration. Compared with treatment group SeNPs and Na_2SeO_3 at the same concentration, there were 8 up-regulated genes and 10 down-regulated genes for ABC transporter (Figure 9).

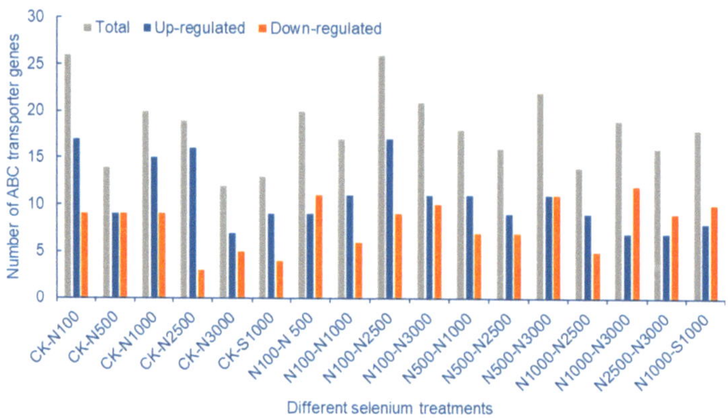

Figure 9. Statistical analysis of differentially expressed cowpea ABC transporter genes under different Se treatments.

2.8. GO Enrichment Analysis of ABC Transporters

The GO enrichment analysis of differentially expressed ABC protein genes showed that all 40 genes were enriched to ATP binding proteins in molecular functions. In cell components, they are mainly concentrated in membrane and membrane components; in the biological process, there are mainly three subtypes of transport, transmembrane transport and transport activities. The main function of ABC transporter is to complete the transmembrane transport of proteins, lipids and other compounds by hydrolyzing ATP to generate energy, which is completely consistent with the significant enrichment analysis of ABC transporter GO function, further indicating that ABC transporter plays an important role in the absorption and transport of Se in cowpea (Figure 10).

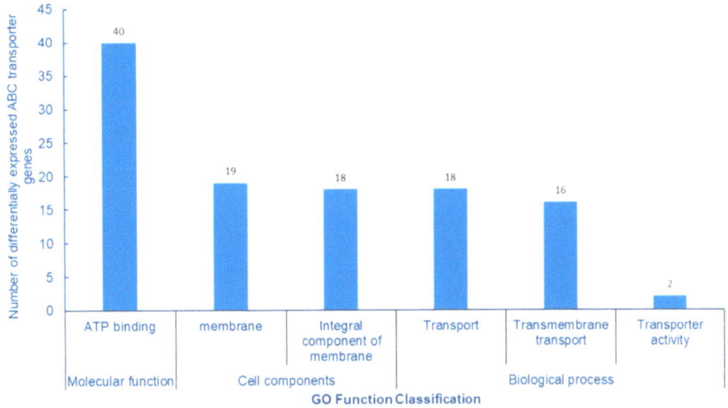

Figure 10. Analysis of GO function enrichment of differentially expressed ABC transporter gene.

2.9. WGCNA Analysis of DEGs

In order to further screen the differentially expressed ABC genes related to the absorption and transport of Se, we conducted correlation analysis between the DEGs and the contents of five Se species in cowpea pods. In the analysis, we set the power value of the module as 10, the similarity parameter as 0.75, and the minimum gene number of the module as 50 (Figure 11a). WGCNA analysis finally obtained 30 modules with similar gene expression patterns, each module contains 76–3332 genes (Figure 11b). Figure 11c shows the correlation analysis of the module characteristic genes and five forms of Se (SeCys2, MeSeCys, Se^{4+}, SeMet, Se^{6+}). It can be seen from the figure that White and Medium purple3 modules are positively correlated with the contents of organic Se, total Se, SeCys2, MeSeCys and SeMet, while Honeydew modules are negatively correlated with the contents of organic Se and total Se, indicating that the characteristic genes of these three modules are highly correlated with the Se accumulation of cowpea (Figure 11c).

In order to display the details of the relevant modules, the first 20 KEGG pathways of the two positive correlation modules are displayed using bubble diagrams (Figure 12). In the White module, N3001 sample has the most up-regulated genes (Figure 12a), and the genes are significantly enriched in the plant hormone signal transport, plant pathogen interaction, and nuclear cellular transport metabolic pathways. It is worth noting that *ABCC11* (Vigun05g292000) protein is enriched in the ABC transporters pathway (Figure 12c). In the Medium purple3 module, N3002 has the most up-regulated genes (Figure 12b), and the genes are significantly enriched in the biosynthesis of secondary metals, biosynthesis of amino acids, nitrogen metabolism, ether lipid metabolism and ABC transporters metabolic pathways. Among them, *ABCC13* (Vigun05g291860) and *ABCC10* (Vigun05g292884) genes are enriched in the metabolic pathway of ABC transporters (Figure 12d).

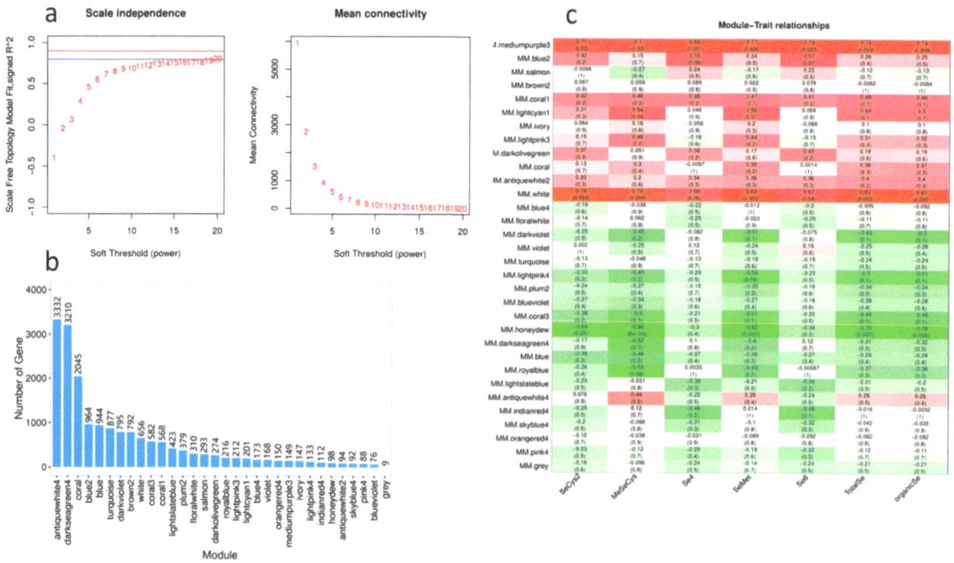

Figure 11. WGCNA analysis of the association between DEGs and different Se forms in cowpea treated with different Se. (**a**) Left: The abscissa represents the power value, the ordinate represents the correlation coefficient, the blue horizontal line represents the correlation coefficient 0.8, and the red horizontal line represents the correlation coefficient 0.9. Right: The abscissa represents the power value, and the ordinate represents the average connectivity of genes; (**b**) Number distribution of genes in different modules, the abscissa represents each module, and the ordinate represents the number of genes; (**c**) Direct correlation between different modules and Se speciation. The abscissa is the character, the ordinate is the module, and Pearson correlation coefficient is used for mapping. Red represents positive correlation and green represents negative correlation. The darker the color is, the stronger the correlation is. The number in the lower brackets represents significant p value. The smaller the value is, the stronger the significance is.

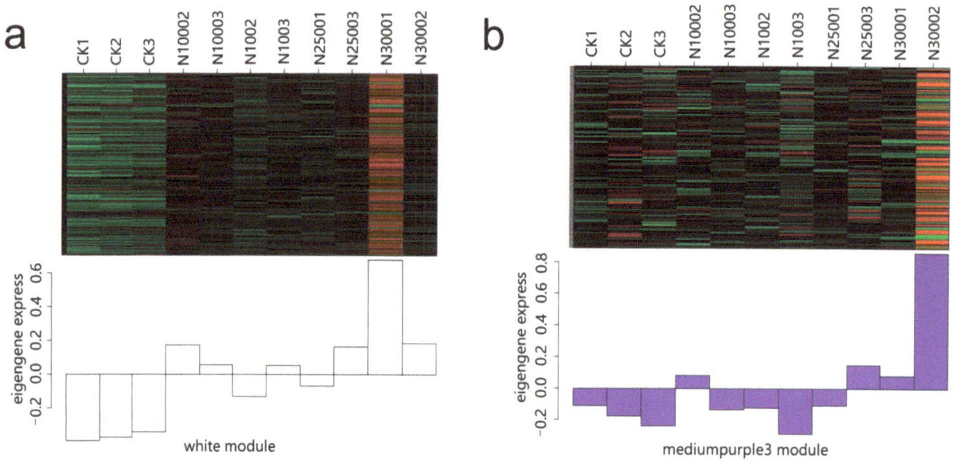

Figure 12. *Cont.*

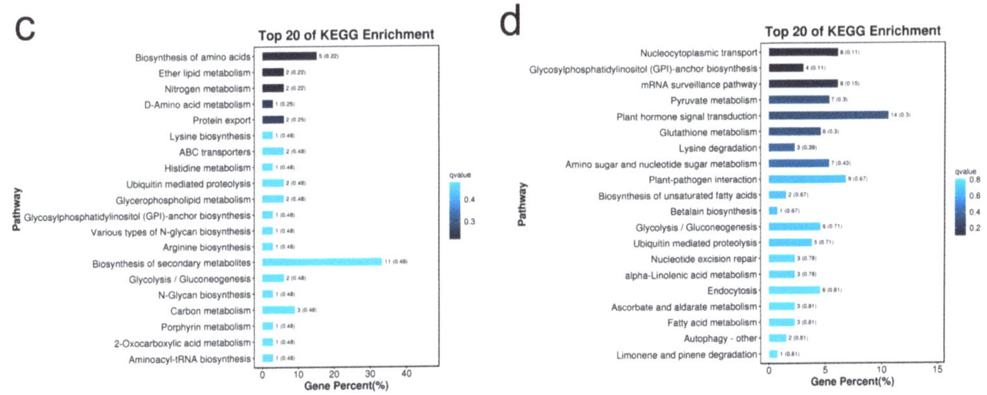

Figure 12. Enrichment analysis of KEGG pathway of white and medium purple3 module genes. (**a**) Heat map of gene expression of white module treated with different Se; (**b**) Heat map of gene expression of medium purple3 module treated with different Se; (**c**) Kegg enrichment bar graph of the first 20 genes with the lowest *p*-value in the white module; (**d**) Kegg enrichment bar graph of the first 20 genes with the lowest *p*-value in the medium purple3 module.

3. Discussion

3.1. Effect of Se on Physiological Indexes of Cowpea Pods

Cowpea pod is rich in protein, soluble sugar, cellulose and polyphenolic compounds, and long-term consumption can effectively supplement the human body needs nutrients [25–28]. Se is an essential trace element for human body and also a beneficial element for plant growth [11,29]. Relevant studies showed that Se application could improve the physiological indicators of leguminous plants, and exogenous Se leaf spraying could effectively increase the content of soluble protein in cowpea pods [30]. The seed soaking with 4 mg/L Na_2SeO_3 could significantly increase the content of soluble protein, soluble sugar and Vc in cowpea by 38%, 28% and 47%, respectively, compared with the control [31]. After selenite treatment of soybean plants at the early flowering and full flowering stages, the fat content in seeds did not change significantly, but the protein content increased significantly [32]. Treatment of peanut leaves with 20 and 40 mg/L SeNPs will increase the total sugar concentration in different branches [33]. Soaking *Vigna radiata* seeds with 0.75 mg/L Na_2SeO_4 will increase the concentration of soluble sugar in *V. radiata* sprouts [34].

This study showed that the two Se sources had effects on the physiological indexes of cowpea pods, and the effects of different Se sources on the nutritional components of cowpea were different. The content of soluble protein in cowpea pods treated with two kinds of Se was higher than that of the control, and the SeNPs treatment showed a significant difference. After two kinds of Se treatment, the cellulose content in cowpea pods increased compared with the control group, and 1.0–2.5 mmol/L SeNPs treatment showed a significant difference. Compared with the control group, the cellulose content in cowpea pods treated with 3.0 mmol/L Na_2SeO_3 was significantly increased. After Na_2SeO_3 treatment, the content of Vc in pods increased, but there was no significant difference. The change of physiological indexes of cowpea pods is an important factor determining the commodity attributes. In the current study, it was found that SeNPs has a greater promotion impact on the nutritional indicators of cowpea than Na_2SeO_3.

3.2. Effect of Exogenous Se Treatment on Total Se Content in Cowpea Pods

After applying exogenous Se (SeNPs or Na_2SeO_3), the absorption and accumulation of total Se in cowpea pods showed a dose-response relationship. The results were consistent with the research results on other leguminous plants such as *Pisum sativum*, *Astragalus sinicus* and *Medicago sativa* [35–37]. The results showed that foliar spraying of 37.5–750.0 mg/L Na_2SeO_3 could effectively increase the contents of macromolecular bound Se and total Se in pea seedlings, which were 8.75–38.75 times and 6.01–55.65 times higher than those in the control, respectively [38]. In the present study, compared with the control treatment, SeNPs and Na_2SeO_3 at the highest concentration (3.0 mmol/L) can increase the Se content in cowpea pods by 538.60% and 800.76%, respectively. In general, the effect of SeNPs on the increase of Se content in cowpea pods is not as good as that of Na_2SeO_3, which may be related to their absorption kinetics, different physical and chemical properties, and different interactions with biological molecules [39]. The inward flow mechanism of Na_2SeO_3 is different from that of SeNPs: Na_2SeO_3 enters plants mainly through the active transport of phosphate transporters, while SeNPs diffuses passively through aquaporins. In wheat and rice, the absorption rate of SeNPs is much lower than that of Na_2SeO_3 [39,40]. It is worth noting that the application of 3.0 mmol/L Na_2SeO_3 has a toxic effect on cowpea plants, which is manifested by a large number of irregular necrotic spots distributed at the edge of leaves and chlorosis between veins. It is speculated that the leaf spraying treatment with high concentration of Na_2SeO_3 may cause the formation of ROS in the mesophyll cells of cowpea, leading to the destruction of cell membrane structure and the decrease of chlorophyll concentration [18]. However, the application of SeNPs with the same concentration did not cause cowpea plants to show obvious symptoms of injury. This phenomenon shows that the plant toxicity of SeNPs is lower than that of selenite, which is consistent with the results observed by Li in the water culture experiment of garlic with different forms of Se [41]. Kovács monitored the Se absorption and accumulation dynamics of *M. sativa* harvested four times in the same growth season. They found that the Se content in the stems and leaves of selenate and selenite treatment groups decreased gradually during the harvest period, and the effect of SeNPs treatment was different from this. In the 10 mg/L SeNPs treatment group, the total Se content of cowpea leaves harvested from the first harvest (14.3 μg/g) to the fourth harvest (37.5 μg/g) gradually increased. In the 50 mg/L SeNPs treatment group, the total Se content decreased in the first three harvests and increased again in the fourth harvest [37].

In this study, cowpea behaves similarly to *M. sativa* under Na_2SeO_3 treatment. The total Se content in the pods harvested in the second harvest is lower than that harvested in the first harvest, but the difference is not significant. However, after SeNPs treatment, the Se content in cowpea pods harvested in the second crop was higher than that in the first crop at a high concentration (2.5 mmol/L). It is likely that the difference will be caused by the different oxidation-reduction, assimilation metabolism and transport modes of SeNPs in plant cells.

3.3. *Effect of Se on Antioxidant Capacity of Cowpea*

Higher plants' antioxidant defense mechanism will be disrupted under adverse environments, resulting in a rise in hydrogen peroxide (H_2O_2), superoxide (O_2^-) free radicals, MDA, hydroxyl anion (OH^-), and other reactive oxygen species (ROS) [42,43]. Excessive levels of hydrogen peroxide and superoxide will harm the plasma membrane and chlorophyll. MDA causes cytotoxic cross-linking and polymerization of protein, nucleic acid, and other living components [44].

Appropriate concentration of Se can reduce the active oxygen produced by plant cell membrane, thus promoting the antioxidant metabolism under stress [44]. Se application can also reduce lipid peroxidation by enhancing the activity of antioxidant enzymes against ROS [45,46]. The results showed that the activities of CAT and glutathione reductase (GR) in cowpea root treated with selenite and selenate were significantly increased, and the increase of CAT activity caused by selenite application was more obvious [44]. A total of 5 μmol/L Na_2SeO_3 increases the tolerance of cucumber seedlings to oxidative stress caused by water shortage by increasing ascorbic acid peroxidase (APX), POD, SOD and CAT activities, and also inhibits the damage of plasma membrane caused by lipid peroxidation [47]. After *V. radiata* sprouts were sprayed with SeNPs leaves, the activities of SOD and POD showed a trend of rising first and then declining. The maximum activities of two enzymes increased by 108.04% and 59.45%, respectively, compared with the control group [48].

The results of this study show that Se application can prevent the damage caused by peroxidation of cowpea plants to a certain extent. Among them, SeNPs treatment can significantly inhibit or reduce the content of MDA in cowpea pods, and improve the antioxidant capacity of cowpea. Se application also affected the antioxidant enzyme activity of cowpea pods. Two kinds of Se treatments could effectively improve the peroxidase activity in cowpea pods, and higher concentrations of SeNPs could significantly improve the peroxidase activity in cowpea pods. The two Se treatments had little effect on the activity of superoxide dismutase in pods and could effectively maintain the activity of SOD. High concentration of SeNPs can effectively improve the CAT activity of cowpea, and CAT may play an important role in antioxidant defense against high concentration of Se stress.

3.4. Effect of Se Treatment on Se Species in Cowpea

The form of Se in plants is the basis for evaluating the effect of Se supplementation. Organic Se forms (Se amino acids) have higher bioavailability than inorganic Se forms (such as selenite and selenate) [29]. The form and content of Se accumulated by plants after applying Se vary with plant species and tissues and organs distributed. For example, SeMet, Se (IV), MeSeCys and γ-Glu-Se-MC can be detected in cowpea grown in natural Se rich soil [49]. Only Se (VI) and SeMet were detected in *M. sativa* after selenate, selenite and SeNPs were applied [37]. After Na_2SeO_3 treatment, the main Se compounds in soybean seeds are SeCys2 and SeMet, accounting for about 74%. In addition, there is less MeSeCys (9%). In this study, the main form of Se detected in cowpea pods after treatment with SeNPs and Na_2SeO_3 of different concentrations was SeMet, followed by MeSeCys, which accounted for 55–100% and 14.5–32.31%, respectively, in cowpea pods. With the increase of the concentration of exogenous Se, the types of Se compounds detected in the pods gradually increased, but organic Se (especially SeMet) was the main form, and inorganic Se was only detected in the high concentration treatment group. Li et al. also found that when the treatment of SeNPs increased from 10 μmol/L to 30 μmol/L in rice, the form of Se accumulated in the shoot of seedlings increased from one (SeMet) to two species (SeMet and MeSeCys), and SeMet was the main form; In the 10 μmol/L Na_2SeO_3 treatment group, the Se forms detected above ground of rice seedlings were SeMet, MeSeCys and Se (VI). It can be seen that the form and content of Se in plant tissues are also closely related to the types of exogenous Se and the amount of Se applied [41].

Se application could maintain or improve the nutritional quality of cowpea, which made it possible to produce Se-rich foods from cowpea. After Se application, the total Se content in cowpea pod, the edible part of cowpea, was significantly increased compared with the control group. Moreover, the Se in the pod was mainly organic Se, which could be better absorbed and utilized by the human body, further indicating the feasibility of Se-rich cowpea production. At present, the research on Se metabolism and transport in leguminous plants is not completely clear. The ABC transporter is a family of transporters with extensive functions. Through the study of the response of ABC transporters to Se in cowpea transcriptome data, we can understand the correlation between ABC transporters and Se transport.

3.5. ABC Transporters and Se Absorption

At present, the mechanism of selenite absorption by plants is not clear. Some studies show that phosphate transporters and silicon transporters participate in the absorption of Na_2SeO_3 by plants [22,23]. *OsPT2*, a rice phosphorus transporter, can promote the effective absorption of Se in plants. *OsPT2* gene is mainly expressed in roots. After absorbing selenite from the outside and transforming it into organic Se, it is transported to the upper part of roots. Overexpression of *OsPT2* gene can improve the Se tolerance of plants [19]. Analysis of transcriptome data of Se treated *Lolium perenne* showed that 28 ABC transporters in ryegrass might be related to the accumulation of selenides [20]. After sodium selenate hydroponic treatment of tea plants, 16 key genes of ABC transporter were found, 12 of which were up-regulated and 4 were down-regulated. It was speculated that the differential expression of ABC transporter in tea plants might be related to the transport of selenate in roots [50]. After transcriptome sequencing of rice seedling leaves treated with SeNPs and Na_2SeO_3, seven differentially expressed ABC transporters were screened, of which five were up-regulated and two were down regulated, indicating that ABC transporters may be involved in Se transport in rice [24].

A total of 131 *ABC* transporters have been identified in the genome of *A. thaliana*, of which 43 are *ABCG* subfamily transporters [51]. A total of 265 *ABC* transporters have been identified in tobacco genome, of which the number of ABCG subfamily transporters is up to 134 [52]. Statistical analysis of cowpea transcriptome showed that there were 107 *ABC* transporter genes, including 27, 14 and 44 ABCB, ABCC and ABCG subfamilies, respectively. It is suggested that ABCB, ABCC and ABCG subfamilies may be the main undertakers of the function of ABC transporter family and play an important role in the transmembrane transport of substances in plants and stress response. The number of ABC subfamily in different plants may be related to the living environment and genetic characteristics of different plants [53]. Compared with animals, the ABC transporters in plants are more diverse in species and function. On the one hand, this may be related to the complex secondary metabolic activities in plants. On the other hand, it plays an important role in plants' resistance to changes in the external environment [54].

Studies have shown that ABCG transporters play an important role in plant organ development, epidermal cuticle formation, hormone transport, secretion of secondary metabolites, resistance to biotic and abiotic stresses, and plant microbial interaction [55]. The function prediction of *GmABCG40* shows that *GmABCG40* gene may respond to the induction of multiple hormones, especially the transportation of abscisic acid in soybean root, and play an important role in the process of soybean resistance to stress and growth and development [56]. When the expression of *GmABCG1* in roots and pods is high, soybean seeds contain more fat. It can be speculated that the expression of *GmABCG1* may be related to the transportation of lipid substances and seed development [57]. Unlike ABCG, ABCB transporters play an important role in the polar auxin transport, plant resistance to heavy metals and aluminum toxicity, etc. [58]. For example, the TAP type transporter OsABCB27 (OsALS1) located on the vacuole membrane is involved in the response of rice to aluminum stress [59]. Studies have shown that AtABCB14 and AtABCB15 transporters play a role in auxin transport in *A. thaliana*. AtABCB25 is considered to be related to cadmium and lead tolerance. AtABCB27 transporters are located in the vacuole membrane and participate in aluminum metabolism [60–62]. It is reported that the transporter OsABCB23 is involved in the transport of iron and sulfur precursors in rice [63].

The number of ABCC subfamilies in plants is only less than that of ABCD and ABCG subfamilies. A total of 15 ABCC subfamilies have been identified in *A. thaliana*, among which AtABCC1 mediates the uptake of radiolabeled folic acid and antifolate [64,65], while AtABCC2 is related to the transport of glucuronic acid conjugates [66]. These two transporters also mediate tolerance to cadmium, mercury and arsenic by transferring phytochelatin into vacuoles [67]. Similarly, in *Sedum alfredii*, the ABCC transporter subfamilies Sa14F190 and Sa18F186 also participate in the tolerance and adaptation of plants to cadmium [68]. In rice, OsABCC1 locates in the vacuole membrane and participates in the accumulation and storage of arsenic in grains [69]. Pan found that plant ABC transporter ABCC8 may act as the plasma membrane transporter of glyphosate, reducing the level of cytoplasmic glyphosate, thus giving plants resistance to glyphosate [70].

In this study, three subfamilies ABCB, ABCC and ABCG obtained from cowpea transcriptome data account for a large proportion in the differential expression of ABC family, indicating that the three subfamilies play a major role in Se stress in leguminous plants. WGCNA analysis showed that the content of organic Se in cowpea materials treated with high concentration of SeNPs was significantly correlated with the expression levels of ABCC11, ABCC13 and ABCC10 transporters. This result suggests that ABCC may be more involved in the transmembrane transport of organic Se in plant cells.

4. Materials and Methods

4.1. Plant Materials

Vigna unguiculata cylindrica was selected as experimental material, and Na_2SeO_3 and selenium nanoparticles (SeNPs) were used as exogenous Se to spray cowpea leaves. The experiment site is the nursery garden of Wuhan Light Industry University (the soil type is loam). The sources of applied Se are, respectively, biological SeNPs and Na_2SeO_3, SeNPs is provided by Se-rich Microorganism Research Group of Wuhan Polytechnic University, and Na_2SeO_3 is purchased from Hubei Cheng Chemical Co., Ltd. (Wuhan, China) [71]. Biosynthesized selenium nanoparticles by *B. subtilis* T5 were characterized systematically using UV-vis spectroscopy, FTIR, Zeta Potential, DLS, and SEM techniques [72].

4.2. Treatment of Cowpea with Se

The experiment began sowing on 31 March 2020. The cowpea seeds with full grains were selected and sown on the flat land without weeds, with the plant spacing of 25 cm and the group spacing of 50 cm. On June 3, exogenous Se was sprayed on the leaves at the early flowering stage. Se concentration gradient is set as 0, 0.1, 0.5, 1.0, 2.5, 3.0 mmol/L. Randomized block design was used in the experiment, which was repeated for 3 times (Table S2, Supplementary Materials). On 11 June 2020, the mature cowpea pods with the same growth status will be picked, washed with ultrapure water, and then subpackaged. They will be placed in the −80 °C refrigerator for the subsequent determination of relevant indicators.

From July to September in 2020, the treatment tests of different harvest time and Se application times were carried out. Sowing on 27 July, applying Se once at the early flowering stage of cowpea (29 August), and sampling twice on 8 September and 25 September, respectively; The second application of Se was carried out in the pod setting period (17 September), and the sampling was carried out on 25 September. The exogenous Se was sprayed on the leaves. The Se source used was SeNPs and Na_2SeO_3. The application concentrations were 0 (control), 0.5 mmol/L (low) and 2.5 mmol/L (high), respectively. The experiment was randomized block design and repeated for three times. During the growing period of cowpea, normal water, fertilizer and pest control management shall be carried out. After sampling, dry it, grind it into powder and sieve it for standby.

4.3. Chlorophyll, Cellulose, Soluble Sugar and Protein, Free Amino Acid, and Vc Contents

Chlorophyll, cellulose, soluble sugar and protein, free amino acid, and Vc content were determined following the method described by Hou [73]. Briefly, the chlorophyll and carotenoids were extracted with 95% ethanol and measured using an ultraviolet-visible (UV-Vis) spectrophotometer at 665, 649, and 470 nm. Soluble sugar and protein contents were determined by anthrone–sulfuric acid colorimetry at 620 nm and Coomassie bright blue G-250 color rendering at 595 nm, respectively. Total free amino acid content was measured via ninhydrin colorimetry at 570 nm. Vc content was measured using the 2,6-dichloroindophenol titration method.

4.4. Determination of Antioxidant Index of Cowpea Pods

The antioxidant index of cowpea pods is determined according to Rao's method [74]. For the determination of hydrogen peroxide and MDA content, refer to the Kit (A003-1-2, A003-3-5) of Nanjing Jiancheng Biological Engineering Co., Ltd. (Nanjing, China).

The activities of CAT, polyphenol oxidase PPO), peroxidase POD, superoxide dismutase SOD and GR in cowpea pods were measured using commercial enzyme activity assay Kits (Nanjing Jiancheng Technology Co., Ltd., Nanjing, China) in accordance with the manufacturer' instructions.

4.5. Total Se and Se Species Concentrations

The total Se content of cowpea pods was determined using hydride generation atomic fluorescence spectrometry (HG-AFS). Briefly, 0.5 g of dry samples were digested with 10 mL of HNO_3 and 2 mL of H_2O_2 in a microwave digestion system. The digested solutions were added with 5 mL of HCl and then heated until they became clear. The solutions were diluted with water to 10 mL for HG-AFS (AFS8510, Haiguang Instrument, Beijing, China) detection. The working parameters for HG-AFS were as follows: negative high voltage, 340 V; lamp current, 100 mA; atomization temperature, 800 °C; carrier gas flow rate, 500 mL/min; injection volume, 1 mL; KBH4 concentration, 3.5%. Se species analysis was performed via liquid chromatography HG-AFS (LC-AFS8510, Haiguang Instrument, Beijing, China). Five standard substances, namely, $SeO3^{2-}$, $SeO4^{2-}$, SeCys2, SeMet, and MeSeCys, purchased from the National Institute of Metrology (Beijing, China), were employed to plot standard curves. Extraction of Se species was performed by placing 0.2 g of dry samples in centrifuge tubes, immersing the tubes in a water bath with ultrasonication at 70 °C for 1 h, and then centrifuging them at 5000 rpm for 10 min. The supernate was measured via an LC–HG-AFS under the following conditions: mobile phase, 40 mmol/L KH_2PO_4 + 20 mmol/L KCl, pH 6.0; flow rate, 1.0 mL/min, chromatographic column, Hamilton PRP-X100 (Hamilton Co., Reno, NV, USA); column temperature, 25 °C; injection volume, 100 μL; cathodic current, 80 mA; carrier gas flow rate, 600 mL/min; and negative high voltage, 350 V. Each treatment included three biological replicates with two technical replicates.

4.6. Transcriptome Sequencing and Analysis

Total RNA of four stages were extracted by the Trizol Kit (Promega, Beijing, China) according to the manufacturer's instructions. The RNA was treated with DNase I (Takara to remove genomic DNA). The RNA integrity and quality were verified by RNase-free agarose gel and NanoDrop 2000 (IMPLEN, Westlake Village, CA, USA).

Total RNA was used as input material for the RNA sample preparations. The clustering of the index-coded samples was performed on a cBot Cluster Generation System using TruSeq PE Cluster Kit v3-cBot-HS (Illumina) according to the manufacturer's instructions. After cluster generation, the library preparations were sequenced on an Illumina Novaseq platform and 150 bp paired-end reads were generated. Raw data (raw reads) of fastq format were firstly processed through in-house Perl scripts. Q20, Q30 and GC content, the clean data, were calculated. All the downstream analyses were based on the clean data with high quality. Reference genome and gene model annotation files were downloaded from genome website directly. The mapped reads of each sample were assembled by StringTie (v1.3.3b, Johns Hopkins University, Baltimore, MD, USA) in a reference-based approach [75].

Transcriptome assembly was then performed using Trinity software (v2.4, the Broad Institute, Cambridge, MA, USA). Gene function was annotated on the basis of the following databases: the National Center for Biotechnology Information non-redundant protein sequences (Nr), Kyoto Encyclopedia of Genes and Genomes (KEGG), Protein family (Pfam), the evolutionary genealogy of genes: Nonsupervised Orthologous Groups (egg-NOG), Swiss-Prot, and Gene Ontology (GO). Fragments per kilobase of transcript per million fragments (FPKM) was used to represent the expression level of unigenes.

To identify key regulatory genes in the absorption and transport of Se corresponding to SeNPs and Na_2SeO_3 treatments, WGCNA was performed by the Novogene online tools (https://magic.novogene.com/customer/ (accessed on 13 September 2022). The module eigengene was defined as the first principal component of a given module, and then used to represent the expression profile of module genes in each sample. The Pearson correlations between the eigengenes of each module and the abundance of flavonoids were performed using R package ggplot2.

4.7. Statistical Analysis

Excel 2021 (v2212, Microsoft Corporation, Redmond, WA, USA) and SPSS (v22.0, IBM Corporation, Amonk, NY, USA) were used for experimental data processing. Duncan's test was used to analyze significant differences ($p \leq 0.05$). All of the experimental data had three biological replicates. Correlation Network was performed using the OmicStudio tools at https://www.omicstudio.cn/tool (accessed on 21 November 2022). The positive correlation threshold is set to be greater than or equal to 0.5, the negative correlation threshold is set to be less than or equal to −0.5, and the *p*-value threshold is less than 0.5, R version 3.6.1, igraph1.2.6.

5. Conclusions

Organic Se is safer than inorganic Se and can be used as a nutritional supplement for people living in Se deficient areas. In the present study, both SeNPs and Na_2SeO_3 treatment can increase the soluble sugar content in cowpea pods. SeNPs treatment increased the content of cellulose in cowpea pods. Na_2SeO_3 treatment increased the content of Vc in cowpea pods. Se application could effectively alleviate the peroxidation of cowpea plants and enhance the stress resistance of cowpea. Compared with SeNPs, Na_2SeO_3 treatment had a better effect on improving the total Se content in cowpea pods. After treatment with SeNPs and Na_2SeO_3, the Se form detected in cowpea pods was mainly SeMet, followed by MeSeCys. ABCB, ABCC and ABCG subfamilies play an important role in the absorption and transport of Se in cowpea. The content of organic Se in cowpea treated with high concentration of SeNPs was significantly and positively correlated with the expression level of three transporters ABCC11, ABCC13 and ABCC10, which means that the ABCC subfamily may be more involved in the transmembrane transport of organic Se. In the future, the functions of these three genes need to be verified, such as the transport and absorption of Se in transgenic yeast. Further research will be beneficial to the molecular assisted breeding of Se-rich legume cultivars.

Supplementary Materials: The following supporting information can be downloaded at: https://www.mdpi.com/article/10.3390/molecules28031398/s1, Figure S1: The heat map of correlation analysis of data among cowpea transcriptome samples; Figure S2: Density distribution of cowpea transcriptome data; Figure S3 GO enrichment analysis of differentially expressed genes in cowpea samples; Table S1: Analysis on the number and proportion of differentially expressed genes in ABC gene family of cowpea in different treatment groups; Table S2: Treatment concentration and sampling tissue of cowpea.

Author Contributions: Methodology, software, formal analysis, funding acquisition, L.L.; conceptualization, resources, writing—original draft, H.C.; resources, writing—review and editing, validation, investigation, Y.W.; research materials, Y.X.; experiment, formal analysis, methodology, software, S.W. (Shuai Wu); software, investigation, visualization, S.W. (Shiyan Wang); software, formal analysis, visualization, C.X.; validation, data curation, S.C. All authors have read and agreed to the published version of the manuscript.

Funding: This research was funded by the Special Projects for Technological Innovation in Hubei Province, Hubei Provincial Department of Science and Technology (No. 2019ABA113).

Institutional Review Board Statement: Not applicable.

Informed Consent Statement: Not applicable.

Data Availability Statement: Not applicable.

Acknowledgments: The research was supported by Horizontal science and technology of Enshi Se-De Bioengineering Co., Ltd., Grant Number se1-202102.

Conflicts of Interest: The authors declare no conflict of interest.

Sample Availability: Samples of the cowpea varieties and nano-selenium synthetic strains are available from the authors.

References

1. Roman, M.; Jitaru, P.; Barbante, C. Selenium biochemistry and its role for human health. *Metallomics* **2014**, *6*, 25–54. [CrossRef] [PubMed]
2. Shu, N.; Cheng, Q.; Arnér, E.S.J.; Davies, M.J. Inhibition and crosslinking of the selenoprotein thioredoxin reductase-1 by p-benzoquinone. *Redox Biol.* **2020**, *28*, 101335. [CrossRef] [PubMed]
3. Cardoso, B.R.; Roberts, B.R.; Bush, A.I.; Hare, D.J. Selenium, selenoproteins and neurodegenerative diseases. *Metallomics* **2015**, *7*, 1213–1228. [CrossRef] [PubMed]

4. Kipp, A.P.; Strohm, D.; Brigelius-Flohé, R.; Schomburg, L.; Bechthold, A.; Leschik-Bonnet, E.; Heseker, H.; German Nutrition, S. Revised reference values for selenium intake. *J. Trace Elem. Med. Biol.* **2015**, *32*, 195–199. [CrossRef] [PubMed]
5. Krinsky, N.I.; Beecher, G.R.; Burk, R.F.; Chan, A.C.; Erdman, J.J.; Jacob, R.A.; Jialal, I.; Kolonel, L.N.; Marshall, J.R.; Taylor Mayne, P.R. Dietary reference intakes for vitamin C, vitamin E, selenium, and carotenoids. *Inst. Med.* **2000**, *19*, 95–185. [CrossRef]
6. White, P.J. Selenium metabolism in plants. *Biochim. Biophys. Acta-Gen. Subj.* **2018**, *1862*, 2333–2342. [CrossRef]
7. Richard, C.T.; Elizabeth, A.H.P.-S. Selenium transport and metabolism in plants: Phytoremediation and biofortification implications. *J. Hazard. Mater.* **2021**, *404*, 124178. [CrossRef]
8. D'Amato, R.; Fontanella, M.C.; Falcinelli, B.; Beone, G.M.; Bravi, E.; Marconi, O.; Benincasa, P.; Businelli, D. Selenium biofortification in rice (*Oryza sativa* L.) sprouting: Effects on Se yield and nutritional traits with focus on phenolic acid profile. *J. Agric. Food Chem.* **2018**, *66*, 4082–4090. [CrossRef]
9. D'Amato, R.; Proietti, P.; Onofri, A.; Regni, L.; Esposto, S.; Servili, M.; Businelli, D.; Selvaggini, R. Biofortification (Se): Does it increase the content of phenolic compounds in virgin olive oil (VOO)? *PLoS ONE* **2017**, *12*, e0176580. [CrossRef]
10. Liao, X.; Rao, S.; Yu, T.; Zhu, Z.; Yang, X.; Xue, H.; Gou, Y.; Cheng, S.; Xu, F. Selenium yeast promoted the Se accumulation, nutrient quality and antioxidant system of cabbage (*Brassica oleracea* var. *capitata L.). Plant Signal. Behav.* **2021**, *16*, 1907042. [CrossRef]
11. Rao, S.; Gou, Y.; Yu, T.; Cong, X.; Gui, J.; Zhu, Z.; Zhang, W.; Liao, Y.; Ye, J.; Cheng, S. Effects of selenate on Se, flavonoid, and glucosinolate in broccoli florets by combined transcriptome and metabolome analyses. *Food Res. Int.* **2021**, *146*, 110463. [CrossRef]
12. Rao, S.; Yu, T.; Cong, X.; Lai, X.; Xiang, J.; Cao, J.; Liao, X.; Gou, Y.; Chao, W.; Xue, H.; et al. Transcriptome, proteome, and metabolome reveal the mechanism of tolerance to selenate toxicity in *Cardamine violifolia* . *J. Hazard. Mater.* **2021**, *406*, 124283. [CrossRef]
13. Silva, V.M.; Boleta, E.H.M.; Lanza, M.G.D.B.; Lavres, J.; Martins, J.T.; Santos, E.F.; dos Santos, F.L.M.; Putti, F.F.; Junior, E.F.; White, P.J.; et al. Physiological, biochemical, and ultrastructural characterization of selenium toxicity in cowpea plants. *Environ. Exp. Bot.* **2018**, *150*, 172–182. [CrossRef]
14. Hall, A.E. Phenotyping cowpeas for adaptation to drought. *Front. Physiol.* **2012**, *3*, 155. [CrossRef]
15. Manzeke, M.G.; Mtambanengwe, F.; Nezomba, H.; Watts, M.J.; Broadley, M.R.; Mapfumo, P. Zinc fertilization increases productivity and grain nutritional quality of cowpea (*Vigna unguiculata* [L.] Walp.) under integrated soil fertility management. *Field Crops Res.* **2017**, *213*, 231–244. [CrossRef]
16. Schiavon, M.; Pilon-Smits, E.A.H. The fascinating facets of plant selenium accumulation–biochemistry, physiology, evolution and ecology. *New Phytol.* **2017**, *213*, 1582–1596. [CrossRef]
17. Dos Reis, A.R.; El-Ramady, H.; Santos, E.F.; Gratão, P.L.; Schomburg, L. Overview of selenium deficiency and toxicity worldwide: Affected areas, selenium-related health issues, and case studies. In *Selenium in Plants*; Springer: Cham, Switzerland, 2017; pp. 209–230.
18. Lanza, M.G.D.B.; Silva, V.M.; Montanha, G.S.; Lavres, J.; de Carvalho, H.W.P.; Dos Reis, A.R. Assessment of selenium spatial distribution using μ-XFR in cowpea (*Vigna unguiculata* (L.) Walp.) plants: Integration of physiological and biochemical responses. *Ecotoxicol. Environ. Saf.* **2021**, *207*, 111216. [CrossRef]
19. Zhang, L.; Hu, B.; Li, W.; Che, R.; Deng, K.; Li, H.; Yu, F.; Ling, H.; Li, Y.; Chu, C. OsPT2, a phosphate transporter, is involved in the active uptake of selenite in rice. *New Phytol.* **2014**, *201*, 1183–1191. [CrossRef]
20. Byrne, S.L.; Durandeau, K.; Nagy, I.; Barth, S. Identification of ABC transporters from Lolium perenne L. that are regulated by toxic levels of selenium. *Planta* **2010**, *231*, 901–911. [CrossRef]
21. Verrier, P.J.; Bird, D.; Burla, B.; Dassa, E.; Forestier, C.; Geisler, M.; Klein, M.; Kolukisaoglu, Ü.; Lee, Y.; Martinoia, E. Plant ABC proteins a unified nomenclature and updated inventory. *Trends Plant Sci.* **2008**, *13*, 151–159. [CrossRef]
22. Zhao, X.Q.; Mitani, N.; Yamaji, N.; Shen, R.F.; Ma, J.F. Involvement of silicon influx transporter OsNIP2; 1 in selenite uptake in rice. *Plant Physiol.* **2010**, *153*, 1871–1877. [CrossRef] [PubMed]
23. Li, H.F.; McGrath, S.P.; Zhao, F.J. Selenium uptake, translocation and speciation in wheat supplied with selenate or selenite. *New Phytol.* **2008**, *178*, 92–102. [CrossRef] [PubMed]
24. Kong, Q.; Li, F.; Qin, L.; Chen, E.; Yang, Y.; Zhang, H.; Guan, Y. Screening and analysis of Se responsive genes in leaves of foxtail millet. *Mol. Plant Breed.* **2021**, *19*, 2798–2810. [CrossRef]
25. Ishikawa, H.; Batieno, B.J.; Fatokun, C.; Boukar, O. A high plant density and the split application of chemical fertilizer increased the grain and protein content of cowpea (*Vigna unguiculata*) in Burkina Faso, West Africa. *Agriculture* **2022**, *12*, 199. [CrossRef]
26. Wang, L.; Lin, C.; Luo, H.; Jin, C.; Yu, Z.; Qian, C. Impact of precooling time on quality and physio-biochemical characteristics of common bean (*Phaseolus vulgaris* L.) and cowpea (*Vigna sinensis*) under controlled atmosphere storage conditions. *Food Sci.* **2018**, *39*, 232–238. [CrossRef]
27. Cai, R.; Hettiarachchy, N.S.; Jalaluddin, M. High-performance liquid chromatography determination of phenolic constituents in 17 varieties of cowpeas. *J. Agric. Food Chem.* **2003**, *51*, 1623–1627. [CrossRef]
28. Cheng, A.; Wu, J.; Qin, H.; Yang, Q.; Liu, C.; Guo, X.; Sun, J. Content and antioxidant activity of polyphenols and flavonoids from 4 food beans. *J. Chin. Cereals Oils Assoc.* **2017**, *32*, 28–32. [CrossRef]
29. D'Amato, R.; Regni, L.; Falcinelli, B.; Mattioli, S.; Benincasa, P.; Dal Bosco, A.; Pacheco, P.; Proietti, P.; Troni, E.; Santi, C.; et al. Current knowledge on selenium biofortification to improve the nutraceutical profile of food: A comprehensive review. *J. Agric. Food Chem.* **2020**, *68*, 4075–4097. [CrossRef]

30. Manaf, H.H. Beneficial effects of exogenous selenium, glycine betaine and seaweed extract on salt stressed cowpea plant. *Ann. Agric. Sci.* **2016**, *61*, 41–48. [CrossRef]
31. Liu, G.; Li, P.; Li, Y.; Ren, X.; Wang, J. Effects of sodium selenite treatment on quality and active oxygen of cowpea. *J. Gansu Agric. Univ.* **2020**, *55*, 85–91. [CrossRef]
32. Ye, J.; Shan, W.; Zhu, G. Effect of foliar application selenium on the yield and quality of soybean. *Xinjiang Agric. Sci.* **2009**, *46*, 506–509.
33. Hussein, H.-A.A.; Darwesh, O.M.; Mekki, B.B. Environmentally friendly nano-selenium to improve antioxidant system and growth of groundnut cultivars under sandy soil conditions. *Biocatal. Agric. Biotechnol.* **2019**, *18*, 101080. [CrossRef]
34. Malik, J.A.; Kumar, S.; Thakur, P.; Sharma, S.; Kaur, N.; Kaur, R.; Pathania, D.; Bhandhari, K.; Kaushal, N.; Singh, K. Promotion of growth in mungbean (*Phaseolus aureus* Roxb.) by selenium is associated with stimulation of carbohydrate metabolism. *Biol. Trace Elem. Res.* **2011**, *143*, 530–539. [CrossRef]
35. Wu, Z.; Guo, D.; Gao, W. Effects of Na_2SeO_3 sprayed at different time on growth and selenium accumulation of pea seedling. *Seed* **2008**, *27*, 40–42. [CrossRef]
36. Liu, F.; Zhou, Q.; Zhou, S.; Huang, Y.; Liu, K.; Zhang, D.; Zhou, H. Effect of application of sodium selenite to soil on the growth and physiological indexes and selenium accumulation of *Astragalus sinicus* . *Chin. J. Soil Sci.* **2016**, *47*, 129–136. [CrossRef]
37. Kovács, Z.; Soós, Á.; Kovács, B.; Kaszás, L.; Elhawat, N.; Bákonyi, N.; Razem, M.; Fári, M.G.; Prokisch, J.; Domokos-Szabolcsy, É. Uptake dynamics of ionic and elemental selenium forms and their metabolism in multiple-harvested alfalfa (*Medicago sativa* L.). *Plants* **2021**, *10*, 1277. [CrossRef]
38. Wu, Z.; Wang, S.; Jiang, Y.; Zhang, J.; Gao, W. Effect of sodium selenite on plant growth and selenium accumulation in pea sprout. *Guangdong Trace Elem. Sci.* **2004**, *11*, 35–38. [CrossRef]
39. Hu, T.; Li, H.; Li, J.; Zhao, G.; Wu, W.; Liu, L.; Wang, Q.; Guo, Y. Absorption and bio-transformation of selenium nanoparticles by wheat seedlings (*Triticum aestivum* L.). *Front. Plant Sci.* **2018**, *9*, 597. [CrossRef]
40. Wang, K.; Wang, Y.; Li, K.; Wan, Y.; Wang, Q.; Zhuang, Z.; Guo, Y.; Li, H. Uptake, translocation and biotransformation of selenium nanoparticles in rice seedlings (*Oryza sativa* L.). *J. Nanobiotechnol.* **2020**, *18*, 1–15. [CrossRef]
41. Li, Y.; Zhu, N.; Liang, X.; Zheng, L.; Zhang, C.; Li, Y.-F.; Zhang, Z.; Gao, Y.; Zhao, J. A comparative study on the accumulation, translocation and transformation of selenite, selenate, and SeNPs in a hydroponic-plant system. *Ecotoxicol. Environ. Saf.* **2020**, *189*, 109955. [CrossRef]
42. Rady, M.M.; Desoky, E.-S.M.; Ahmed, S.M.; Majrashi, A.; Ali, E.F.; Arnaout, S.M.A.I.; Selem, E. Foliar nourishment with nano-selenium dioxide promotes physiology, biochemistry, antioxidant defenses, and salt tolerance in phaseolus vulgaris. *Plants* **2021**, *10*, 1189. [CrossRef]
43. Qados, A.M.S.A. Effects of salicylic acid on growth, yield and chemical contents of pepper (*Capsicum annuum* L) plants grown under salt stress conditions. *Int. J. Agric. Crop Sci.* **2015**, *8*, 107–113.
44. Silva, V.M.; Tavanti, R.F.R.; Gratão, P.L.; Alcock, T.D.; Dos Reis, A.R. Selenate and selenite affect photosynthetic pigments and ROS scavenging through distinct mechanisms in cowpea (*Vigna unguiculata* (L.) walp) plants. *Ecotoxicol. Environ. Saf.* **2020**, *201*, 110777. [CrossRef] [PubMed]
45. Reis, H.P.G.; de Queiroz Barcelos, J.P.; Junior, E.F.; Santos, E.F.; Silva, V.M.; Moraes, M.F.; Putti, F.F.; dos Reis, A.R. Agronomic biofortification of upland rice with selenium and nitrogen and its relation to grain quality. *J. Cereal Sci.* **2018**, *79*, 508–515. [CrossRef]
46. Yin, H.; Qi, Z.; Li, M.; Ahammed, G.J.; Chu, X.; Zhou, J. Selenium forms and methods of application differentially modulate plant growth, photosynthesis, stress tolerance, selenium content and speciation in *Oryza sativa* L. *Ecotoxicol. Environ. Saf.* **2019**, *169*, 911–917. [CrossRef]
47. Jóźwiak, W.; Politycka, B. Effect of selenium on alleviating oxidative stress caused by a water deficit in cucumber roots. *Plants* **2019**, *8*, 217. [CrossRef]
48. Wang, Y.; Zhao, T.; Chang, S.; Li, S.; Pang, X.; Zheng, L.; Kang, Y. Effect of nano-selenium spraying on the growth characteristics, nutritional quality, phenolics content and antioxidant activity of mung bean sprouts. *J. China Agric. Univ.* **2019**, *24*, 39–46. [CrossRef]
49. Shao, S.; Dernovics, M.; Long, S. Study on the forms of selenium in Enshi selenium rich beans. *J. Miner.* **2015**, *35*, 671. [CrossRef]
50. Hu, Y. *Identification and Analysis of Genes Related to Selenium Assimilation and Metabolism in Tea Plant Roots*; Chinese Academy of Agricultural Sciences: Hangzhou, China, 2016.
51. Zhu, L.; Xu, J.; Zhang, D. Molecular evolution, expression and functional network prediction analysis of ABC transporter gene family in *Arabidopsis Thaliana*. *Plant Physiol. J.* **2012**, *48*, 1151–1166. [CrossRef]
52. Xie, X. *Identification and Functional Characterization of Nicotiana Tabacum Abc Transporters Involved in The Transport of Secondary Metabolites*; Chongqing University: Chongqing, China, 2014.
53. Wang, X.; Sun, W.; Ma, Y.; Han, E.; Han, L.; Sun, L.; Peng, Z.; Wang, B. Research progress of ABC transporters in *Arabidopsis thaliana* . *Plant Physiol. J.* **2017**, *53*, 133–144. [CrossRef]
54. Niu, L.; Li, H.; Song, Z.; Dong, B.; Cao, H.; Liu, T.; Du, T.; Yang, W.; Amin, R.; Wang, L. The functional analysis of ABCG transporters in the adaptation of pigeon pea (*Cajanus cajan*) to abiotic stresses. *PeerJ* **2021**, *9*, e10688. [CrossRef]
55. Do, T.H.T.; Martinoia, E.; Lee, Y. Functions of ABC transporters in plant growth and development. *Curr. Opin. Plant Biol.* **2018**, *41*, 32–38. [CrossRef]

56. Li, Y.; Zhang, Y.; Li, D.; Wang, T.; Chen, W.; Wang, X.; Li, Y.; Li, W. Functional prediction and expression analysis of GmABCG40. *Soybean Sci.* **2017**, *36*, 502–507. [CrossRef]
57. Han, E.; Wang, H.; Han, L.; Li, J.; Li, Y.; Wang, B. Cloning and functional analysis of *GmABCG1* gene in *Glycine max* . *Mol. Plant Breed.* **2018**, *16*, 7925–7932. [CrossRef]
58. Xu, Y.; Zhang, S.; Guo, H.; Wang, S.; Xu, L.; Li, C.; Qian, Q.; Chen, F.; Geisler, M.; Qi, Y.; et al. OsABCB14 functions in auxin transport and iron homeostasis in rice (*Oryza sativa* L.). *Plant J.* **2014**, *79*, 106–117. [CrossRef]
59. Huang, C.F.; Yamaji, N.; Mitani, N.; Yano, M.; Nagamura, Y.; Ma, J.F. A bacterial-type ABC transporter is involved in aluminum tolerance in rice. *Plant Cell* **2009**, *21*, 655–667. [CrossRef]
60. Kaneda, M.; Schuetz, M.; Lin, B.S.P.; Chanis, C.; Hamberger, B.; Western, T.L.; Ehlting, J.; Samuels, A.L. ABC transporters coordinately expressed during lignification of Arabidopsis stems include a set of ABCBs associated with auxin transport. *J. Exp. Bot.* **2011**, *62*, 2063–2077. [CrossRef]
61. Kim, D.; Bovet, L.; Kushnir, S.; Noh, E.W.; Martinoia, E.; Lee, Y. AtATM3 is involved in heavy metal resistance in Arabidopsis. *Plant Physiol.* **2006**, *140*, 922–932. [CrossRef]
62. Larsen, P.B.; Cancel, J.; Rounds, M.; Ochoa, V. Arabidopsis ALS1 encodes a root tip and stele localized half type ABC transporter required for root growth in an aluminum toxic environment. *Planta* **2007**, *225*, 1447–1458. [CrossRef]
63. Zuo, J.; Wu, Z.; Li, Y.; Shen, Z.; Feng, X.; Zhang, M.; Ye, H. Mitochondrial ABC transporter ATM3 is essential for cytosolic iron-sulfur cluster assembly. *Plant Physiol.* **2017**, *173*, 2096–2109. [CrossRef]
64. Lefèvre, F.; Boutry, M. Towards identification of the substrates of ATP-binding cassette transporters. *Plant Physiol.* **2018**, *178*, 18–39. [CrossRef] [PubMed]
65. Raichaudhuri, A.; Peng, M.; Naponelli, V.; Chen, S.; Sánchez-Fernández, R.; Gu, H.; Gregory, J.F.; Hanson, A.D.; Rea, P.A. Plant vacuolar ATP-binding cassette transporters that translocate folates and antifolates in vitro and contribute to antifolate tolerance in vivo. *J. Biol. Chem.* **2009**, *284*, 8449–8460. [CrossRef] [PubMed]
66. Liu, G.; Sánchez-Fernández, R.; Li, Z.-S.; Rea, P.A. Enhanced multispecificity of Arabidopsis vacuolar multidrug resistance-associated protein-type ATP-binding cassette transporter, AtMRP2. *J. Biol. Chem.* **2001**, *276*, 8648–8656. [CrossRef] [PubMed]
67. Park, J.; Song, W.Y.; Ko, D.; Eom, Y.; Hansen, T.H.; Schiller, M.; Lee, T.G.; Martinoia, E.; Lee, Y. The phytochelatin transporters AtABCC1 and AtABCC2 mediate tolerance to cadmium and mercury. *Plant J.* **2012**, *69*, 278–288. [CrossRef]
68. Feng, T.; He, X.; Zhuo, R.; Qiao, G.; Han, X.; Qiu, W.; Chi, L.; Zhang, D.; Liu, M. Identification and functional characterization of ABCC transporters for Cd tolerance and accumulation in *Sedum alfredii* Hance. *Sci. Rep.* **2020**, *10*, 20928. [CrossRef]
69. Song, W.; Yamaki, T.; Yamaji, N.; Ko, D.; Jung, K.; Fujii-Kashino, M.; An, G.; Martinoia, E.; Lee, Y.; Ma, J.F. A rice ABC transporter, OsABCC1, reduces arsenic accumulation in the grain. *Proc. Natl. Acad. Sci. USA* **2014**, *111*, 15699–15704. [CrossRef]
70. Pan, L.; Yu, Q.; Wang, J.; Han, H.; Mao, L.; Nyporko, A.; Maguza, A.; Fan, L.; Bai, L.; Powles, S. An ABCC-type transporter endowing glyphosate resistance in plants. *Proc. Natl. Acad. Sci. USA* **2021**, *118*, e2100136118. [CrossRef]
71. Deng, K.; Li, L.; Li, L.; Xu, F.; Yuan, H.; Zha, S.; Xiao, X.; Yu, J.; Cheng, S.; Cheng, H. Molecular mechanism of selenium affecting the synthesis of flavonoids in *G. biloba* leaves. *Plant Mol. Biol. Rep.* **2022**, *40*, 232–246. [CrossRef]
72. Duan, Y.; Li, M.; Zhang, S.; Wang, Y.; Deng, J.; Wang, Q.; Yi, T.; Dong, X.; Cheng, S.; He, Y.; et al. Highly efficient biotransformation and production of selenium nanoparticles and polysaccharides using potential probiotic *Bacillus subtilis* T5. *Metabolites* **2022**, *12*, 1204. [CrossRef]
73. Hou, F. *Experiments in Plant Physiology*; Science Press: Beijing, China, 2022; pp. 23–28.
74. Gui, J.Y.; Rao, S.; Gou, Y.; Xu, F.; Cheng, S. Comparative study of the effects of selenium yeast and sodium selenite on selenium content and nutrient quality in broccoli florets (*Brassica oleracea* L. var. *italica*). *J. Sci. Food Agric.* **2022**, *102*, 1707–1718. [CrossRef]
75. Pertea, M.; Pertea, G.M.; Antonescu, C.M.; Chang, T.C.; Mendell, J.T.; Salzberg, S.L. StringTie enables improved reconstruction of a transcriptome from RNA-seq reads. *Nat. Biotechnol.* **2015**, *33*, 290–295. [CrossRef]

molecules

MDPI

Article

Lignin Degradation by *Klebsiella aerogenes* TL3 under Anaerobic Conditions

Zhuowei Tu [1], Alei Geng [1,2,*], Yuhua Xiang [1], Anaiza Zayas-Garriga [1], Hao Guo [1], Daochen Zhu [1], Rongrong Xie [1] and Jianzhong Sun [1,*]

1 Biofuels Institute, School of the Environment and Safety Engineering, Jiangsu University, Zhenjiang 212013, China; zwtu99@hotmail.com (Z.T.); mr.linyue@outlook.com (Y.X.); anaiza.zayas10@gmail.com (A.Z.-G.); gh15036951363@hotmail.com (H.G.); dczhucn@ujs.edu.cn (D.Z.); rrxie@ujs.edu.cn (R.X.)

2 Changzhou Engineering and Technology Institute, Jiangsu University, Changzhou 214153, China

* Correspondence: galxj@ujs.edu.cn (A.G.); jzsun1002@hotmail.com (J.S.)

Abstract: Lignin, the largest non-carbohydrate component of lignocellulosic biomass, is also a recalcitrant component of the plant cell wall. While the aerobic degradation mechanism of lignin has been well-documented, the anaerobic degradation mechanism is still largely elusive. In this work, a versatile facultative anaerobic lignin-degrading bacterium, *Klebsiella aerogenes* TL3, was isolated from a termite gut, and was found to metabolize a variety of carbon sources and produce a single kind or multiple kinds of acids. The percent degradation of alkali lignin reached 14.8% under anaerobic conditions, and could reach 17.4% in the presence of glucose within 72 h. Based on the results of infrared spectroscopy and 2D nuclear magnetic resonance analysis, it can be inferred that the anaerobic degradation of lignin may undergo the cleavage of the C-O bond (β-O-4), as well as the C-C bond (β-5 and β-β), and involve the oxidation of the side chain, demethylation, and the destruction of the aromatic ring skeleton. Although the anaerobic degradation of lignin by TL3 was slightly weaker than that under aerobic conditions, it could be further enhanced by adding glucose as an electron donor. These results may shed new light on the mechanisms of anaerobic lignin degradation.

Keywords: lignin linkages; *Klebsiella aerogenes*; lignocellulose; catalytic mechanism

Citation: Tu, Z.; Geng, A.; Xiang, Y.; Zayas-Garriga, A.; Guo, H.; Zhu, D.; Xie, R.; Sun, J. Lignin Degradation by *Klebsiella aerogenes* TL3 under Anaerobic Conditions. *Molecules* **2024**, *29*, 2177. https://doi.org/10.3390/molecules29102177

Academic Editor: Alejandro Rodríguez Pascual

Received: 6 April 2024
Revised: 28 April 2024
Accepted: 3 May 2024
Published: 7 May 2024

1. Introduction

Lignocellulose represents the most abundant renewable source of carbon on the planet. Every year, more than 200 billion tons of non-food lignocellulosic biomass is produced [1], which is a valuable source of organic carbon that may substitute petroleum in the synthesis of various organic chemicals. Lignocellulosic biomass is generally composed of lignin (26–31%), hemicellulose (25–32%), and cellulose (41–46%). As the largest non-carbohydrate fraction of lignocellulose, the utilization of lignin has attracted increasing attention. Lignin is a complex racemic aromatic heteropolymer, which is made up of three differently methylated hydroxycinnamic alcohol monomers: *p*-coumaryl alcohol (H), coniferyl alcohol (G), and sinapyl alcohol (S) linked by various C-O and C-C bonds [2,3]. The high phenyl-ring structure endows lignin with unique physiochemical characteristics, and could be converted into a series of value-added chemicals [4,5].

However, lignin is highly resistant to chemical and biological degradation. It not only confers mechanical resistance to wood, but also leads to the non-specific adsorption of cellulolytic as well as hemicellulolytic enzymes during the saccharification of lignocellulosic biomass, and is associated with enzyme inhibition [6]. This recalcitrant polymer is usually found in the secondary cell wall of plant cells, where it combines with hemicellulose to create an amorphous matrix that embeds cellulose protofibrils and inhibits their biodegradation [7]. Lignin's recalcitrance has hindered the conversion of biomass into bioproducts; nevertheless, exploring strategies for lignin degradation may allow new industrial technologies to improve the economics of lignocellulosic biorefineries [8].

In the natural world, a series of fungi and bacteria are able to break down lignin, with bacteria being the versatile microbes for this process, which is also an essential part of global carbon recycling. Compared to fungi, bacteria possess faster growth rates and more diverse and efficient metabolic pathways for the breakdown of lignin aromatic compounds [4,9,10]. Plenty of research focuses on the aerobic lignin transformation by micro-organisms, which primarily use free radicals and enzymes such as peroxidase as well as laccase to oxidatively tear lignin apart. Nevertheless, the oxygen supply is a major cost in the fermentation industry, which results in oxygen-free anaerobic fermentation becoming of great potential in industrial application. There are relatively few studies on the anaerobic degradation of lignin by individual bacterial strains, whose specific mechanism is still ambiguous. While some anaerobic microbes just use the methyl groups in lignin [11], others may need electron donors.

The anaerobic lignin degradation process is substantial and may involve an array of enzymes as well as pathways. A series of anaerobic lignin-degrading bacteria have been isolated. *Acetoanaerobium* sp. WJDL-Y2, which was isolated from pulp and paper mill sludge, demonstrated kraft lignin degradation and modification capabilities [12]. Rashid et al. validated the lignin-degrading and gas-producing abilities in four anaerobic bacterial strains isolated from municipal-solid-waste-containing soil [13]. The genomic profiles of some anaerobic lignin-degrading bacteria have also been revealed. *Enterobacter lignolyticus* SCF1, a facultative anaerobic bacterium, utilizes the 4-hydroxybenzoate degradation pathway, peroxidase/peroxygenase system, glutathione biosynthesis, and upregulation of glutathione S-transferase (GST) proteins for lignin degradation [14]. *Klebsiella* sp. BRL6-2 possesses genes associated with a complete protocatechuate pathway, six putative peroxidase genes, two lactate dehydrogenase genes, and two catalase genes [15]. *Tolumonas lignolytica* BRL6-1T harbors various lignin degradation pathways; hence, it could enzymatically cleave the β-aryl ether bonds, depolymerizes lignin through extracellular peroxidases, and breaks down lignin-derived aromatic compounds [16]. Chaput et al. isolated a novel anaerobic lignin-degrading species, *Sodalis ligni*, from forest soil, and also found a series of lignin and aromatic compound degradation/metabolism enzymes in its genome [17]. Despite the knowledge of the genomic information of these bacteria, the anaerobic degrading mechanism of lignin is still not fully understood.

In the present study, a facultative anaerobic lignin-degrading bacterium, *Klebsiella aerogenes* TL3, was isolated from a termite gut, the latter of which is considered to be the most efficient bioreactor of lignocellulose, as well as a reservoir of lignocellulolytic microbes. TL3 is capable of anaerobically degrading alkali lignin as well as several other carbon sources to produce a single kind or multiple kinds of acids under anaerobic or aerobic conditions. Specific studies on the changes of the physicochemical properties of lignin and the degradation capacity of TL3 revealed a unique degradation pattern, which might shed new light on the mechanisms of the anaerobic decomposition of lignin.

2. Results

2.1. Growth of TL3 on Various Carbon Sources

A variety of substrates can be effectively utilized by TL3 to produce multiple compounds (Table 1). Surprisingly, the aerobic metabolism of glucose led to an exclusive product of propanoic acid (3.11 g/L), while the aerobic metabolism of xylose gave a single product of butyric acid (1.81 g/L). In addition, no accumulation of acids was found in the degradation of avicel, lignin, and corn cob by TL3 under aerobic condition, but the bacterial growth was clearly observed. By comparison, the degradation of different carbon sources by TL3 under anaerobic conditions produced various types of acids. The richest variety of acids was generated by glucose degradation, of which propionic acid was the most abundant at 2.7 g/L. Of particular note, TL3 was able to degrade corn cob and alkali lignin under anaerobic conditions to produce lactic acid and succinic acid.

Table 1. Fermentation products by TL3 on different carbon sources.

Methods	Fermentation Products (g/L)	Carbon Sources				
		Avicel	Glucose	Xylose	Corn Cob	Lignin
Anaerobic	Lactic Acid	ND [a]	0.73 ± 0.01	0.85 ± 0.02	0.23 ± 0.01	0.13 ± 0.02
	Acetic Acid	ND	1.25 ± 0.02	0.52 ± 0.01	ND	ND
	Succinic Acid	0.64 ± 0.03	1.07 ± 0.00	1.69 ± 0.01	0.82 ± 0.02	1.43 ± 0.00
	Propanoic acid	0.26 ± 0.02	2.70 ± 0.03	0.32 ± 0.03	ND	0.26 ± 0.01
	Butyric Acid	ND	1.38 ± 0.02	ND	ND	ND
Aerobic	Lactic Acid	ND	ND	ND	ND	ND
	Acetic Acid	ND	ND	ND	ND	ND
	Succinic Acid	ND	ND	ND	ND	ND
	Propanoic acid	ND	3.11 ± 0.01	ND	ND	ND
	Butyric Acid	ND	ND	1.81 ± 0.02	ND	ND

[a] "ND", not detected.

2.2. The Lignin Degradation Curve

To determine the growth characteristics and the capability of lignin degradation by strain TL3 under different conditions, cultures were grown for 3 days in mMRS or MSM media with lignin as the carbon source, with or without the addition of glucose. A typical lignin-degrading strain, *Sphingobium* sp. SYK-6, was used as a control strain. Optical densities (OD_{600} and OD_{310}) were measured every 12 h. Lignin contents correlated with the absorbance at 310 nm, where a decrease in absorbance indicates a reduction in the concentration of soluble phenolic and polymeric lignin [16]. As shown in Figure 1a, without the addition of glucose, TL3 was able to grow in the mMRS medium (OD_{600} reached 0.401), but there was almost no change in OD_{310}, indicating that TL3 could grow in this complex medium with minimal lignin consumption. By comparison, TL3 was indeed able to grow in the mMRS medium in the absence of lignin or glucose, reaching a maximum OD_{600} of 0.198 (Supplementary Materials: Figure S1). From Figure 1b, it can be observed that TL3 grew rapidly in the mMRS medium supplemented with glucose, reaching a maximum OD_{600} of approximately 0.616 within 12 h, with the percent degradation of lignin reaching approximately 17.8%. Interestingly, as shown in Figure 1c, TL3 was able to grow in the MSM medium with lignin as the sole carbon source, reaching a maximum OD_{600} of 0.486 within 72 h, and achieving a percent degradation of 14.8%. At this point, the color of the medium changed from brown to black-green. In Figure 1d, TL3 reached its highest OD_{600} (0.142) in the MSM medium supplemented with glucose at 36 h, and the percent degradation of lignin was approximately 17.4%. In most cases, the control strain, SYK-6, did not grow well under anaerobic conditions, except for that in the rich medium supplemented with glucose (Figure 1b).

2.3. FTIR Plots of Lignin Degradation by TL3 under Various Conditions

In order to understand the structural changes in lignin by bacterial treatment, FTIR spectroscopy was employed to analyze lignin samples before and after degradation. Figure 2 depicts the FTIR spectra of alkali lignin before and after treatment with strain TL3 under different conditions. Clear differences in the characteristic peaks of lignin were observed before and after degradation, primarily ranging between 1800 cm^{-1} and 800 cm^{-1}, with the most significant variations observed when a little glucose was added into the lignin-mineral medium as an electron donor. Across all conditions, the absorption peaks at 1596 cm^{-1}, 1425 cm^{-1}, 1374 cm^{-1}, and 1142 cm^{-1} exhibited a decrease, suggesting potential side-chain oxidation, methyl removal, and aromatic-ring-framework modification (Table S1). However, the mMRS-HL and mMRS-YL spectra were quite similar to CK, suggesting very little modification in the lignin structure has occurred under these conditions, and the growth of TL3 might primarily rely on the complex nutrition factors in the medium, which is consistent with the results of Figures 1a and S1. By comparison, in the MSM medium, significant changes occurred in the lignin absorption peaks between 1270 cm^{-1}

and 816 cm^{-1} under anaerobic conditions, particularly when glucose was added, where the peaks at 1218 cm^{-1}, 1142 cm^{-1}, and 1084 cm^{-1} almost disappeared, indicating the extensive disruption of the β-O-4 linkage. The increased absorption at 1033 cm^{-1} indicated a higher C_α-ether content after TL3 treatment. The peaks at 854 cm^{-1} and 816 cm^{-1} nearly disappeared, while those at 3434 cm^{-1} and 2938 cm^{-1} increased, suggesting the modification of lignin side-chains by strain TL3. Overall, lignin degradation was more effective in the MSM medium compared to the mMRS medium, especially when glucose was supplemented as an electron donor in the MSM medium. Hence, studies were primarily carried out using the MSM medium hereafter.

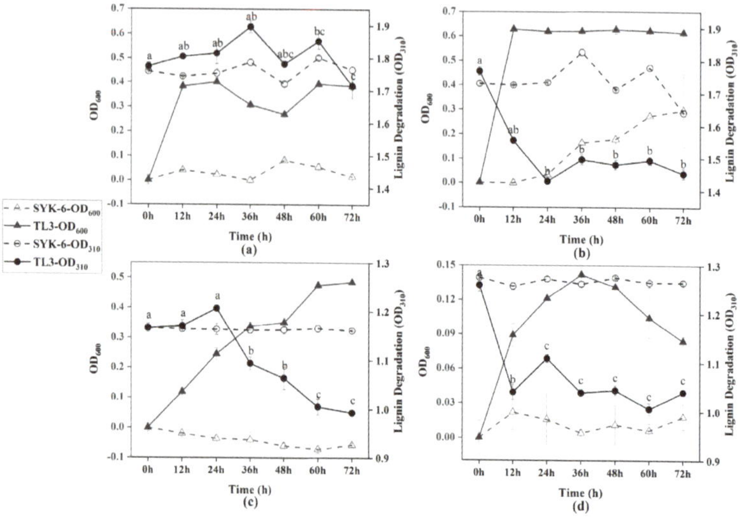

Figure 1. Growth curve of TL3 and anaerobic lignin degradation under different conditions: (**a**) mMRS medium supplemented with 1 g/L lignin; (**b**) mMRS medium supplemented with 1 g/L lignin and glucose; (**c**) MSM medium supplemented with 1 g/L lignin; and (**d**) MSM medium supplemented with 1 g/L lignin and glucose. Error bars showing the standard error of triplicate samples. Different letters indicate statistically significant differences among groups at $p < 0.05$.

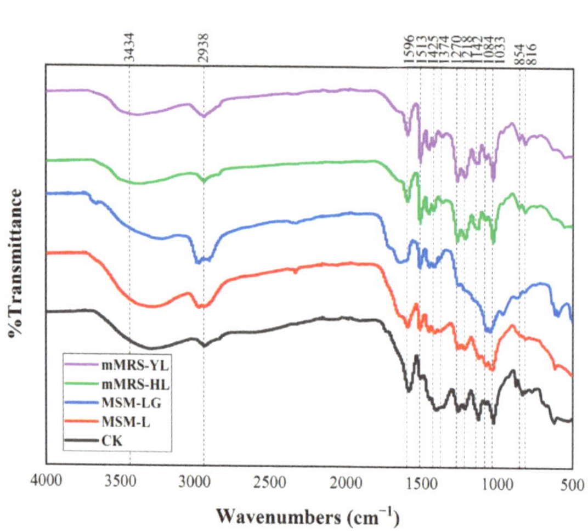

Figure 2. The FTIR spectra of lignin before and after treatment by TL3. CK, untreated alkali lignin; MSM-L, lignin was anaerobically treated with TL3 in MSM media; MSM-LG, lignin was anaerobically

treated with TL3 in MSM media supplemented with 1 g/L glucose; mMRS-HL, lignin was aerobically treated with TL3 in mMRS media; and mMRS-YL, lignin was anaerobically treated with TL3 in mMRS media.

2.4. GC-MS Analysis of the Metabolites

In order to understand the metabolic profiles of lignin by TL3, we conducted a GC-MS analysis to identify the products of lignin decomposition in the MSM medium. Untreated lignin served as a control group. The chromatographic peaks were identified and assigned according to the National Institute of Standards and Technology (NIST) library. The peak areas from the chromatograms were used to determine the amount of each aromatic degradation product. As shown in Table 2, the control sample had quite a few extractive chemicals, while a series of new chemicals showed up in the TL3-treated sample, suggesting a substantial decomposition of lignin. These newly appeared compounds included aliphatic hydrocarbon, aromatic hydrocarbons, aliphatic acid, phenolic acid, phenol, ester, etc. Furthermore, with the addition of glucose as an electron donor (TL3 + Glucose treatment), lignin degradation became more thorough, and an array of end metabolites with a small molecular weight accumulated, such as propanoic acid and formic acid. It is noteworthy that there were almost no intermediates such as ferulic acid and *p*-coumaric acid in the degradation products, but, rather, mainly several kinds of end products.

Table 2. Products of lignin degradation in different treatments.

NO.	RT (min)	Compounds	Treatment		
			Control	TL3	TL3 + Glucose
1	1.344	Acetic acid, (propylthio)-	−	+	−
2	2.879	3-Ethylheptanoic acid	−	−	+
3	3.256	Propanoic acid, ethyl ester	−	+	+
4	3.400	Formic acid, butyl ester	−	+	+
5	6.400	*p*-Xylene	−	+	−
6	9.766	L-(+)-Lactic acid	−	+	−
7	11.921	1,3-Butanediol, diacetate	−	+	−
8	13.398	Benzaldehyde, 3,4-dimethyl-	+	+	−
9	13.944	Oxalic acid, 6-ethyloct-3-yl isobutyl ester	−	+	−
10	14.132	Heptadecane, 8-methyl-	+	+	−
11	15.287	Butyric acid, 4-pentadecyl ester	−	+	−
12	15.587	Butanoic acid, octyl ester	−	+	−
13	17.032	Octacosane	+	+	−
14	17.288	Phenol, 2,5-bis(1,1-dimethylethyl)-	−	+	+
15	18.943	Pentadecane, 2,6,10-trimethyl-	−	+	−
16	19.532	Hexane, 2,4-dimethyl-	−	+	−
17	19.543	Heneicosane	+	+	−
18	20.786	3-Vanilpropanol, bis(trimethylsilyl)-	−	+	−
19	21.286	Phthalic acid, isobutyl undecyl ester	−	+	+
20	21.798	Heptacosane	−	+	−
21	21.863	7,9-Di-tert-butyl-1-oxaspiro(4,5)deca-6,9-diene-2,8-dione	+	−	−
22	21.942	Methoxyacetic acid, 3-tridecyl ester	+	+	−
23	22.020	Benzenepropanoic acid, 3,5-bis(1,1-dimethylethyl)-4-hydroxy-, methyl ester	−	+	−
24	22.287	n-Hexadecanoic acid	+	+	−
25	22.565	Benzoic acid, 3-[(2,2-dimethyl-1-oxopropyl)amino]-	−	+	−
26	24.187	Hexadecanoic acid, 2-hydroxy-, methyl ester	−	+	−
27	24.508	Octadecanoic acid	+	+	−
28	26.109	Hexanedioic acid, bis(2-ethylhexyl) ester	−	+	−
29	26.309	Phenol, 2,2′-methylenebis[6-(1,1-dimethylethyl)-4-methyl-	−	+	+

Table 2. *Cont.*

NO.	RT (min)	Compounds	Treatment		
			Control	TL3	TL3 + Glucose
30	26.686	2-Nonenoic acid, methyl ester	−	+	−
31	26.799	Benzenepropanoic acid, 3,5-bis(1,1-dimethylethyl)-4-hydroxy-, octadecyl ester	−	−	+
32	27.142	Hexadecanoic acid, 2-hydroxy-1-(hydroxymethyl)ethyl ester	−	+	−
33	27.253	1,2-Benzenedicarboxylic acid, dioctyl ester	−	+	−
34	27.797	Benzoic acid, 2,5-bis(trimethylsiloxy)-, trimethylsilyl ester	+	+	−
35	28.508	Butanedioic acid, 2-hydroxy-2-methyl-, dimethyl ester	−	+	−
36	28.819	Octadecanoic acid, 2-hydroxy-1-(hydroxymethyl)ethyl ester	−	+	−
37	28.897	Terephthalic acid, di(2-ethylhexyl) ester	−	+	−
38	28.842	Fumaric acid, dodecyl nonyl ester	−	+	−

Note: RT, retention time; "+" present; "−" absent.

2.5. 2D-HSQC NMR Analysis of the Residual Lignin

The cleavage of the lignin C-O-C bond is crucial in the bacterial lignin degradation process. In this study, lignin linkages such as β-O-4, β-β, and β-5 showed varying degrees of cleavage after treatment with TL3 (Figures 3 and S2 and Table S2). The contours in the NMR spectra are used to determine changes in lignin linkages; a weakening or strengthening of the contours indicates a change in bond energy. As depicted in Figure 3a–c, the β-O-4 aryl ether linkage (A_α, $A_{\beta(G)}$, and $A_{\beta(S)}$) was almost completely degraded under aerobic conditions, and was more pronounced than that under the optimized anaerobic degradation conditions (with the addition of glucose as an electron donor). The partial cleavage of the β-β (B_α, B_β, and B_γ) and β-5 (C_β, and C_γ) linkages was also observed, with a similar effectiveness both under the aerobic and the anaerobic conditions. In Figure 3d–f, it can be observed that the syringyl ($S_{2/6}$) units were completely degraded both under aerobic and anaerobic conditions, with oxidation occurring under aerobic conditions; while the *p*-coumaric acid ($pCA_{3/5}$, pCA_β) and the ferulic acid (FA_2, FA_6, FA_β) structure showed almost no change before and after treatment. Altogether, these results suggested that the lignin degradation level under aerobic conditions was slightly higher than that under the optimized anaerobic conditions (with an electron donor).

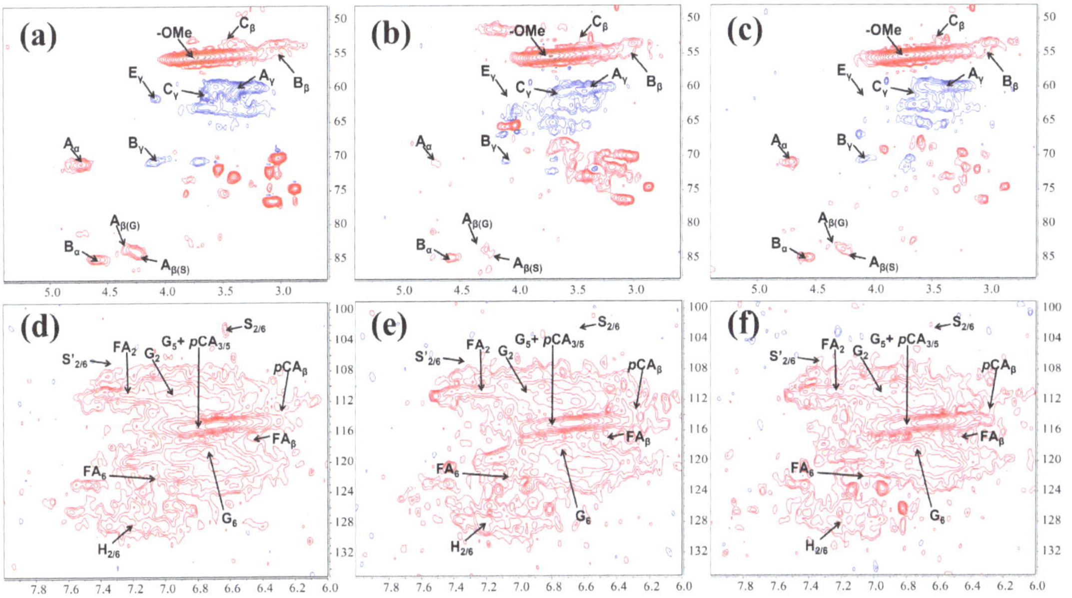

Figure 3. 2D ^{1}H-^{13}C HSQC NMR spectra of lignin before and after treatment by TL3: (**a–c**) oxygenated aliphatic region; and (**d–f**) aromatic region. (**a**,**d**) Untreated lignin; (**b**,**e**) lignin was treated by TL3 in MSM media under aerobic condition; and (**c**,**f**) lignin was treated by TL3 in MSM media supplemented with 1 g/L glucose under anaerobic condition.

3. Discussion

This study presents the profiles of lignin degradation both under aerobic and anaerobic conditions by a newly isolated bacterium, *Klebusiella aerogenes* TL3, which was isolated from the gut of a termite, *Odontotermes formosanus*. The degradation process led to the breakdown of key lignin linkages such as β-O-4, β-β, and β-5, resulting in the production of low-molecular-weight aromatic compounds and organic acids. TL3 was also capable of degrading some monosugars as well as some polysaccharides, including cellulose and natural corn cob. These findings suggested that TL3 is a versatile bacterium, which may shed light on the knowledge of lignocellulose decomposition by facultative anaerobic bacteria.

The versatile degradation capability of TL3 is unique among known bacteria. Previously, it has been reported that *Klebsiella* sp. strain BRL6-2 was capable of utilizing various substrates under aerobic and anaerobic conditions without discrimination, including D-cellobiose, D-mannose, Tween 20, D-glucose, and D-fructose [15]. Many anaerobic cellulolytic bacteria, such as *Thermoanaerobacter* spp., *Clostridium* spp., and *Caldicellulosiruptor* spp., are good at decomposing cellulose, but some of them do not utilize xylan as well as xylose, and few of them can assimilate lignin [18]. In contrast, TL3 can utilize cellulose, lignin, and some monosugars, suggesting its powerful metabolic capability (Table 1). Known potential lignin-degrading bacteria belong to the phyla *Proteobacteria*, *Actinobacteria*, *Firmicutes*, and *Bacteroidetes* [19], and most of them decompose lignin through extracellular oxidases, which require the availability of oxygen [20], while many anaerobic lignin-degrading bacteria do not metabolize cellulose or xylan [14,21]. In comparison to the aforementioned studies, TL3 exhibits growth on lignin both under aerobic and anaerobic conditions, making it a versatile and unique strain for the degradation of lignocellulose.

The anaerobic lignin degradation process is featured by a quick yet incomplete decomposition, and could probably be enhanced by the addition of specific electron acceptors. In the presence of 0.2% glucose, the concentration of lignin decreased from 0.14% to approximately 0.092% after anaerobic treatment by *Tolumonas lignolytica* BRL6-1 [16]. *Ace-*

toanaerobium sp. WJDL-Y2 can degrade approximately 24.9% of kraft lignin under anaerobic conditions [12]. *Enterobacter lignolyticus* SCF1 can anaerobically degrade 56% of lignin when supplemented with xylose (with 0.8% xylose and 0.05% lignin in the culture medium), but it cannot grow on liquid media with lignin as the sole carbon source [14]. The degradation of lignin in the aforementioned results requires the presence of electron acceptors (such as glucose, xylose, etc.) to achieve efficient lignin degradation. Similarly, in this study, the addition of glucose to the culture medium promoted lignin degradation from 14.8% to 17.4% (Figure 1c,d). In the aerobic degradation of lignin with different concentrations by *Pseudomonas* sp. Hu109A, the highest percent degradation of 88.13% was achieved at a lignin concentration of 0.5 g/L, while the percent degradation was approximately 60% at a lignin concentration of 1 g/L [22]. It is known that the degradation rate decreases with increasing lignin concentration, partly due to the biotoxicity of lignin to bacteria and the inhibitory effect of the degradation products on bacterial growth. SCF1 and WJDL-Y2 have a high anaerobic degradation efficiency and can reach maximum conversion within approximately 48 h [12,14], while the aerobes degraded lignin continuously at a slow pace for 7–28 days [22]. The higher conversion of lignin under aerobic conditions may correlate with its longer time span of degradation than that under anaerobic conditions.

The cleavage of various linkages is crucial for the process of lignin decomposition. The predominant linkages in lignin are β-O-4 aryl ether linkages, which account for 45–60% of the total linkages, while other linkages in natural lignin include β-5 phenylcoumaran (6–12%), 5-5′ biphenyl (9–22%), β-β′ resinol (2–4%), β-1 dibenzodioxocin (1–9%), and 4-O-5′ biphenyl ether (1–7%) linkages [23]. The aryl ring cleavage, carboxylation, and demethylation happened during the anaerobic degradation of kraft lignin by strain WJDL-Y1 [24]. Thomas et al. treated various natural lignocellulosic fibers with several anaerobic fungi, revealing the loss of both β-O-4 and β-5 linkages, the significant hemicellulose reshaping, and the removal of dangling units such as ferulic acid, *p*-coumaric acid, and *p*-hydroxybenzoic acid [25]. Apparently, TL3 treatment led to more kinds of linkage reduction (β-O-4, β-5, and β-β linkages), with the syringyl units being completely removed, and the presence of an electron acceptor promotes the linkage cleavage (Figure 3). In addition, the extent in the cleavage of lignin linkages by TL3 was significantly higher under aerobic conditions than that under anaerobic conditions (Figure 3b,c). This was also evidenced by the fact that more types of lignin linkages were disrupted by some aerobic lignin degraders. For example, an NMR analysis revealed that the β-O-4, β-β, β-5, and β-1 linkages were broken during the aerobic degradation of kraft lignin by the secretome of *P. putida* KT2440 [26]. In addition, an introduction of H_2O_2 into the secretome activated peroxidases, resulting in the increased cleavage of β-O-4 and C-C (β-β and β-1) linkages, with a further enhancement on the cleavage of these linkages upon the addition of Mn^{2+}, suggesting the power of hydroxy radicals in linkage cleavage, which might be absent in the anaerobic degradation process. However, the removal of aromatic rings by TL3 is less pronounced (Figure 3d–f), possibly due to the lack of relevant degrading enzymes.

The metabolic intermediates and end products were also different from other known lignin degraders. Compared to fungi, bacteria exhibit lower lignin-degrading activity, but their versatile metabolic capabilities enable them to degrade various aromatic compounds derived from lignin [27]. Previous studies have identified compounds such as benzoic acid, vanillin, acetosyringone, 2-methylbenzoic acid, coumaric acid, and vanillic aldehyde in the extractives of kraft lignin [24,28]. Vanillin and coumaric acid are present in almost all lignin degradation products by aerobic microbes, with the main products of aerobic lignin degradation by Hu109A also including dibutyl phthalate, acetosyringone, 2,4-di-tert-butylphenol, and eugenol [22]. This differs significantly from our results, where lignin monomer substances such as syringic acid and guaiacol, as well as other O-demethylated derivatives such as gallic acid and protocatechuic acid, could not be detected in the treated samples (Table 2), perhaps because of their different metabolic pathways. The lignin degradation products by TL3 include derivatives of various lignin monomeric structures such as *p*-xylene, phthalic acid, and phenylpropionic acid, which is similar to the results of a previous study on lignin degradation by the anaerobic lignin-degrading bacterium, *Trabulsiella* sp. [29], but different from another one, *Acetoanaerobium* sp. WJDL-Y2 [12]. This indicates the extreme diversity in microbial selectivity and the preference for lignin bond cleavage and aromatic compound metabolism, resulting in vastly different degradation products among different bacteria [30].

In conclusion, it is evident that *K. aerogenes* TL3 is a versatile facultative anaerobic lignin-degrading bacterium capable of simultaneously degrading the main components of lignocellulose. Its efficiency in anaerobic lignin degradation was a little weaker than that under aerobic conditions, with a percent degradation of up to 17.4% in the presence of glucose. TL3 primarily disrupts lignin side chains and cleavages the β-O-4, β-5 and β-β′ linkages, which, accordingly, leads to the production of small-molecule organic acids such as propionic acid, lactic acid, and succinic acid. The difference in lignin degradation by TL3 between aerobic and anaerobic conditions may help better understand the process and mechanisms of lignocellulose decomposition in nature. Moreover, this study may shed new light on the knowledge of the mechanism of anaerobic lignin degradation, and pave a way toward the efficient anaerobic conversion of biomass.

4. Materials and Methods

4.1. Strain and Media

TL3 was isolated using a modified deMan Rogosa Sharpe (mMRS) medium [31] supplemented with 1 g/L lignin instead of glucose from the gut of a termite (*Odontotermes formosanus*), which was collected from the campus of Jiangsu University by us. TL3 was deposited at China General Microbiological Culture Collection Center (CGMCC 1.61912), and identified as *Klebusiella aerogenes* because it shared 99.8% 16S rDNA similarity with the known members of this species (GenBank accession: PP577623). Luria-Bertani (LB) medium contains (g/L) tryptone 10 g, Yeast extract 5 g, and NaCl 5 g. The mMRS medium contains (g/L) tryptone 1 g, YE 2 g, beef extract 2 g, glucose 5 g, KH_2PO_4 1.5 g, $K_2HPO_4 \bullet 3H_2O$ 2.9 g, sodium acetate 1 g, triammonium citrate 0.5 g, $MgCl_2 \bullet 6H_2O$ 0.2 g, $CaCl_2 \bullet 2H_2O$ 75 mg, and $MnSO_4$ 0.05 g. After boiling and nitrogen-flushing for 10 min, L-cysteine was added into the mMRS medium as a reducing reagent at a final concentration of 0.5 g/L. The mineral salt medium (MSM) contains 100 mM KH_2PO_4, 75 mM KOH, 15 mM $(NH_4)_2SO_4$, 1 mM $MgSO_4$, 3.9 μM $FeSO_4$, and 1 g/L alkaline lignin (CAS8068-05-1, Catalog number 370959, Sigma–Aldrich, St. Louis, MO, USA). One gram per liter of glucose was added into the MSM medium as an electron donor where necessary.

4.2. Metabolic Characterization of Different Carbon Sources

In order to determine the fermentation profiles of strain TL3, fresh TL3 cultures were firstly transferred to 15 mL of mMRS medium at 1% inoculum and incubated anaerobically at 37 °C and 180 rpm for 24 h. Then, the TL3 culture was inoculated into mMRS medium supplemented with 10 g/L various carbon sources such as glucose, avicel, xylose, corn cob (60 mesh), lignin, etc., and grew anaerobically or aerobically at 37 °C and 180 rpm for 24 h. The metabolites were determined using high-performance liquid chromatography (HPLC) as described below: The fermentation broth was centrifuged at 13,000 rpm, and the supernatant was filtered and acidified with 0.2 M sulphuric acid. The HPLC was equipped with an Aminex HPX-87H column and a refractive index detector, utilizing 4 mM sulphuric acid as the mobile phase at a flow rate of 0.4 mL/min, and the temperature of the column was maintained at 50 °C.

4.3. Time Courses of Bacterial Growth and Lignin Degradation under Anaerobic Conditions

For the determination of bacteria growth and lignin degradation under anaerobic conditions, TL3 and the control strain were grown in the MSM medium or the mMRS medium, with or without the supplement of 1 g/L glucose. One milliliter of fermentation broth was taken every 12 h, and the optical density was measured at 600 nm, deionized water was used as blank control, and the above experiments were conducted in three parallel experiments. Lignin concentration was measured by retrieving 1 mL of culture from anaerobic septum bottles, diluting 1:2 in distilled water, filtering out cells, and measuring the absorbance at 310 nm as described by Billings et al., 2015 [16]. For statistical analysis, Turkey's test was carried out using Origin™ Pro 9.0 software.

4.4. Fourier-Transform Infrared Spectroscopy (FTIR) Analysis

After three days of aerobic or anaerobic incubation, the fermentation broth was centrifuged at 13,000 rpm to remove the cells. Subsequently, the supernatant was acidified with 0.5 M sulfuric acid to precipitate the alkali lignin, which is water-insoluble under acid condition and water-soluble at weak acid condition or at even higher pHs. Samples were washed repeatedly with acidified pure water and dried in an oven at 60 °C. The lignin FTIR spectra in attenuated total reflection (ATR) mode were collected by the potassium bromide tablet method by taking a certain amount of dried, treated, or untreated lignin samples, respectively. The lignin samples were mixed with potassium bromide at a mass ratio of 1:100, ground to 200 mesh powder, and prepared into thin slices of uniform thickness. The FTIR spectra were collected, with pure potassium bromide as the background. The wavenumber range was 4000–400 cm^{-1}, and the number of scans was 64 with a resolution of 1 cm^{-1}.

4.5. Gas Chromatography–Mass Spectrometry (GC-MS) Analysis

Lignin was anaerobically degraded by TL3 in 50 mL of MSM medium with or without the supplement of glucose for 3 days; while the control sample was not inoculated with bacteria. Samples were centrifuged at 13,000 rpm for 15 min to remove biomass. Then, the supernatants were acidified to pH 2–3 with 6 M HCl and thoroughly extracted afterward with 150 mL of ethyl acetate. The organic layer of the extraction mixture was collected and reduced to 10 mL by rotary evaporation at 37 °C and dewatered by anhydrous Na_2SO_4 to remove moisture. Then, 100 μL of the organic layer was silylated after the evaporation of the solvent under nitrogen stream. For the silylation procedure, 100 μL dioxane and 10 μL pyridine were added into samples and vortexed in glass tubes followed by silylation, which was performed with 50 μL of N,O-bis(trimethylsilyl)trifluoroacetamide (BSTFA; Sigma-Aldrich, St. Louis, MO, USA). The mixture was heated in a water bath at 80 °C for 45 min with periodic shaking to dissolve the residues. An aliquot of 1 mL of silylated mixture was injected into a gas chromatography–mass spectrometry (GC-MS) system (AntoSystem XL GC-TurboMass; PerkinElmer, Waltham, MA, USA). The analytical column was connected to the system was a PE-5MS capillary column (20 m × 0.18 mm internal diameter, 0.18 mm film thickness). A solvent delay of 3.0 min was selected. In the full-scan mode, electron ionization mass spectra in the range of 30–550 (m/z) were recorded at electron energy of 70 eV. All standard monomeric phenolics compounds (1 mg) were derivatized and chromatographed as above. In order to identify the low-molecular-weight lignin-related compounds as trimethylsilyl (TMS) derivatives derived from bacterial treatment, their mass spectra were compared with that of the data of GC-MS spectral library (Wiley, NIST) available in the instrument.

4.6. Nuclear Magnetic Resonance (NMR) Analysis

The 2D-^{1}H-^{13}C heteronuclear single-quantum coherence (HSQC) NMR spectra were obtained using a Bruker Avance III HD 500 MHz spectrometer operating at a frequency of 125.12 MHz for the 13C nucleus as described by Ji et al. (2021) [32]. Thirty to fifty milligrams of the dry lignin samples were dissolved in 0.6 mL deuterated dimethylsulfoxide (DMSO)-d6 and the spectra were collected at 298 K. A standard Bruker adiabatic HSQC pulse sequence (hsqcetgpsisp2.2) was used with the following spectra acquisition condition: 1.0 s pulse delay, 64 scans, 1024 data points for ^{1}H, 256 increments for ^{13}C, and a 1JC–H of 145 Hz. The ^{1}H and ^{13}C spectral widths are 13.0 and 220.0 ppm, respectively. The central DMSO solvent peak ($\delta^{13}C/\delta^1H$ = 39.5/2.49 ppm) was used for chemical shifts calibration. HSQC spectra were processed and analyzed with Mestrenova (version 12.0.2) with a matched cosinebell apodization and 2 × zero filling in both dimensions.

Supplementary Materials: The following supporting information can be downloaded at: https://www.mdpi.com/article/10.3390/molecules29102177/s1, Table S1: Assignment of FTIR spectra of lignin samples; Table S2: Main lignin 2D ^{1}H-^{13}C cross-peak assignments in the HSQC spectra. Figure S1: Growth of TL3 and SYK-6 in mMRS medium without additional carbon source. Figure S2: Main detected lignin linkages.

Author Contributions: Conceptualization, Z.T. and A.G.; data tabulation, Z.T., A.Z.-G., H.G. and Y.X.; formal analysis, Z.T., D.Z. and R.X.; supervision, J.S.; writing—original draft preparation, Z.T. and A.G.; writing—review and editing, Z.T. and A.G.; funding acquisition, A.G. All authors have read and agreed to the published version of the manuscript.

Funding: This work was financially supported by the Zhenjiang Key Research and Development Project (CN2023001) and Changzhou Application Basic Research Project (CJ20230015). R.X. acknowledges the support of the National Key R&D Program of China (2023YFC3403600).

Institutional Review Board Statement: Not applicable.

Informed Consent Statement: Not applicable.

Data Availability Statement: All data sets used and analyzed are available upon reasonable request.

Conflicts of Interest: The authors declare no conflicts of interest.

References

1. Nanda, S.; Azargohar, R.; Dalai, A.K.; Kozinski, J.A. An assessment on the sustainability of lignocellulosic biomass for biorefining. *Renew. Sustain. Energy Rev.* **2015**, *50*, 925–941. [CrossRef]
2. Upton, B.M.; Kasko, A.M. Strategies for the conversion of lignin to high-value polymeric materials: Review and perspective. *Chem. Rev.* **2016**, *116*, 2275–2306. [CrossRef] [PubMed]
3. Andlar, M.; Rezic, T.; Mardetko, N.; Kracher, D.; Ludwig, R.; Santek, B. Lignocellulose degradation: An overview of fungi and fungal enzymes involved in lignocellulose degradation. *Eng. Life Sci.* **2018**, *18*, 768–778. [CrossRef] [PubMed]
4. Weng, C.; Peng, X.; Han, Y. Depolymerization and conversion of lignin to value-added bioproducts by microbial and enzymatic catalysis. *Biotechnol. Biofuels* **2021**, *14*, 84. [CrossRef] [PubMed]
5. Jin, X.; Li, X.; Zou, L.; Zheng, Z.; Ouyang, J. Biological Valorization of Lignin-Derived Aromatics in Hydrolysate to Protocatechuic Acid by Engineered *Pseudomonas putida* KT2440. *Molecules* **2024**, *29*, 1555. [CrossRef] [PubMed]
6. Mate, D.M.; Alcalde, M. Laccase engineering: From rational design to directed evolution. *Biotechnol. Adv.* **2015**, *33*, 25–40. [CrossRef] [PubMed]
7. Blaschek, L.; Murozuka, E.; Serk, H.; Ménard, D.; Pesquet, E. Different combinations of laccase paralogs nonredundantly control the amount and composition of lignin in specific cell types and cell wall layers in Arabidopsis. *Plant Cell* **2023**, *35*, 889–909. [CrossRef] [PubMed]
8. Ragauskas, A.J.; Beckham, G.T.; Biddy, M.J.; Chandra, R.; Chen, F.; Davis, M.F.; Davison, B.H.; Dixon, R.A.; Gilna, P.; Keller, M. Lignin valorization: Improving lignin processing in the biorefinery. *Science* **2014**, *344*, 1246843. [CrossRef] [PubMed]
9. Janusz, G.; Pawlik, A.; Sulej, J.; Świderska-Burek, U.; Jarosz-Wilkołazka, A.; Paszczyński, A. Lignin degradation: Microorganisms, enzymes involved, genomes analysis and evolution. *FEMS Microbiol. Rev.* **2017**, *41*, 941–962. [CrossRef]
10. De Gonzalo, G.; Colpa, D.I.; Habib, M.H.M.; Fraaije, M.W. Bacterial enzymes involved in lignin degradation. *J. Biotechnol.* **2016**, *236*, 110–119. [CrossRef]
11. Neumann, G.T.; Pimentel, B.R.; Rensel, D.J.; Hicks, J.C. Correlating lignin structure to aromatic products in the catalytic fast pyrolysis of lignin model compounds containing β–O–4 linkages. *Catal. Sci. Technol.* **2014**, *4*, 3953–3963. [CrossRef]
12. Duan, J.; Huo, X.; Du, W.J.; Liang, J.D.; Wang, D.Q.; Yang, S.C. Biodegradation of kraft lignin by a newly isolated anaerobic bacterial strain, *Acetoanaerobium* sp. WJDL-Y2. *Lett. Appl. Microbiol.* **2016**, *62*, 55–62. [CrossRef]
13. Rashid, G.M.M.; Duran-Pena, M.J.; Rahmanpour, R.; Sapsford, D.; Bugg, T.D.H. Delignification and enhanced gas release from soil containing lignocellulose by treatment with bacterial lignin degraders. *J. Appl. Microbiol.* **2017**, *123*, 159–171. [CrossRef]
14. Deangelis, K.M.; Sharma, D.; Varney, R.; Simmons, B.; Isern, N.G.; Markilllie, L.M.; Nicora, C.; Norbeck, A.D.; Taylor, R.C.; Aldrich, J.T.; et al. Evidence supporting dissimilatory and assimilatory lignin degradation in *Enterobacter lignolyticus* SCF1. *Front. Microbiol.* **2013**, *4*, 280. [CrossRef] [PubMed]
15. Woo, H.L.; Ballor, N.R.; Hazen, T.C.; Fortney, J.L.; Simmons, B.; Davenport, K.W.; Goodwin, L.; Ivanova, N.; Kyrpides, N.C.; Mavromatis, K. Complete genome sequence of the lignin-degrading bacterium *Klebsiella* sp. strain BRL6-2. *Stand. Genom. Sci.* **2014**, *9*, 19. [CrossRef]

16. Billings, A.F.; Fortney, J.L.; Hazen, T.C.; Simmons, B.; Davenport, K.W.; Goodwin, L.; Ivanova, N.; Kyrpides, N.C.; Mavromatis, K.; Woyke, T.; et al. Genome sequence and description of the anaerobic lignin-degrading bacterium *Tolumonas lignolytica* sp. nov. *Stand. Genom. Sci.* **2015**, *10*, 106. [CrossRef] [PubMed]
17. Chaput, G.; Ford, J.; DeDiego, L.; Narayanan, A.; Tam, W.Y.; Whalen, M.; Huntemann, M.; Clum, A.; Spunde, A.; Pillay, M. *Sodalis ligni* strain 159R isolated from an anaerobic lignin-degrading consortium. *Microbiol. Spectr.* **2022**, *10*, e02346-21. [CrossRef]
18. Vishnivetskaya, T.A.; Raman, B.; Phelps, T.J.; Podar, M.; Elkins, J.G. Cellulolytic microorganisms from thermal environments. In *Extremophiles: Microbiology and Biotechnology*, 1st ed.; Anitori, R.P., Ed.; Caister Academic Press: Norfolk, UK, 2012; pp. 131–158.
19. Silva, J.P.; Ticona, A.R.P.; Hamann, P.R.V.; Quirino, B.F.; Noronha, E.F. Deconstruction of lignin: From enzymes to microorganisms. *Molecules* **2021**, *26*, 2299. [CrossRef] [PubMed]
20. Bugg, T.D.; Ahmad, M.; Hardiman, E.M.; Singh, R. The emerging role for bacteria in lignin degradation and bio-product formation. *Curr. Opin. Biotechnol.* **2011**, *22*, 394–400. [CrossRef]
21. Khan, M.U.; Ahring, B.K. Lignin degradation under anaerobic digestion: Influence of lignin modifications—A review. *Biomass Bioenerg.* **2019**, *128*, 105325. [CrossRef]
22. Nawaz, M.Z.; Shang, H.; Sun, J.; Geng, A.; Ali, S.S.; Zhu, D. Genomic insights into the metabolic potential of a novel lignin-degrading and polyhydroxyalkanoates producing bacterium *Pseudomonas* sp. Hu109A. *Chemosphere* **2023**, *310*, 136754. [CrossRef] [PubMed]
23. Bugg, T.D.H.; Ahmad, M.; Hardiman, E.M.; Rahmanpour, R. Pathways for degradation of lignin in bacteria and fungi. *Nat. Prod. Rep.* **2011**, *28*, 1883–1896. [CrossRef] [PubMed]
24. Duan, J.; Liang, J.; Wang, Y.; Du, W.; Wang, D. Kraft Lignin Biodegradation by *Dysgonomonas* sp. WJDL-Y1, a new anaerobic bacterial strain isolated from sludge of a pulp and paper mill. *J. Microbiol. Biotechnol.* **2016**, *26*, 1765–1773. [CrossRef] [PubMed]
25. Lankiewicz, T.S.; Choudhary, H.; Gao, Y.; Amer, B.; Lillington, S.P.; Leggieri, P.A.; Brown, J.L.; Swift, C.L.; Lipzen, A.; Na, H.; et al. Lignin deconstruction by anaerobic fungi. *Nat. Microbiol.* **2023**, *8*, 596–610. [CrossRef] [PubMed]
26. Xu, Z.; Peng, B.; Kitata, R.B.; Nicora, C.D.; Weitz, K.K.; Pu, Y.; Shi, T.; Cort, J.R.; Ragauskas, A.J.; Yang, B. Understanding of bacterial lignin extracellular degradation mechanisms by *Pseudomonas putida* KT2440 via secretomic analysis. *Biotechnol. Biofuels Bioprod.* **2022**, *15*, 117. [CrossRef] [PubMed]
27. Masai, E.; Katayama, Y.; Fukuda, M. Genetic and biochemical investigations on bacterial catabolic pathways for lignin-derived aromatic compounds. *Biosci. Biotech. Bioch.* **2007**, *71*, 1–15. [CrossRef] [PubMed]
28. An, X.; Cheng, Y.; Zang, H.; Li, C. Biodegradation characteristics of lignin in pulping wastewater by the thermophilic *Serratia* sp. AXJ-M: Performance, genetic background, metabolic pathway and toxicity assessment. *Environ. Pollut.* **2023**, *322*, 121230. [CrossRef] [PubMed]
29. Suman, S.K.; Dhawaria, M.; Tripathi, D.; Raturi, V.; Adhikari, D.K.; Kanaujia, P.K. Investigation of lignin biodegradation by *Trabulsiella* sp. isolated from termite gut. *Int. Biodeter. Biodegr.* **2016**, *112*, 12–17. [CrossRef]
30. Lee, S.; Kang, M.; Bae, J.-H.; Sohn, J.-H.; Sung, B.H. Bacterial valorization of lignin: Strains, enzymes, conversion pathways, biosensors, and perspectives. *Front. Bioeng. Biotechnol.* **2019**, *7*, 209. [CrossRef]
31. Li, N.; Geng, A.; Tu, Z.; Fan, Y.; Xie, R.; Li, X.; Sun, J. Isolation of *Lactococcus* sp. X1 from termite gut, and its application in lactic acid production. *Fermentation* **2023**, *9*, 85. [CrossRef]
32. Ji, Q.; Yu, X.; Wu, P.; Yagoub, A.E.-G.A.; Chen, L.; Taiye, M.A.; Zhou, C. Pretreatment of sugarcane bagasse with deep eutectic solvents affect the structure and morphology of lignin. *Ind. Crop Prod.* **2021**, *173*, 114108. [CrossRef]

MDPI AG
Grosspeteranlage 5
4052 Basel
Switzerland
Tel.: +41 61 683 77 34

Molecules Editorial Office
E-mail: molecules@mdpi.com
www.mdpi.com/journal/molecules